T0269182

CAMBRIDGE LIBRARY COLLECTION

Books of enduring scholarly value

Polar Exploration

This series includes accounts, by eye-witnesses and contemporaries, of early expeditions to the Arctic and the Antarctic. Huge resources were invested in such endeavours, particularly the search for the North-West Passage, which, if successful, promised enormous strategic and commercial rewards. Cartographers and scientists travelled with many of the expeditions, and their work made important contributions to earth sciences, climatology, botany and zoology. They also brought back anthropological information about the indigenous peoples of the Arctic region and the southern fringes of the American continent. The series further includes dramatic and poignant accounts of the harsh realities of working in extreme conditions and utter isolation in bygone centuries.

Notes by a Naturalist on the Challenger

The *Challenger* Expedition of 1872–6 was conceived to examine the deep sea floor worldwide and disprove the theory of a 'dead zone' in the oceans below a certain depth. Using a modified Royal Navy ship, the expedition sailed nearly 70,000 nautical miles across the globe, collecting oceanographic data and marine specimens, and laying the foundations for the science of oceanography by later publishing fifty volumes of reports. The naturalist Henry Moseley (1844–91) recounts the voyage in this 1879 work, covering visits to many remote islands and the taking of samples at hundreds of locations. The voyage's achievements included the collection of over 4,000 new marine species and the discovery of the world's deepest ocean trench (Challenger Deep). Moseley's observations on native peoples also proved important as traditional cultures were changing rapidly at the time. Illustrated with numerous woodcuts, this narrative illuminates an adventure of great scientific significance.

Cambridge University Press has long been a pioneer in the reissuing of out-of-print titles from its own backlist, producing digital reprints of books that are still sought after by scholars and students but could not be reprinted economically using traditional technology. The Cambridge Library Collection extends this activity to a wider range of books which are still of importance to researchers and professionals, either for the source material they contain, or as landmarks in the history of their academic discipline.

Drawing from the world-renowned collections in the Cambridge University Library and other partner libraries, and guided by the advice of experts in each subject area, Cambridge University Press is using state-of-the-art scanning machines in its own Printing House to capture the content of each book selected for inclusion. The files are processed to give a consistently clear, crisp image, and the books finished to the high quality standard for which the Press is recognised around the world. The latest print-on-demand technology ensures that the books will remain available indefinitely, and that orders for single or multiple copies can quickly be supplied.

The Cambridge Library Collection brings back to life books of enduring scholarly value (including out-of-copyright works originally issued by other publishers) across a wide range of disciplines in the humanities and social sciences and in science and technology.

Notes by a Naturalist on the *Challenger*

Being an Account of Various Observations
Made During the Voyage
of H.M.S. Challenger Round the World,
in the Years 1872–1876,
Under the Commands of Capt. Sir G.S. Nares,
and Capt. F.T. Thomson

HENRY NOTTIDGE MOSELEY

CAMBRIDGE
UNIVERSITY PRESS

CAMBRIDGE
UNIVERSITY PRESS

University Printing House, Cambridge, CB2 8BS, United Kingdom

Cambridge University Press is part of the University of Cambridge.

It furthers the University's mission by disseminating knowledge in the pursuit of education, learning and research at the highest international levels of excellence.

www.cambridge.org
Information on this title: www.cambridge.org/9781108074834

© in this compilation Cambridge University Press 2014

This edition first published 1879
This digitally printed version 2014

ISBN 978-1-108-07483-4 Paperback

ANTARCTIC ICEBERGS
From Sketches by the Author.

C.F.Kell, Lith. London E.C

NOTES

BY

A NATURALIST ON THE "CHALLENGER,"

BEING

AN ACCOUNT OF VARIOUS OBSERVATIONS

MADE DURING THE

VOYAGE OF H.M.S. "CHALLENGER" ROUND THE WORLD, IN THE YEARS 1872—1876,

UNDER THE

Commands of Capt. Sir G. S. NARES, R.N., K.C.B., F.R.S., and Capt. F. T. THOMSON, R.N.

By

H. N. MOSELEY, M.A., F.R.S.,

FELLOW OF EXETER COLLEGE, OXFORD, MEMBER OF THE SCIENTIFIC STAFF OF
H.M.S. "CHALLENGER."

With a Map, Two Coloured Plates, and numerous Woodcuts.

London:
MACMILLAN AND CO.

1879.

LONDON :

HARRISON AND SONS, PRINTERS IN ORDINARY TO HER MAJESTY,
ST. MARTIN'S LANE.

ERRATA.

PAGE	LINE	FOR	READ
50	9	*Myrmelion*	*Myrmeleon.*
56	36	*Halcyon erythroryncha*	*Halcyon erythrogastra.*
86	24	Garde de l'eau	Gare à l'eau.
105	29	which	whom.
121	37	*Neospiza*	*Nesospiza.*
122	note 1 and 4	ibid.	ibid.
125	11	Inaccesible	Inaccessible.
128	32	acts . . . assists	act . . . assist.
163	9	North or West	north of west.
169	5	*Hastia*	*Haastia.*
181	13	*Halabæna*	*Halobæna.*
257	10	*Phalangister*	*Phalanjista.*
276	27	Musschenbrook	Musschenbroek.
293	9	Chorophyly	Chlorophyll.
295	13	*Solitarias*	*Solitaria.*
301	8	Lats. 16° and 20° E.	Lats. 16° and 20° S.
352	14	*alerrimum*	*aterrimum.*
385	Heading	Amboina	Banda.
499	11, 24	Hawai	Hawaii.
504	33 and in note	ibid.	ibid.
514	3	*Chionider*	*Clionider.*

PREFACE.

THE contents of this book were mainly written on board H.M.S. "Challenger," and sent home from the various ports touched at, in the form of a journal. Much of the book has been printed directly from the sheets of foreign note paper on which the journal was transmitted. Since, however, very much of it was intended for family reading, a good deal of matter descriptive of various well-known animals and phenomena has been struck out, as well as the accounts of the long voyages and various dredgings.

A considerable amount of the less technical matter, even though treating of matters often before described, contained in the original letters, has, however, been retained in the hopes that it may interest general readers. The whole has been revised and corrected, and the scientific names of birds and animals have been corrected, as far as possible, by means of the notices of the "Challenger" collections published by various specialists since the arrival of the ship in England.

I venture here to make an appeal to specialists engaged in working at "Challenger" material to spare me copies of their papers, and to assure them that, as I hope this book will show, I take an interest in the collections of all kinds made during the "Challenger's" voyage, that I took a large share myself in the bringing together of the collections, and that such papers

will not be thrown away upon me.   It is almost impossible to
collect them unaided from the various Journals in which they
are published.

I have given in the form of foot-notes, and also in some
instances, at the ends of the chapters, references to various works
relating to the subjects treated of.   No attempt has, however,
been made to afford a complete bibliography ; I have merely
noted down those works which I have happened to consult
myself, in the hopes that such references may be found of some
assistance.   At the end of the book I have given a list of books
and papers published, relating to the " Challenger " Expedition.
This is no doubt far from perfect ; I have only made it as full
as my information allows.

I should have wished to have been able to illustrate this
book more fully ; the reason I have not done so is simply that
of expense.   I have introduced amongst the figures, which are
otherwise new, about twelve which are printed from clichés,
and which may no doubt be familiar to some readers as occurring in other works.   I make no scruple to use them as illustrating the subject, and as being better than no figures at all
of the objects referred to.

Since it was not considered expedient to attach a Botanist
to the " Challenger " Expedition, because the special work of
the ship lay in deep-sea exploration, I undertook the collection
of plants during the voyage.   I received instructions at Kew
before starting.   My best thanks are due to my friend Sir
Joseph Hooker for the constant encouragement in my work
which he conveyed to me by letter throughout the voyage,
and for the care and trouble bestowed on my collections.   I
have to thank further my friend Prof. Oliver for the prompt-
ness with which he examined the collections and named them,
and made arrangements for the description and enumeration
by various authors of all the cryptogams as well as for kindly

correcting and arranging for publication of my own notes on plants collected, and for presenting them to the Linnean Society.

I have to thank my friend Mr. W. T. Thistleton Dyer, for much kind assistance and information conveyed to me throughout the cruise. I have also to thank the various authors, whose names will be found in the list of papers at the end of this book, for having undertaken the description of my collections of cryptogams and other plants.

To my friend Prof. G. Rolleston I am indebted for various kind offices performed during the course of my voyage, and for having seen through the press several scientific papers sent home by me for publication.

I would here express my obligations to Sir C. Wyville Thomson for having selected me as a member of the Scientific Staff of H.M.S. "Challenger." The fact that the "Challenger" Expedition started at all, is principally due to the energy and perseverance of Sir C. Wyville Thomson and Dr. Carpenter. I sincerely hope that before very long another scientific exploring expedition may be despatched from England under Government auspices.

I would also return my best thanks to Captain Sir G. S. Nares and Captain F. T. Thomson for their invariable kindness and courtesy to me personally, and for the many occasions on which they gave me special assistance in my work during the voyage.

I am further bound to express my gratitude to all my messmates and many friends on board the ship, who constantly helped me in various ways. The interests of a Commander and a First Lieutenant on board a man-of-war are as directly opposed to all scientific operations, especially those of Botany and Zoology, which are necessarily more or less connected with dirt and untidiness, as they can possibly be. I have to thank Captain J. F. L. P. Maclear and Lieuts. P. Aldrich and A. C. B. Bromley, for having so often good-naturedly put up with

my various messes and disfigurements of their decks. My best
thanks are further due to Staff-Commander T. H. Tizard, Navigating Officer of the "Challenger," who piloted us so safely
amongst so many reefs and into so many little-frequented
harbours. He was always ready to afford information during
the voyage, and has also done me much most generous service
in this way subsequently, whilst I have been preparing the
present work for the press. A reference to his valuable papers
on deep-sea temperatures will be found in the list at the end of
this book.

To Mr. R. Richards, our Paymaster, we were all indebted for
the careful planning of many pleasant excursions on shore and
various acts of kindness.

My indebtedness to my colleagues, Mr. J. Y. Buchanan,
Mr. J. Murray, Dr. J. J. Wild, and the late R. Von Willemoes
Suhm, I have expressed in several places in the text of this
work.

It is perhaps somewhat out of place in a private work of
the present kind, to express my gratitude to the actual pro-
moters of the "Challenger" Expedition—The Lords of the
Admiralty. I cannot, however, refrain from saying, as has so
often been done before by others, that all honour is due to
them for having promoted this memorable Expedition, and for
the completeness with which it was furnished in every respect.
The thanks of all scientific men are due to them, and I cannot
but feel personally thankful in consideration of the extreme
pleasure which I derived from the voyage.

Thanks are no less due to the two successive hydrographers
to the Admiralty, Vice-Admiral Sir G. H. Richards, Knt., C.B.,
F.R.S., &c., and Captain F. J. Owen Evans, R.N., C.B., F.R.S.,
for the skill with which the Expedition was planned and carried
out. I have to thank both of them for many advantages derived,
and also for their personal kindness to me on all occasions.

To Mr. W. B. Blakeney, R.N., I am personally indebted for various arrangements made at home by him, during the voyage, for my benefit.

I have, finally, to thank my friend Prof. E. Ray Lankester, Fellow of Exeter College, for assistance and advice received during the progress of this book; and also my friend the Rev. Thomas Sheppard, B.D., Fellow of Exeter College, for having kindly read through all the proof sheets, and assisted in their correction.

EXETER COLLEGE, OXFORD,
December, 1878.

CONTENTS.

CHAPTER I.

TENERIFFE, ST. THOMAS, BERMUDA.

Circumstances of the Voyage, p. 1. Teneriffe, 2. Cochineal Plantations, 2. Excursion up the Peak. Tradewind Cloud, 3. Zones of Vegetation, 4. Sunset seen above the Clouds, 5. Rabbits and other Animals on the Peak, 6. Peculiar Spider's-web, 8. Catching Sharks off Sombrero Island, West Indies, 8. Appearance and habits of Remora, 9. Pilot fish, 9. Island of St. Thomas, 11. Calcareous Seaweeds, 12. Sea Urchins with poisoned Spines, 12. Burrowing Spider, 13. Nest of Termites, 14. Pelicans edible, 15. Sand-box tree, 15. Defensive colouring of Spines of Cacti, 16. Beach Conglomerate, 17. Sea Beans, 17. Bermuda, 18. Calcareous Sand-rock, 18. Caves, 22. Vegetation, 23. Peat, 23. Boatswain Birds, 25. Land Nemertine, 26. Corals in Caves, 27.

CHAPTER II.

AZORES, MADEIRA, CAPE VERDES.

Fayal Island, Azores, 29. Porpoises on the Feed, 30. Town of Horta, 30. Peculiar Dress of the Women, 31. Island of Pico, 32. St. Michael's Island, 32. Native Ferns and Australian introduced Trees, 33. The Threshing-floor and Women at the Mill, 33. Vegetation of the Azores, 34. Hot Springs at Furnas, 35. Plants Growing in the Hot Water, 36. Caldeira des Sette Cidades, 37. Madeira, 38. Grand Cural, 39. Curious Caps worn by the Men, 40. The Island at Sunset, 41. St. Vincent Island, Cape Verdes, 41. Vegetation of the Island, 42. Ascent of Green Mountain, 43. Different Causes of Variation of Vegetation with Altitude, 45. Structure of Basaltic Dykes, 46. Calcareous Seaweeds on Bird Island, 46. Habits of Crabs, 48.

Miniature Oasis, 51. Flying Gurnet Hooked, 51. Mode of Catching Bonito, 53. Island of Fogo, 54. Porto Praya, St. Jago Island, 55. Use of Foot in Feeding by Kites, 55. Kingfisher and Galinis, 56. Hauling the Sein, 57. A Large Shark, 57. San Domingo Valley, 59. Monkeys, 60. Remarkable Freshwater Crustacean, 60. Limestone Band in the Cliff of the Harbour, 64.

CHAPTER III.

ST. PAUL'S ROCKS AND FERNANDO DO NORHONA.

St. Paul's Rocks, 67. Equatorial Current, 68. Nests of Noddies, 69. Predatory Habits of Grapsus strigosus, 70. Fishing off the Rocks, 71. Nests of Boobies, 72. Pugnacity of the Young Birds, 72. Other Inhabitants of the Rocks, 73. Fishing for Cavalli with Salmon tackle, 74. Geological Structure of the Rocks, 75. Seaweeds Growing on the Rocks, 76. Fernando do Norhona, 77. Calcareous Sand-rock containing Volcanic Intermixture, 78. Tree Shedding Leaves in Dry Season, 78. Jatropha urens, 79. Birds, 79. Brazilian Convicts, 80. St. Michael's Mount, 82. Frigate Birds Nesting, 83. Pigeons Nesting with Sea Birds, 83. Lizards of the Islands, 84.

CHAPTER IV.

BAHIA.

Harbour and Town of Bahia, 85. Religious Procession, 86. Black Angels, 87. Land Planarians, 89. Clicking Butterfly, 89. Primeval Forest, 90. Shooting Humming Birds and Toucans, 91. Caxoeira, 93. Mewing Toads, 93. Excursion to Feira St. Anna, 93. Mule Riding, 94. Former Highway Robbers, 95. Inn at Feira St. Anna and its Guests, 96. The Fair, 97. Anteaters Eaten as Medicine, 97. Vaqueiros, 98. Tailing Cattle, 99. Horse Dealing, 100.

xii CONTENTS.

German Settler in the Country, 100.
Driving Cattle in the Bush, 101. Farm
Slaves, 102. Preparation of Cassava,
102. Overburdened Ant, 104. Three-
toed Sloth, 104. Slavery in Brazil, 105.

CHAPTER V.

TRISTAN DA CUNHA, INACCES-
SIBLE ISLAND, NIGHTINGALE
ISLAND.

Settlement of the Island, 108. Geological
Structure, 109. Vegetation, 110. Tem-
perature of Fresh Water, 111. Phylica
arborea, 111. Rigorous Climate, 112.
Condition of the Settlers, 113. In-
accessible Island, 114. Rock-hopper
Penguins, 117. Tussock grass. 117.
Penguin Rookeries, 119. Peculiar Land
Birds, 121. Noddies and other Sea
Birds, 123. Southern Skuas, 123.
Wild Swine, 124. Change of Habits
of Penguins, 125. Nightingale Island,
126. Vast Penguin Rookery, 127.
Seal Caves, 127. Rocks Worn by the
Feet of the Penguins, 128. Molly-
mauks and their Nests, 130. Deriva-
tion of Seamen's Names for Southern
Animals, 129. Dogs run Wild in a
Penguin Rookery, 132. Migrations of
Penguins and Seals, 133. Insects, &c.,
of the Group, 134. Flowering Seasons,
134. Sea Beans, 135. Relations of the
Flora, 135.

CHAPTER VI.

CAPE OF GOOD HOPE.

Aspect and Formation of the Country,
138. Simon's Bay, 139. Appearance
of the Vegetation, 140. The Road to
Cape Town, 140. The Silver Tree,
142. Habits of Baboons, 143. The
Rock Rabbit, 144. Habits of Rodent
Moles, 145. Kitchen Middens, 147.
Burial Places of Natives, 149. Antelopes, 150. An Ostrich Farm, 151.
Tracks of Animals in the Sand, 152.
Great Variety of Flowering Plants, 153.
Clawless Otter, 154. Land Planarians,
154. Chameleon, 154. Jackass Penguins, 155. Bdellostoma, 156. Rare
Whale with Long Tusks, 157. Peripatus Capensis, the Ancestor of Insects,
159. The Turacou, 161.

CHAPTER VII.

PRINCE EDWARD ISLAND. THE
CROZET ISLANDS.

Appearance and Formation of Marion
Island, 163. Vegetation of the Island,

165. Azorella selago, 165. Limit of
Vegetation in Altitude, 168. Relations
of the Flora, 169. Former Extension
of Land in this Region, 169. Nesting
of the Great Albatross, 172. Mode
of Courtship, 174. Skuas, 174.
"Johnny" Penguins, 175. Rock-
hoppers, 175. Rookeries of King Penguins, 176. Absurd Appearance of the
Young Birds, 177. Singular Mode of
Incubation, 178. Habits of Sheath-
bills, 179. Appearance of the Crozet
Islands, 181. Tree-trunks found in the
Islands by former Voyagers, 182.

CHAPTER VIII.

KERGUELEN'S LAND.

Position of the Island, 184. Its Mountains and Fjords, 185. Active Volcano,
186. Christmas Harbour, 186. Sea
Elephants and Fur Seals, 187. Shooting Teal, 190. The Kerguelen Cabbage, 191. Wingless Flies and Gnats,
192. Vegetation at Successive Heights,
193. Fossil Wood, 195. Rookeries of
Rock-hopper and Macaroni Penguins,
195. Penguins Inhabiting a Cave, 196.
Betsy Cove, 196. Glaciation of the
Land Surface, 197. Iceborne Rocks,
198. Excavation of the Fjords, 199.
Beds of Burnt Coal, 199. The Sea
Leopard, 200. Killing Sea Elephants,
201. Nature of the Trunk of the Sea
Elephant, 202. Carrion Birds, 206.
The Giant Petrel, 206. Habits of
Several Burrowing Petrels, 207. The
Diving Petrel, 208. Habits of Sheath-
bills, 209. Struggle for Existence
amongst the Birds. 213. Whaling
amongst the Kelp, 213.

CHAPTER IX.

HEARD ISLAND.

Diatoms on the Sea Surface, 216. Macdonald Island, 216 Whisky Bay,
Heard Island, 217. Coast-line composed of Glaciers, 219. Structure of
the Glaciers, 219. Terminal and
Lateral Moraines, 220. Glacier Stream,
221. Rocks Cut by Natural Sand
Blast, 222. Lava Flow and Denuded
Crater, 222. Scanty Vegetation, 224.
Range in Elevation of Arctic and
Southern Plants Compared, 225. Mode
of Hunting Sea Elephants, 227. Habits
of these Animals, 228. Sealers Inhabiting Heard Island, 229. Birds of the
Island, 229.

CHAPTER X.

AMONGST THE SOUTHERN ICE.

First Iceberg Sighted, 232. Typical Forms of Southern Bergs, 233. Preservation of Equilibrium, 234. Washlines, 234. Caverns, 235. Bi-tabular Bergs, How Formed, 236. Weathering of Bergs, 238. Stratification of Ice in Bergs, 239. Cleavage, 240. Scarcity of Rocks on Bergs, 242. Discoloured Bands in the Ice, 243. Rev. Canon Moseley on the Motion of Glaciers, 244. Colouring of Bergs, 245. Blue Bergs, 246. Surf on the Coasts of Bergs, 246. Scenic effects of Icebergs, 246. Appearance of the Packice, 248. Discolouration of Ice by Diatoms, 249. Gales of Wind amongst the Icebergs, 250. Snow Bow, 252. Whales Blowing, 252. Grampuses, 253. Birds amongst the Ice, 253. Antarctic Climate in Summer, 254.

CHAPTER XI.

VICTORIA. NEW SOUTH WALES.

Excursions into the Bush near Melbourne, 256. Opossum Snare, 257. Tracks of the Aborigines on Tree-trunks, 258. Town of Sandhurst, 259. The Highest Tree in the World, 260. Aborigines on a Government Reserve, 261. Ornithorynchus paradoxus, 262. Leaves of Australian Trees, why Vertically Disposed, 264. Fur Seal in the Open Sea, 265. Sydney Harbour, 266. The Blue Mountains, 266. Excavations in the Ground caused by Rain, 267. Shooting Opossums by Moonlight, 267. Fruit-eating Bats, 268. Hunting Bandicoots, 269. Browera Creek, 270. Intimate Relation of Land and Sea Animals, 271. Geological Import of this, 272. Medusæ in Fresh Water, 272. Kitchen Middens, 273. Drawings by Aborigines, 273. Handmarks, 275. Trigonia and Oestracion, 276.

CHAPTER XII.

NEW ZEALAND. THE FRIENDLY ISLANDS. MATUKU ISLAND.

Wellington, New Zealand, 277. The Rata Tree, 278. Kingfisher with Littoral Habits, 278. Peripatus, 279. Egg Capsules of Land Planarians, 279. The Vegetation of the Kermadec Islands, 280. Red coloured Muscles of the Shark, 281. Island of Eua, 282. General Appearance of the Island of Tongatabu, 282. Tongan Natives, 283.

Mode of Hairdressing, 284. Facial expression of the Natives, 284. A Pea-Jacket a Badge of Distinction, 285. Town of Nukualofa, 286. Dress of Tongan Women, 287. Getting Fire by Friction, 287. Deserted Plantations, 290. Fruit-bats Feeding on Flowers, 291. Herons, Tree-swifts, and other Birds, 291. Parasitic Algæ in Foraminifera, 292. Matuku Island, Fiji Group, 293. The Island an Ancient Crater, 293. Its Vegetation, 294. Encircling Reef, 294. Flocks of Lories, 295. Periophthalmus, a Fish Living on Land, 295. Living Pearly Nautilus, 297. Its Mode of Swimming, 297. Account of the Nautilus, by Rumphius, 299.

CHAPTER XIII.

FIJI ISLANDS.

Position and Area of the Islands of the Group, 301. Kandavu Island, 302. Grindstones for Stone Adzes, 302. Shooting Birds in the Woods, 303. Terrestrial Hermit Crabs, 304. Visit to a Barrier Reef, 306. Ovalau Island, 308. Excursion to Livoni, 308. Fijian Convicts, 309. Log Drum, 309. Native Hairdressing, 310. Kaava Drinking, 311. Buying Stone Adzes, 313. Excursion to Mbau Island, 314. Structure of the Island, 315. Na vatani tawaki, 316. Relics of Cannibalism, 318. Interview with King Thackombau, 319. Connection of Wooden Drums and Bells, 321. Excursion up the Wai Levu, 322. Sugar Plantations at Viti, 323. Freshwater Sharks, 325. Joe the Pilot, 325. Fijian Fortifications and Tombs, 326. A Chief's House and his Children, 328. A Missionary Meeting, 329. Various Modes of Painting the Body, 331. Grand Dancing Performances, 331. Primitive Origin of Music, Poetry, and the Drama, 333. Wesleyan Missionary, 335. Albino Native, 335. Congregation of Races at Levuka, 336. Fijian Modes of Expression, 336. Laughter, 337. Cicatrisation, 338. The Ula, 338. Particulars concerning Cannibalism, 339.

CHAPTER XIV.

NEW HEBRIDES. CAPE YORK. TORRES STRAITS.

Api Island, New Hebrides, 342. Fringing Reefs, 343. Proofs of Elevation, 344. Coral Living Detached, 344. Natives of Api, their Ornaments and Weapons, 345. Condition of Returned

Labourers, 345. Expression of the Emotions, 346. Raine Island, 347. Its Geological Structure, 347. Its Vegetation, 348. Nesting of Wide-awakes, Gannets and Frigate Birds, 349. Dead Turtles, 350. Somerset, Cape York, 350. Nests of White Ants, 350. Combination of Indian and Australian Features in the Vegetation, 351. Various Birds, 351. Habits of the Rifle Bird, 352. Birds Fertilizing Plants, 353. Camp of the Blacks, 354. Habits of these Natives, 354. Curious mode of Smoking, 356. Food of the Blacks, 357. They cannot Count Higher than Three, 358. Absolute Nudity of the Men, 359. Coral Flats, 360. Collection of Savage Weapons at Cape York, 361. Wednesday Island, Torres Straits, 361. Structure of Coral Flats, 362. Giant Clam, 362. Native Graves, 363. Booby Island, 363. A Halting Place for Birds during Migration, 364. Many Land Birds on an Almost Bare Rock, 364.

CHAPTER XV.

ARU. KE. BANDA. AMBOINA. TERNATE.

Appearance of the Aru Islands, 366. Trees Transplanted by the Waves, 368. Masses of Drift Wood, 368. Malay Language, 369. Ballasting a Guide, 369. Management of Clothes during Rain, 369. Back Country Natives, 370. Great Height of the Trees, 371. Nests of the Metallic Starling, 372. Parrots and Cockatoos, 372. Bird-winged Butterflies, 373. Shooting Birds of Paradise at Wanumbai, 375. Deposit of Lime in Streams, 378. Boat Crews from the Ke Islands, 379. Fungus Skin Disease, 379. Ke Island Dancing, 380. Houses at Ke Dulan, 381. Leaf Arrows, 381. Bird caught in a Spider's Web, 382. Ascent of the Volcano of Banda, 382. Algæ growing in the Hot Steam Jets, 383. Numerous Insects at the Summit, 384. Alteration in Sea Level, Marked on Living Corals, 385. Nutmeg Plantations, 386. Transportation of Seeds by Fruit-Pigeons, 386. Saluting at Amboina, 387. Danger to the Eyes in Diving for Corals, 389. Raised Reefs, 389. Myrmecodia and Hydnophytum, 389. Moluccan Deer, 390. Ternate Island, 390. Chinese and their Graves, 391. Sale of Birds of Paradise, 391. Ascent of the Volcano, 392. The Mountain Vegetation, 392. The Terminal Cone, 393. View from the Summit, 394.

CHAPTER XVI.

THE PHILIPPINE ISLANDS.

Zamboanga, Mindonao Island, 395. Paddy Fields and Buffalos, 395. The Lutaos and their Pile-Dwellings, 396. Pile-Dwellings on Dry Land, 398. The Ground Floor a late addition to the First Story, 399. Wide Distribution of Pile-Dwellings, 399. Their Possible Origin, 400. Dances Performed by the Lutaos, 401. Bamboo Jews Harp, 401. Lutao Canoe and Weapons, 402. Search for Birgus Latro, 403. Birds' Eggs hatched in the Sea Sand, 403. Alcyonarian Corals. Basilan Island, 404. Cart-wheels cut from Living Planks, 405. Galeopithecus and Flying Lizard, 406. Cebu Island, 407. Mode of Dredging up Euplectella, 407. Mactan Island, Raised Reef, 408. Large Cerianthus, 408. Trachytic Volcano at Camiguin Island, 409. Temperature at which Plants can Grow in Hot Mineral Water, 410. Manila-hemp Plantations, 411. Manila, 411. Shirt worn over Trousers, 411. Clothes Originally Ornamental only, 412. Half-hatched Ducks' Eggs Eaten, 412. Cock Fighting, 412. Sale of Indulgences, 414.

CHAPTER XVII.

CHINA. NEW GUINEA.

Hong Kong, 415. Pigeon English, 415. Chinese Method of Writing compared with European Methods, 417. Development of Chinese and Japanese Books from Rolls, 417. Plants colonizing a Pagoda, 419. Sights of Canton, 419. Chinese and English Examinations, and their subjects compared, 420. The Honam Monastery, 421. Chinese Floral Decorations, 421. A Chinese Dinner, 422. Dragons' Bones and Teeth, 423. Origin of Mythical Animals, 423. Chinese Account of the Dragon, 425. The last Dragon seen in England, 426. Use of Unicorn's Horn as Medicine in Europe, 426. Chinese and English Medicine compared, 428. Chinese Accounts of the Pigmies and of Monkeys, 428. English Mythical Animals, 430. The Sea Serpent, 430. Owls living with Ground Squirrel in China, 431. Off the Talaur Islands, 432. Driftwood off the Ambernoh River, New Guinea, 432. Animals Inhabiting it, 434. Humboldt Bay, 435. Signal Fires of the Natives, 435. Bartering at Night, 436. Numbers of Canoes, 437. Relative Prices of Native Property, 439.

Attempts at Thieving, 439. Modes of Expression, 440. Mode of Threatening Death by Signs, 441. Armed Boat Robbed, 442. Villages of Pile-Dwellings, 445.

CHAPTER XVIII.

THE ADMIRALTY ISLANDS.

History of Visits to the Islands, 449. Eagerness of the Natives for Iron, 451. Trade Gear, 451. Trading with the Natives, 452. Geological Structure of the Islands, 455. Orchids and Ferns overhanging the Sea, 455. Fern resembling a Liverwort, 455. Difficulties in Collecting Words of their Language from the Natives, 456. Their Methods of Counting, 457. Curious Mode of Expressing Negation, 457. Physical Characteristics of the Natives, 458. Hairiness of Races Compared, 459. Possible Signification of Moles, 459. Clothes, Hair Dressing and Ornaments of the Natives, 460. Tattooing and Painting, 463. Betel-Chewing and Food, 464. Houses, Temples, and Canoes of the Natives, 465. Their Implements and Weapons, 467. Artistic Skill of the Natives, 469. Their Musical Instruments, Dancing and Singing, 471. Their Polygamy, 472. Fortification of their Villages, 472. Wooden Gods, 473. Skulls and Hair in their Temples, 474. Their Religion, 474. Disposition of the Natives, 477. Their Fear of Goats and Toys, 477. Population of the Islands, 478. Domestic Animals, Birds and other Animals at the Islands, 478. Habits of Gar-Fish, 479.

CHAPTER XIX.

JAPAN. THE SANDWICH ISLANDS.

Tedious Voyage to Japan, 481. Jinriksha Coolies, 482. Worship of the White Horse, 482. Japanese Sight-Seers, 483. Consulting the Oracle, 483. Japanese Pilgrims, 484. Book Shops and Religious Shops, 484. River Embankments, 485. Rice Fields, 485. Houses of Wood and Paper, 485. English Bed-room Exhibited at the Exhibition, 486. Money Boxes, 487. Pilgrims and Priests, 487. Interest taken by the People in Tojins, 488. Cold Water Cure, 488. Painting of the Face in China and Japan, 489. Japanese Tattooing, 491. Japanese Modes of Expression, 482. Japanese Pictures and Theatres, 493. Barren

Appearance of the Sandwich Islands, 495. Honolulu, 495. Supremacy of American over Native Productions, 496. Principal Trees of Oahu Island, 497. King Kalakaua, 497. Hawaian Burials, 498. Visit to the Crater of Kilauea, 499. Ponds of Fluid Lava, 501. Mode of Formation of Pele's Hair, 502. Lava Fountains and Cascades, 502. Recent Eruptions, 503. Hawaian Hook Ornament, 504. Its Probable Religious Signification, 505. Hawaian Stone Club, 510. Affinities between New Zealand and Hawaian Art, 510. Inter-breeding on Isolated Islands, 512.

CHAPTER XX.

TAHITI. JUAN FERNANDEZ.

Death of Rudolph Von Willemoes Suhm, 513. Scientific Papers and Journals left by Him, 513. Papeete, 514. Excursion into the Mountains, 516. Fly-Fishing in a Mountain Stream, 516. Uses of the Wild Banana, 517. Vegetation Composed mainly of Ferns, 518. Camping at Night, 519. Tahitian Mountain Map, 520. Ascent to 4,000 feet Altitude, 521. Petrels Nesting at this Height, 521. Their Possible Influence in Distribution of Plants, 522. Ignorance of the Natives Concerning the Mountains, 523. Mode of Alternation of Generations in the Mushroom Coral, 524. Structure of the Millepora, 525. Structure of the Stylasteridæ, 528. Catching Land-Crabs, 535. Tahitian National Air, 536. Juan Fernandez, 537. Preponderance of Ferns, 537. Destruction of Trees, 538. Gunnera Chilensis, 538. Conspicuous Flowers, 539. Humming Birds of the Island, 539. Their Fertilization of Flowers, 539. Smallness of the Island Compared with the Number of Endemic Forms, 541. Endemic Palm, 541. Dendroseris, 542.

CHAPTER XXI.

CHILE. MAGELLAN'S STRAITS. FALKLAND ISLANDS. ASCENSION.

Valparaiso, 543. The Andes not Conspicuous, 543. Cattle lassoed in the Streets, 544. Excursion up the Uspallata Pass, 544. Leafless Mistletoe on the Leafless Cactus, 545. An Equestrian Hair Cutter, 546. Dead and Disabled Animals on the Pass, 547. Use of the Lasso in Robbery and Flirtation, 548. Cleverness of a Horse on a Mountain Path, 548. Fjords of

the Western Coast of Patagonia, 549.
Density of the Forest, 549. An Anchor
Broken, 550. Fuegians, 550. Wild
Geese at Elizabeth Island, 551.
Kitchen Middens, 552. The Falkland
Islands, 553. Visit to Port Darwin,
553. Scotchmen turned Gauchos, 554.
Chapinas and Tropijes, 554. Wild
Horses and their Habits, 555. Various
Modes of Handling Cattle in Different
Parts of the World, 557. Goose-Bolas
made of Knuckle Bones, 558. Flies and
Gnats with Rudimentary Wings, 558.
Skeleton of Ziphioid Whale, 559.
Fuegian Arrow-heads Scattered in the
Islands, 560. Habits of Jackass Pen-
guin, 560. Ascension Island, 561.
Land Crabs, 561. The Hatching of
Turtles' Eggs, 561. Shooting at Fly-
ing Fish, 562. Birds at Boatswain
Bird Island, 563.

CHAPTER XXII.

LIFE ON THE OCEAN SURFACE
AND IN THE DEEP SEA. ZOO-
LOGY AND BOTANY OF THE
SHIP. CONCLUSION.

Plants of the Ocean Surface, 566. Fauna
of the Sargasso Sea, 567. Protective
Colouring of Pelagic Animals, 568.
Variety of Pelagic Animals, 569.
Flight of the Albatross, 569. Flight
of Flying Fish, 570. A Pelagic Insect,
571. Pelagonemertes described, 572.
Phosphorescence of Pelagic Animals,
574. Giant Pyrosoma, 574. Uncer-
tainty as to Range in Depth of Pelagic
Animals, 575. The Depth of the
Oceans and Depressions on the Earth's
Surface, 576. Deep-Sea Dredging, 578.
Vast Pressure existing in the Deep
Sea, 579. Experiment showing this
made by Mr. Buchanan, 579. Condi-
tions under which Life Exists in the
Deep Sea, 580. Range of Plants in
Depth, 581. Food of Deep-Sea Ani-
mals, 581. Experiment on Rate of
Sinking of a Salpa, 582. Vegetable
and Animal Débris Dredged from
Great Depths, 583. The Deep Sea a
High Road for Distribution of Animals,
583. Deep-Sea Faunas and Alpine
Floras Compared, 585. Nature of
Deep-Sea Fauna a source of Disap-
pointment, 586. Remarkable Deep-
Sea Ascidian, 587. Localities specially
Rich in Deep-Sea Forms, 589. Rela-
tions of Deep-Sea Animals to One
Another, 590. Phosphorescent Light
in the Deep-Sea, 590. Colours of
Deep-Sea Animals, 591. Cockroaches,
Moths, Mosquitos, House-flies, Cric-
kets, Centipedes and Rats on board the
" Challenger," 592. Plants on board
the Ship, 594. Pet Parrot, Casso-
wary, Ostriches, Tortoises, Spiders,
Fur Seal, and Goat on Board, 594.
Adaptation to Sea Life, 596. Small-
ness of the Earth's Surface, 597. Slow
Rate of Travelling, 597. Man and
possibly Protoplasm existent on the
Earth alone, 598. Necessity for im-
mediate Scientific Investigation of
Oceanic Islands, 599.

LIST OF BOOKS AND PAPERS RELATING TO THE " CHALLENGER "
EXPEDITION, 601–606.

GENERAL INDEX.

LIST OF COLOURED PLATES.

ANTARCTIC ICEBERGS, to face Title-page.

VIEW OF PACK ICE FROM FORETOP, to face page 248.

TRACK CHART, WITH CONTOUR OF THE BOTTOM OF THE OCEAN, to be inserted at
the end of the book.

A NATURALIST ON THE "CHALLENGER."

CHAPTER I.

TENERIFFE, ST. THOMAS, BERMUDA.

Circumstances of the Voyage. Teneriffe. Cochineal Plantations. Excursion up the Peak. Trade-wind Cloud. Zones of Vegetation. Sunset seen above the Clouds. Rabbits and other Animals on the Peak. Peculiar Spider's Web. Catching Sharks off Sombrero Island, West Indies. Appearance and habits of Remora. Pilot Fish. Island of St. Thomas. Calcareous Seaweeds. Sea Urchins with Poisoned Spines. Burrowing Spider. Nest of Termites. Pelicans edible. Sand-box Tree. Defensive colouring of Spines of Cacti. Beach Conglomerate. Sea-beans. Bermuda. Calcareous Sand-rock. Caves. Vegetation. Peat. Boatswain Birds. Land Nemertine. Corals in Caves.

Circumstances of the Voyage.—H.M.S. "Challenger," a main-deck corvette, with auxiliary steam power, left Portsmouth on December 21st, 1872, for a voyage of three years and a half round the world. The object of her cruise was to investigate scientifically the physical conditions and natural history of the deep sea all over the world. The ship was with that aim specially fitted with sounding and dredging apparatus, and carried a scientific staff, appointed by the Lords of the Admiralty, and placed by them under the direction of Sir Charles Wyville Thomson, F.R.S., &c. I accompanied the expedition as one of the naturalists on this staff.

In consequence of the special nature of the mission, the sea voyages were tedious and protracted, the ship being constantly stopped on its course to sound and dredge. Since the results obtained by deep-sea dredging, even in most widely distant localities, were very similar and somewhat monotonous, all reference to them will be deferred to the end of this narrative ; where their natural-history aspects will be discussed shortly as a whole, and where oceanic animals and plants will also be treated of to some extent.

The voyage of the "Challenger" occupied three years and

B

155 days, and out of this period about 520 days, or portions of these, were available for excursions on shore. A very large proportion of the time in harbour was necessarily spent at places where dockyards and workshops were available for repairs to the ship. The stays made at less-frequented places of especial interest to the naturalist were comparatively short. This circumstance should be borne in mind by the reader.

After stopping at Lisbon, Gibraltar, and Madeira, which latter island was afterwards visited a second time, and will be referred to in the sequel, the ship reached Teneriffe, one of the Canary Islands, and anchored off Santa Cruz, the chief town of the island, on February 7th, 1873.

Teneriffe, Canary Islands, February 7th to 14th, 1873.—The most striking feature in the natural vegetation of Teneriffe is the *Euphorbia canariensis.* The fleshy prismatic branches of this plant are devoid of leaves, have a blueish-green colour, and are perfectly straight and perpendicular, being disposed side by side, and 10 or 15 feet in height. The plant is abundant all over the rocks at a low elevation, and resembles a cactus in appearance. It has an abundant milky juice, which is very acrid and poisonous. Of the introduced vegetation, the plantations of the broad-lobed cactus (*Opuntia*), employed for the raising of the Cochineal insect, are curious. The crop of insects was, in the month of February, just being started on the plant, that is to say, the female insects were being placed upon the leaf-shaped lobes of the plant to lay their eggs, and start a fresh brood. The females are, when thus put out at the beginning of the season, held on to the plants by means of white rags tied round the lobes. Hence the fields, when seen at a distance, look as if they contained some crop bearing a continuous sheet of large white blossoms. I was greatly puzzled by them when looking at them as the ship was approaching the island. The island is so steep and rocky that it has been terraced for purposes of cultivation, and nearly every available spot has been treated in this manner.

I accompanied a party on an excursion up the Peak. The way led from Santa Cruz, through the Cochineal fields, and up a steep but well-engineered road, planted with tamarisk trees to the summit of the central ridge of the island. Here was passed a dilapidated town, thoroughly Spanish in its architecture, with some fine houses in it in a ruinous condition. The central square of the town was overgrown with weeds, and its streets mostly covered with grass ; but so are many in the capital, Santa Cruz, itself. On the way, droves of mules, ponies, and donkeys were passed, laden with country produce. The countrymen wear a peculiar dress, black trousers reaching only to the knee, and an ordinary blanket of the natural colour of the wool, drawn into pleats at one end to go round the neck, and worn over the shoulders as a cloak. If the blanket were dyed of some dark or bright tint the dress would not look very remarkable ; but its dirty-white colour has a strange appearance. The countrywomen have very fine figures and are most of them very handsome. We passed through another town where a private collector has a museum containing a number of mummies, skulls, and relics of the Guanches, the ancient inhabitants of the Canaries. The " gabinète," the owner of which was absent, was in a somewhat decayed condition, and was a sort of general collection of curiosities, a survival of the old Raritätenkammer, which is the parent of modern more select collections, just as the West African fetisch house may be regarded as the primitive and savage representative of the Raritätenkammer. Man seems to be almost the only mammal that collects and stores uneatable objects. Amongst birds, on the other hand, the collecting instinct is widely spread, as witness magpies and Bower-birds,* and even Penguins, one of which collects variously-coloured pebbles. It will be a great pity if the Guanche remains, contained in the Teneriffe Gabinète, do not reach some good European museum.

From the neighbourhood of this second town was obtained the first view of the far-famed Peak, " Pico de Teyde." The middle part of the mountain was concealed by a dense bank of white clouds, the condensed vapour of the trade wind. Beneath, a broad valley stretching down to the bright blue sea, with its snow-white edging of surf, was thrown partly into deep shadow by the cloud-bank, partly lit up by the bright hot sun. The sun

---

* O. Beccari, " Le Capanne ed i Giardini del *Amblyornis inornata.*" Ann. del Mus. Civ. di St. Nat. di Genova, Vol. IX, 1876–7.

shone brilliantly upon the snowy peak of the mountain, high up
in the sky above the clouds.   On the shore lay the town of

PEAK OF TENERIFFE FROM THE ROAD ABOVE OROTAVA.
(From a sketch by the Author.)

Orotava, from which the ascent was to be made.   The English
vice-consul at Orotava, who kindly made arrangements for the
trip, told me that the growth of the vine in Teneriffe was fast
being supplanted by the cultivation of Cochineal ; 2,000 pipes
only were being produced around Orotava, whereas 200,000
were formerly made.   He expected, however, that since Cochineal was falling in price, the wine trade would revive.   The
Canary wine is certainly of most excellent flavour.

The route up the mountain lay up a long sloping ridge,
which leads to the base of the actual cone of the Peak.   This
ridge is bounded by a precipice on the side facing Orotava.   The
villagers tried to dissuade the party from going farther after we
had ascended about 2,000 feet, saying that we should be frozen
to death.

The well-known zones of vegetation of the Peak of Teneriffe
are not very well defined on the route which we adopted.   The
limit of cultivation was reached at about 3,000 feet, at which
height corn of some kind was just springing up, and we passed

above this into a zone covered with a tree-like heath (*Erica arborea*). This heath continued for about 2,000 feet, and then ceased abruptly, and we came, higher up, amongst large blueish-green bushes of a sort of broom (*Spartocytisus nubigenus*), called by the natives " Retama," amongst which we pitched our tent, at an elevation of 6,500 feet. Above the Retama, a small violet (*Viola teydeana*) is said to extend up to 10,000 feet, and above this all is barren. The pine (*Pinus canariensis*) which grows on some parts of the mountain is not seen on the usual track of ascent. A halt was made amongst the heath for lunch, and plenty of water-cresses were found growing in a spring. We had to carry water up with us from this spring, since there is no water to be obtained above, except by melting snow. The porous volcanic ashes soak up all the water yielded by the natural melting of the snow above, and there is no place where any can be gathered.

At about 4,000 feet elevation we went through a dense bank of cloud, formed by the trade wind, a similar one to that which was seen from below on the day before, and which had hidden the middle of the mountain from our view, but not the same, for in the early morning there had not been a cloud in the sky. The bank formed at about mid-day. At our camp, far above this cloud-bank, the sun shone brightly, until about six o'clock in the evening, when it began to disappear, and the air, which had been almost too hot, became suddenly cold, the temperature going down almost to freezing point.

We enjoyed a very extraordinary sunset effect. The upper surface of the cloud-bank stretched away like a snow-white billowy sea beneath us in every direction, hiding the actual sea from our sight entirely, but just allowing us a glimpse of the far-off island of Palma, which appeared as a purple streak at the edge of the cloud horizon. As the sun went down the clear sky beyond the white motionless cloud-bank became tinged of a brilliant orange colour, and over it there shot out from the descending sun a fan of pale crimson streamers deeply tinted at their base, and gradually fading off into the dark blue sky above but visible nearly to the zenith. Beyond the great cloud-bank more distant streaky clouds, lit up of a brilliant violet, formed a sort of background to the scene. Some amongst these little distant clouds from time to time assumed fantastic shapes, and once we were almost persuaded that we were looking upon the sea in the distance with two very far-off ships upon it, but it was merely a delusion. The sea was entirely shut out from our view, except once for a few instants when a small rift in the cloud-bank occurred and gave us a momentary glimpse of the rippling surface far below, a sort of vista view dimmed by the misty frame through which it was seen.

All the while the snowy peak itself was perfectly cloudless and stood out clear and sharp against a deep blue arctic-looking sky. Soon the sunlight faded and the moon came out bright, and the peak glistened in its light, which was strong enough for me to read by easily. The view of our tent and camp fire amongst the dark broom bushes with the moonlit snowy peak in the background, fronted by some dark ridges of lava, was most picturesque.*

We set fire to some of the large Retama bushes and soon had a tremendous blaze, the bushes fizzing and crackling loudly in the flare, the flames shooting high up into the air so that they were seen at Orotava, and even at Santa Cruz. The ground froze on the surface around our tent during the night, the thermometer standing at 30° F. just before sunrise.

I walked from the camp to the Canadas—a remarkable plain covered with scoriæ, and shut in on nearly all sides by a perpendicular wall of basaltic cliff. From this plain of vast extent the present terminal cone of the mountain rises. The Canadas represents an ancient and much larger crater in the centre of the remnant of which the more modern smaller peak has been thrown up. The bottom of the Canadas is dotted over with the Retama. The ground was devoid of any other vegetation. I was surprised to find that rabbits were tolerably abundant in the Canadas. I saw several but could not manage to get a shot as they were wary. They feed on the Retama. They have no holes, but live in any chance crack or hole in the rock

---

* For an account of the Peak of Teneriffe and its cloud phenomena, see C. Piazzi Smyth, F.R.S., &c., "Teneriffe : an Astronomer's Experiment." London, Reeve, 1858.

or under the bushes; hence I could not trap them, though I took traps with me for the purpose. They are small. I obtained in Orotava a stuffed specimen of a black variety with a white spot on the forehead, which is occasionally found. Of birds in the Canadas I saw only a lark and a warbler (*Sylvia*), and of lower animals I found only a Lepisma and a Centipede (*Scolopendra*) which were very abundant under the blocks of pumice.

The radiant heat of the sun was extremely powerful on the arid plain of the Canadas. We had no guides, and our mule drivers had left us. All refused to accompany us at this season of the year to the top of the peak. We therefore ascended only to a height of about 9,000 feet, the last 200 feet of which was climbed over snow. Here we watched the often described struggles of the opposing winds, the trades and anti-trades, as shown by the eddying and twisting of the wreaths of cloud.

In the neighbourhood of the camp at 6,500 feet, winter was evidently still in force as far as the animals were concerned. All the spiders and beetles I could find there were under stones, apparently hybernating. I was astonished to find at this altitude a Gecko (*Tarentola?*) also hybernating, coiled up in a hole under a stone. This lizard has a long range in altitude, since I found another specimen close to sea level.

After two nights we moved our camp to a spring at about 3,500 feet altitude amongst the Arboreal Heath, on the verge of the precipice bounding the ridge by which we had ascended. Here it was much warmer at night, and at daybreak the temperature was only as low as 45° F. But we had descended within the cloud-bank and had heavy rain, and should not have succeeded in lighting a fire for cooking had we not been helped by a mountain shepherd who was evidently well accustomed to setting a fire going in the rain, and soon got our kettle to boil. He was a fine powerful man and very honest and obliging, as were all the peasants with whom we came in contact. Stimulated with a shilling he turned collector, and soon returned with boxes full of snails and beetles. The steep side of the ridge overlooking Orotava is covered with a luxuriant vegetation of laurels, heaths, and ferns, and is very different in this respect from the comparatively barren surface of the slope above. A finch (*Fringilla teydeana*) peculiar to the island of Teneriffe, is to be obtained only in some pine woods near Orotava, and is rare.

In the Cochineal plantations a spider (I believe an *Epeira*) is very common, which makes a horizontally extended web, composed of fine square meshes. The web is supported by suspending threads in the midst of a globular labyrinth of irregularly disposed fibres. In the centre the horizontal net is drawn upwards into a short conical tube, at the end of which is an opening. The female always occupies a position immediately over this hole, which is apparently intended to allow of easy access to either side of the net. The egg bags are suspended in a vertical line immediately over the opening, and are often as many as four in number. In those I examined, the uppermost bag always contained fresh eggs, the lower fully developed young, and the others two intermediate stages. The male lives in the lower part of the irregular globular mass, and is very much smaller than the female, but is marked with brilliant silver patches on the abdomen.

In one of the churches at Santa Cruz is a flag taken by the Spanish from Nelson, and there preserved as a trophy. The ship left Teneriffe on February 14th, and reaching the trade winds on February 20th, sailed pleasantly before them across the Atlantic to the Virgin Islands.

Off Sombrero Island, March 15th, 1873.—Whilst dredging was proceeding off the Island of Sombrero, on the approach to St. Thomas, two sharks (*Carcharias brachiurus*) were caught with a hook and line. One of these had the greater portion of one of its pectoral fins bitten off, there being a clean semicircular cut surface, where the jaws of another shark had closed and nipped it through. Attached to the sharks were several "Sucker-fish" (*Echineis remora*), as commonly is the case. Sometimes these "Suckers" drop off as the shark is hauled on board. Sometimes they remain adherent, and are secured with their companion. In this case four out of six "Suckers" were obtained with the two sharks. They were seen to shift their position on the sharks frequently as these struggled in the water fast hooked.

The Remora is a fish provided, as a means of attachment, with an oval sucker divided into a series of vacuum chambers by transverse pleats. The sucker is placed on the back of the fish's head. The animal thus constantly applies to the surfaces to which it attaches itself, such as the shark's skin, its back. Hence the back being always less exposed to light is light-coloured, whereas the belly, which is constantly outermost and exposed, is of a dark chocolate colour. The familiar distribution of colour existing in most other fish is thus reversed. No doubt the object of the arrangement is to render the fish less conspicuous on the brown back of the shark. Were its belly light-coloured as usual, the adherent fish would be visible from a great distance against the dark background. The result is that when the fish is seen alive it is difficult to persuade oneself at first that the sucker is not on the animal's belly, and that the dark exposed surface is not its back. The form of the fish, which has the back flattened and the belly raised and rounded, strengthens the illusion. When the fish is preserved in spirits the colour becomes of a uniform chocolate and this curious effect is lost. When one of these fish, a foot in length, has its wet sucker applied to a table and is allowed time to lay hold, it adheres so tightly that it is impossible to pull it off by a fair vertical strain.

Fishing for sharks was a constant sport on board the ship when a halt was made to dredge anywhere within a hundred miles or so of land in the tropics. Sharks were not met with in mid-ocean. Mr. Murray* examined these sharks thus caught, and reports that they all, whether obtained in the Atlantic or Pacific Ocean, belonged to one widely distributed species, excepting one other kind obtained off the coasts of Japan. The hammer-headed shark (*Zygœna malleus*) was taken by us only with a net on the coasts.

The sharks were often seen attended by one or more Pilot-fish (*Naucrates sp.*) as well as bearing the "Suckers" attached to them. I often watched with astonishment from the deck this curious association of three so widely different fish as it glided round the ship like a single compound organism.

* J. Murray, "Proc. R. Soc.," No. 170, 1876, p. 540.

The sharks, as a rule, were not by any means so easily caught as I had expected. Frequently they were shy and would not take a bait near the ship, though they never failed to bite if it was floated some distance astern by means of a wooden float. It is always worth while for naturalists to take what sharks they can at sea, since their stomachs may contain rare cuttle fish which may not be procured by any other means. The sharks caught were always suspended over the screw well of the ship. It was amusing on the first occasion on which one was got on board, sprawling and lashing about on the deck, to see two spaniels belonging to officers on board put their bristles up and growl, ready to fly at the fish. The dogs would probably have lost their heads in its mouth if not driven back.

Sometimes the sharks were bold enough and would bite at a bit of pork hung over the ship's side on the regulation shark hook which is supplied to ships in the navy, and which is an iron crook as thick as one's little finger, and mounted on a heavy chain. No shark was hooked during the voyage which was large enough to require such a hook. Nearly all the sharks caught and seen were very small, from five to seven feet in length. The largest obtained was, I think, one netted at San Jago, Cape Verde Islands, which was 14 feet in length. Large sharks seem scarce. I was disappointed, and had expected to meet with much larger ones on so long a voyage. The largest shark known seems to be *Carcharodon rondelettii* of Australia. There are in the British Museum the jaws of a specimen of this species which was 36½ feet in length. (Günther, " Catalogue of Fishes.") The "Challenger" dredged in the Pacific Ocean in deep water numerous teeth of what must be an immensely large species of this genus. The great Basking-shark (*Selache maxima*), a harmless beast with very minute teeth, ranging from the Arctic seas to the coast of Portugal, has been known to attain a length of more than 30 feet.

Sharks occasionally seize the patent logs, which being of bright brass and constantly towed, twirling behind ships, no doubt appear to them like spinning baits intended for their use. The pilot fish often mistakes a ship for a large shark, and swims for days just before the bows, which it takes for the shark's snout. After a time the fish becomes wiser and departs, no doubt thinking it has got hold of a very stupid shark, and hungrily wondering why its large companion does not seize some food and drop it some morsels. The "Suckers" often make the same mistake and cling to a ship for days when they have lost their shark. I fancy that porpoises and whales, when they accompany a ship for several days, think they are attending a large whale. A Hump-back whale followed the "Challenger" for several days in the South Pacific.

Island of St. Thomas, March 16th to 24th, 1873.—The island of St. Thomas, one of the Virgin Islands, or Danish West Indies, was reached on March 16th. As the ship steamed in towards the harbour, Frigate birds soared high over-head with their long tail feathers stretched widely out. A number of brown pelicans (*Pelicanus fuscus*) were flying at a moderate height near the shore, and every now and then dashing down with closed wings into the water on their prey like gannets, their close allies. Often several of the birds dashed down together at the same instant.

The island of St. Thomas itself, as well as its outliers, is covered with a wild bush growth, which at first sight might perhaps be taken for original vegetation, but which is composed of plants which have overrun deserted sugar plantations. It is only in a few remote parts of the island that any original forest exists and in small streaks of broken ground bordering the watercourses. The whole of the country in the island of St. Thomas and in all the immediately adjoining islands was cropped with sugar-cane until the emancipation of the slaves in 1848. Since that time the ground has been allowed to run wild. There was only one estate partly under cultivation at the time of the ship's visit, and the owner of it, Mr. Wyman, told me that he made no sugar, but found sufficient sale for his canes in the raw state to be cut up and sold for chewing. Mr. Wyman was nearly ruined by the emancipation, and said that the planters received only 50 dollars per head compensation for the loss of their slaves, and that after the lapse of three years' time.

All about the shores in every small bay were to be seen wrecks of vessels of all kinds, and in various stages of dilapidation, which had been wrecked by the hurricanes, for which

St. Thomas is notorious, and close to our anchorage was a portion of a large iron dock which had been sunk before ever it could be used. Behind the town of St. Thomas are hills rising to a height of 1,400 feet at their highest point.

I landed at one of the many wooden jetties amongst numerous negroes of both sexes lolling about and chewing sugar-cane, their constant occupation. The shore is covered with corals bleached white by the sun, and amongst these lay quantities of calcareous seaweeds (*Halimeda opuntia* and *H. tridens*), branching masses composed of leaf-shaped joints of hard calcareous matter articulated together. These were all quite dry and bleached white, and hard and stiff, like corals. Seaweeds belonging to two very different groups of algæ thus secrete a calcareous skeleton, Halimeda and its allies belonging to the Siphonaceæ, green algæ, and Lithothamnion and allied genera belonging to the Corallinaceæ, which are red coloured algæ. These lime-secreting algæ are of great importance from a geological point of view as supplying a large part of the material of which calcareous reefs and sand rocks are built up. At St. Thomas the Siphonaceæ are especially abundant, whereas at other places, as at St. Vincent, Cape Verde Islands, the Corallinaceæ appear to supply most of the calcareous matter separated from the sea water by plants.

The rise and fall of the tide at St. Thomas is only about a foot; yet along the very margin of the water I found plenty of animals living, some of them only just awash. Sea urchins (*Diadema antillarum*), with extremely long sharp spines, were very common. The spines penetrate a bather's foot or hand with the greatest facility, and breaking off leave a very unpleasant wound. In gathering specimens I got wounded in the finger, though I took great care ; so well are the animals protected. The animals keep their long spines in constant motion, so that it is very difficult to avoid being pricked if one tries to handle one. The wound produced by the spines is apt to fester, but there appears to be no poison on the spine. In the case, however, of another genus of sea urchins which I dredged in abundance in shallow water on the Philippine coast, and in which the short spines are hollow and tubular at their extremities, a definite poison certainly exists. Probably there is a poison gland in the tube. A sharp stinging pain, like that produced by the sting of a wasp, but not quite so intense, is felt at the instant when one of these spines pierces the flesh, and the pain lasts for about five minutes. These urchins are peculiar, because they have a perfectly flexible test or shell, and are, I believe, of the genus *Asthenosoma* (Grube). Allied forms are common in great depths, but in these I never experienced so marked a stinging effect as in the case of the shallow-water ones.

Large Chitons, three inches in length, were abundant along the shore of St. Thomas, and a very large Annelid with glistening yellow setæ (*Eunice*), was a constant feature about the water's edge, crawling over the rocks. In dredging in shallow water most of the seaweeds obtained were of a brilliant green colour,* and amongst these lived a crab and a Squilla which were of exactly the same shade of green, evidently for protection and concealment.

There is only one kind of Humming-bird at St. Thomas. It is very common, and constantly to be seen hanging poised in the air in front of a blossom or darting across the roads. It is remarkable how closely Humming-birds resemble in their flight that of Sphinx moths, such as our common Humming-bird Sphinx, named from this resemblance. There are in their flight exactly the same rapid darts, sudden pauses, and quick turns, the same prolonged hovering over flowers. The most conspicuous bird is called commonly in the island " Black-witch " (*Crotophaga ani ?*). These birds are usually to be seen in flocks of three or four, in constant motion amongst the bushes, and screaming harshly when they apprehend danger. The birds behave very much like magpies. They are somewhat smaller than the English magpie and black all over. They belong structurally to the family of the cuckoos (*Cuculidæ*).

A large ground spider (*Lycosa*) is very abundant in the island, inhabiting a hole in the ground about six inches in depth, and from half an inch to an inch in diameter, and with a right-angled turn at the bottom to form a resting chamber for the spider. Some negro boys dug the spiders out for me. They

* *Udotea cyathiformis, U. conglutinata,* and *U. flabellata,* and others.

said that their bite was poisonous, and that they fed on lizards, leaving their holes at night to search for them. The boys soon grubbed one out with a knife, a great heavy venomous-looking brute about three inches across. It bit savagely at my forceps. The holes of these spiders were so common, that on one tolerably clear patch of about an acre in extent, they were dotted over the entire area at about one or two feet distance from one another. I noticed the holes at once, and was astonished when the boys told me they were spiders' holes.

A species of White-ant (*Termite*) is very common, which makes large globular nests as much as two feet in diameter, and which are perched high up in the fork of a tree. The nests are made of a hard brown comb. From the bottom of the tree covered galleries about half an inch in breadth lead up on the surface of the bark to the nest, looking like long narrow brown streaks upon the trunk of the tree. The galleries usually follow a some-what irregular course up the trunk to the nest, reminding one of the curious deviations which are always to be seen in footpaths, cut out by people walking across fields, in their endeavours to go straight from one point to another. The galleries, or rather tubular ways, for they have bottoms to them, are made of the same tough brown substance as the nests, and are cemented firmly to the bark. Though they are so broad in order to allow numerous ants to pass and repass, they are only high enough for the ants to walk under. I broke one of these galleries, and a number of soldier Termites came out and began biting my hands, hardly making themselves felt, but as brave as if they had a sting. I had to break a considerable length of the gallery before I got to any of the working Termites, as they had retired from the scene of danger. A species of Peripatus* is found in St. Thomas, but I did not succeed in meeting with any. An Agouti, a species of rodent (*Dasyprocta*), occurs in the island, and Mr. Wyman told me that it was common in the gullies near his sugar plantation.†

* See Chapter VI.

† Mr. Wallace, "The Geographical Distribution of Animals," London, Macmillan, 1876, Vol. II, p. 63, in the account of the mammals of the West Indies, says an Agouti inhabits "*perhaps* St. Thomas." There seems to have been doubt about the matter.

I went out on a shooting excursion to the opposite side of the island in pursuit of wild goats. The only game we brought back was a wild common fowl which I had shot in the bushes. Goats, pigs, guinea fowl, and the domestic fowl breed in the wild condition in various parts of the island, being sprung, as I was told, in most instances, from stock which has escaped and been scattered during hurricanes. The ferine fowls are very wary like their progenitors, the Indian Jungle-fowl, and are not at all easy to shoot. We sat down to lunch on the shore. Flights of the brown pelicans (*Pelicanus fuscus*), kept passing over our heads, flying always almost exactly over the same spot on their way from one feeding ground to another. We shot a number of them as they flew over at the desire of the German overseer of the farm where we had left our horses, who wanted the birds for eating. I should have thought a pelican to have been, next to a vulture, almost the least palatable of birds, but the man said they were very good. There were about 300 tame goats at the farm, and a few cows. The milk was sent into the town every morning in wine bottles and fetched about eighteen pence a bottle.

Large silk cotton trees (*Eriodendron*) are common, growing along the road-sides in St. Thomas. These trees are shaped something like walnut trees, but have a rough bark. They bear large green pods full of a substance like cotton. Perched in the forks and all over their branches are numerous epiphytes of the pine-apple order (*Bromeliaceæ*). On the far side of the island I saw several "Sand-box" trees (*Hura crepitans*). The tree is one of the *Euphorbiaceæ*, allied to our Spurges, and has a poisonous irritant juice ; but its most remarkable peculiarity is its fruit. A number of seed capsules, shaped like the quarters of an orange, are arranged together side by side as in an orange, so as to form a globular fruit. When the fruit has become quite ripe and dry, suddenly all the capsules split up the back, opening with a strong spring, and the whole fruit flies asunder, scattering its seeds for a distance of several yards, and making a noise like the report of a pistol. I gathered one of the fruits which is called commonly "Sand-box," because it was formerly used for holding sand to sift over writing instead of blotting-paper. It was boiled in oil when gathered and this prevented its flying asunder. The fruit I gathered went off with considerable violence when I touched it one day on board ship after it was dry, but it did not make much noise.

Another Euphorbiaceous tree, the Manchineel, grows in St. Thomas, and its juice is almost as poisonous as that of the "Sand-box" tree. The fable ran that if a man allowed rain to drop off its leaves on to his skin, his skin would be burned and inflamed by it.

I landed one day on one of the small outliers of St. Thomas, Little Saba Island, about a mile and a half distant from the main island. A puffin (*Puffinus sp.*) was nesting in holes amongst the grass, laying a single large white egg. The birds allowed themselves to be caught in the nest with the hand. Our spaniels kept bringing them to us, retrieving them with great delight. The island was covered with thorny cactuses. It was impossible to avoid their prickles, and I got covered with them when in pursuit of wild goats and pigeons. There were four kinds of cactuses, a prickly-pear (*Opuntia*) with spines three-quarters of an inch long ; a quadrangular stemmed cactus, like the most familiar one in green-houses ; a cactus with rounded ribbed stem, growing in candelabra-like form (*Cereus*), and a large dome-shaped cactus, a foot and a half high and bearing a crown of small red flowers (*Melocactus*).

The spines must be a most efficient protection to the cactus from being devoured by large animals. I have often noticed that if one approaches one's hand slowly towards some of the forms with closely set long spines, doing it with especial care to try and touch the end of one of the spines lightly without getting pricked, one's hand always does receive a sharp prick before such is expected, the distance having been miscalculated. There seems to be a special arrangement in the colour of the spines in some cases, possibly intended directly to bring about an illusion, and cause animals likely to injure the plant to get pricked severely before they expect it, and thus to learn to shun the plant. Whilst the greater length of the spines next the surface of the plant is white, the tips are dark-coloured or black. The black tips are almost invisible as viewed at a good many angles against the general mass as a background. The spines look as if they ended where the white colouring ends, and the hand is advanced as if the prickles began there, and is pricked suddenly by some unseen black tip. The experiment is easily tried in any cactus house at home.

In the beach of Little Saba Island there was being formed a reddish sandstone conglomerate rock composed of the *débris* of the rock of which the higher parts of the island consist, cemented together by calcareous matter derived from the corals, and calcareous sand. This rock, which was hard and compact, contained embedded in it plenty of the various corals from the beach and large Turbo shells (*T. pica*) with their nacre quite fresh in lustre, and their bright greenish colour unimpaired.

Large examples of these Turbo shells, as much as two inches in diameter at the base, are in St. Thomas carried up far inland by terrestrial Hermit-crabs. I saw a large number of them amongst the bush at an elevation of 1,000 feet, some of them with the crabs in them, many empty. These large heavy sea shells occurring in abundance at great heights, puzzled geologists until it was found that they were carried up by the crabs.

On the shore at Little Saba Island grew a number of plants of *Guilandina bonduc*. This plant bears a pod covered with prickles which contains nearly spherical beans of about the size of a hazel nut, which have a perfectly smooth, as it were, enamelled surface, and are flinty hard. These seeds float, and are carried by ocean currents to distant shores, and are in Tristan da Cunha and Bermuda known as "Sea-beans," and supposed to grow at the bottom of the sea. Don Jose de Canto showed me one found in the Azores.

The coral reefs of St. Thomas are remarkable for the large size and luxuriant growth of certain corals upon them, especially two species of the genus madrepora named from their resemblance to antlers, *Madrepora cervicornis* and *M. alcicornis*. I saw at Little Saba Island, a Brain-coral which measured four feet in diameter of the base and three feet in height.

A list of the flowering plants of St. Thomas, and other information, is given in " A Historical Account of St. Thomas, W.I." By J. P. Knox. New York, Charles Scribner, 1852.

Bermuda, April 5th to 21st, and May 27th to June 12th, 1873.
—Bermuda is entirely a coral island, that is to say, the complete
mass of the island now above water, and that below sea level, as
far at least as excavations which have been made have extended,
has been brought together by the agency of lime-secreting animals
and plants, aided by the winds and waves, and alterations in the
height of the sea-bed. It is the most distant coral island from
the equator, lying about 9° of latitude north of the Tropic of
Cancer, in about the same latitude as Madeira, which island has,
however, no coral reefs. It is distant from Cape Hatteras, the
nearest point of the American coast, about 600 miles.

Bermuda consists of a series of islands, some very small
indeed, others several miles in length, there being, it is said, an
island for every day in the year. The islands are disposed in an
irregular semicircle, and the larger ones of the chain are narrow
and elongate in form. This semicircle or rather semiellipse is
completed below water, or made into an entire atoll shape by a
series of coral reefs, as may be seen by a glance at the chart. A
few narrow and winding passages lead in through the reefs to the
harbours of St. George's, Ireland Island, and Hamilton the
capital town. The highest point is only about 300 feet above
the level of the sea.

The islands are almost entirely composed of blown calcareous sand, more or less consolidated into hard rock. In
several places, and especially at Tuckers-town and Elbow Bay,
there exist considerable tracts covered with modern sand dunes,
some of which are encroaching inland upon cultivated ground,
and have overwhelmed at Elbow Bay a cottage, the chimney of
which only is now to be seen above the sand. The constant
encroachment of the dunes is prevented by the growth upon
them of several binding plants, amongst which a hard prickly
grass (*Cenchrus*) with long, deeply-penetrating root fibres, is the
most efficient, assisted by the trailing *Ipomœa pes capræ*
When these binding plants are artificially removed, the sand at
once begins to shift, and the burying of the house and the
present encroachment at Elbow Bay are said to have originated
from the cutting through of some ancient sand-hills for military
purposes.

The sand is entirely calcareous and dazzling white when seen in masses. When examined closely, in small quantities, it is seen to consist of various-sized particles of broken shells. By gathering samples from the shores where the material of which the sand is formed is first thrown up, and selecting portions where eddies of the wind have left the heavier particles together, a sand full of large fragments of shell, and containing even many whole shells of smaller species, may be obtained, and from the examination of these an accurate conclusion may be arrived at as to the main constituents of the finer more comminuted sand, which is driven inland by the wind, blown up into the dunes, and from which the whole island above water has been formed.

The sand may be seen to be made up in by far its greater part of the shells of Mollusca. Species of *Tellina*, *Cardium*, and *Arca* contribute most largely to compose the mass, together with large quantities of pink-coloured fragments derived from a *Spondylus*, which is common about the islands. A few Gasteropodous shells contribute fragments, and a considerable number of Foraminiferous shells occur in the sand, and no doubt careful examination would reveal the presence of fragments of tubes of *Serpulæ*, corals, calcareous algæ, *Bryozoa*, and *Cirrhipedc* shells; but there can be no doubt that by far the greater mass is derived from the shells of Mollusca.* Thus, although the foundations of Bermuda, and its natural breakwaters and protections, without which it would not exist, are formed by corals, the part above water is mostly derived from another source, and even below the water the same is the case for some distance, for the same beds of sandstone were met with in an excavation carried to a depth of 50 feet.

The shells, more or less broken, are thrown up upon the beach, and there pounded by the surf. As the tide recedes, the resulting calcareous sand is rapidly dried by the sun, and the

* It would be of great interest to determine by careful microscopic examination, what are the relative percentages of the very various calcareous structures composing the calcareous sands of coral islands in different parts of the world. I collected specimens of all the calcareous sands accessible during the voyage of the "Challenger" with that object. They vary very much in composition, some being mainly Foraminiferous.

finer particles are borne off inland by the wind, to be heaped up into the dome-shaped dunes. The rain, charged with carbonic acid, percolates through the dunes, and taking lime into solution, re-deposits it as a cement, binding the sand grains together.* Successive showers of rain, occurring at irregular intervals, some charged more, some less highly, with carbonic acid, and forming each a crust on the surface of the dune of varying thickness, produce a series of very thin, hard layers in the mass of sand, alternating with seams of less consolidated and sometimes quite loose sand. Crusts of consolidated sand are to be observed commonly on the surfaces of fresh sand dunes. These layers or strata of the hardened sand follow in form the contour of the dunes, and thus, where these have been perfect domes or mounds, dip outwards in all directions, with curved surfaces from a central vertical axis. Such an arrangement is constantly to be seen where sections of the older rocks are exposed. I saw especially good instances of it in a small island, near Castle Island in Harrington Sound. Where banks or long rounded ridges of sand have been formed, strata following the surfaces of these in inclination are produced.

All kinds of curious irregularities in arrangement are to be found in the bedding of the strata, resulting evidently from the encroachment of one dune upon the edge of another, or the action of various eddies of wind, or the burying of a small dune in the edge of a larger one. In some cases, an already hardened

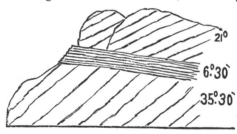

STRATA OF SAND ROCK, CASTLE HARBOUR, BERMUDA.

dune, after having suffered denudation by the action of the waves, has become buried in a more recent sand mound, and

* The process is described by Jukes in his account of Raines Islet. "Voyage of the 'Fly,'" p. 339, and elsewhere.

this process may have been repeated several times, as the accompanying diagram, showing the arrangement of bedding in some rocks at Castle Harbour, will show. I saw no rock in Bermuda with an inclination in its bedding of more than 35° 30', which is not much more than the slope of some of the sand-hills.

Dana terms this calcareous sand-rock, " Drift sand-rock."* Nelson terms it "Æolian formation" in his account of the geology of the Bermudas.† Jukes observed that in Heron Island the main strata of calcareous rock composing the island dipped outwards from the longitudinal axis of the island towards the shore, north and south, with an inclination of from 8° to 10°, and Nelson observed similar dispositions of the strata at Bermuda.

The rock of Bermuda presents all degrees of consolidation, from beds of mere unagglutinated friable sand to extremely hard and compact stone. The main component rock is a good deal softer than Bath stone. A much harder rock occurs at two places in the islands only, and is quarried for the construction of forts. The red fragments of *Spondylus* shell are especially well preserved in it. A bed of lignite was found at a depth of 40 feet below sea level in excavating for dockyard purposes, being evidently an ancient peat bed, such as those which now occur in the islands, overwhelmed by the sand. Besides these primary sand rocks, a conglomerate is being formed on the shore in some places, composed of beach fragments cemented together, as usually occurs in coral islands. The sand rock contains various fossils, most abundantly a land snail (*Helix*) now abundant in the islands, and a much larger one, now extinct, but closely resembling the present species in other respects than size. The bones of turtles and birds are also found in the rock, and all the common marine shells of the islands. The rock, when exposed, is honeycombed by the action of the rain, and that of sea water, and on the coast its surface has a remarkable corroded appearance. It is eaten into cup-like hollows all over, separated from

* Dana, "Corals and Coral Islands." Sampson Low & Co. London, 1875, p. 182.

† Major-Gen. Nelson, R.E., "On the Geology of the Bermudas." Trans. Geol. Soc. London, Vol. V, 1840.

one another by extremely sharp projecting points and edges of thin laminæ, which break with a crackling noise under the feet. In some places on the coast the rock has been left by denudation projecting in isolated pinnacles and peaks of fantastic form.

The surface of the rock is not only honeycombed by the action of rain, but hardened by re-deposit of carbonate of lime ; and a fresh surface exposed to the weather soon becomes covered with a hard film. Extensive caverns exist all over the islands, undermining the rock in all directions, and filled at the bottom with water, which, in caves near the sea, rises and falls with the tide and is salt. At Paynter's Vale Cave the water is only brackish, so that the communication underground with the sea must be slight. Such caves must necessarily result from the consolidation of masses of loose sands by means of the percolation of rain water. The carbonate of lime taken up must leave cavities unless the whole mass were to shrink gradually ; but as the outer or upper layers receive the water first, they become consolidated, and hardened more thoroughly than the inner. Subsequently, these outer layers being hardened, the water ceases to take up so much lime from them, but passes through cracks and chinks, to dissolve away the softer interior, which sinks and falls in. A cave is the result, on the roof of which stalactites form at once.

The falling in of the roofs of ancient caves gives rise to many peculiar features in the landscape of Bermuda. The stalagmites at Walsingham Cave are far under water, proving a sinking of the floor of the cave which might possibly be supposed to be local, due to the giving way of some hollow beneath ; but since the same condition is to be seen in nearly all the caves, and there is the further evidence of the sunken bed of lignite, there seems no doubt that there has been a general sinking of the island in comparatively recent times. In some places on the coast of Bermuda are reefs composed by Serpulæ, which were called by Nelson Serpuline reefs. These often form regular circles or tiny atolls, as it were, about 20 to 30 feet in diameter. The form evidently results from the fact that the most externally placed animals have a great advantage in procuring food over those placed behind them or in the centre of the area.

The scenery of Bermuda is in some respects not unlike that of northern lake districts, for the numerous small islands which are dotted over the sounds and land-locked sheets of water are covered with vegetation down to the water's edge. The dark colour of the juniper trees (*Juniperus barbadensis*), called in the island "cedar," the prevailing foliage, not unlike that of pines in appearance, gives the landscape a northern aspect, and on cloudy days, the island as viewed from the sea, looks cold and bleak. Only the extreme lowness of all the land is characteristic and distinctive. Next conspicuous to the juniper as a general feature in the vegetation, is probably the oleander, which having been introduced, flourishes everywhere. A large portion of the uncultivated land is covered with a dense growth of another introduced plant, *Lantana camera*, a most troublesome weed.

The most refreshing and beautiful vegetation in Bermuda is that growing in the marshes and caves. The marshes or peat bogs lie in the inland hollows between two ranges of hills. These bogs are covered with a tall luxuriant growth of ferns, especially two species of Osmunda (*O. cinnamomea* and *O. regalis*). Some ferns are restricted to particular marshes. In some *Acrostichum aureum* grows densely to a height of from 4 to 5 feet. Together with the ferns grow the juniper which thrives in the marshes, and a Palmetto, which gives a pleasing variety to the foliage.

The peat of these marshes is mainly composed of the *débris* of the rhizomes of the ferns and roots and bases of the sedges, especially of one very large species of Cladium. A bog moss grows in the marshes, but is not abundant enough to take much share in the peat formation. The peat burns well and has very much the appearance of ordinary home peat. The stems of junipers are occasionally found in it in good preservation, and of larger size than any now growing on the island. The formation of peat at sea level in so warm a climate, seems very unusual. Darwin has dwelt on the peculiar conditions of climate necessary to the formation of peat. In South America and the Falkland Islands, as here, the peat is formed by the slow decomposition of plants other than mosses.*

* Darwin, "Journal of Researches." 2nd Ed. London, J. Murray, 1845, p. 287.

I have referred to the falling in of the roofs of caves. At the mouths of nearly all the caves are hollows with steep rocky sides, produced by the falling in of former extensions of the caves. One of the largest of these is at the mouth of Paynter's Vale Cave. This hollow is sheltered from the sun by its steep walls, and is hence constantly shady and moist. It is a natural fernery, fifteen species of ferns being found within its small compass, two of them occurring nowhere else in the islands. Wild coffee trees thrive amongst the ferns in the hollow. The plants of Bermuda, which are of West Indian origin, were transported thither, probably, as Grisebach* states, by the Gulf Stream, or general drift of heated surface water in this direction. Others may have travelled with the cyclones which pass constantly from the West Indies in the direction of Bermuda, and sometimes reach the island. There are no winds blowing directly from the American coast which would be likely to carry seeds, the anticyclones taking a different direction. It is, however, probable that the occurrence of American plants in the islands is connected with the fact that the islands are visited from time to time by immense numbers of migratory birds from that continent, especially during their great southern migration.

Of these the American Golden Plover (*Charadrius marmoratus*) seems to visit Bermuda in the greatest numbers, but various other birds, frequenting marshes, Gallinules, Rails and Snipes, arrive in no small quantities every year. These birds have probably brought a good many plants to Bermuda, as seeds attached to their feet or feathers, or in their crops. The seed used for the onion crops in Bermuda is all imported yearly, mostly from Madeira, and the potato seed is brought from the United States. Various seeds cannot fail to reach the island with these imports, and the constant importation of hay must have led to the introduction of many more.

Shipwrecks furnish additions to the flora occasionally. A vessel laden with grapes was wrecked on the coast a short time ago. The boxes of grapes were washed ashore, and the grape seeds germinated in abundance, so that Sir J. H. Lefroy was able to gather a number of small plants for his garden.

* A. Grisebach, "Die Vegetation der Erde." Leipzig, 1872. 2te Bd. II, s. 454.

The only export of the Bermudas is vegetables; potatoes, onions and tomatos. These are said to be best in the world, and they reach New York very early in the season and command a very high price. The "Mudians" are, however, so lazy that they do not grow enough potatoes for home consumption, and at the time of our visit to the islands, at the same time that new potatoes were being exported to New York, large quantities of the former year's American crop were being imported in the returning steamers.

Some of the most conspicuous of the present land-birds of Bermuda, such as the "Red bird," or Cardinal, have been introduced for ornamental effect. The birds most interesting to us were the "Boatswain birds" (*Phaethon flavirostris*), since we now met them in numbers for the first time, though we afterwards became so familiar with them amongst the Pacific Islands and elsewhere. The birds are white, a little smaller than our commonest English gull, and shaped more like a sea swallow or tern, though allied to the gannets and cormorants; in their tails are two long narrow feathers of a reddish tint, which as the bird flies, are kept extended behind, and give it a curious appearance.

BOATSWAIN BIRD.

The birds breed, more or less gregariously, in holes in the rock formed by the weathering out of softer layers. It is easy to secure them in the hole by clapping a cap over its mouth and often both male and female can be caught together. It is however quite a different matter to get hold of them for stuffing: their bills are very sharp and strong, and they fight furiously, screaming all the while. Only one egg is laid, and it is of a dark red colour like that of the Kestrel. Rats abound in the islands, and I saw one hunting about the holes evidently on the look-out for eggs or young. These must be the only enemies the birds have except man, and they would find no difficulty in driving the rats off, but I saw several eggs broken and sucked, no doubt in their absence.

On one of the islands I saw a pair of crows, but they were very scarce, since blood-money to the extent of two-shillings a head had been put upon their heads by the Government.

Crabs abound at Bermuda: a species of Grapsus, a crab which will be frequently referred to by me, climbs the mangrove trees with the greatest ease. A white Sand-crab (*Ocypoda*), burrows deep in the sand-hills, and is very difficult to dig out, and a huge ugly Land-crab (*Cardisoma*) is common further inland. A small White-crab (*Remipes*) lives in the sand on the shore just below the verge of the water; it burrows rapidly in the sand until covered, and then by ejecting a small jet of water from its gills clears a small passage for respiration, remaining concealed.

A land Nemertine worm was discovered by Von Willemoes Suhm, living in moist earth. Only one other terrestrial Nemertine was known hitherto, and that was discovered by Semper in the Philippine Islands; this worm Von Suhm named *Tetrastemma agricola*, placing it in the same genus with certain aquatic species.* When irritated it darts out its armed proboscis with great rapidity in defence. It also uses the proboscis as an aid in progression, shooting it out and fixing its tip to a distant point and then drawing the body up to the point by contracting the protruded organ. The animal is ciliated all over, and has two pairs of eyes. The earth in which it lives contains a good deal of salt. The animal was found to live for hours in salt water, but to die at once when placed in fresh water.

The corals of Bermuda may be seen growing to great advantage by the use of a water glass. The species are very few in number, there being only about ten species of Anthozoan corals, and two of Hydrozoan. The latter two species of *Millepora* are

---

* A. Von Willemoes Suhm, Ph.D., "On a Land Nemertine found in the Bermudas." Ann. and Mag. Nat. Hist. 1874, XIII, p. 409.

very abundant, and contribute largely to the reef formation.
While some species such as the great " Brain coral," (*Diploria*

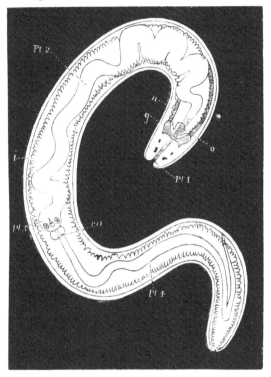

LAND NEMERTINE, TETRASTEMMA AGRICOLA. (YOUNG MALE.)

Pt 1—4 Successive portions of the proboscis; 1 entrance; 2 papillary portion; 3 pouch
of stylets; 4 glandular portion; *ca* muscular entrance of glandular portion;
o mouth; *i* intestine; *g* ganglion; *n* lateral nerves.

(After a figure by Von Willemoes Suhm.)

cerebriformis), which is conspicuous at the bottom as a bright
yellow mass appear to prefer to grow where the water is
lighted up by the sunshine; other species, such as *Millepora
ramosa* and *Symphyllia dipsacea*, seem to thrive best in the
shade. One species, *Mycedium fragile*, which forms very thin
and fragile plate-like laminæ, which are, when bleached white,
almost the most beautiful of corals, occurs growing in colonies
in great abundance, in water from a foot to a fathom in depth
inside small caverns.

All around the Bermuda coast, wherever it is at all sheltered, large black Holothurians, are excessively numerous. They are to be seen covering the white sandy bottom all over, lying a few feet only apart.

I was greatly indebted during my stay at the Bermudas to General Sir J. H. Lefroy, C.B., F.R.S., then governor of the islands, both for his kind hospitality and constant information and assistance in scientific matters.

For a further account of the geology of the Bermudas, see "Nautical Magazine," 1868, p. 486, and also J. M. Jones, F.L.S. on the "Geological Features of the Bermudas." Trans. Nova Scotian Institute of Nat. Hist., May, 10, 1869.

For the Mollusca, Rev. H. B. Tristram, Proc. Zool. Soc., 1861, p. 403. For the Birds of Bermuda, Lieut. Reid, R.E., F.Z.S., "Zoologist," 1877. Repr. from the "Field" newspaper.

For a general account of the Natural History of the islands, see "The Naturalist in Bermuda," by J. M. Jones, F.L.S. London, 1859.

For the Vegetation, see Dr. Rhein, "Ueber die Vegetations-Verhält- nisse der Bermudas-Inseln." Vortrag gehalten beim Jahresfeste der S.N.G., 25. Mai, 1873. Also papers on collections made by me in the Journal of the Linnean Society, for which see the list of papers at the end of this work.

CHAPTER II.

AZORES, MADEIRA, CAPE VERDES.

Fayal Island, Azores. Porpoises on the Feed. Town of Horta. Peculiar
Dress of the Women. Island of Pico. St. Michael's Island. Native
Ferns and Australian-introduced Trees. The Threshing floor and
Women at the Mill. Vegetation of the Azores. Hot·Springs at
Furnas. Plants Growing in the Hot Water. Caldeira des Sette
Cidades. Madeira. Grand Cural. Curious Caps worn by the Men.
The Island at Sunset. St. Vincent Island, Cape Verdes. Vegetation
of the Island. Ascent of Green Mountain. Different Causes of
Variation of Vegetation with Altitude. Structure of Basaltic Dykes.
Calcareous Seaweeds on Bird Island. Habits of Crabs. Miniature
Oasis. Flying Gurnet Hooked. Mode of Catching Bonito. Island
of Fogo. Porto Praya, St. Jago Island. Use of Foot in Feeding by
Kites. Kingfisher and Galinis. Hauling the Sein. A Large Shark.
San Domingo Valley. Monkeys. Remarkable Freshwater Crus-
tacean. Limestone Band in the Cliff of the Harbour.

Azores, July 1st to 10th, 1873.—After a voyage of 19 days
from Bermuda on July 1st, the "Challenger" steamed in towards
the island of Fayal, which was soon sighted as a blue haze in the
far distance which mingled with the clouds and showed a faint
outline only here and there. The haze became darker and
darker as the island was approached and the outline more
distinct, and at last we began to make out the shape of the
island clearly with our glasses, and to see the great belt of
cultivation on the lower region, with its thickly set rectangular
patches of ripe corn. The highest point of the island is only a
little over 3,000 feet above sea level; this part of the structure
was not sighted at all by us, for it remained always covered
with clouds.

The whole of the Azores are volcanic, only on Sta. Maria
Island is there a small deposit of limestone containing marine
shells, of miocene date. The islands are composed of beds of lava,
basaltic and trachytic, and cones of scoriæ and pumice. As we

approached Fayal numerous craters became visible, of the usual truncated conical form, but in all stages of decay, and as usual of all sizes. Some huge volcanic masses form the main ridge of the island, and from the slopes and bases of these numerous baby volcanoes rise up, and are seen clustering together in irregular groups.

One crater close to the shore and partly cut into by the waves was very conspicuous. In its loose pumice walls the sea had made an excavation, and had exposed vertical columns of harder trachyte. The lip of the crater facing the sea is partly broken down, and a view is thus obtained right into the conical hollow inside, which is now partly under cultivation. The crater is called Castello Branco by the inhabitants.

The whole lower part of the island, which has a more gradual slope than the steep cones above, is closely cultivated, and showed as seen from seawards a series of intermingled bright green and yellow fields interspersed with glistening white villages, and numerous churches and monasteries.

As we neared shore, a large shoal of porpoises was seen close by, going at great speed in full chase after fish, the whole shoal skipping together, four or five feet out of water for several successive bounds in hot pursuit. The shoal was closely attended by a flock of gulls which follow in order to pick up the fish which are bitten or wounded by the porpoises, but which the porpoises have no time to stop to pick up. In the Arafura sea, I have seen frigate birds hanging over a shoal of porpoises with the same object, and in just the same manner in the tropics terns and noddies follow the shoals of large predatory fish (*Caranx*) to pick up the crumbs. The demeanour of a shoal of porpoises on the feed is a very different thing from their lazy rolling motion which one more commonly sees.

We rounded a promontory formed of two old craters, one of them with its seaward half entirely demolished by the waves, and its hollow inner slope terraced for cultivation, and came in sight of Horta, the capital town of Fayal. It is almost the most beautifully situated town I have ever seen. It is built along the shore of a wide bay, the white houses being crowded together on a very narrow, almost flat belt of land. Immediately behind the main body of houses, rises a series of steep hills covered with the most brilliantly green gardens, orange trees, and magnolias, with houses dotted amongst them at various heights, and here and there churches and monasteries. The lower hills are backed by the main mountain mass, the summit of which was hidden in the clouds. In full view of Horta is the island of Pico with its towering cone.

The town is thoroughly Portuguese in appearance. The houses are whitewashed as at Lisbon, with green Venetian blinds and window frames and balconies. The women are better looking than at Lisbon. They dress in remarkable dark blue cloth cloaks with enormous long coal-scuttle shaped hoods to them, so that one has to look down a sort of tunnel to see a pretty face at the end of it, and it is impossible to get any but a full face view of a beauty, or to steal a sly glance at all. The girls save up their money most carefully, in order to become possessed of one of these fashionable cloaks. They cost about six pounds, and a girl has to work two years and a half to get one. Horta has many primitive ways. The old women sit at their doors and spin with the spindle and distaff.

COSTUME OF WOMEN AT HORTA.
(From a photograph.)

The gardens are all surrounded by high walls to protect them from the furious gales which blow here in winter, and which would else destroy all the fruit trees. Fruit was abundant, apricots were bought at 20 for a penny. The prevalence of small pox in the town prevented our making any stay. I slipped on shore in a fruit boat, or I should not have been allowed to land at all.

The sea beach has a most peculiar appearance to an eye not

accustomed to volcanic shores, being composed of fine volcanic sand which is absolutely black. The sand is made up of ground-up lava and ejected dust, and is full of crystals of olivine, augite, hornblende, and quartz, with abundance of magnetic iron particles, which cling to a magnet when it is brought near.

The ship was off Pico in the evening of July 2nd. The clouds gradually cleared off the island, at first hovering about its summit, then remaining as a belt some way below the top of the cone, and finally disappearing altogether, and leaving the majestic peak in full view, lit up by a splendid red sunset glow. The peak is a steep cone, rising abruptly to 7,613 feet above sea level from a more gently sloping base, on which are numerous secondary craters which look like little pimples on the surface of their huge parent. The top of the cone is cut off horizontally and out of the huge crater on the top arises towards one side of it a little secondary cone which forms the highest point of the whole.

St. Michael's Island, July 4th to July 9th, 1873.—We neared the island of San Miguel. The island has mountains of from 2,300 to 3,500 feet altitude at either end, and a lower range of hills joins these together. Ponta Delgada, the capital of San Miguel, lies on the sea shore opposite, about the middle of the lower range of land.

The volcanic cones and slopes leading from these to the sea are formed of light pumice and ash soil, very friable and easily cut into by the action of water. Hence, water-courses have cut their way deep into the surface of the country, and as San Miguel is viewed from seawards, its most striking feature is formed by the numerous deep gullies which are seen running parallel to one another, and with almost straight courses from the high land down to the sea. Pouta Delgada is composed of houses similar to those of Fayal, but it is not nearly so pretty as the latter town, the land behind not being steep, and there being no bay shut in by hills. A breakwater is required to form a harbour.

I formed one of a large party which paid a visit to the valley of Furnas and its hot springs, distant about 30 miles from the port town. We travelled in carriages, each drawn by four mules. From the nature of the country already described, we had to cross numerous water-worn gullies, and our road led constantly up and down steep hills. We crawled up one side of the ridges, and made fearful dashes down the other, the mules going with great spirit. We passed between fields of maize and corn, with tall hedges of reeds (*Arundo donax*), planted round them to break the force of the wind, and a kind of lupine planted in geometrical patterns amongst the corn to be ploughed in after the crop was reaped, as manure.

We passed many fine flower gardens, planted with a large variety of Australian, New Zealand, and South American plants, and went by numerous hills, small volcanic cones, planted with firs and various timber trees with great care. The appearance of the island has been wonderfully modified by careful plantation, most of the work having been done by a Mr. Brown, a gardener from Kew, who was brought to the island 30 years ago by Don Jose de Canto, to superintend the laying out of his garden.

We halted for luncheon at a small stream under a clump of Australian blue gum trees, beneath which on the margins of the stream grew a profusion of ferns. Here flourished the cosmopolitan break fern, and another *Pteris* (*P. arguta*); *Woodwardia radicans*, not so long ago discovered to occur in Great Britain, a splendid bright green fern, with large fronds, the tips of which bend over to meet the soil, and then take root, whence the name; *Asplenium monanthemum*, hardly to be distinguished in appearance from our home *A. trichomanes*; *Asplenium marinum*, *Adiantum nigrum*,—the lady fern, the hart's tongue, the male fern, and the common polypody. With these was *Osmunda regalis*, and abundance of the Maiden hair.

We crossed the lower central ridge of the island, and looked down upon the bright blue sea on the other side. We passed a threshing floor where threshing was going on in the old biblical style, as all over the Azores, where primitive customs are maintained to an extraordinary degree. The threshing floor is a circular flat space, usually near a house in the home corn-field, about 40 feet in diameter, and with a bottom of cement or some hard mortar. On this the corn is laid and pairs of oxen are driven round and round over it, yoked to a heavy wooden sledge-like machine, like that used for dragging casks on in England.

A man often sits or stands on the drag, and the girls ride on it for fun. Usually two yoke of oxen are employed. At the floor we halted at, the oxen were not muzzled, and were feeding freely, but they often are so, as we saw at other floors.

A little further on we came upon two women grinding at the mill. A pair of circular stones, one placed on the top of the other, are used; the upper fitted with a straight upright handle, the thing being in fact a simple quern. Two women standing facing one another, catch hold of the handle, one at the top, the other lower down, and they send the upper stone round at a good pace, each exerting her strength when the handle is furthest off from her, and thus pulling to the best advantage.

We next passed a small town, Ribeira Grande, where there were numerous churches and a monastery, and a pretty patch of public garden laid out by Mr. Brown, and planted principally with Australian shrubs, *Banksias* and *Melaleucas*. At a road-side inn, at which we pulled up to water the mules and refresh the drivers, the church choir was singing remarkably well, practising an ancient chant in a room overhead, with a piano as an accompaniment. None of the poorer houses in the town, or indeed all over the island, have any glass in the windows, but only shutters. Glazed windows are scarce; only the priests, shopkeepers, and merchants have them.

We turned up inland from the sea, and mounted the high land, making across the island again in a zigzag direction. At last we gained the summit and came out upon a moor covered with bog myrtle (*Myrica faya*), break fern, *Woodwardia radicans*, heath (*Erica azorica*), and a splendid fern (*Dicksonia culcita*), which almost forms a tree, and which has a beautiful golden brown silky substance covering its young shoots, which is gathered and used for stuffing cushions. Several tree ferns have a similar substance developed on them. The moor looked very like a Scotch moor, and stretched away far over the flat hill tops.

There are 40 flowering plants found in the Azores, which grow nowhere else in the world, *Erica azorica*, the heath, is one of them. The rest of the plants are either South European, or belong to the Atlantic flora, a name given to a series of plants which grow on the Azores, Canaries, and Madeira, and nowhere else. Of these Atlantic plants 36 are found in the Azores.* Examples of them are the laurel (*Laurus canariensis*) and the juniper (*Juniperus brevifolia*). One little plant, a Campanula (*C. vidalii*), is found only on one small rock on the east coast of Flores (one of the Azores), and nowhere else in the world. Nearly all the shrubs and trees of the Atlantic group of islands are evergreens.

We crossed a stretch of the plateau, and suddenly looked down on the other side of it into an immense deep, nearly circular crater, beautifully green. Its undulating bottom was dotted over with white houses amongst gardens and corn-fields, and in the distance was seen a small column of steam hovering over the hot springs. We drove down a steep incline for at least a couple of miles, and at last reached the village of Furnas. The road hence to the hot springs led across a small stream fed by them, deeply stained red, and smelling strongly of sulphuretted hydrogen. Thence the path went up a little valley, cut out in the low ridge of very fine light whitish ashes which separates the main Furnas valley from that part of it in which the Furnas lake is situate. It is a beautiful tiny glen, with dark evergreen foliage on its steep banks, and on the swamp borders of its narrow bed were masses of the brilliant green leaves of the eatable Arum (*Caladium esculentum*), one of the staple foods of the Polynesians, their "*taro.*" The *taro* is cultivated all over the islands, but thrives here especially in the warm mineral water.

The Furnas lake is about three miles in circumference. There are two groups of boiling springs, the one at the margin of the lake, the other close to the town of Furnas.

The boiling springs near the lake are scattered over an area of about 40 yards square, covered with a greyish clayey deposit; a geyser or hot-spring formation, being composed of matter deposited by the hot water. No doubt the present hot springs are the dwindled remains of former fully developed geysers. The principal spring consists of a basin about 12 feet in diameter, full up to within about 2 feet of the brim of a blueish water, which in the centre is in constant and most violent ebullition,

* A. Grisebach, "Die Vegetation der Erde." Leipzig, 1872, 2ter Bd. s. 503.

the water being thrown up a foot in height as it boils forth. A constant column of steam rises from the basin. Near by is a sort of fissure, from which issue at short irregular intervals jets or splashes of boiling water mingled with steam and sulphuretted hydrogen in abundance. This spring makes a gurgling, churning sort of noise; the large basin, a sort of roar.

In the sides of the fissure grow, in the area splashed by the hot water, some green lowly organized algæ (*Botryococus*), which form a thick crust upon the rock surface. Similar growths of lowly organized plants in the water of hot springs have been observed in various parts of the world.* At a couple of feet distance from this hot spring rushes up a perfectly cold iron spring with a considerable stream of water.

All around are small openings, from which sulphuretted hydrogen and other gases issue with a fizzing noise, and coat the openings with bright yellow crystals of sulphur. The ground around is hot, too hot in many places for the hand to rest upon, and it is somewhat dangerous to approach the pools of hot water at all closely, since the hard crust on the surface may give way and one may be let fall into the boiling mud.

Just above these hot springs is a beautiful mountain stream, which forms little cascades as it tumbles down to the lake valley from the fern-clad moor above.

At the town of Furnas is an inn kept for families who come in the season to drink the waters and bathe. There is a free bath house built by the Government, with marble baths and hot and cold mineral water laid on to each. The whereabouts of the springs near the town are marked by clouds of steam. The springs are scattered over a larger area than at the lake springs, and the grey geyser formation is piled into irregular hillocks around them, instead of presenting a nearly flat surface as at the other springs. Here the principal spring is like that at the lake, but the amount of hot steam rushing up is much greater, and the noise is almost deafening. The water is thrown up about

* For further account of the vegetable growths in the hot spring of Furnas, see Linn. Journ. Bot., Vol. XIV, p. 321. Also papers on the same by Mr. W. T. Thiselton Dyer and Mr. W. Archer, ibid., pp. 326–328.

two or three feet in a constant hot fountain. Close by are sulphur springs with hot water issuing in violent intermittent splashes; and there is also one deep chasm, from the depths of which boiling hot blue mud is jerked out in similar splashes. The mud hardens on the sides of the cavity into a crust made up of successive laminæ.

The natives use the natural hot water to heat sticks or planks in, in order to bend them. They also sometimes dig holes in the mud and set their kettles in them to boil. As at the other springs, there are cold springs issuing from the ground, close to the boiling ones. One spring has its water charged with carbonic acid and effervescing. All the springs empty into one small stream, which then runs down to the sea, with a complex mixture of mineral flavours in its water, and retains its heat for several miles.

In the shores of the lake there are large extents of geyser deposit, forming strata 40 or 50 feet in thickness, and evidently resulting from hot springs, now worked out, but with a few small discharge pipes of heated gas remaining active here and there. Near the seaward end of the lake is a hole, where, as in the Grotto del Cane, an animal, when put into it, becomes stupefied by inhaling the carbonic acid gas discharged.

I made an excursion from Ponta Delgada to the Caldeira des Sette Cidades, or Cauldron of the Seven Cities. It is a marvellous hollow of enormous size, with two lakes at its bottom and a number of villages in it. One slowly climbs the mountains from the sea and suddenly looks down from the crater edge upon the lakes, 1,500 feet below. On the flat bottom of the crater, which is covered with verdure and cultivated fields, are several small secondary craters, the whole reminding one of a crater in the moon. One of these small craters has been so cut up by deep water-courses, that between them only a series of sharp radiating ridges is left standing, and the crater has thus a very fantastic appearance.

San Miguel was suffering from a drought, at the time of our visit, which had been of long duration. A grand procession therefore took place in order to procure rain, in which a miraculous image the "Santo Christo," the jewels presented at the shrine of which are reputed amongst the people to be worth one million sterling, was carried round the town. The figure is apparently of wood and is in a squatting posture with the legs crossed. It was borne in a litter, with a canopy over it, on men's shoulders. Next day, from seawards, we saw clouds hanging low over the island, and it seemed as if the image had been again miraculously successful.

The most complete account of the geology of the Azores is that of G. Hartung, "Die Azoren." Leipzig, Engelmann, 1860. See also F. Du Cane Godman, "Nat. Hist. of the Azores." London, Van Voorst 1870. Also T. Vernon Wollaston, "Testacea Atlantica." London, Reeve and Co. On the Coleoptera Crotch, P.Z.S., 1860, p. 359.

Madeira, February 3rd to 5th, July 15th to 17th, 1873.—Madeira is a mass of mountainous rocks, rising to 6,000 feet in height. The town of Funchal nestles close to the water's edge and straggles up the side of the valley in which it lies. In the early morning the island, viewed in clear weather from seawards, is of a beautiful hazy violet, whilst the sea is of the deepest blue.

The beach at the landing-place near the town is formed of large pebbles of basalt and is very steep. In landing, boats provided underneath with runners like those of a sledge are used on account of the surf. They are backed in stern first and are hauled up directly they ground by men stationed on shore. The main part of the town lies close to the beach and is very like the old part of Lisbon.

The fish market yields many rare fish to naturalists. Deep-sea fish every now and then find their way, for some reason or other, to the surface at Madeira and get picked up, and several very rare fish are known from here only; as for example, a curious small fish,* allied to the Angler, described by Dr. Günther from a single specimen. The "Challenger" dredgings yielded several close allies, and showed that the fish in question was undoubtedly a deep-sea form, as had been surmised. Huge Tunnies, weighing some of them from 60 to 100 lbs., are sold in the market. Their flesh is quite red, like beef, and they are cut up and sold just like butchers' meat. The great beauty of Funchal lies in its gardens, where plants of tropical and tem-

* *Melanocetus*, "Proc. Zool. Soc.," 1864, p. 301.

perate climates thrive together. Bananas, pine-apples, aloes,
vines, prickly pears, guavas, mangoes, oranges, grow together,
with a profusion of flowers.

The island being resorted to by so many invalids, the
cemetery forms a conspicuous feature in the scenery. The coffin-
makers have the unfeeling habit of manufacturing their wares in
front of their shops in the public streets. The roads are narrow
and run directly up and down the steep slopes. They are paved
with small pieces of basalt, three or four inches long. The
stone pavement has become, by constant use, polished and
slippery, and the traffic is carried on by means of sledges on
runners instead of with wheels. These come down the steep
hills at a very rapid pace.

I made an excursion to the Grand Cural. We rode ponies
which trotted or galloped up the steepest hills. A native went
with each pony and hung on to its tail to help himself along
when the pace was fast. We passed through the gardens on the
outskirts of the town; then higher, through fields of sugar-cane
and corn, up amongst the vineyards, terraced on the hill sides, and
with the vines trained on horizontal trellis work; then past the
hovel-like cottages of the country people, till we reached the
district of pine and sweet chestnut trees.

The pine woods were deliciously cool. We passed them and
came out upon open grass slopes with occasional patches of basalt
rock sticking up out of them, the slopes themselves being composed of disintegrated scoriæ. We climbed the slopes on foot
and reached a height of about 5,000 feet. From thence we had
a commanding view of the Grand Cural, a huge gorge or rent in
the mountain mass, precipitous on one side and almost so on the
other. The precipitous side opposite us was in the deepest
shadow, so much so, that we could hardly trace the details upon
its surface, but we could yet see that every available ledge had
been terraced and brought into cultivation. The sun shone
brightly on the dark red and purple scoriæ and lava, and on its
clothing of chestnuts and pines, on our side of the chasm, which
being thus in high light contrasted forcibly with the deep gloom
of the opposite wall. A magnificent panorama of the south side
of the island was visible from our position, with its volcanic

cones and white houses scattered amidst the green. After we had enjoyed the scene but a few moments, a thick mist shut it from our view and we descended.

It is only in the highest parts of the island of Madeira, that anything is to be seen of the true indigenous vegetation. Below, cultivation has destroyed the native plants. On the upper slopes the common furze and broom and the brake fern grow in abundance.

The countrymen of Madeira wear, on gala days, curious pointed blue cloth caps, very small, and resting only on the back of the head. The point is a long pointed cylinder, which sticks out stiffly from the back of the head. It seems to be a curious abnormal development, due to insular isolation, of the pointed bag which hangs down from the knitted worsted nightcap-like head covering of Mediterranean and Spanish seamen, and English yachting men. The point seems to be a sort of rudimentary

CAP WORN BY PEASANTS OF MADEIRA.

organ which has undergone subsequent modifications for the sake of ornament. A minute tag of the red lining of the cap is turned up in front and behind with great care, and no doubt is also a rudiment of some former appendage of the head dress. There seems to be a curious general tendency in the Atlantic islands, amongst the inhabitants, to develop strange head dresses. The hoods of the women of the Azores have been described. Besides these, the men wear, or wore, in some of the islands, a curious cap, in which a pair of side flaps have been developed into a regular pair of horns, projecting vertically above the head.

I was told that Madeira wine is sometimes manufactured in the island out of red wine, the colour being taken out with animal charcoal. I knew that red wine was constantly made out of white wine, but had not suspected the opposite manufacture.

July 16th, 1873.—On our second visit to Madeira we were unable to land owing to the prevalence of small pox on shore.

I visited a steamboat which came into the harbour for coals and which was running between the Bight of Benin and Liverpool. The whole ship was covered with cages full of grey parrots ; even in the forecastle, in the seamen's sleeping place, every available nook was full of parrots. The deck was covered with various African monkeys, and there was a large wild cat in a den, and some large snakes (Pythons) in a box. All these animals were intended for sale in Liverpool.

We left Madeira in the evening. The ship passed quickly out of the lee of the land and into the trade wind and was soon driving along before it, dashing a sheet of foam from under the bows. There was a splendid sunset. The sky was lighted up with brilliant golden and red tints, behind and to the west of the hazy blue mountains of Madeira, in front of which floated here and there small filmy clouds. Beneath the higher mountains, were the green lower ranges, half lighted up by the evening light, half in intense black shade. Lower down again, on the shore, lay the glistening white town with its dark black cliffs on either hand.

As it grew darker, the lower ranges and details of the view became gradually lost, and at last all that was to be seen was the dark outline of the mountains against the sky, with the twinkling lights of Funchal far below, and a few lights dotted about on the hill-side above. At last we lost sight of the island altogether and sped south before the breeze, not to return so far north of the equator again for nearly two years, when we reached Yeddo, in Japan, in nearly the same latitude.

For a list of works and papers relating to the Zoology of Madeira, see "Preussische Expedition nach Ost-Asien." Zoologie, 1tes Kap. Madeira, pp. 1–25.

Cape Verde Islands, July 27th to August 9th, 1873.—The ship was off the island of St. Vincent of the Cape Verde group on July 27th, and the islands of Sta. Lucia and St. Antonio were in sight ; a heavy mist hanging over the high mountains of the latter. We anchored at Porto Grande, the harbour of St. Vincent.

The island is about 12 miles long by six broad. It has an irregularly oval form, and consists of a flat central tract more or less broken by low hills surrounded by a range of high land. The low central district is evidently the bottom of an ancient crater, of the wall of which the high surrounding range is the remains. The range is composed of strata dipping outwards from the ancient centre of eruption. It is cut up by a series of deep valleys having a general radiate arrangement, into ridges of various heights, which are again cut up by secondary transverse valleys so as to culminate in a series of irregular peaks.

Some of the ridges are of considerable altitude. The Green Mountain is 2,483 feet in height, and one other mountain to the extreme south of the island, 2,218 feet. A break in the encircling range to the north-west forms the harbour or Porto Grande, in the entrance to which lies a small island, called Bird Rock, a fragment of the range, once continuous in this direction.

More barren and desolate-looking spots than St. Antonio and St. Vincent appear as approached from seawards, after they have been suffering from their usual prolonged droughts, it is impossible to conceive of. Their general aspect reminded me of that of Aden or of some of the volcanic islands in the Red Sea. At the time of our visit, no rain had fallen for a year at St. Vincent. Sometimes it does not rain for three years.

The mountains are of black volcanic rock terminating seawards in precipices, in which the numerous dikes, which traverse them in all directions, stand out conspicuously, often projecting far through weathering of the matrix. Between the hill ranges, stretches a flat sandy plain covered with sand dunes and with ranges of low rounded hills of a bright red ochre tint. The white sandy plain terminates at the head of the harbour in a sandy shore, where is a miserable town, composed mostly of mere hovels, and a black coaling jetty.

The whole was glaring in a fierce sun, and appeared almost devoid of vegetation, but from the anchorage some black tufts could be made out with a telescope, which consisted of small bushes of lavender (*Lavandula rotundifolia*), the most abundant plant in the island, and on the summits of the higher hills a few Euphorbia bushes (*E. tuckeyana*) could be made out in the same way. On the sandy plain at one spot is a thick growth of low tamarisk bushes which stretches from the shore inland, amongst which at about half a mile from shore is a group of half a dozen small trees. These are a Tamarind (*Tamarindus indica*), some thorny acacias (*A. albida*), and *Terminalis catappa*. They stand in an old enclosure in front of the ruins of a house, and are green and flourishing, and show that much might be done by cultivation, even for St. Vincent.

From a statement in Horsburg's Directory, in the description of St. Vincent, that "as much wood may be cut here in a short time as can be stowed away," I was led to suppose that possibly in old times there was much more vegetation in the island and hence more rain, and that the trees had been destroyed as at San Jago, according to Darwin,* but I find that in accounts of the island published in 1676,† the vegetation is described as having almost exactly the same appearance and range as at the present day. The firewood is mentioned, but described as a bush, evidently the tamarisk, and said to be scanty and very bad. The island is described as being as barren as it is now.

The plains I found covered all over with the spiny fruit of a small creeping plant (*Tribulus cistoides*). Almost the only plants retaining any living and green leaves were the lavenders, on the bushes of which were to be found here and there a green sprout put forth apparently in anticipation of the wet season. Many of the plants were so chip dry, that I had to gather specimens in boxes, as they would not stand pressing.

The plains were covered with grass seeds. The island is said to become green as if by magic after rain, and at St. Jago, where the rain had been earlier, the plains at about 500 feet elevation were covered at the time of our visit with a bright green coat of seedlings; but a day's moderate rain which occurred on July 30th at St. Vincent had not produced any visible effect by August 5th, the day on which we sailed. The bottoms of the valleys and hill-slopes to the southward, are covered with a dry hay-like grass; but the goats and cattle kept in this part of the island were dying in numbers from starvation.

On June 30th, I made an excursion with a small party, up

* "Journal of Researches." London, J. Murray, 1845, p. 2.

† Dapper's "Africa." Amsterdam, 1676. "Eilanden van Africa" p. 83.

Green Mountain. It was raining, and the coal contractor on shore, who arranged matters for our trip, warned us that we should all catch a terrible fever if we went and got wet. We went however, and did not suffer, and I cannot help thinking that it is to some extent the extremely rare occurrence of rain which inspires dread of it in St. Vincent. Our party of three started on two ponies and a donkey, over the latter of which Murray soon broke a pet walking-stick of mine of Bermuda juniper, in trying to urge him into the right path. A strapping negress, one of the coaling gang, started on foot for the mountain with the lunch on her head.

The road led over the bottom of the old crater, and then up the steeper end of the mountain by a zigzag path in places built up in steps and in others hewn out of the rock. The soft friable soil of the plain was in many places already converted into tenacious mud by the rain.

As the hill-slopes are ascended from the plains, the plants become greener and more abundant. In a narrow gorge at the commencement of the ascent of the mountain, some small gardens were passed, at an elevation of about 200 feet above sea level. They contained sugar-cane, pumpkins, and a small date palm ; and maize was just being planted in them. There were a few cotton bushes growing near. At 700 feet, Euphorbias and woody Composites commenced, and the hill-side was covered with coarse dry grass. At 1,000 feet, small *Boraginaceous* bushes with pink flowers (*Echium stenosiphon*) commenced. At 1,300 feet I found the first patch of moss and *Marchantia*, with a fern and a live snail. At 1,700 feet a Statice (*S. Jovis barba*) was abundant on the cliff.

The lavender grows right up to the top of the mountain, but is there entirely fresh and green instead of black and withered as below. A leafless trailing Asclepiad (*Sarcostemma daltoni*) commenced at 900 feet. All the plants on Green Mountain appear to extend their range of growth to the summit. On the summit, the land is all more or less under cultivation, and maize, potatoes, tomatos, and pumpkins grow there. There are several cottages on the summit, and near one is a double circle of large Agaves.

In the Green Mountain, the appearance of the several plants at successive heights is due mainly to the gradual increase in amount of moisture received by the soil as a higher and higher zone is reached. Closely similar conditions determine the distribution of plants on many other mountains, such as on Green Mountain in the Island of Ascension.

The distribution of plants in successive zones on mountains which is most familiar, is that brought about by a successive decrease in temperature with increase of altitude, the Alpine flora being that which withstands a prolonged covering of snow. In Kerguelen's Land thus, a rapid decrease of vegetation is encountered as the mountains are ascended, and at 1,000 feet most of it ceases.

On some active volcanoes, however, as at the Banda Group near the Moluccas, a gradual decrease in the vegetation in correspondence with increased altitude is brought about by exactly opposite conditions, namely, gradual increase of heat. Here, close to the crater at the summit, the soil is excessively hot, yet one or two plants grow in it where it is almost too hot for the botanist's hand, and these straggle upwards, beyond distanced more sensitive competitors, till a region is reached which is barren of all but lowly organized algæ, which grow around the mouths of natural steam jets, as about the hot springs in the Azores and elsewhere.

In very high latitudes only, apparently, is the vegetation not influenced by altitude. On the mountains in East Greenland, the same plants extend from sea level up to as high as 7,000 feet altitude. This circumstance is accounted for by the fact that here the sun never rising far above the horizon, its rays strike the mountain-slopes nearly or quite vertically, and hence by their greater power compensate for the larger amount of heat lost by radiation at great elevations. The flat land receives the rays on the other hand very obliquely, and hence with much less force.[*]

The combination of effects due to difference of aspect with regard to the trade wind and sun produces a marked difference

---

[*] "Die Zweite Deutsche Nordpolarfahrt in den Jahren 1869 und 1870," 2ter Bd. Wissenschaftliche Ergebnisse. Leipzig, F. A. Brockhaus. " Klima und Pflanzenleben auf Ostgrönland," von Adolph Pansch in Kiel.

in the altitudes at which plants can grow at various aspects in St. Vincent. Thus *Aizoon canariense*, a Malvaceous plant which on Bird Rock grows close to the sea level on its windward side, does not commence on the leeward side of the hills of the main island till 700 or 800 feet. On the mountains on the southern side of the island, the vegetation does not come so far down the windward slopes, since the wind is heated and dried before reaching them, by passing over the hot central plain.

Vertical dikes of basalt are very numerous all over the island, penetrating the main component rocks, by the disintegration of which they are often weathered out so as to project as walls. They usually show a columnar structure, the columns being as usual at right angles to the cooling surfaces. I saw several in which the cleavage in the centres of the masses was laminar and parallel to the lateral surfaces, whilst on either side the dikes were composed of very regular small columns disposed at right angles to these surfaces. In the Auvergne district, I have observed dikes in which laminar cleavage parallel to the surfaces occurred at the sides of the dikes and the columnar cleavage in the centre just the opposite condition.

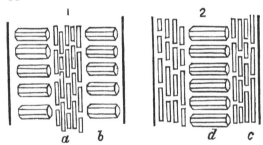

DIAGRAMS OF THE CLEAVAGE STRUCTURE OF BASALTIC DYKES.

1 In S. Vincente; 2 in the Auvergne, near M. Dore; *a* central portion with laminar cleavage; *b* lateral regions with horizontal columnar cleavage; *c* lateral regions with laminar cleavage; *d* central mass with horizontal columnar cleavage.

On Bird Island, the rocks about tide mark, are covered with a broad band of a dense incrustation composed of Coralli-naceæ, which forms a striking feature in the appearance of the island as seen from the sea, and is more marked here than on the main island. The Corallinaceæ are seaweeds which secrete a dense skeleton of carbonate of lime. The incrustation on Bird

Island is of several colours, white, bright pink or cream colour, and is mainly composed of two species, of *Lithothammion, L. polymorphum* and *L. mamillare.* The incrustation assumes very varied forms, being simply incrusting, and following the form of the rock surface on which it rests, or forming smooth rounded convex masses, or being covered with a close set series of projections, sometimes of considerable length, and with a sinuous arrangement.

I broke off specimens from the mass with my geological hammer. It is bored in all directions by Mollusks, such as *Lithodomus caudigerus*—a Senegambian species with two curious little tails at the hinder extremities of the valves so cut out as to lap over one another when the shells are closed. On the whole, plant-life seems to play a far more important rôle than do corals in accumulating carbonate of lime around the Cape Verdes. The principal rôle in this respect is however played by the larger Foraminifera, of the shells of which the calcareous sand of St. Vincent is mainly composed.

I made excursions every day along the shore or over the hot sandy plains or over the sharp and rugged lava, in search of plants and animals. So desolate is the place that a naval schoolmaster, who had come to St. Vincent to join the "Challenger," got lost on one of the mountains just before the arrival of the ship, and died of exposure. His body was found only after the lapse of several months.

On a visit to Bird Rock, I found that the sea birds' dung forms there, as at St. Paul's Rocks, pendent stalactite-like masses. The rock is composed of volcanic conglomerate and tuff, traversed in all directions by dikes of hard almost obsidian-like lava. Small rock pools at a short distance above the waves were filled with solid salt evaporated out from the spray. On the main island, on the windward side, the shore rocks are covered high up with an incrustation of salt dried out from the spray blown up by the trade wind. Men-of-war use Bird Rock occasionally as a target, and there were plenty of broken shot and shell upon it.

At low tide, along the shore of the main island, numerous rock pools were exposed at low tide. These are inhabited by vast numbers of sea urchins (*Echinometra*) which rest within

rounded cavities in the rock excavated by the urchins for themselves, both in the calcareous sand rock and volcanic conglomerate. With these was a coral (*Porites*) which forms small rounded
masses, bright yellow or whitish pink in colour, and a grey Palythoa, a compound sea anemone, that is a colony composed of
sea anemones closely joined together, and here forming sheet-
like masses often a foot in diameter, encrusting the rock.   An
Aplysia, or sea slug, with a pair of large skin folds continued up
from the sides of the body, and lapping together over the back
of the animal, was common, and is probably the one referred to
by Darwin, as seen at St. Jago.*

A Rock-crab (*Grapsus strigosus cf*), was very abundant, running about all over the rocks, and making off into clefts on one's
approach. I was astonished at the keen and long sight of this crab.
I noticed some make off at full pace to their hiding places at the
instant that my head showed above a rock fifty yards distant.
The crab often makes for the under side of a ledge of rock when
escaping from danger, and may then be caught resting in fancied
security by the hand brought suddenly over it from above.   The
dry rocks were covered with the dung of the crab, which is in
the form of small brittle white sticks about an inch in length,
very puzzling objects at first sight.   The cast shells of the crab,
which are bright red and very conspicuous, were lying all over
the rocks.

At Still Bay, on the sandy beach on which, although it is on
the leeward side of the island and the sea surface was smooth,
a heavy rolling surf was breaking, I encountered a Sand-crab
(*Ocypoda ippeus*), which was walking about, and got between it and
its hole in the dry sand above the beach.   The crab was a large
one, at least three inches in breadth of its carapace.   In this
species of crab, the eyestalks are very long.   The eyes are on the
side of the stalks which are longer than eyes, and projecting
above them are terminated by a tuft of hairs.   When the
animal is on the alert, these long eyestalks are erected and
stand up vertically side by side far above the level of the
animal's back.

With its curious long column-like eyes erect the crab bolted

* Darwin, "Journal of Researches," p. 6.

down towards the surf as the only escape, and as it saw a wave
rushing up the shelving shore dug itself tight into the sand and

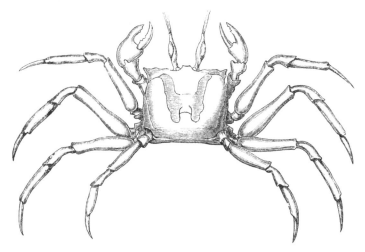

OCYPODA IPPEUS.
(About half natural size.)

held on to prevent the undertow from carrying it down into the
sea.   As soon as the wave had retreated, it made off full speed
along the shore.   I gave chase, and whenever a wave approached,
the crab repeated the manœuvre.   I once touched it with my
hand whilst it was buried and blinded by the sandy water, but
the surf compelled me to retreat, and I could not snatch hold of
it for fear of its powerful claws.   At last I chased it, hard
pressed, into the surf in a hurry, and being unable to get proper
hold in time it was washed down into the sea.

The crab evidently dreaded going into the sea.   These sand-
crabs breathe air through an aperture placed between the bases
of the third and fourth pairs of walking legs, and leading to the
gill chamber.   They soon die when kept for a short time
beneath the water, as shown by Fritz Muller's experiments.*

A lizard or gecko is very common both at St. Vincent and
San Jago.   It appears to be the *Tarentola Delalandii* of Madeira,
or closely allied to this.

* " Facts and Arguments for Darwin," p. 33.   London, John Murray,
1869.

E

A beetle, a species of Cicindela, is very common on the dry sand along the seashore, and is very difficult to catch. The beetles sit five or six together on the sand, and fly off before the wind directly they are approached. They are so quick that I could not catch them with my net. I found, however, that if a handful of sand were thrown at them, they seemed paralyzed for a few moments, and could be picked up with the hand.

Most of the insects on the island are to be found amongst the clumps of tamarisk. An Ant-lion (*Myrmelion*) is very common, making pitfalls for the ants under the lee of all the tamarisk bushes. Spiders are abundant. A large and handsome yellow spider (*Nephila*), makes large webs of yellow silk everywhere amongst the bushes. The silk is remarkably strong, and the supporting threads of the web often bend the tips of the tamarisk twigs, to which they are fastened, right down. Either the spider drags on the thread and bends the twig, or the twig becomes bent in growing, after being made fast to. The result is that the thread is kept tense, although yielding to the wind.

I ascended one day one of the steep slopes on the north-east side of the town, on the leeward side of the encircling range of the island. It was terribly hot and parchingly dry, but the instant the summit was reached, the refreshing trade wind was felt in full force, and its influence was everywhere seen in the increased vegetation, and wherever it lapped over the crest, or crept through a gully, green tufts marked its range.

I climbed a peak about 850 feet in altitude, from which there was a comprehensive view of the island, showing well the general outward dip of the strata composing the encircling range. In the distance was the irregular mountainous outline of the island of St. Antonio, which was blue and hazy-looking, with a line of white clouds hanging against it at a height of about 2,000 feet. How I longed to be at the summit of the principal mountain, 7,000 feet high, to see the European wild thyme growing there far above the Atlantic and African plants! A sheer precipice led down from my feet to the surf and the sea driven into white crested waves by the trade wind, which was blowing with more than ordinary violence, so that it was difficult to stand on the edge of the cliff.

I found a chasm in the cliff where it was possible to descend. At about 200 feet from the bottom of the cliff, where the stratified volcanic rock was intersected in all directions by dikes, was a very small spring, from which issued perhaps a quarter of a pint of water in an hour. It was the only natural spring I saw in the islands, although a few others exist. There was green slimy matter round the spring composed of diatoms and other low algæ, and a small mass of vegetable mould, in which grew two plants which I had not met with elsewhere in the island, a yellow flowered crucifer (*Sinapidendron Vogelli*) and *Samolus Valerandi*.

This miniature oasis was only about four feet in circumference, and absorbed the whole of the water yielded by the scanty spring. A number of wood-lice sheltered in it. I suppose the seeds of these two plants must have been carried to the spring by birds coming to drink.

On returning to the town down the leeward slopes, I passed the principal wells of the town ; they are dug in a now dry stream bed, and are about 15 feet in diameter, and 25 to 30 feet in depth. There was plenty of water in them, but it was slightly brackish, and probably partly derived from the sea.

The trammel net was set nightly in the harbour by Mr. Cox, the boatswain, and yielded some fine fish ; amongst these were some large flying gurnets which evidently, from their being caught in the trammel, frequent the bottom a good deal like our wingless gurnets. One was caught with a line at the bottom. I hooked one, however, near the surface, when fishing with a rod and trout tackle for small mackerel and silver fish. This was quite a novel experience in fishing. The flying fish darted about like a trout and then took a good long fly in the air, and in an instant was down in the water again and out again into the air, and being beyond my skill in playing with such light tackle, soon shook itself loose and got free.

A species of Balistes, called the trigger-fish, because it has a stout trigger-like spine on the back and the belly, which can be erected as a defence, was caught in the net. The living fish when held in the hand makes a curious metallic clicking noise by grating its teeth ; similarly *Diodon antennatus* makes a curious noise by the movement of its jaws, as noticed by Darwin.* I have heard the sound in the case of a *Diodon hystrix* caught at St. Thomas ; it is a sort of grunting sound. A large hammer-headed shark (*Zygœna malleus*), about 12 feet long, was also netted and put an end to the net fishing for some time by tearing the net to pieces.

We left St. Vincent on August 15th. I went on that day with Captain Nares on a boat excursion to collect corals in a small bay with a westerly aspect, not far from Porto Grande. On our way we passed under a rocky mountain, 1,594 feet in height, which has an outline remarkably like that of a man ; the nose, mouth, and chin, are well marked, and the entire range in connection looks like a giant lying on his back.

The small bay we visited was bounded by steep cliffs. On the rocks beneath was the usual zone of calcareous seaweeds. A coral (*Cœnopsammia Ehrenbergiana*), composed of bundles of delicate tubes fused together side by side, covered the rocks profusely just below tide level, forming bright vermilion and bright yellow masses, which showed out conspicuously as the swell fell now and then and exposed the rock surface lower down than usual. The coral appears to vary in colour in an irregular manner, some clusters of the coral being red, with the exception of one or two tubes at one corner of the mass, which were yellow, and I saw a young yellow bud given off from a red parent tube. Some masses were entirely yellow, and in some places only yellow corals were to be seen, but on the whole the red predominated.

At the north point at the mouth of the bay was a regular fishing station, where two young Africans were fishing, and where the whole rock was reeking of dead and decaying fish, and a small cave was full of *débris*, having evidently been made use of by fishermen for many years.

The two young negroes at first occupied themselves in catching small fish with a short bamboo rod, baiting with pounded fish, and catching various little rock fish and a Scarus. They then began pounding and breaking up the small fish and

* Darwin, "Journal of Researches," p. 14.

throwing largish pieces of the mass into the verge of the surf off the point to attract large fish.

They watched until they saw a large fish taking these baits on the top of the water, and then they threw a bait on a hook attached to a long cod line. They thus caught a large Cavalli (*Caranx*), of the mackerel tribe, which they had to play for some time and finish with a spear. Large Garfish (*Belone*) sometimes came within reach, and were easily caught, being very ravenous.

One fish, a kind of Bonito or tunny (*Thynnus Argentivittatus*), of about 25 lbs. in weight, was attracted by the baits, and coming close in swam backwards and forwards in front of the stand on the rock, taking every bait thrown on to the top of the water. The negroes kept feeding the fish for some time to give it confidence. A very strong piece of cord with a hook like a salmon gaff made fast to it, was then baited with a small bit of fish, just enough to cover the point of the hook, and a stout bamboo was used as a rod. The cord was hitched tight round one end of it, with about a foot of it left dangling with the hook. One negro held the rod and the other the cord.

The bait was held just touching the surface of the water. The fish swam up directly and took it, the negro holding the bamboo struck sharply and drove the big hook right through the fish's upper jaw, and both men caught hold of the line and pulled the fish straight out on to the rock. The negroes evidently felt quite certain of their fish directly they saw it swimming backwards and forwards in front of the rock. I was astonished that so large a fish could be caught in so absurd a manner. The negro holding the pole was not six feet from the fish when it took the bait.

The inhabitants of St. Vincent are mostly negroes from the adjacent coast. In the town at Porto Grande there was an albino negress, who was exhibited to visitors.

Of birds the most conspicuous at St. Vincent are the scavenger vultures (*Kathartes pernicopterus*), the same which are to be seen in great numbers about the native town at Aden, and about all the towns of Egypt and northern Africa, and which even follow caravans across the desert as gulls follow ships. The birds were always to be seen about the waste land close to the town where garbage was thrown, and were often to be seen hunting over the refuse heaps in company with ravens and crows. Some small finches were common in flocks on the hills and some small hawks.

At the periods of migration, quails are extremely abundant on the island, as at St. Jago, and often afford good sport to naval officers; they are, however, mere birds of passage here, and there were none at the time of our visit. Of sea birds I saw a cormorant and a bird which looked in the distance like a Merganser. Gulls and terns were absent entirely.

I was told that the goats which are wild on the island, have all attained a red colour resembling that of the rocks, and that they were hence very difficult to find and shoot; I, however, saw none myself.

August 6th.—The island of Fogo was in sight; it appeared to our view as two truncated cones, showing out against the sky above a bank of clouds. One of the cones, which is 9,000 feet in height, is much higher than the other, and has a tiny secondary cone at one edge of its main terminal crater, just like Pico in the Azores. The volcano is active, but had no smoke issuing from it as we passed. The peaks showed out against the sky far above the horizon.

I was constantly astonished at the great height above the horizon to which high mountainous islands seem to rise when viewed from a long distance at sea. This appearance was especially marked in the case of the Peak of Teneriffe. One is apt to scan the region of the horizon, when the Peak is just in sight far too low down, being accustomed to search for much less elevated objects which become visible directly they rise above the horizon. The line of sight traversing in that direction, clearer air allows the summit of the high distant mountain to be visible long before the base.

When we were approaching the Azores, we sighted the island of Corvo at a distance of sixty miles. The island appeared remarkably near, being thrown up high above the horizon probably by atmospheric refraction. The distance of the island was guessed from its appearance at from seven to twenty-five miles. The island disappeared from view before mid-day by a change in the condition of the atmosphere, which nevertheless appeared clear.

St. Jago Island, August 7th, 8th, and 9th, 1873.—The ship anchored at Porto Praya, the port town of San Jago, Cape Verde Islands, on August 7th. The harbour is exposed to the south-west, and, during the rainy season, from August to October, when south-west gales are frequent, is unsafe. The harbour is bounded by black basaltic cliffs, in which, in several places, a fossiliferous limestone bed, which is described by Darwin, shows out as a conspicuous white streak.

The town is placed on an isolated mass of a flat, elevated plain, which terminates abruptly seawards in the cliffs above described. A deep valley, with a flourishing grove of cocoanut trees at its bottom, separates this mass from the main table-land on the east side. On the west side, at the base of the mass, lies a sandy plain which extends far back into the country and terminates seawards in a sandy bay, admirably adapted for the use of the sein net. On this plain, behind the town, is a large plantation of date-palms, with artificially irrigated gardens beneath their shade. The dates were hanging thick upon the trees, but were as yet yellow and unripe; in ripening they turn first red and then deep purple or black.

There is a large Baobob tree near the town, which has been mentioned by travellers: its stem is irregular in transverse section and short; it measured 42 feet in circumference at the time of our visit. The tree was then in full flower, with no fruit as yet of any size.

The country rises inland in a succession of terrace-like steps often remarkably flat at the tops, and formed by successive flows of lava. The flat table-land nearest the sea was parched and had very little green upon it. Behind rises a succession of small conical hills and higher table-lands, which were brilliantly green.

As the ship came to anchor, a flock of kites (*Milvus korschum*) came wheeling round the stern, just as do gulls ordinarily, and keep swooping down after garbage from the ship. Instead of seizing the morsels with their beaks, like gulls, they did so with their claws, putting out one foot for the purpose as they swooped down, and seizing the food with it with wonderful precision. As they rose they bent down their heads and ate the food at once on the wing from their claws. Some large fish came round the ship, and amongst them some sharks, one of which was seen to seize one of the kites as it put its foot down to the water and carry it down after a short struggle.

I landed with a party in search of quail shooting. We landed at a small stone jetty under the cliff beneath the town, and mounted by a zigzag path and steps to the top; here just above the landing-place are the barracks, one-storied, with iron-grated unglazed windows, a conspicuous feature in the view of the town from the anchorage. The town consists of about two dozen two-storied houses, mostly surrounding a public square, and a number of one-storied hovels and low wooden houses, disposed in three or four parallel streets, along the ridge on which the town stands. The inhabitants are nearly all negroes, the remainder being Portuguese and half-castes. Attempts were being made to improve the place, and there was a fountain in the middle of the square with young trees planted round it and good water is laid on to the town from a distance of several miles.

As soon as we landed we were beset by a crowd of negro boys, wanting to carry our cartridge bags and show us where plenty of quails and galinis were to be found. We each selected our boy and made for the high flat plain across the valley to the west. The plain was covered with tufts of short dry grass, and scanty patches of young seedling grasses just coming up. Scattered about were patches of the darker green of the abundant trailing Convolvulus (*Ipomœa pes capræ*). The elevated plains are intersected in all directions by deep gorges cut out by watercourses which were now quite dry; the gorges have usually steeply sloping sides which terminate above in a range of cliffs.

Quails were not at all plentiful, being only migratory visitors to the island, and not having as yet arrived. The entire party shot only about twenty. The Kingfisher mentioned by Darwin (*Halcyon Erythroryncha*), is common. The bird is peculiar to the island, though very closely allied to an African species. It is a beautiful bird of a brilliant blue and white with a red beak.

Like many other kingfishers it is not aquatic in its habits, but feeds mainly on locusts and other small terrestrial animals. It has a terribly harsh laughing cry, a feeble imitation of that of its congener of Australia, the laughing jackass.

We met with several flocks of wild galinis, which are abundant on the island, but are very difficult to approach. The birds inhabit the slopes of the gorges which are covered with a thick growth of oil trees (*Jatropha curcas*) which have very much the habit and general appearance of castor-oil plants. The flocks of galinis station sentries to keep a look-out from some rocky eminence, and these, when once they have discovered an enemy, never lose sight of him, but carefully watch the stalking operations of a sportsman and give warning as soon as he gets too near to their comrades and is just expecting to get a shot.

We returned to the town in the afternoon in order to join a seining party. All English men-of-war on foreign service are provided with a sein net, and a seining party is regarded as a sort of lark or picnic by the Blue-jackets. There are always plenty of volunteers eager to go, and a good many officers are ready to join.

With us, Mr. Cox, the boatswain, was the great man on such occasions, and he enjoyed the sport as much as anyone in the ship. The party of volunteers, of perhaps thirty men besides the officers, goes ashore in the afternoon at about four o'clock in one of the cutters with the net in the dingey, the smallest ship's boat. Then the net is payed out, and everyone is dressed and prepared for going into the water up to his neck and hauling on the lines. At last in comes the bag of the net, or "cod" as Mr. Cox calls it. It is run up the beach with a final spurt, and then comes the fun of handing out the fish and looking at the many unfamiliar forms, for which the Blue-jackets have all sorts of extraordinary names.

At one haul on the present occasion there was a large shark (*Carcharias sp.*), 14 feet long in the net. Mr. Cox in the dingey following the net as usual as it was drawn in, in order to free it if it should hitch on the bottom, sighted the shark swimming round within the rapidly decreasing circle, and making bolts at

the net to try and break through. And the beast would have
burst through had not Mr. Cox hammered it on the head with
a boat-hook whenever it turned at the net, whilst the men
belaboured it with anything they could get hold of as it got
drawn into shallow water.

There was great excitement, and it seemed very uncertain
whether the shark would not break the net and let out not only
itself but all the other fish. At last we ran the brute up high
and dry, and then it suffered instant punishment.

The sailor has absolutely no pity upon a shark. I have
heard one of our men say to a shark which he had just hauled
on to the forecastle with a line, " Ah, thou beggar, thou'd hurt I
if I was in the water and now I'll hurt thee," whereupon
he caught it a vicious kick and proceeded to gouge it. When a
big shark like the present one is landed it is regarded as a
general enemy, against whom everyone has an old score to
pay off. Mr. Cox shoves the boat-hook about five feet into its
mouth and down its throat. The others job the beast in the
eyes with sticks and knives and make a deep slash across
the tail to prevent its lashing out, and proceed to open the belly,
where the usual miscellaneous collection is found ; lots of ships'
beef bones, a two pound lead sinker of a fishing line, with chop-
stick and hooks complete, &c., &c.

We caught plenty of fish. Gray and red mullet, a Gar fish
or Greenbone, with long slender beak-like jaws (*Belone*), and
another fish closely like the Greenbone, but with a long beak-
like lower jaw only, the upper jaw appearing as if cut off close
to the snout (*Hemiramphus*). With these were other curious fish
with deformed-looking heads (*Argyreiosus setipinnis, Galeoides
polydactylus*).

A fire had been lighted on the shore and we had a ship's
boat's cooking stove with us. We fried some of the fish, and
with bread and preserved meats and plenty of beer made a
good supper, and set to work again hauling the net till it had
long been dark. Then we had hot tea and grog, and packed our
net and fish into the boats and pulled on board.

We did not reach the ship until past 11 P.M., and at 3 A.M.
I was, by arrangement, to start on a trip to try and ascend the

high mountain of the island called San Antonio, 7,400 feet
in height, in search of the European plants which grow there.

I had a very short sleep and landed at 3 A.M. I found two
horses ready at the landing-place but my guide was not there,
and it was a long time before I could make the men with
the horses, who spoke only Portuguese, understand what I
wanted. At last a negro, who was sleeping on the pier, agreed to
find the guide, John Antonio, for a shilling, and I sat down on
the pier wall to listen to the surf and watch the crabs (*Grapsus
strigosus*) running about, for nearly an hour.

The parapet of the jetty had a capping upon it projecting
some distance and with a rounded edge. I saw a crab running
on the jetty, and I thought I could catch it, but to my astonishment it ran with readiness over the edge of the parapet, round
the projection and down the flat face of the wall, with all the
ease of a fly under similar circumstances.

At last my guide, John Antonio, a negro who spoke English,
arrived. He was to have been at the rendezvous at 3 A.M., but
said he was too sleepy. We mounted and rode off inland;
after about an hour's ride day began to break. As we ascended
successive terraces the hills became greener and greener, being
covered with a continuous carpet of seedling grass, and other
herbs, as yet only two or three inches in height. John said that
it would be a foot or eighteen inches high later on, and that then
the quails would abound and the galinis breed, so that the
breeding season of these birds appears here to occur in autumn,
determined by the rainy season. Numbers of the galinis are
taken when quite young, and their eggs are also sought after.

The quantity of birds of prey in San Jago is remarkable.
We passed numerous large falcons at rest on dead trees and
several hawks, and an owl flew across the road just at daybreak.
I saw also two eagles in San Domingo Valley. Ravens and
crows are abundant.

The valley of San Domingo, into which our road at length
led, is deep, with precipitous cliffs and steep mountains on either
side, rising from 1,000 to 2,500 feet above sea level. The valley
is broken here and there by lateral offsets and backed towards its
head by irregular mountain masses. The view up the valley is

very beautiful. Beneath the cliffs, which are encrusted with lichens and stained of various colours, often of a deep black, are steep talus slopes covered with oil trees, with a few other shrubs sparingly intermingled. At the bottom of the valley is a strip of comparatively level land, on which are cultivated all sorts of tropical fruits, pineapples, bananas, oranges, lemons, guavas and cocoanuts : with cassava, sweet potatoes and sugar-cane as field crops.

All along the valley a little way up the slopes are small huts, where boys are stationed, whose duty it is to keep off the monkeys, which abound amongst the rocks, and the wild blue rock pigeons (*Columba livea*) which are very numerous, and were seen flying about in flocks and alighting in the road as we went along.

John Antonio said that the monkeys used their tails to pull up the sugar cane and cassava with, an unlikely story, since the monkeys must be some imported African species run wild. I was astonished to hear that there were monkeys at all in the island, and have not seen the fact mentioned in any account of the place. John said that the monkeys never came out in wet weather. I did not see any of them. The boys kept up a constant shouting, which resounded through the valley.

At the bottom of the valley is a small stream running rapidly over the stones, like a trout stream, and everywhere very shallow. In this stream grow watercresses and several familiar English water plants, and I found two ferns on the banks. Two kinds of freshwater shrimps live in the stream under the stones, and are very abundant, notwithstanding the shallowness of the water. One is a Palæmon, a large prawn, as big as the largest specimens of our common river crayfish, and with long and slender biting claws.

The other kind is a very different animal, somewhat smaller, and of the genus Atya, which is distinguished by having no nippers on the larger pairs of walking legs, but only simple spine-like ends to them. The two front pairs of walking legs have, however, most extraordinarily shaped claws at their extremities ; quite unlike any occurring in other Crustacea, except the Atyidæ, as will be ·seen from the figure. These claws or nippers have slender arms of equal length and dimensions, which are linked together so as to open and shut like a pair of forceps, closing flat against one another.

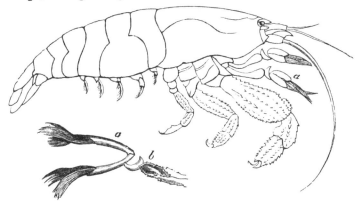

ATYA SULCATIPES. (Natural size.)
a One of the front pairs of walking legs. Beneath; the same pair enlarged; *a* the nippers widely open; *b* the crescent-shaped joint to which they are hinged.

At their extremities these forceps arms are provided with thickly-set brushes of long hairs, as long as the arms themselves. These hairs expand in the water when the forceps are opened, and evidently form a widely-sweeping grasping surface, by which small particles of food or minute animals can be caught. No doubt these forceps catch the food of the Atya, and the larger legs with simple pointed ends enable it to hold on to the stones in the rapid stream.

The pair of forceps is not attached directly at its hinge joint to the end of the limb, but at a point on the side of one of the arms. Here it is hinged on to a crescent-shaped joint, into the crescent of which the rounded end of the forceps is received when the apparatus is retracted and at rest. The complicated manner of jointing gives a very wide sweep and great mobility to these very curious prehensile organs.

The genus Atya must, from its very wide distribution, be a very ancient one. Species of the genus occur in the West Indies, in the Philippines, in Samoa, and in Mexico, besides in the Cape Verde Islands. The Cape Verde species * is possibly

* Atya sulcatipes (Newport)? A. scabra (Leach). "Ann. and Mag. Nat. Hist.," 1847, p. 158, where is a list of species. Upolu is in it placed by

identical with one occurring in Mexico. In Mexico and the West Indies the animal occurs in the sea : elsewhere in fresh water.

I am greatly indebted to Sr. Jose M. Quirino Chaves, U. S. Vice-Consul of Porta Praya, who most kindly sent me specimens of the above described crustacea, on my writing to him, when preparing this journal for the press. The only specimen which I secured on my visit was lost by accident on board the "Challenger." The Palæmon is called in the island "Christao," The Atya, " Mouro."

John Antonio said there were no fish in the San Domingo stream, " cos river fresh water." He evidently thought that fish were to be found only in the sea.

We passed the village of San Domingo, which consists of scattered thatched stone houses, and the road became worse and worse, being sometimes knee-deep in mud. The ponies, small fine-built bays, began to show signs of giving in, and soon spurring would not make mine move further. I had to dismount and flounder back to a cottage, where we had a rest, and fed the ponies with grass. The excursion up the mountain is evidently too long for one day, although John Antonio had declared before-hand that it was an easy matter. I had been riding five hours, and we were still a long way from the place where the actual ascent commences. The ponies went very badly, at little more than a foot-pace. It was raining more or less during the whole time that we were in the valley.

The Portuguese, at whose house we stopped, said that it was impossible to ascend the mountain in the rainy season, because of the falls of stones, or stone avalanches, which were common and dangerous. All this I failed to find out before leaving the town, the natives of the island there knowing nothing of the mountain. At the house I got some coffee, which was grown in the valley just below.

I ascended the steep side of the valley, to a ridge about 1,500 feet above sea level, but did not find anything in the plant

mistake in New Zealand instead of Samoa. M. Edwards places Atya with Alpheus. Dana, (U.S. Exp. Ex. Crustacea), places Atya, Atyoides and Caridina, in a special family Atyidæ, next the Astacidæ.

way to reward me, the plants being the same as lower down the slope. The oil tree (*Jatropha curcas*) grew up to the top of the slope. There were none of the mountain plants which occur at St. Vincent at this height. There were a good many fungi. They apparently spring up luxuriantly during the wet season. Plants generally grow at a lower level at San Jago than at St. Vincent. Thus, *Sarcostemma Daltoni* in San Jago grows abundantly almost at sea level on the cliffs near the harbour. In St. Vincent I found none lower than 900 feet. The plant was in full bloom at San Jago. In St. Vincent I found only a single blossom, though the plant was very abundant.

I exchanged a drink of ship's rum with my Portuguese host for his cup of coffee. He had a very pretty young yellow wife, who on my return to the house was pounding maize in a large wooden mortar, assisted by a very black servant girl, each of them wielding a heavy pestle, and striking alternately, like blacksmiths on an anvil. A little water was sprinkled on the maize to assist the process.

John Antonio was well known all along the road, and most elaborate courtesies passed between him and every one we met, or whose house we passed by, sometimes a Creole, sometimes a Portuguese. He explained that the Creole greeting which he used meant, " What you feel ? " In Portuguese he always addressed everyone as Sir, and after mutual congratulation on the subject of health, he entered into a lengthy explanation of who I was, which wasted a great deal of our time. John was a thin, spare man, with a very ragged coat and trousers, which had evidently once been respectable on a previous owner. He was perpetually hungry and thirsty.

As soon as the horses were rested we started back. I shifted my single spur, for John and I wore a pair between us, to my left foot, and managed to reach the town by 3 P.M., in time to join a second seining party. The seining was suddenly brought to a conclusion, for a south-west gale being expected, we were hurried on board. A heavy swell had set in by the time we reached the ship so that there was some difficulty in getting up the ship's side. We found all the boats hoisted, and steam up, ready for sea at a moment's notice.

San Domingo Valley, with its succession of mountain ridges and peaks becoming bluer and bluer in the distance, is one of the finest mountain valleys I have ever seen, and the tropical vegetation gives it an especial charm. The sight of such a place is particularly delightful to a man who has for weeks been trudging the arid hills and plains of St. Vincent, and who has just ascended to it from the almost equally sterile plains about the coast of San Jago.

The gale did not come as was expected, and another opportunity of landing being afforded, I went with Buchanan to look at the peculiar limestone bed described by Darwin.* On our way we passed through the grove of cocoanut trees ; at the foot of these trees are numerous holes of a large land crab (*Cardisoma*) ; the female of this land crab was found by Von Willemoes Suhm to carry its eggs and newly hatched young under its abdomen ; the young emerge from the eggs in the larval zoëa condition,† and are found in that state attached to the abdominal legs of the mother.

As we made our way along the cliff we disturbed a flock of rock pigeons which breed abundantly in the cliff, and also a wild cat, which was no doubt watching them. The cat was of a reddish tabby colour; they are very abundant on the island, and it is not easy to understand how so many animals of prey, cats, hawks, crows, &c., manage to subsist here. In the quail season no doubt they have abundance, but in the dry season they must often be nearly starved.

The limestone band exposed in the cliff around the harbour is topped by a thick mass of basaltic lava, which as it flowed over the limestone baked and heated it, and altered its structure. The limestone band crumbles and weathers away, and thus leaves a hollow all along the cliff about half way up its height, which forms a convenient path for men and goats. By the cropping out of the limestone the under surface of the lava-

---

* C. Darwin, " Journal of Researches," pp. 5, 6.

† R. Von Willemoes Suhm, "On some Atlantic Crustacea from the 'Challenger' Expedition." " On the Development of a Land Crab." Trans. Linn. Soc., 2 Ser. Zoology, Pt. I, 1875, p. 46. Proc. R. Soc., No. 170, 1876, p. 582.

flow is exposed to view and in many places ripple marks can be seen in it.

The limestone bed, where exposed to the air, is of a dazzling white; it is full of rounded nodules of a calcareous alga as described by Darwin,* a species of *Lithothammion.*† I dredged closely similar nodules to these in ten fathoms off the Philippine Islands, in bushelsfull. These nodules were living masses of *Corallinaceæ*, but loose rounded and unattached, yet covering and composing the sea bottom. The basalt, undermined by the cropping out of the limestone, falls in large masses and splitting off with great regularity leaves the cliff with a remarkably smooth vertical surface.

Red or precious Coral occurs at San Jago, and also at St. Vincent. There are four or five Spanish boats, and seven belonging to Italians, engaged in the fishery for it at San Jago It occurs in about 100 to 120 fathoms, and is dragged for with swabs as in the Mediterranean: the strands of the swabs are made up into a net with about a four-inch mesh. A duty of a dollar a kilogram is paid to the Government on the coral.

A pair of huge fish came round the ship whilst at anchor in the harbour during the afternoon; one, supposed to be the male, was struck with a harpoon, but after some time managed to draw it out by its struggles; it twisted up the harpoon and was said even to have moved the ship in its throes. I did not see the fish, but from the description, coupled with the fact that there were a pair of the fish, it seemed probable that the fish were the huge ray *Cephaloptera*, the "Devil fish," which has curious horn-like projections sticking out in front on either side of the mouth. The fish were described as "as big as an ordinary dining-room table." ‡

The voyage from San Jago to St. Paul's Rocks occupied nineteen days. When we were two days out some swallows paid us a visit, flying behind the ship. We ran at first parallel with the African coast, and then stretched over westwards to St. Paul's

---

* C. Darwin, "Volcanic Islands," p. 3. Smith and Elder, London, 1866.
† Prof. G. Dickie, "Journ. of Linn. Soc.," Vol. XIV, p. 346.
‡ For an account of a visit to Porto Praya, see G. Bennett, "Wanderings in New South Wales," Vol. I, p. 15. London, R. Bentley, 1834.

Rocks.   We passed first through a region where we had a pretty steady south-west wind, an African land breeze or monsoon. Here we had occasional heavy showers, but not so much rain as was to be expected, since we were passing a region where it rains on an average for seven hours out of every twenty-four, all the year round.   We next steamed through the belt of equatorial calms to reach the south-east trade winds, and left the Guinea current, which was running at the rate of 21 miles in 24 hours. We entered the trade wind on August 21st, and the air became damp and cooler than before, and we were soon running before the wind at the rate of seven or eight knots.

CHAPTER III.

ST. PAUL'S ROCKS AND FERNANDO DO NORHONA.

St. Paul's Rocks. Equatorial Current. Nests of Noddies. Predatory
Habits of *Grapsus strigosus.* Fishing off the Rocks. Nests of
Boobies. Pugnacity of the Young Birds. Other Inhabitants of the
Rocks. Fishing for Cavalli with Salmon tackle. Geological Structure
of the Rocks. Seaweeds growing on the Rocks. Fernando do Nor-
hona. Calcareous Sandrock containing Volcanic Intermixture. Tree
Shedding Leaves in dry season. *Japtropha urens.* Birds. Brazilian
Convicts. St. Michael's Mount. Frigate Birds Nesting. Pigeons
Nesting with Sea Birds. Lizards of the Islands.

St. Paul's Rocks, August 28th and 29th, 1873.—The ship
arrived at St. Paul's Rocks, on August 25th. The rocks are
about 540 miles distant from the coast of South America, and
350 miles from the island of Fernando do Norhona. The group
of rocks is scarcely more than half a mile in circumference, and
their highest point is only 64 feet above sea level.

At 5 P.M., the rocks were about half a mile from the ship.
Their smallness is the striking feature in their appearance as
they are approached. They show themselves as five small projecting peaks, which are black at their bases, and white with
birds' dung on their summits. A yellowish-white band shows

ST. PAUL'S ROCKS.

out about tide mark. The sea was dashing up in foam at the
south-east end of the rocks, and a long line of breakers stretching
from the opposite end marked the course of the equatorial
current.

The birds were to be seen hovering over the island in
thousands. Only three kinds inhabit it. Two noddies and the

F 2

booby. The noddies (*Anous stolidus* and *A. melanogenys*) are small terns or sea swallows, black all over, with the exception of a small white patch on the head. The booby (*Sula leucogaster*) is a kind of gannet. The full-grown birds are white on the belly, with a black head and throat; the black ending on the neck, where it joins the white in a straight conspicuous line. The back is dark. The younger birds are brown all over. Some few of both birds soon came off to have a look at the ship.

We moved gradually up to the islands, sounding as we went; the Captain and Lieutenant Tizard mounted into the foretop, and steered the vessel from thence, looking out for rocks. The water is deep right up to the rocks, and a hawser was sent on shore in a boat, and made fast round a projecting lump of rock, and the ship was moored by means of it in about 100 fathoms of water, although not more than 100 yards distant from shore. Such an arrangement is only possible under the peculiar circumstances which occur here. The wind and current are constantly in the same direction, and keep a ship fastened to the rock always as far off from it as the rope will allow.

I never properly realized the strength of an oceanic current until I saw the equatorial current running past St. Paul's Rocks. Ordinarily at sea the current of course does not make itself visible in any way; one merely has its existence brought to one's notice by finding at mid-day, when the position of the ship is made known, that the ship is 20 miles or so nearer or farther off from port than dead reckoning had led one to suppose she would be, and one is correspondingly elated or depressed. But St. Paul's Rocks is a small fixed point in the midst of a great ocean current, which is to be seen rushing past the rocks like a mill-race, and a ship's boat is seen to be baffled in its attempts to pull against the stream.

Between the two extremities of the main body of rocks, is a bay, enclosed by a somewhat semicircular arrangement of the rock masses. We landed on the eastward side of this bay. Landing from a boat is a little difficult. There is a perpetual swell running in the bay, although it is on the sheltered side of the rocks, and one has to jump as the boat rises, and cling to the rocks as best one may.

I landed in the first boat. The rock was covered with noddies, and their nests, some containing eggs, whitish in colour, with red spots at the larger end, and others with young in them, little round balls of black down. The air was full of noddies and boobies, circling about, and screaming in disgust at the invasion of their home.

The noddies' nests are made of a green seaweed (*Caulerpa clavifera*) which grows on the bottom in the bay and around the rocks, and which getting loosened by the surf, floats, and is picked up by the birds on the surface. The weed is cemented together by the birds' dung, and the nests having been used for ages, are now solid masses, with a circular platform at the summit, beneath which hang down a number of tails of dried seaweed. The older nests pro-

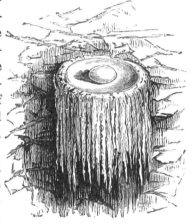

NEST OF NODDY AT ST. PAUL'S ROCKS.

ject from the cliffs on the shel-tered sides of the rocks, like brackets, having been origin-ally commenced, as may be seen by the complete gradua-tions existing, by a pair of birds laying an egg on a small projecting ledge of rock and adding a few stalks of weed.

It is only the stronger and more vigorous noddies that are able to occupy and hold posses-sion of a nest of this descrip-tion. There are only a limited number of such on the island, there not being cliffs enough to accommodate more.* The island being somewhat over-populated, a great many noddies have to put up with the bare flat rocks as breeding-places, and there they lay their eggs in any slight

---

* The two species of noddy occurring at the rocks are so nearly alike, that I did not notice at the time that there was more than one species present ; a fact which I have since learnt from Mr. Howard Sanders' paper—"On the Laridæ of the Expedition," Proc. Zool. Soc., 1877, pp. 797, 798. Possibly the birds, which make bracket-like nests, are of one species only, and those which build on the ground, of the other.

hollow or chink. They are plucky birds, and the old ones sometimes make dashes at the head of an intruder who goes too near their nest. They had so little fear of man, from want of experience of his cruelty, that we could have caught any number of them with our hands.

In vast abundance, all over the rocks, crawls about a crab (*Grapsus strigosus*), the same as that already noticed at the Cape Verde Islands. This crab has been referred to by nearly all visitors to the rocks. It is far more wide-awake than the birds, and keeps well out of reach, being thus of some difficulty to catch. The crabs are all over the rocks, every crevice has several in it.

You are fishing, and you have put down at your feet a nice bait, cut with some care and difficulty from a fish sacrificed for the purpose. You are absorbed in the sport. A fish carries off your bait; you look down and see two crabs just disappearing into an impracticable crevice, carrying your choice morsel between them. You catch a fish and throw it down beside you. Before long you find a swarm of crabs round it, tearing morsels off the gills, using both claws alternately to carry them to their mouths; and a big old crab digging away at the skin of the fish, and trying to bite through it.

If a bird dies the crabs soon pick its bones, and I saw one old crab profiting by our having driven off all the old birds, and carrying off a young bird just hatched. The older crabs are richly coloured, with bright red legs. The crabs have odd ways, and curious habits of expressing anger, astonishment, suspicion, and fear, by the attitude of their claws. When two old crabs meet unsuspectingly in a crevice they dodge one another in an amusing way, and drawing their legs together strut on tiptoe.

In the tropics one becomes accustomed to watch the habits of various species of crabs, which there live so commonly an aerial life. The more I have seen of them the more I have been astonished at their sagacity. I had, I do not know why, always considered them as of low intelligence.

Admiral Fitzroy gives an account of the large numbers of fish caught off the rocks by his men, and states that they hauled the fish up from the bottom with difficulty because they were always rushed at by voracious sharks.

In the evening volunteers for fishing were called for, and I went in the jolly-boat with about six officers and four or five men. A cutter full of men also put off. We made fast to the line across the bay, and for a long time got nothing, till at last, when we were getting tired, one man caught a shark, about three feet long, and we all got good bait from him.

Then we caught more sharks, and it was at last discovered that we ought to have been fishing at the surface, and not at the bottom. As soon as we took the sinkers off our lines and allowed the baits to float we began to haul in large fish, some of them 20 lbs. in weight, as 'fast as possible. The fish were " Cavalli " (= seahorse ?)—a species of *Caranx*, which is allied to our mackerel, and very good to eat.

The fish were very game, and pulled hard, making phosphorescent flashes as they dashed about in the water under the boat, it being now dark. Every now and then someone hooked a shark (*Carcharias sp.*), and then there was a tremendous fight, and all the lines in the boat were tangled and fouled as the big fish rushed around. At last it either broke the line, or was hauled on board. When the latter was the case everyone stood clear, whilst the shark hammered in its flurry the thwarts and bottom of the boat, till they resounded. At last its tail was cut, and it was then soon slit up into bait pieces.

Sometimes, a tremendous sudden pull was felt at one's line, and it went fizzing through one's fingers without possibility of checking it. The only thing to be done was to take a turn round a belaying pin. Then came a check, and the line broke right off, without even a momentary struggle, and some big shark went off with hook and bait, without probably noticing anything the matter. We returned to the ship at 12 P.M., with enough fish to give the whole ship's company a breakfast.

In the morning I went to a white peak on the western side of the bay. This rock forms the home of the boobies, which are not nearly so numerous as the noddies, and seem to be almost restricted to this one peak out of the five of which the islands are made up.

The whiteness of the rock is caused by the birds' dung, which in some places forms on the rocks, as described by Darwin, an enamel-like crust, which is hard enough to scratch glass. I found some of this at about 45 feet above sea level.

The rock is 50 feet in height, steep on the sheltered sides, and there hung all over with the bracket-like nests of the noddies; the weather-side slopes more gently; and all over it, on every little flat space, are the boobies' nests, mere hollows, some containing two eggs, but mostly with one only. The eggs are as large as a fowl's, sometimes dirty-white all over, sometimes blotched with brown.

In many of the nests were young, which were of all ages; some just out of the egg, ugly big-bellied black lumps, without a particle of down or feathers; then larger ones, as big as one's fist, covered with white down; then others as large as a fowl, thickly clothed with down; then larger ones again, with brown wing feathers and brown feathers on the breast, the white down remaining only in patches, about the head especially. Then birds with brown feathers all over, full-sized and just beginning to fly.

Two almost full-grown birds, as big nearly as geese, were having a desperate fight at the bottom of the slope as I came up. They evidently thought each other the cause of the whole disturbance. They fought furiously with their sharp bills, flapping their wings, and half screaming, half croaking, with anger. They fought till they were quite exhausted, and could not stand, but went at it again after they had rested awhile and recovered their breath.

Some old boobies were sitting on their young on the top of the peak. They would not move until actually pushed off the nest. The young, both of boobies and noddies, are very brave, and scream and strike out hard at anything put near them. Our spaniels could not tackle the young boobies, but after one or two pecks fought quite shy of them; and even the little noddies kept the dogs pretty well at bay, twisting round in the nests and always showing front. Natural selection has no doubt brought about this bravery in the young, to protect them from their constant enemies, the crabs.

Around all the nests were small flying fish, which are brought by the old birds in their crops, and ejected for food for the young or for the females whilst sitting. Fitzroy visited St. Paul's Rocks on February 16th ; Ross on May 29th ; we on August 29th ; on all these occasions eggs and young birds were found. Hence, breeding goes on all the year round.

The only other terrestrial inhabitants of the rocks besides the birds are insects and spiders which prey on them. They are most of them to be found by breaking up the nests of the noddies. Darwin* mentions the following :—A pupiparous fly (*Olfersia*), living on the booby as a parasite. This fly belongs to the same group as the curious *Nycteribia*, so common on the bodies of fruit-eating bats. The group is remarkable for the fact that the female, instead of laying, like most insects, eggs which produce grubs, produces a chrysalis, from which the fly in a short time emerges.

A Staphylinid beetle (*Quedius*), a tick, a small brown moth, belonging to a genus which feeds on feathers, and a wood-louse, living beneath the guano, and spiders, complete Darwin's list. We found two species of spiders, which cover the rock in some places with their web, and in addition to the insects noted by Darwin, the larva of a moth, apparently a *Tortrix*, and a small *Dipter*. Von Willemoes Suhm also found a *Chelifer*, but could not find either the beetle or wood-louse.

Besides these there are of course to be reckoned the lice, parasites usual upon the two birds, and the list of air-breathing inhabitants seems then complete.

St. Paul's Rocks being close on the equator, the sun was extremely powerful, and the white guano-covered rocks reflected the radiant heat-rays with the same effect as does a snow surface in Switzerland. Our faces were severely sunburnt. At the base of the "Booby's hill" is a flat expanse of rock with tide pools upon it, in which were shoals of small fish, a black and yellow banded *Chœtodon* and numerous small gobies. The sides of the pools were covered with a grey *Palythoa*, a sea anemone, forming colonies of the same species apparently as that at St. Vincent,

* Darwin, "Journal of Researches," p. 10.

Cape Verde Islands. The only seaweeds, however, growing in these pools were encrusting nullipores (*Corallinaceæ*).

Numerous Cavalli had been caught by the men fishing from off the rocks in the morning. Lieutenant Aldrich started fishing for them with a salmon rod and tackle. The fish fought for the bait, racing after it as it was drawn along the top of the water in the small bay. One could pick out the largest fish in the shoal and manœuvre the bait with the rod, so as to prevent any but that one taking it. The fish showed fine sport, and I broke my salmon rod over one of them in trying how hard I could give him the butt; we played them until tired out, and then gaffed them.

The Cavalli bite best in the early morning and at night; at noon and in the afternoon they seem to cease feeding, and as soon as they leave the field open, shoals of trigger-fish (*Balistes*), a species of a sooty black colour with a blue streak along the base of the anal and posterior dorsal fins, appear on the scene, and rush at the baits and soon clear the hooks, being nearly safe from being hooked because of the smallness of their mouths. These fish are quite fearless and are small, weighing only about one pound, and of no use for food.

With these fish appears a bright red and green Wrasse (*Labrus*), and a small blue Chœtodon with dark stripes. Three other fish which I saw caught were a Barracuda pike (*Sphyræna barracuda*), a yellow eel with black spots (*Muræna*), and a red Beryx. A Rock-lobster, a small Palinurus, is very common about the rocks, and is to be seen clinging to the rock, having crawled just above the reach of the waves. I caught some of these in lobster pots which I set for them.

Late in the afternoon I had to procure three boobies for stuffing. They are by no means so foolish as their name would imply. They had learnt by experience, even in a day, and I now had considerable difficulty in getting within shot of the old birds.

I climbed the highest peak, which is 64 feet above sea level; the top affords only just standing room ; from it one sees the whole of the rocks, and their smallness in size is most striking; here is an island group 540 miles distant from the nearest mainland, and yet not nearly so large as, say, the Holmes in the Bristol Channel.

The group consists of five peaks of rock, disposed in four principal masses which are separated by three narrow channels, through which the surf perpetually roars and boils ; over one of these channels it is possible to cross at low water, the tide rising and falling here about five feet. The rocks are disposed in a sort of horse-shoe round the bay ; they are composed of hard black rock, and another yellowish rock with black laminæ in it, " full of variously coloured pseudo fragments," according to Darwin a variety of the former black rock.

There are in places bands of a green stone resembling Serpentine. The whole is intersected by various veins, mostly nearly vertical and running in all directions, consisting of various rocks, viz. : brown ferruginous laminæ, a coarse conglomerate of beach pebbles, and a finer conglomerate which contains fragments of sea shells and nullipores, and which are considered by Darwin as evidently of later origin than the main mass of the rocks. These seams of conglomerates have the appearance of having been formed of beach fragments washed into fissures in the rock and consolidated there. Each face of the containing fissure is covered by a peculiar dense and hard black layer of about a quarter of an inch in thickness. This black layer is mentioned by Mr. M'Cormick in " Ross's Voyage"; Mr. Buchanan found it to be composed of "phosphate of lime, peroxide of manganese, a little carbonate of lime and magnesia, with traces of copper and iron."* He considers that the rocks as a whole may be classed as Serpentine.

Mr. Darwin has dwelt on the importance of the fact that the rocks are not volcanic, like nearly all other oceanic islands. The depth to the eastward of St. Paul's Rocks is irregular, and a depth of only 1,500 fathoms was obtained shortly before we approached them, succeeded by deeper water. There is no connecting ridge between the rocks and Fernando do Norhona. No doubt the rocks are the remnants of a much larger tract of land now submerged, probably once continuous with these irregular masses in their neighbourhood, and which may have had a vegetation of its own.

* J. Y. Buchanan, " Proc. R. Soc.," No. 170, 1876, p. 613.

With regard to the present vegetation, as stated by Darwin and Ross, there are no aerial plants on the rocks, not even a lichen; I found however a microscopic alga (*Protococcus affinis*), growing on the guano in sheltered places and colouring it of a dull green. In the stagnant pools on the rocks grow two low green algæ, *Prasiola minuta* and *Oscillaria sordida*, and a few diatoms.

The rocks are poorly supplied with the larger species of seaweeds, apparently because these are unable to endure the constant heavy surf. The high-tide mark is formed by a band of a pinkish white nullipore (*Lithothammion polymorphum*); its calcareous masses form an incrustation on the rocks, in places two inches in thickness, and which is bored in all directions by tubicolous annelids, and has its surface thus pierced all over by small round holes. This band is referred to by M'Cormick as the work of coral insects; there are no corals at all about the rocks, except in deep water.

Above the band of *Lithothammion* is a band of dark red staining on the rocks, caused by an encrusting alga (*Hildenbrandtia expansa*), and from the region of the tide mark depends a filamentous brown seaweed (*Chonospora atlantica*). The green weed (*Caulerpa clavifera*), of which the noddies build their nests, grows in from two to twenty fathoms about the rocks.

Of the whole of the eleven species of non-microscopic algæ belonging to the rocks, two are peculiar, and the remainder are known to occur at widely different localities at the Cape of Good Hope, east coast of Australia, Venezuela, &c.*

I went out for a second night's fishing. The fish for some reason did not bite so well as before, having possibly, like the birds, profited by experience; but the men in one of the cutters alongside us, kept up a succession of songs with hearty choruses, and with the aid of rum and beer and the moonlight, and an occasional bite, the time soon passed away until midnight, when our boat returned to the ship with a party which had been stationed on the rocks to observe stars for determination of longitude.

* Prof. G. Dickie, "Algæ collected at St. Paul's Rocks." Linn. Jour, Botany, Vol. XIV, p. 311.

Accounts of St. Paul's Rocks are to be found in C. Darwin, "Journal of Researches," 2 Ed., p. 8. "Volcanic Islands." Smith and Elder, London, 1844, pp. 31, 32. Fitzroy, "Voyage of 'Adventure' and 'Beagle.'" Ross, "Voyage to the Antarctic and Southern Regions," Vol. I, pp.14–18 ; with extracts from the Journal of Mr. M'Cormick, Surgeon to the "Erebus."

Island of Fernando do Norhona, September 1st and 2nd, 1873.— The ship reached the island of Fernando do Norhona on September 1st. The island is in lat. 3° 50′ S., and is about 200 miles distant from Cape San Roque, the nearest point of South America. The main island of Fernando do Norhona is about four miles in length, and nowhere more than four and a-half broad, and the length of the group formed by it and its outliers is seven geographical miles. The main island is long and narrow, and stretches about N.E. and S.W.

At the eastern extremity is a series of islets known as Platform Island, St. Michael's Mount, Booby Island, Egg Island, and Rat Island. On the southern side of the main island are several outlying rocks, one of which, called Les Clochers, or Grand Père, appears as a tall pinnacle with a rounded mass of rock balanced on its summit.

At about the middle of the northern coast of the main island is a remarkable column-like mass of bare rock, which projects to a height of 2,000 feet, and is known as the Peak. The south-western extremity of the island runs out into a long narrow promontory, which is composed of a narrow wall of rock.

In this, at one spot near sea level, the sea has broken a quadrangular opening through which the sea dashes in a cascade. This opening, known as the "Hole in the Wall," is visible from a considerable distance at sea. At the opposite extremity the island terminates in a low sandy point with sand dunes upon it, beyond which stretch out the outlying islets already referred to.

The Peak forms a most remarkable feature in the aspect of the island as viewed from the sea, and appears to overhang somewhat on one side. One other hill in the island is 300 feet in height. The island is volcanic, but has evidently undergone a vast amount of denudation, so as to obliterate all traces of the centres of eruption. The Peak is composed of phonolith, or clinkstone, as is also St. Michael's Mount, which is a conical mass 300 feet in height.

Rat Island and Booby Island are formed of a calcareous sandstone, an Æolian formation like that of Bermuda, but here containing volcanic particles intermixed. This rock is weathered in a closely similar manner to that at Bermuda, the exposed surface being covered with irregular projecting pinnacles with excessively sharp honeycombed surfaces, in places on Rat Island as much as two feet in height.

On the western side of Rat Island, close to the shore, a beach of large oval pebbles of phonolith is embedded in this sand rock. In Platform Island the sand rock overlies columnar volcanic rock. The main island is thickly wooded and appears beautifully green from the sea.

The principal trees are what Webster, who visited the island in 1828, calls the Laurelled *Bara*, which has dark green laurel-like leaves, and an abundant milky juice, but the exact nature of which is unknown, since I did not succeed in procuring a specimen, and a Euphorbiaceous tree, or rather tall shrub, called by Webster, Jatropha or Pinhao (*Japhopha gossypifolia*).

It has a pink flower, and at the time of our visit had only single tufts of young leaves immediately beneath the inflorescence, although in full flower. Its bare stems and branches render it a striking object amongst the green of the creepers when the forest is viewed from the sea. Webster says that it casts its leaves in July and August, that is, at the commencement of the dry season. It is evidently the tree mentioned by Darwin as occurring on the Peak.

There is a dry and a rainy season on the islands. The rainy season is from January to July, and the dry from July to December. In the dry season there is occasionally want of water, but it often falls heavily during this season, as it did during our stay, on September 2nd.

Fernando do Norhona is used by the Brazilians as a convict settlement. Close to the base of the Peak is the citadel or small fort, on which the Brazilian flag was seen flying as we approached the shore, and beneath this are the convict buildings, a group of low huts, with the governor's house, a small church, and a long low building in which some of the convicts are locked up at night. Farther to the eastward on some low-lying land close to the beach is an old ruined fort, off which we anchored at about 4 P.M.

Captain Nares landed at once and paid a visit to the governor of the island to ask permission for our parties to land and explore, and I availed myself of permission to follow him on shore and hear the result of the interview. The surf was heavy on the sandy beach ; one of our boats was upset in it, and I got a sea round me in landing, up to my neck.

I found the littoral blue flowered convolvulus (*Ipomœa pes capræ*), so common in the West Indies and Cape Verde Islands, abundant on the shore. It was beset by a Dodder (*Cuscuta*), which parasite was seen twining round it everywhere in masses.

A horrible pest, a stinging plant, *Jatropha urens*, one of the *Euphorbiaceœ*, was very common. The plant has a thick green stem, and leaves resembling those of our common garden geraniums in shape, and a small white flower. The plant is covered with fine sharp white bristles, which sting most abominably. I lassoed a specimen with my knife. lanyard and kicked it up by the roots and carried it on board carefully slung on a stick, but I got stung as I was putting it in paper to dry, though handling it with forceps, and the stinging sensation lasted for more than two days. The pain is like that produced by the nettle, but far more intense.

The path to the settlement led through the woods. The ground was covered with innumerable large black crickets (*Gryllus*). These are most astonishingly abundant, especially around the cultivated fields. The woods were also full of flocks of reddish brown doves (*Peristera geoffroyi*), a species which occurs in Brazil, and has possibly been introduced into the island. They are in vast numbers, and, being scarcely ever shot at, are so tame that we had to throw stones at them to make them take wing. Many of them had nests and eggs, and they probably breed all the year round.

I saw also a small warbler (*Sylvia*), with greenish brown plumage, and a bird which, from its appearance and song, I took to be a thrush of some kind. Mice are extraordinarily abundant,

running about everywhere amongst the bushes. Large butterflies seemed to be absent. I saw only a small blue butterfly
(*Polyommatus*). A tomtom was being beaten as a call for the
convicts, which reminded me of the exactly similar drumming
which wearies one on coffee estates in Ceylon.

On the slope of a hill opposite the fort is a square of open
space, roughly pitched with stones, at the top of which is the
governor's house, with a row of bread-fruit trees planted in front
of it. A black sentry was lolling in front of the house.

I was told that there was a garrison of about 120 men on the
island, and that these, with a few officials, constituted the entire
non-convict population. There were said to be 1,400 convicts on
the island. They are all let loose during the day-time, the blacks
being locked up at night whilst the whites are allowed to live in
their huts with their families, if they have any. They have to
answer a roll-call daily, and are flogged if they fail.

They are all criminals, political prisoners not being confined
here; many of them are murderers, capital punishment not being
exacted in Brazil. They have as a rule a horribly ruffianly
appearance, especially the blacks, and being mostly half naked
they appear especially savage.

All are, however, not of this bestial type. Some few are
well educated, and convicts do duty as waiters and interpreters
to the governor. The interpreter of the time being was a most
gentlemanly looking fellow and well dressed. He was well
informed and spoke English and French well; he was most polite,
and on the governor's producing coffee and cake, took a cup with
the rest.

He told us that the ordinary punishment for a convict was
50 lashes, but that troublesome ones got as many as 500 lashes
delivered with a rod cut from one of the native trees. No one
had ever stood to receive more than 250 cuts. After that they
were supported by means of rests placed under the arms until
the flogging was complete. Then they were taken to the
hospital and never seen again. He had known a man receive
700 lashes. Two-thirds of the convicts had been flogged during
the last seven months. He said he himself had had a misfortune
and had got 64 years' imprisonment. He had bought off 20

of these. He would like a bible and some newspapers. He would sooner die than be flogged. His statements must be taken for what they are likely to be worth.

The convicts receive a small pay, and are obliged to find their own living. The black ones are obliged to work for ten hours daily on Government plantations. Some of these convicts go out fishing on small rafts made of three or four logs lashed together, provided with a small stool for a seat. A basket for the fish is placed on the raft in front of the seat, and a small fishing-rod is stuck up behind.

CONVICT ON FISHING EXPEDITION.
(From a sketch by Lieutenant H. Swire, R.N.)

The men steer these rafts with great dexterity through the surf with a paddle, usually standing up to paddle, and sitting down to fish. At a distance, the raft being almost entirely under water, the men look as if walking on the water. These rafts were termed "catamarans" by the naval officers. Sailors are apt to apply this term to any out-of-the-way canoe or boat for which they have no other name. I believe the word is of South American origin. No boats of any kind are allowed on Fernando do Norhona, for fear the convicts should use them to escape with.

The huts of the convicts form a sort of small town round the square. They have most of them a bit of garden enclosed. I saw several women and children. There are plantations of sugar-cane, maize, cassava, sweet potatoes, pumpkins, bananas, and melons. The latter are remarkably fine in size and flavour, both water and marsh melons; we paid about three pence each for them.

We had to wade in up to our middle, to reach our boats

on account of the surf. A large shoal of dolphins (*Delphinus*) was feeding in the bay close to the shore.

The governor having first given full permission for exploration subsequently retracted it, and sent off a message to say that he would allow no surveying or collecting. This was most unfortunate, since very little is known of the fauna and flora of Fernando do Norhona.

September 2nd.—I landed with Captain Nares on St. Michael's Mount, a conical outlying mass of phonolith, 300 feet in height. It is comparatively inaccessible, and owing to its steepness has never been cultivated; hence it seemed likely to yield a fair sample of the indigenous flora of the group. Most of the plants collected proved, when examined at Kew, to be common Brazilian forms, but a fig tree (*Ficus norhonæ*) with pendent aerial roots like the banyan, which grew all over the upper parts of the rock, and which in favourable spots forms a tree 30 feet in height, proved to be of a new species and peculiar to the island, as far as is yet known.*

The only land birds which I saw on the island were the doves, but I saw a nest, probably that of a finch. The principal

FRIGATE BIRD. TACHYPETES AQUILA.

bird inhabitants of the island were boobies and noddies of the same species as at St. Paul's Rocks, but far shyer here than there,

---

* *Ficus norhonæ.* D. Oliver, F.R.S., "Icones Plantarum," Vol. III, 3rd Ser., p. 18, p. 1222.

and boatswain birds and frigate birds (*Tachypetes aquila*). These latter soared high overhead, looking, with their forked tails, like large kites.

All these birds nest on the rock. They circled round our heads in vast numbers as we stood on the top of the rock. The frigate birds put their nests here well out of harm's way, on the very verge of a precipice which was quite inaccessible. I could look down and see the nests, five or six of which were built close together, almost touching one another, and each containing a single egg.

On the low cliffs of Booby Island, the noddies and boobies nest on all the available ledges, and sat on their nests quite undisturbed as we rowed past them. It was curious to see the doves nesting together with these two sea birds on the same ledges and with their nests intermingled with theirs. The utmost harmony seemed to prevail on the breeding ground. A similar association of land and sea birds occurs in Great Britain. In caves on the coast of Harris, in the Hebrides, starlings and rock pigeons nest together with cormorants.*

Progression on Rat Island is by no means pleasant. The calcareous sand rock of which the island is composed, is, as has been before described, weathered on the surface in the same curious manner as at Bermuda. The surface is here so deeply excavated by pluvial action as to leave projecting a series of sharp edged honeycombed pinnacles, often two feet in height, and separated from one another by intervening jagged holes and crevices. Into these, as they are in many places overgrown by creepers, one's foot and leg readily slip and may easily get badly bruised and cut; whilst in putting out one's hand to save a fall it is not at all improbable that one lays hold of a vigorous plant of *Jatropha urens*, which can show no quarter even if it had the will.

A small Gar-fish (*Belone*) was caught in abundance at the foot of St. Michael's Mount. A Grapsus (*G. strigosus*), the same species as that at St. Paul's Rocks, occurred on the shore rocks, but as far as I saw, Land-crabs and Sand-crabs (*Ocypoda*) are absent from Fernando do Norhona.

* Macgillivray, "British Water Birds," Vol. II, p. 397.

Two lizards occur in the islands, which are South American in their affinities.* One, *Thysanodactylus bilineatus*, is one of the *Iguanidæ*. The genus is distinguished by a scaly projection on the outer side of the hinder toes. The species occurs also in South America. We did not meet with this lizard, which was obtained in the island by the officers of H.M.S. " Chanticleer."

The other lizard, *Euprepes punctatus*, belongs to the *Scincidæ*. The species is peculiar to Fernando do Norhona, its nearest ally, *E. maculatus*, inhabiting Demerara. This lizard is very abundant on the main island, and especially so on St. Michael's Mount, where it is remarkably tame. Some specimens are more than a foot in length. I did not see the Gecko mentioned by Webster.

I could find no fern on any of the islands, nor any moss or Liver-wort. These may, however, no doubt occur on the moister parts of the main island. Fernando do Norhona is in its fauna and flora closely allied to South America. It has however, a peculiar species of fig and a peculiar lizard. Possibly amongst the three land birds noted, other than the dove, peculiar species may occur, but it seems unlikely that it will hereafter yield either in fauna or flora any very remarkable endemic forms. The seaweeds of the island are found by Professor Dickie to be related chiefly to those of the Mexican Gulf.

Accounts of Fernando do Norhona are to be found in Webster's narrative of Capt. Foster's voyage. " Voyage of the Chanticleer." London, 1834. See also Appendix for Webster's notes on the Geology and Natural History of the Island.

Darwin's " Journal of Researches," p. 11.

Darwin's " Volcanic Islands," p. 23.

"Report and Charts of the U.S. brig 'Dolphin,'" edited by Lieut. S. P. Leet. Washington, 1845, p. 75.

Snow's " Voyage to Tierra del Fuego and the South Seas." London, Longmans, 1857, p. 32.

* Gray, " British Museum Catalogue of Lizards," p. 193.

CHAPTER IV.

BAHIA.

Harbour and Town of Bahia. Religious Procession. Black Angels. Land Planarians. Clicking Butterfly. Primæval Forest. Shooting Humming Birds and Toucans. Caxoeira. Mewing Toads. Excursion to Feira St. Anna. Mule Riding. Former Highway Robbers. Inn at Feira St. Anna and its Guests. The Fair. Anteaters Eaten as Medicine. Vaqueiros. Tailing Cattle. Horse Dealing. German Settler in the Country. Driving Cattle in the Bush. Farm Slaves. Preparation of Cassava. Overburdened Ant. Three-toed Sloth. Slavery in Brazil.

Bahia, Brazil, September 14th to 25th, 1873.—The ship approached Bahia under steam and sail, on September 14th. It was all the morning almost a dead calm, and at noon the stock of coal came to an end, with the exception of a few bushels which had to be reserved for steaming to anchorage amongst the shipping in the harbour. The ship crept slowly towards shore in the afternoon, under sail, at the rate of about a mile an hour.

As the shore was approached, swarms of a butterfly, a Heliconia (*H. narcea*), filled the air, and settling on the ship, alighted everywhere and penetrated even into the ward-room. With these a few beetles, flies, and a Hymenopterous insect came on board, whilst a land bird settled in the rigging.

The anchor was dropped in the harbour at about half-a-mile from shore. The city of Bahia or San Salvados lies on the north side of a wide and extensive bay, the Bahia de todos os Santos, or Bay of All Saints. On the north side of the bay is a slightly elevated ridge, stretching east and west, on which the town is built, and under the lee of which is the anchorage for shipping.

The town resembles Lisbon in the general appearance of its buildings. These are mostly whitewashed, with very numerous windows. They rise one above another on the hill-side, with a large number of convents and churches interspersed amongst the houses. The churches have all two towers at the west ends, as at Lisbon, and usually an open plateau or square in front. The architecture is thoroughly Portuguese.

The bright green tropical vegetation, the palms and banana plants, interspersed between the buildings, give the town in reality a different look from that of a home Portuguese town. A small strip of flat land intervening between the foot of the ridge occupied by the main town, and the harbour, affords space for wharves and warehouses for the mail steamers and general shipping. There were a large number of small trading vessels at anchor in the harbour, and two Brazilian vessels of war, a gun brig, and a small iron ram, which had conspicuous shot marks on its hull, received in the Paraguayan war.

The usual mode of ascent from the lower shipping district to the higher town is by means of sedan chairs of the old European pattern, which are painted black, with yellow beading, and are carried up the hill, each by a pair of negroes. A mechanical lift was being constructed to take the place of this primitive arrangement.

I preferred walking, and made my way through steep narrow stinking streets, where slops were being constantly emptied from upper stories without any warning or " Garde de l'eau." After a stiff climb, I reached the main street of the town, which runs all along the top of the ridge, and was just in time to see a religious procession, held in commemoration of the day of the saint of one of the churches.

The bells of the church were clanging and tinkling, sounding something like Swiss cow-bells, a regular jangle, "tinkle, tinkle, tinkle, cling, cling, clang," and the procession was pouring itself from the church door. First came men in blue cassocks with white surplices over them, carrying lighted paper lanterns on poles. They marched on and then formed line on each side of the street for the rest of the procession to pass.

Then came men with white cassocks and black surplice-like vestments, also bearing lanterns, and at intervals amongst them were borne silver crosses with bunches of artificial flowers on

silver-mounted poles, carried on either side of each of them.
Amongst these also walked here and there a priest, in the usual
cassock and alb, and one or two old monks with hooded robes
and double chins, with a well-nourished appearance.

A crowd of acolytes succeeded, dressed nearly like the
priests, and, like them, mostly white-skinned or but slightly
yellow. All the remainder of the procession had deep yellow-
brown or almost black faces. A body of priests came next, and
then the saint, carried on a silvered platform on the shoulders
of eight bearers.

The saint was a wooden figure, of life-size, with a Vandyke-
like countenance, black hair, moustache, and beard. He was
dressed in a stiff crimson velvet cape, worked with gold lace,
crimson trunk hose, and flesh tights over very thin and shaky
legs, and had a curious sort of plume or cockade of feathers and
tinsel sticking up at the back of his head.

In front of the saint, skipped along two little girls, one of
them with a dark yellow complexion, the other jet black. They
were dressed as angels, with wings of feathers and tinsel.
Around the saint marched a guard of soldiers with fixed
bayonets, and immediately behind came a military brass band
in full bray, but playing well. Another body of soldiers followed with fixed bayonets and led by their officers with drawn
swords.

Behind the procession followed a crowd of negro women,
crushing through the street. The negro women of Bahia are
strapping females and apt to become very stout. The balconies
in the narrow street were crowded with the wives and daughters
of the townspeople, who pelted the saint as he passed with
bouquets of flowers.

Vespers were going on at the churches. I entered one, an
oblong building with a small apse for a chancel, and a row of
rectangular pillars on either side, shutting off the aisles. There
were three or four clerestory windows, but no others. The
interior was profusely ornamented with bright colour and gilt
tracery in relief. The chancel and altar, which had an elaborate
gilt reredos, were brilliantly lighted up by candles, whilst the
body of the church was comparatively dark, having no light but

that which reached it from the chancel. The air was full of incense, and the whole effect was fine and impressive.

The floor of the church was crowded with negro women, kneeling and singing at intervals a simple chant in response to a choir which could not be distinguished in the gloom. There were a few white women in the church, but they appeared to go into the aisles and not to mix with the blacks.

After the procession was over, fireworks, rockets full of crackers and blue lights, were let off, and the soldiers marched to their barracks. They were small dark-skinned dwarfed-looking men. Fireworks are as invariable concomitants of religious ceremonies in Bahia as in China, and as they are let off before as well as after the ceremonies, occasionally wake one up at 4 A.M.

There are tramways in Bahia leading to the railway station, the Campo Grande, and out into the country. The Campo Grande is a large open space, turfed and surrounded by trees. It is here that the best residences are, and there are several hotels, including a Swiss one, and a German one with a Kegel-bahn, and where dinner is served in regular German style. There are large numbers of Germans in Bahia, and a great part of the trade is in their hands.

There are public gardens in Bahia, and a theatre, and at certain seasons an opera troupe comes from Rio de Janeiro to perform. At the distance of a mile or two from the town, where the country tramway ends, the roads degenerate at once into mere green lanes, and lead between a succession of small mud-built cottages, each with its fenced garden, and numerous intervals of neglected land, often planted with coffee bushes but overgrown with weeds.

The principal features of the vegetation are made up of banana plants and large mango and Jack-fruit trees. The Jack-fruit is a huge sort of bread-fruit, as large as a man's head, and grows on a large tree with dark green laurel-like foliage. These three trees are no more indigenous than are the people with whose well-being they are so closely bound up, but are of Asiatic origin, as the people are of European and African extraction.

At a short distance from the town the country is covered

with a thick wild growth, but with numerous scattered cottages. The inhabitants of these are mostly black, but there are many whites amongst them, and white and black children are to be seen playing together on almost every doorstep.

I frequently visited these suburbs to search for Land-planarian worms,* which I found resting beneath the sheathing leaf stalks of the banana plants, just as I had found them in Ceylon, and accompanied, curiously enough, as in Ceylon, by a peculiar slug (*Vaginulus*).

A butterfly which makes a clicking sound whilst flying, a fact first observed by Darwin, is common near Bahia.† I only heard the sound when pairs were flying together in courtship. I do not know whether the butterfly in question at Bahia is *Papilio feronia*, the species which Darwin met with at Rio de Janeiro. It has, however, the peculiar drum with a spiral diaphragm with it at the base of its wings, as described by Doubleday. This organ of sound is large and conspicuous.

I made an excursion with one of the sub-lieutenants about 20 miles inland, along the railway intended to reach Pernambuco, but at the time of our visit, completed only for about 60 miles to the Rio Francisco. Free passes were given by the railway company to all the officers of the " Challenger," and the officials of the line, who were Englishmen, were extremely hospitable and gave us every possible assistance.

Leaving Bahia, the railroad led along the shores of the bay, fringed with gardens and houses. Further on the land was covered with wild vegetation, with occasional sugar plantations and frequent cottages. Almost the whole of the land has been cleared at some time or other of the dense forest which once covered it.

On a sugar plantation, ground is cleared in patches. The patches are planted and cultivated for about fifteen years and are then allowed to run waste, or sleep, as the Brazilians put it. A fresh piece of land is then cleared, and so the whole estate is

---

* See H. N. Moseley. " Notes on the Structure of several Forms of Land Planarians." Quart. Journ. Micro. Sci. Vol. XVII, New Ser., p. 273.

† C. Darwin, " Journal of Researches," p. 33.

gradually gone over, and the original clearing eventually reached again. The forest land on the banks of the Lower Moselle is cultivated in much the same way.

There were no large trees to be seen along the route, but rather a dense growth of large shrubs and small trees, bound together by creepers and loaded with epiphytic plants, amongst which the *Bromeliaceæ*, plants allied to the pineapple, were most conspicuous, especially one with a bright scarlet and blue inflorescence.

Near the station where we stopped there was a small river and a patch of primæval forest, which was what we had come to see. A guide led us a short distance into the forest. The most striking feature about it was the immense height of the trees, their close packing and great variety. At home we are accustomed to forests composed mainly of one gregarious species of tree. Here the trunks are covered with parasites and climbers. Mistletoes of various kinds, some of them with scarlet flowers, grow amongst the upper branches, from which also hang down the stems of various creepers in festoons, often sweeping the ground. In the forks of the great branches repose the large green masses of the Bromeliaceous plants, and up the trunks climb numerous aroids with their huge sagittate leaves. The ground is covered with decaying branches, and here and there dead trunks, on which grow fungi in abundance.

The forest was so thick as to be quite gloomy and dark, and as we passed along the path we heard no sound and saw no living animal, except a few butterflies (*Heliconiæ*), some small fish in a little stream, along which the path led, and an Oven-bird gathering mud for its curious nest. There were two deserted armadillo holes close to the path, but we saw no mammal of any kind, nor did I see a single wild mammal during my short stay in Brazil, notwithstanding the abundance of forms which exist in the country. The abundant vegetation hides them from the casual view, and they are not conspicuous, as in an open country, such as California.

We returned to the railway station, where we found beds made up for us in the waiting room. Thanks to the energy of the English railway officials, Bass's ale is to be had at all the stations on the line at 2s. 2d. a bottle.

As soon as it was dark, numbers of fireflies came out. The small negro boys of the village lighted a bonfire and sat round it, making horrible squealing noises by blowing through short conical tubes, made by rolling up strips of palm leaf spirally, and so arranged that at the mouth-piece there are two pieces placed flat against one another, as in the reed of a hautboy. Such excruciating sounds seem to be as pleasing to the youthful African ear as to that of the London street boy.

Next morning at daybreak, we started off to a part of the forest where the negro guide said there were Toucans. We passed a tree covered with white blossoms, over which about a dozen Humming-birds of three species were hovering. We shot some, but it is not an easy matter to obtain them in good condition. They are of so light weight that they often hang amongst the leaves when killed, and even when they do fall it is almost impossible to watch them and distinguish them from the falling leaves knocked off by the shot.

Then the ground beneath the bushes is frequently covered with thorny plants and sharply cutting grasses, amongst which it is not pleasant to force one's way, and where search is almost hopeless. The negroes who make it their business to collect Humming-birds for sale can afford to wait till they get their birds in good position.

The birds did not care at all for the sound of a gun but went on buzzing like sphinx moths over the flowers quite unconcernedly, whilst their companions at the same bush are being shot one after another. They can even often be caught with a butterfly net, or knocked down with a hat. I saw five species on the wing whilst in the neighbourhood of Bahia.

We turned into the gloomy forest and for some time saw nothing but a huge brown moth, which looked almost like a bat on the wing. All of a sudden, we heard, high upon the trees, a short shrieking sort of noise ending in a hiss, and our guide became excited and said " Toucān." The birds were very wary and made off. They are much in request and often shot at. At last we got a sight of a pair, but they were at the top of such a very high tree that they were out of range.

At last, when I was giving up hope, I heard loud calls, and

three birds came and settled in a low bush in the middle of the path. I shot one, and it proved to be a very large toucan (*Ramphastos ariel*). The bird was not quite dead when I picked it up, and it bit me severely with its huge bill. Most of the plumage of the bird is of a jet black colour, but the throat is of a brilliant orange, and the breast has a bright scarlet patch. The bill is brightly coloured yellow at its base, and has a light blue streak along its upper crest, but these colours soon fade after the bird is skinned. The skin round the eye is coloured scarlet.

Into the wide bay of Bahia, which is twenty miles across in the broadest part, open several navigable rivers, on two of which steamers ply regularly. The Peruaguacu is the largest of these rivers, and it is navigable for 54 miles up to a town called Caxoeira. At Caxoeira a railway was in process of construction. The English engineer of the line, a Mr. Watson, most hospitably provided me with a free pass by the steamer to Caxoeira, and one of his own mules, and a guide for a trip up country thence.

The river steamers are small paddle-boats, old and dirty. The Caxoeira boat was crowded with passengers, mostly Brazilians and negroes, but amongst them several German Jews going up to buy diamonds.

The bay has all the appearance of an inland lake, there being several islands scattered about in it covered with green to the water's edge. Near its mouth the banks of the river are somewhat low but backed by hills, and here and there are mangrove swamps. As the river was ascended the hills and cliffs on either hand soon became higher. They are thickly covered with vegetation, but with cliffs and occasional rock masses showing out bare amongst it.

The scenery on the whole is not so unlike that of the Rhine, excepting that there are no castles : but the white buildings of sugar estates perched here and there on the tops of the lower hills take their place. The far-off hills appear of the usual blueish green due to distance, and successive ranges become gradually yellower as they lie nearer to the eye of the observer and show more and more plainly the forms of the vegetation clothing them; only in the actual foreground do the palms and feathering bamboos, planted in long lines as boundaries, distinguish the scenery as tropical. The bamboos are especially conspicuous, from the bright yellow green of their foliage. The steamer left Bahia at 10 A.M. and reached Caxoeira at 4 P.M.

Caxoeira.—There are two towns at Caxoeira, one on each side of the river. These consist of the usual whitewashed houses and two or three churches, one broad street and several narrow ones, with mostly dirty dilapidated two-storied houses tailing off towards the country into one-storied hovels. On the river, canoes hollowed out of a single tree trunk, simple and trough-like in form and pointed at both ends, ply between the town and its suburb. They are large enough to contain six persons.

The hotel at which we stayed consisted of a restaurant below and a long barn-like chamber above, with a passage down the middle, and a series of small bed chambers on either hand, enclosed by partitions about twelve feet in height. As one lay in bed one looked up at the bare rafters and tiles, and was apt to receive unpleasant remembrances from the bats. I have seen sleeping places arranged in the same manner in the hotel at Point de Galle, Ceylon, and it is closely similar in all Japanese houses; the great disadvantage is that you have to put up with the snorings and conversations of all the guests in the hotel.

In the evening, just outside the town, in a small pond, a number of small toads were making a perfectly deafening noise. The sound is like a very loud harsh cat's mew, and I could not at first believe that it would come from so small an animal. It is however not unlike the extraordinary moan made by the fire-bellied toad of Europe (*Bombinator igneus*), but much louder and with more distinct intervals between the sounds. The frog tribe made a horrible noise at night at Caxoeira, a bull frog shouting the loudest with a deep bass voice.

Trip inland.—I started on my trip in the morning. I was to go to Feira St. Anna, about 28 miles from Caxoeira, to see the great fair held there every Monday and from thence go down to St. Amaro, a town on another river running into the bay, whence I could take steamer for Bahia. Caxoeira, Feira St. Anna, and

St. Amaro, form with each other roughly an equilateral triangle, being each distant from the other about eight leagues.

My guide was a German, who acted as interpreter on the railroad. He spoke English, French, Italian, Spanish and Portuguese, and had been in Brazil about twelve years. He was a wild sort of young fellow, and had undergone various vicissitudes of fortune, having been once reduced to selling jerked beef, and once having been a dancing-master. He was a capital merry companion, knowing everyone on the road and having a joke for all.

We rode extremely well-broken mules of large size that ambled along, rendering it no labour to ride. Mine much preferred his natural rough trot to ambling, and tried to make me put up with it, finding that I was a tyro at mule riding. But I was told that I was ruining the beast by letting him get into bad habits, and was told to dig in my spurs and jerk back his head with the bit at the same time. This receipt never failed to make the poor brute so thoroughly uncomfortable that he ambled as softly as possible at once.

The road led up the steep side of the river valley on to the table land above. From the top of the hill there is a fine view of the river and its valleys, and the white town below. Some trees, the leaves of which turn scarlet before dropping, set off the green of the rest of the landscape. In their action on foliage and plant life generally, the wet and dry seasons take the place of summer and winter at home, and many plants become bare of their leaves at the dry season, and only burst out again into leaf at the commencement of the wet season. This condition is far more marked in other regions of South America. Humboldt observed that certain trees anticipated the coming wet season, and put out their leaves some weeks before there was any appearance of its approach.

The road was very much like a green lane. In places a regular slough of mud, in others dry and sandy; it was broad, but usually more or less overgrown with grass and weeds, with a narrow track picked out along the best ground by the mules. There were numerous cottages along the road, and fields of tobacco, maize, and cassava; every now and then a bit of wood was passed with beautiful flowers growing about it, and amongst them numerous forms of *Melastomaceæ* with their characteristic three-veined leaves.

I saw here most of the plants which I had collected at Fernando do Norhona growing as road-side weeds. As we rode on, a splendid Iguana, about three feet in length, ran across the road. I was astonished at the brilliant dark green and bright yellow-green colouring of the animal, and have never seen any other lizard so bright.

Every now and then a village was passed; in the first, as it was Sunday, the villagers were enjoying a cock-fight. Every villager keeps a fighting-cock. Good Lisbon wine is sold along the road; the drinking-places consist of a hole about a yard square in the gable-end of the usual mud-walled cottage, placed at such a height as to be convenient to a man on horseback, who thus gets his drink without dismounting. Ladies travel along the road either in the saddle or in a sedan chair slung between two horses or mules by means of a long pole.

A thick growth of myrtles and shrubs which was passed, was pointed out as having been the hiding-place of a notorious highway robber, a negro named Lucas, who used to lay in wait for merchants on their way to the fair at St. Anna; he was the terror of the district, and committed several murders and worse atrocities. Though he was caught and executed in 1859, stories about him are already beginning to assume a mythical dress, and I was told that miraculous flowers grew out from a tree to which he bound one of his victims, a white girl, leaving her to die of exposure.

We took seven and a half hours over the 28 miles to Feira St. Anna.

Feira St. Anna.—The town consists of about three long parallel streets, with a broad cross street, or rather open oblong space on which the small dealers erect their booths on fair day. We rode into the town at about five o'clock in the evening.

The girls were all dressed in their best, expecting home their various sweethearts who are away all the week in search of cattle, and only come to town on Sundays in time for the fair on Monday. Several of them greeted my guide as an old friend, as we rode up a long street to the other end of the town. Here is an open common-like space surrounded by houses, which serves as tobacco and cattle market. We stopped at an inn close to the market.

The inn was a one-storied house, consisting of an eating room fronting the street, and two sleeping rooms, and a kitchen behind. The eating room had large windows with jalousies, but no glass, looking out upon the market. It had a cement floor, a trestle table at one end for eating on, a small table opposite with a red curtained box upon it, containing the household gods, the Virgin in plaster, and Sta. Antoinetta in china, and a half round table with an inkstand for the use of those customers who could write.

The host, an old Brazilian, greeted us with great politeness, and we bowed according to custom to the assembled guests. The company consisted of about half-a-dozen cattle dealers, who were in animated discussion concerning the prices of stock. One of them, who was quite black, was evidently the sharpest of the lot, and a wag. Presently there came in a dirty coarse-looking grey-haired man with a black skull-cap on; he wore a dilapidated black garment something like an Inverness cape. He was chief vicar of the town; he was in considerable excitement, and addressed himself to the black cattle dealer, who produced a letter for him.

The reverend gentleman had not got his spectacles with him, so the host proceeded to spell out the letter aloud. It appeared that the vicar did a bit of general trading, and had sent some horses, mules, and slaves to a neighbouring fair, in hopes of a good price. The letter was to inform him that he had made a bad speculation, and that no buyer had been found. The vicar was in a great rage, and made an excited oration about the hardships of his position and terrible depreciation in the value of slaves, and left. He was said to receive £60 per annum as stipend and fees in addition.

We had some excellent fresh beef for dinner, fried in small pieces with garlic and potatoes and carrots, and with it farinha, the coarse meal made from cassava root, the fine siftings from which are tapioca. The farinha is universally used here, and is very good with gravy.

The sleeping apartment was a space of about eight feet square, separated from the front room by a low partition: in it were three light cane-bottomed sofas, one at each end, and one opposite the door; they were packed so close together as to touch one another. A neatly folded small coverlet and a pillow were placed in the middle of each.

Here we turned in; the third bed being occupied by a very dirty dealer in tobacco. Rendered sleepless by the fleas, I lay awake most of the night listening to the mingled crying of children, barking of dogs, croaking of frogs in the marsh below, and squeaking and groaning of the axles of the ox-carts bringing merchandise to the fair.

Though other charges were comparatively cheap, we had each to pay two shillings for our beds, as did also some of the cattle dealers who slept in a small house over the way, rented by the host for that purpose, and to keep the guests' saddles and bridles in.

At 6 A.M. there was no bustle or signs of the fair, and not till 9 or 10 o'clock did strings of mules, laden each with a pair of bales of tobacco, arrive opposite the inn. The mules carry about seven or eight arrobas (arroba = 25 lbs.). The tobacco comes to the market compressed and cut into neat rectangular bundles; the merchants test it by pulling some from the bundle and rolling a rough cigar.

In the broad open street in the middle of the town were rows of small booths, at which farinha, fruit, vegetables, and jerked beef, imported largely from Buenos Ayres, were for sale; the dried beef varies in price from six to two milreis = 2s. an arroba. It seemed singular that it should pay to bring it to a place where fresh meat was so abundant.

Other stalls offered needles and thread, sweet stuff for children, &c.; but most trying to a naturalist's eye, were stalls where various Rodents and other small native animals were for sale, spitted on wooden skewers, roasted and dried for eating. Amongst these I saw at least a dozen of the tree-climbing ant-eater, the Tamandua, and many Three-toed Sloths: the skulls of all were split open, and they were utterly lost to science. The flesh is supposed to cure various diseases.

H

Makers of the long riding boots so fashionable here wandered about the fair trying to sell their handiwork, and I bought from a similar wanderer one of the vaqueiro's leather hats, which did me the best of service in thick and thorny forests throughout the remainder of the cruise; with this on my head I could butt my way head first into any bush with impunity.

Close by the market-place was the church of the vicar already mentioned, which had a mosque-like dome ornamented with variously coloured dinner and tea plates set in patterns in cement, a very original form of decoration.

In the leather market quantities of skins of leather were exposed for sale, and also tanned puma skins used for saddle-cloths, and boa-constrictor skins, also tanned, used to make boots and said to be remarkably waterproof.

But the great sight of the fair is the cattle market, the situation of which has already been described; the cattle are bred at estates far up the country, where they run wild in the bush and are caught and branded, and drafted for market every two years.

The men who look after and drive the cattle are termed "vaqueiros" in Portuguese. They are of all shades of colour, from black to white; they are dressed when at work from head to foot in undyed red brown leather; they wear leather breeches, high leather boots with huge spurs, a leather coat like a longish jacket, and a leather hat with rounded close-fitting crown and broad brim: they ride small rough horses, which are worth at Feira St. Anna from £4 to £5. They ride in saddles of the form commonly called Mexican or Spanish.

The vaqueiros receive as payment from the owners every tenth head of cattle brought to market. They are, of course, extremely expert riders, and it is marvellous what work they get out of their small horses.

The breeders rarely bring the cattle to market on their own account, but sell them to dealers, who take them to Feira St. Anna, and hand them over to other dealers again, who sell them in Bahia or Caxoeira.

The cattle are driven by the vaqueiros, who use a short leather thong to strike them with. Bands of from 20 to 50 head of cattle were being driven into the market as we approached. A vaqueiro rides in front of each herd, one on each side, and one or more behind. They keep up a constant shouting, and bring the animals along at a fair pace.

Every now and then, a beast wilder than the rest, or less exhausted by the long journey from the interior, breaks away, and goes off at full gallop over the open market-place or up the street. Off gallop two or three vaquieros, in full chase, with outstretched arms, spurring their horses to the utmost. They try to drive the beast back into the herd, and often succeed forthwith; but often it gets in amongst another herd, and then it is wonderful to see how rapidly they manage to single it out, get it on the outside of the herd, and start it afresh.

Sometimes the animals are very fresh and wild, and make off at full pace, and cannot be headed. The vaqueiros then strain every effort to come up behind them, catch hold of their tails, and spurring their horses forward so as to get up alongside the beasts, give a sudden violent pull, which twists the animals round, and throws them sprawling on their sides.

The cattle, though they fall so heavily that this expedient is resorted to as little as possible at the fair, because it bruises the meat, are often up after a fall and off again in an instant; but two or three falls knock the breath out of them, and they are then driven back to the herd quietly. Sometimes, even this treatment does not subdue them, and then they are lassoed round the horns and dragged back.

The various herds were driven in compact bodies against the walls bounding the market, and some of the vaqueiros dismounted, and kept the cattle together by the use of their thongs and shouting, but one at least at every herd remained mounted ready to chase any animal which might break away. The scene was most exciting. Often three or four cattle were loose at once and careering madly in all directions, jumping over obstacles like deer, and with two or three vaqueiros after each, at full gallop, spurring their little horses to the utmost, twisting and turning with wonderful dexterity.

One wild cow went right up the main street. She was very fast, and five vaqueiros had a sort of race after her; now one gained a little, now another, and it appeared as if the beast were going to make off altogether; but at last a big black vaqueiro shot ahead, and threw her sprawling in the road. I kept close to a sheltering corner, ready to retreat round it when a beast came in my direction.

The cattle dealers rode round from herd to herd, on their mules and horses, and most of the dealing was done on horseback. As soon as a herd was sold, it was driven off, one or more vaqueiros accompanying the drovers, according to the wildness of the cattle.

In the middle of the open space, horses and mules were being sold. The sellers of the horses were mounted on them, and were showing off their paces in an open lane formed amongst a crowd of buyers and lookers-on. The sellers made their horses amble full pace up the lane, turn sharp round, and return: and on reaching the starting-point, stop suddenly, without slacking pace in the least beforehand, in doing which the animals were thrown almost back upon their haunches. The being able to stop thus suddenly when in full pace is one of the points most admired in horses by Brazilians.

The horses are small, but well made. Good well-trained horses cost about £40. Good riding mules are worth as much or even more. The Brazilians of the better class ride their ambling horses, with their legs straight and stiff and carried right forward, with the toes turned up and the tips of the toes only resting on the stirrup irons. The vaqueiros, however, ride much in the usual English fashion.

Sheep are used as beasts of burden in a small way in Feira St. Anna. I saw three or four laden with small barrels of water slung across their backs. They were driven by children, who were thus taking water from the well outside the town round to the various houses. The sheep seemed perfectly trained, and went along at a smart pace. Sheep are used as beasts of burden in Ladak to transport goods over the mountains of Little Thibet, and carry from 20 to 30 lbs.; * but their use for such purpose is very uncommon.

In the crowd we met with a German farmer, who was a

* "The Middle Kingdom," Williams, Vol. I, p. 204.

friend of my companion, and he invited us to pass the night at his house, his farm lying on the road to St. Amaro, by which we were to travel. We had our mules brought up to the inn door, and there gave them a feed of maize to make sure that they got it. We saddled them ourselves in front of the inn, and after much ceremonious shaking of hands with the host, and polite speeches, rode off.

On the road we passed several herds of cattle, which were being driven towards Bahia. In one of these some of the cattle were very wild. There were three vaqueiros in charge of it, a man, and two lads of from 16 to 18 years of age. There was thick bush on either side of the road, and every now and then the cattle broke away into this. The use of the rough lurcher-like dogs which follow the vaqueiros now appeared. In the thick scrub the vaqueiro could do nothing without his dog. The cattle are out of sight in an instant, and go off dashing full pace through the bushes. The dogs are after them at their heels at once, and drive them to the vaqueiros, who dash off into the thick of the bushes in pursuit, bending right forward in the saddle, and stooping till their heads are beside their horses' necks, to avoid the branches.

One cow came full charge down the road behind me, and I had only just time to back my mule into the bush out of the way. One of the lads was after her. He seized her tail just as he was opposite to me, held on for about twenty yards, and then digging in his spurs and shooting forwards, turned her over with a thud. She was up, however, again, and off into the bush in an instant, and he after her with the dog in full pursuit, and I saw him disappear under the branch of a tree with his body laid right back on his horse's rump to avoid it.

We passed about sunset through a village, where there is a hospital, a very substantial building, erected by the vicar, who diligently collected subscriptions for that purpose for many years. The church was lighted up and the people were going to vespers. One of the villagers was pointed out to me by the German farmer as being the hereditary owner of a large estate worth several thousand pounds, and a number of slaves. He was quite black and dressed in tatters, and looked like a slave himself, and was driving cows along the road. He could neither read nor write.

Our host was an emigrant from the Hartz District. He had been out in Brazil about 14 years, and had a farm of several hundred acres, most of which was grass land; the grass growing where sugar had once been planted. He bought cattle and sheep at Feira St. Anna, kept them some time on his farm, and then killed them and sold the meat in St. Amaro and the district. He also grew a large patch of sugar-cane, which was ground at a large mill close by, he receiving half the sugar produced as his share. He had bought one slave: all foreigners, except English, being allowed to possess slaves in Brazil. The slave was married to a girl, who was principal servant in the house. The farmer had assisted the girl to buy her freedom.

Frau Wilkens, his wife, who had no children, described the girl as most trustworthy, honest, and deeply attached. Her small child, a chubby little negro, was a great pet in the house. The greater part of the work on the farm was done by slaves hired from the owners of neighbouring plantations. There was a row of about thirty very small wooden houses or huts on a neighbouring hill, where the slaves belonging to the owner of the sugar mill lived.

Cassava or Mandiocea, which is a Euphorbiaceous plant, allied to our common spurge, was also grown on the estate, and there was a small manufactory of farinha. The Cassava (*Jatropha manihot*) is an indigenous South American plant, though now widely spread in the tropics, and was cultivated in Brazil by the original inhabitants, before they were molested by Europeans. The plant is not unlike the castor-oil plant in appearance, and is planted in rows slightly banked up.

The tubers are long and spindle shaped. The preparation of them was conducted in a small hut. A large fly-wheel was turned by a negro, and drove, by means of a band, at a rapid rate, a small grinding wheel provided with iron cutting teeth. The cassava root, which had been peeled and washed by a negress, was reduced to a coarse meal by means of the grinding wheel. The meal was then put into a wooden trough, and a board was tightly pressed upon it by means of a lever, heavily weighted with stones. The cassava was thus left in the press for twelve hours, in order that the poisonous juice which it contains should be expressed. The meal was then taken out and dried on a smooth stone surface, beneath which a wood fire was burning.

The resulting chalky-white meal, when sifted, yields samples of three degrees of fineness. The finest, a white flour-like powder, is tapioca, *i.e.*, true, original tapioca, an imitation of which, made from potato starch, is commonly sold in England. The intermediate sample is used in starching clothes and in cooking; and the coarsest substance, which is coarser than oatmeal, and consists of irregularly-shaped dried chips of the roots, is called farinha, and is, as before described, commonly eaten with gravy at dinner, taking the place of bread, and forming a staple article of food.

Our host was well to do, having thrived best of all the emigrants who came out with him, and, having no family to provide for, talked of going home soon. An old German was staying in the house, an idler, whose real occupation was gardening, his father having been Imperial gardener, as he informed us with great pride. He had landed, more than twenty years before, at Rio, and had reached Bahia on foot. He was now travelling from estate to estate, and staying at each as long as he could, under pretence of doing up the garden, but although he had been two months at the farm, the few square yards of garden were as yet untouched.

He had been too lazy to learn Portuguese, and understood very little. He did a little trade in the way of peddling books. He seemed, however, a favourite at the farm, and was well taken care of, tea being made as a special luxury for him, and he had many stories to tell, and quaint sayings, and had amusingly strong Prussian sympathies.

The farmer guided us to a large tract of primitive forest close by, which was extremely difficult to penetrate. Here I caught a curious bat (*Saccopteryx canina*). This bat has remarkable glandular pouches on the under sides of the wings, at the elbow-joints; these pouches are well developed only in the males, rudimentary in the females, and secrete a red-coloured strongly-smelling substance, supposed to act as a sexual attraction. The bat was resting on a bare tree-trunk, asleep, the dense forest growth overhead making this exposed situation quite dark enough for it. I caught it with a butterfly net.

On our way back to the farm, we watched some ants carrying off bits of cassava leaves to their holes. One cannot walk anywhere in the neighbourhood of Bahia without seeing these Leaf-cutting Ants (*Œcodoma*) at work. Their habits have been described by many observers, and recently by Mr. Belt * at great length.

One soldier-ant was carrying a piece of young cassava root, two inches in length. It held the stick by one end thrown over its back, but not touching it, the other end projecting far behind the insect. There was just a balance. The slightest extra weight on the hinder tip of the stick would have upset the bearer backwards. The ant staggered from side to side under its burden, like a heavily-laden porter, and got along very slowly.

I pulled the burden away and then put it back again. The ant struggled a long while to get it back into its old position, but could not. Then it tried to balance it crossways by the middle, but one end always tilted up, and the other stuck against the ground. So at last the ant cut the stick in two, and carried off one half, a worker hoisting the other. The further road to St. Amaro lay through sugar estates all the way. I left St. Amaro early next morning by steamer, and reached Bahia at 10 A.M.

Bahia.—On the quay I bought a living full-grown Three-toed Sloth (*Bradypus tridactylus*) from a countryman for two shillings. We kept the animal alive in our work-room for some days, where it hung on to the book shelves and bottle racks, and crawled about. As I could not get it to feed, I had to kill it.

The beast was the most inane-looking animal I ever saw, and never attempted to bite or scratch; none of us could look at its face without laughing. It merely hung tight on to anything within reach. It showed, however, one sign of intelligence.

* "The Naturalist in Nicaragua," by Thos. Belt, p. 71, *et seq.* London, John Murray, 1874.

I hung it on a brass rod used for suspending a lamp beneath one of the skylights in our room. It remained there half a day, hanging head downward, and constantly endeavouring to reach the book shelves near by, but without success. At last it found out an arrangement of its limbs by which this was possible, and got away from the lamp rod, and in future whenever I hung it up on the rod it climbed to the book shelves within five minutes or so.

When I reached the ship I found that a case of yellow fever had occurred on board. This determined our immediate departure, and we sailed for Tristan da Cunha direct, being obliged to hasten to cold weather, for fear of other cases breaking out. We thus missed our intended visit to the islands of Trinidad and Martin Vas, to which I had looked forward with the greatest interest, since they are the only islands in the Atlantic, the flora and fauna of which are absolutely unknown.

A word or two about slavery in Brazil. A law is now in force by which every child born in the country is free, and further, a master is obliged to free a slave if the slave can raise a sufficient sum to buy himself off. The value to be paid is fixed by a Government valuer, and the sum is always fixed as low as possible by him. Slaves commonly buy themselves off, and a Society exists which assists them to do so, advancing the money on loan, and receiving it back by instalments. Slaves also go round and collect money from charitable people to assist them in the matter. The fact that the children become free, and that the slaves can buy themselves off so cheaply, has made them fall very much in value. A female slave's time is much taken up with her children, which a master has to feed, although after all they do not belong to him. Hence a strong young man was worth, at the time of our visit, only about £120, and a young woman about £70 to £80.

The slaves, however, do not often change hands. Old families pride themselves on the numbers of their hereditary slaves, and often having fallen in'the world and being poorly off, have nevertheless a dozen slaves, for whom they find hardly any work, and whom they can scarcely afford to keep. These slaves are much attached to their masters, and often their masters to them. A member of the House of Assembly has been known to refuse to speak on an occasion of importance, because his foster brother, a slave, had just died.

The slaves are hired out as servants: and foreign residents, especially English, who cannot hold slaves, hire them as domestic servants. They make the engagements with the slaves themselves, and pay them the wages, and the slaves carry these wages to their owners, who, if kind ones, give them a fourth part or so as a present. Other slaves are hired from the owners, but not the best ones. At the best hotel in Bahia, kept by a German, most of the servants were thus hired slaves. The proprietor said that was much better than buying slaves, since when they were ill you sent them back to their owners and got fresh ones.

Owners also employ their slaves as sellers of various goods in the streets. The slaves are usually well treated, but in some rare cases owners are cruel and beat them. At Caxoeira, a pretty girl was collecting money to buy herself off because, according to her story, her master beat her constantly. There is no slave market in Bahia. The slaves that have not been born in the country, but were brought from the coast, have marks cut on their cheeks, the marks of the tribes to which they belong, and of which they are proud. There are many of these to be seen in the streets ; but there is no means of distinguishing a slave from a freed man. The following slave statistics are taken from the *Anglo-Brazilian Times* :—

Brazil Slave Statistics.—" In the province of Goyaz the 8,903 slaves registered in 1872, had on the 31st of December, 1875, become reduced to 7,888 by 357 deaths, 222 liberations, and 436 removals. At the same date there existed 921 freeborn children of slaves. In the province of Pernambuco, during the same four years, the 106,201 slaves diminished 3,386 by death, and 1,049 by emancipations. From September 28th, 1871, to the end of December last, the number of children of slaves born free under the law of 1871 was 12,312, of whom 2,802 died, leaving 9,510. In the province of San Paulo there died, from April, 1872, to the end of 1875, of the 147,746 slaves registered, 8,561 and 3,410 were emancipated. In 111 of the 151 parishes the freeborn births were 18,176, of whom 5,861 had died.

We left Bahia on September 25th. The voyage to Tristan da Cunha was not very eventful. A suspicious case of fever appeared on board, and we were for some time in anxious suspense as to whether we were not going to suffer from an epidemic of yellow fever, but all turned out well. We crossed the track of sailing vessels bound round the Cape, and sighted two English vessels bound for Chittagong and Point de Galle. There is some doubt as to when the first Albatross was met with; but a bird, either an Albatross or the Giant Petrel (*Ossifraga*) was seen on October 4th, in lat. 27° 43'. We arrived at Tristan da Cunha on October 15th.

108

CHAPTER V.

TRISTAN DA CUNHA, INACCESSIBLE ISLAND, NIGHTINGALE ISLAND.

Settlement of the Island. Geological Structure. Vegetation. Temperature of Fresh Water. *Phylica arborea.* Rigorous Climate. Condition of the Settlers. Inaccessible Island. Rock hopper Penguins. Tussock Grass. Penguin Rookeries. Peculiar Land Birds. Noddies and other Sea Birds. Southern Skuas. Wild Swine. Change of Habits of Penguins. Nightingale Island. Vast Penguin Rookery. Seal Caves. Rocks Worn by the Feet of the Penguins. Mollymauks and their Nests. Derivation of Seamen's Names for Southern Animals. Dogs run Wild in a Penguin Rookery. Migrations of Penguins and Seals. Insects, &c., of the Group. Flowering Seasons. Sea Beans. Relations of the Flora.

Tristan da Cunha, Oct. 15th, 1873.—The ship arrived at Tristan da Cunha on October 15th. The island of Tristan da Cunha is one of a group composed of three, the other two being called Nightingale and Inaccessible Islands. Besides these, another small island, Gough Island, lies about 200 miles to the south and somewhat to the east of Tristan da Cunha, and from its vegetation would seem to be naturally included in the group.

Tristan da Cunha itself lies in Lat. 37° 2′ 48″ S., Long. 12° 18′ 20″ W., distant westward from the Cape of Good Hope, 1,550 miles, and about one-third farther from Cape Horn, lying nearly on a line drawn between the two Capes; it lies 1,320 miles south of St. Helena. The island is about 16 square miles in area,* it is nearly circular in form, its highest point is 8,326 feet above sea level.

The latest information concerning the inhabitants of the island, extant at the time of our visit, is to be found in the

---

* I regret exceedingly, that owing to ignorance of the nature of a German geographical square mile, I concluded that Grisebach had, in his " Veg. der Erde," made an error in describing the area of Tristan as two geographical square miles, and that I stated this in "Journ. Linn. Soc." Bot., Vol. XIV, p. 328.

"Cruize of H.M.S. 'Galatea,' " p. 28 (London, Allen and Co., 1869). In this account reference is made to the various mentions of the place in books of travel. The visit of the Dutch brig " Dourga " in about 1827 is omitted.* Before the time of the second exile of Napoleon, the island had been settled by some American agriculturists ; but their adventure failed, and the place was but scantily inhabited until the date at which Napoleon was sent to St. Helena.

A corps of Artillery was then sent to Tristan, and batteries were begun to be constructed. A corporal named Glass received permission to stay on the island when the men were withdrawn, and a small colony sprang up which has lasted till the present time, Glass having been for many years regarded as a sort of governor. The numbers were at one time over 200, but were at the time of our visit about 90 ; the younger members of the settlement constantly migrate to the Cape.

We anchored at early morning on the north-west side of the island of Tristan da Cunha, nearly opposite to the settlement. The island here rises in a long black cliff range ; above this stretches a plateau about 2,000 feet above sea level, on which can be discerned from below two or three small secondary craters; above the plateau rises the Peak, a conical mountain with rounded summit, which at the time of our visit and throughout the year, excepting in the middle of summer, is covered over with a smooth shining cap of snow, its lower slopes being dotted over with irregular patches of snow, between which the dark rocks showed out in relief. The whole island has a peculiar cold barren uninhabitable appearance, which seems to be character-istic of the islands of the Southern Ocean.

The cliffs show a very regular stratification, and are com-posed throughout of a series of beds lying nearly horizontally, but dipping slightly towards the shores, at least they appear to do so east and west of the anchorage. The beds, which are con-spicuously marked, are alternately of hard basalt and looser scoriaceous lava, with occasional beds of a red tuff. The whole section is traversed by numerous dykes, mostly vertical and

* " Voyage of the Dutch Brig of War, ' Dourga,'" p. 2. Trans. by W. Earle, London, John Madden and Co., 1840.

usually narrow, and is not unlike that exposed in the Grand Cural at Madeira in appearance.

Streams, or rather cascades, which come dashing down to the sea during the constant heavy rains, have eaten their way into the cliffs, and their beds form conspicuous features in the view as narrow gullies, descending the rocks in a series of irregular steps. At the foot of the cliffs, immediately opposite the anchorage, are débris slopes and irregular rocky and sandy ground, forming a narrow strip of low shore land.

The settlement lies on a broader and more even stretch of low land which extends westwards. In the margin of this lower tract a small low secondary cliff has been formed by the waves. Steep débris slopes lead from the cliffs above to the settlement tract, and the cliffs are here and there broken into ledges and deep gullies, by which ascent to the summit is easy.

At the landing-place the beach is formed of black volcanic sand, but elsewhere in the neighbourhood, of coarse basaltic boulders. At the summit of the Peak, as the inhabitants told us, is a crater basin with a lake at the bottom of it. From their description given, it appears that there is something like the Cānadas of the Peak of Teneriffe, around the terminal crater.

The cliffs have a scanty covering of green, derived mainly from grasses, sedges, mosses, and ferns, with darker patches of the peculiar trees of the island (*Phylica arborea*), and the crowberry (*Empetrum nigrum var. rubrum*). These dark patches become more and more marked towards the summit. Conspicuous patches of bright green are formed under the cliffs at the foot of the watercourses by a dock (*Rumex*). Further dotted about amongst the other herbage are rounded tufts of pale blueish-green, consisting of the tall reed-like grass (*Spartina arundinacea*), which is peculiar to the Tristan da Cunha group and Amsterdam Island.

On nearer inspection the damp foot of the cliff is found to be covered with mosses and liverworts, which latter form, in favourable situations, continuous green sheets covering the earth beneath the grass.

Two ferns, an Asplenium (*A. obtusatum, Forst.*), growing in the clefts of the rocks just as does our home *A. marinum*, and

Lomaria alpina are most abundant under the cliffs. The *Lomaria* plants where situate on stony slopes, and comparatively starved, were all provided with fertile fronds, whilst when growing in rich vegetable mould, they were commonly without fructification.

The commonest flowering plants under the cliffs are *Apium australe*, wild celery, almost the same as the common garden plant abundant here, in Tierro del Fuego, and in the Falkland Islands : the crowberry: the common sow-thistle, a cosmopolitan weed : and a plant with strongly scented leaves (*Chenopodium tomentosum*), which is used as tea by the islanders, a decoction of the leaves being drunk with milk and sugar. The islanders call it " tea."

Creeping amongst the damp moss, is a small narrow-leaved plant with small bright red berries (*Nertera depressa*).

The streams which run down the cliffs, and which vary from violent dashing cascades in rain time, to narrow rills fed only by the melting of the snow above in dry weather, were small at the time of our visit ; their water soaks into the banks of sand at the foot of the cliffs and on the shores, and is mostly lost, but in some places reappears in the shape of shallow freshwater ponds close to the sea beach.

The water of the streams had a temperature of 50° F., whilst the ponds were warmer, 54° F. The temperature of the lower regions of the island is no doubt constantly reduced by the descent of the cold water from the snow far above ; in the gully above the settlement, shrubs of *Phylica arborea* commence at about 400 feet elevation.

The trees have in this locality all been cut down for firewood, but there is still plenty of wood on the island : *Phylica arborea* is the only tree occurring in the islands ; it is a species found only in the Tristan da Cunha group, in Gough Island, and in the far-off island of Amsterdam, 3,000 miles distant. Other species of the genus occur at the Cape of Good Hope, but they are low and shrubby. It belongs to the natural order of the Buckthorns (*Rhamnaceæ*).

The foliage of the tree is of a dark glossy green, with the under sides of the narrow, almost needle-like leaves, white and downy. Hence the tree, which in habit is very like a yew, presents as a whole a mixture of glaucous grey, and dark olive-green shades; it bears berries of about the size of sweet-peas which are eaten by the finch which lives in the islands.

The constant heavy gales do not permit the tree to grow erect; the trunk is usually procumbent at its origin for several feet, and then rises again often at a right-angle. It is always more or less twisted or gnarled. In sheltered places, as under the cliffs on the north-east of Inaccessible Island, the tree is as high as 25 feet, but it is not nearly so high on the summit of the island, though the trunks are said there to reach a length of 30 feet or more.

The largest trunk I saw was about one foot in diameter, but they are said to grow to eighteen inches. The wood of the tree is brittle, and when exposed, rapidly decays, but is serviceable when dried carefully with the bark on. The German settlers in Inaccessible Island, used it even for handles to their axes and other tools.

The Tristan da Cunha group has a terrible climate. For nine months in the year there is constant storm and rain, with snow. It is only in the three summer months that the weather is at all fine. In October the "bad season," as the islanders called it, was just beginning to pass away, but the weather was so uncertain that the ship might have had to leave her anchorage at a moment's notice, and only a steamer dared anchor at all. Hence no one of our party was allowed to go for more than half an hour out of sight of the ship, nor for a distance of more than an hour's walk from the settlement.

I botanized under the cliffs on the lowland in the morning, and intended to reserve the upper plateau and cliff ascent for the afternoon, but as I was making my way up the steep slope above the settlement in the afternoon at about 3 or 4 o'clock, suddenly a dark squall came scudding over the sea, and rapidly reaching us, and climbing the hill-side, chilled us to the bone. My guide, a small boy, born and bred in the island, crouched down instantly under the tall grass and fern, lying on his side, drawing up his legs, tucking in his head, and screwing himself down into the grass like a hare into her form. We followed his example, and found that the perfection of the shelter to be thus obtained from such scanty herbage was astonishing.

The squall being felt at the anchorage, up went the recall

flag on board the ship, and as soon as the hail ceased, I had to hurry down to the shore, without having ascended the mountain side for more than 500 feet. I was only able to secure a specimen of the tree fern (*Lomaria Boryana*), which grows in the islands, and is common also in the Falkland Islands and Fuegia, and at the Cape of Good Hope.

The boy was peculiarly taciturn, and, like all the islanders, extremely curt in his language, and very independent. Like most of the others he showed a strong Yankee twang in the little I got him to say, and he seemed to have considerable difficulty in understanding what I said to him in ordinary English, and indeed often not to be able to understand at all.

Having heard that there were penguins in the island, but at some distance, and not to be approached without wading, I had offered a reward of £1 for a pair, with their eggs. I found them ready for me in one of the huts, and I paid for them. Had I known what countless numbers I was so soon to be amongst I should not have made such an offer, but I have found in the long run, that on a voyage like this, where there is so much uncertainty, it is always best to take the very first opportunity, and I always landed on the places we visited with the very first boat, even if it were only for an hour in the evening. It may come on to blow, and another chance may never occur. I strongly advise any naturalist similarly situated to do the same.

The cottages of the Tristan people are built of huge blocks of a soft red tuff, fitted together without mortar, and are thatched with tussock grass. They are all low one-storied houses, with small enclosures formed with low stone walls about them, in which a few vegetables are grown, and pigs and geese roam about. The potato fields are all walled for shelter from the wind. A large quantity of potatoes are grown, and form the principal source of food.

The islanders had about 400 or 500 head of cattle and about as many sheep. They often lose cattle in the very cold weather from exposure. There is no horse on the island. Formerly there were numbers of wild rabbits, but they are now almost, if not quite, extinct, as are certainly the wild goats and pigs, which have been entirely killed off.

I

The Sea Elephants (*Morunga elephantina*) have almost entirely deserted the island. The last was seen two years before our visit on the beach, just below the settlement. Seals are seldom seen on the island. The islanders make yearly visits to Inaccessible and Nightingale Islands in pursuit of seals, but these are becoming scarcer every year.

A mouse lives about the houses in the settlement, but there is no rat on the island.

This I gathered from conversation with some of the islanders in one of the cottages; the walls of which were decorated all over with pictures from illustrated newspapers. Several of the women were dark, of mixed race, from the Cape of Good Hope.

On the way down to the beach I saw two willow bushes growing in the stream running down from the settlement. The stream has cut deeply into the alluvial soil, and the willows, here entirely sheltered from the wind, thrive well. They could only grow in such a place.

We got geese, sheep, beef, and potatoes from the Tristan people, who knew well how to charge the full value for everything. They are all sharp at a bargain, and as on an average twelve ships visit them each year, or one a month, they manage to live pretty comfortably without working very hard.

Four or five of them who came on board to receive the money for the provisions, stayed as long as ever they could, till the ship was well under way, begging for all sorts of things, such as matches and copybooks for their children, and putting down all the drink they could get. They never have any store of strong drinks on shore, because when any spirits are landed the liquor is cleared out at once in a single bout. At last the men went over the side, and we made off for Inaccessible Island, where, as we heard from the Tristan people, there were two Germans, who might be in distress.

The appearance of Tristan da Cunha, as seen in the distance, is very remarkable. The snowy peak up in the clouds shows out far above the high dark plateau, with its precipitous cliffs everywhere leading down to the sea.

Inaccessible Island, October 16th, 1873.—The ship moved over to Inaccessible Island and kept close under its high cliffs all night.

Inaccessible Island lies W. by S. ½ S. of Tristan, distant about 23 miles, *i.e.*, from the Peak of Tristan to the centre of Inaccessible Island. The island is about 4½ miles in length, from east to west, and about 2 miles broad, 4 square miles in area. The highest point of the island is 1,840 feet in altitude. We anchored on the north-east side.

All night the penguins were to be heard screaming on shore and about the ship, and as parties of them passed by, they left vivid phosphorescent tracks behind them as they dived through the water alongside.

In the morning we had a view of the island. It presented on this side a range of abrupt cliffs, about 1,000 feet in height, of much the same structure as those of Tristan, viz., successive layers of basalt, traversed by vertical or oblique dykes, but mostly by narrow vertical ones. At the foot of the cliffs are some very steep débris slopes extending in one place a long way up the cliff, but not so as to render the ascent possible.

In front of these stretches a strip of narrow uneven ground, formed of large detached rocks and detritus from the cliffs above, which terminates seawards in a beach of black boulders and large pebbles. In one place, where the cliff is somewhat lower than elsewhere, there is a waterfall, which at the time of our visit was scantily supplied with water, but from the marks left by it on the rocks and vegetation, evidently attains much greater dimensions in rainy weather. The cascade pours right down from the high cliff above into a dark pool of peaty water on the beach below. The rocks about its course are covered with mosses and green incrusting plants.

The face of the cliff generally is sprinkled over with green, the vegetation consisting principally of tussock grass (*Spartina arundinacea*), *Apium graveolens* (a small sedge), *Sonchus olera-ceus* (Sow thistle), *Rumex* (Dock), and ferns: with dark green patches of *Phylica arborea* on the débris slopes and ledges. The strip of accessible lower shore land is mostly covered with a dense growth of tall grass, called by the Tristan people "tussock," but quite different in structure from the well-known tussock of the Falklands, though in outward habit resembling it very closely.

Amongst the grass are several patches or small coppices of *Phylica arborea* trees, which keep the ground beneath them free from tussock, it being covered instead with a thick growth of sedges, ferns, and mosses, which form an elastic carpet on the dark peaty soil. Amongst the moss creeps *Nertera depressa*, with its bright red berries, and the Potentilla-like *Acæna ascendens* grows here and there together with the "tea-plant" of the islanders.

The stems and branches of the Phylica trees are covered with lichens in tufts and variously coloured crusts, and the branches of the trees meeting overhead these little islands as it were, in the seas of tall grass, afford most pleasant shady retreats, which seem a perfect paradise after the terrible struggle and fight through the penguin rookery, which it is necessary to endure in order to reach them.

In the early morning, we made out with a glass two men standing on the shore gazing at the ship. The Captain went on shore first, and brought off the men, who proved to be the two Germans we had heard of at Tristan da Cunha. They were overjoyed at the chance of escape from the island; we gave them breakfast, and heard something of their story.

They both spoke English, one of them remarkably well. They were brothers; one of them had been an officer in the German army during the war, the other one a sailor. They had got landed at Inaccessible Island by a whaling vessel, in the hopes that they would be able to make a considerable sum by killing fur seals, and taking their skins. They had been bitterly disappointed.*

After breakfast, I landed with one of the Germans as guide with a large party. We passed through a broad belt of water, covered with the floating leaves of the wonderful seaweed *Macrocystis pirifera*, which here, as at Tristan and Nightingale Island, forms a sort of zone around the greater part of the

---

* For an account of the sojourn of the Germans in the island, and valuable particulars as to the habits of the various birds, see an article by Mr. R. Richards, Paymaster, H.M.S. "Challenger," "Two Years on Inaccessible," in the "Cape Monthly Magazine," Dec., 1873. Cape Town, J. C. Juta.

island, and of which we afterwards saw so much at Kerguelen's Land.

As we approached the shore, I was astonished at seeing a shoal of what looked like extremely active very small porpoises or dolphins. I could not imagine what the things could be, unless they were indeed some most marvellously small Cetaceans; they showed black above and white beneath, and came along in a shoal of fifty or more, from seawards towards the shore at a rapid pace, by a series of successive leaps out of the water, and splashes into it again, describing short curves in the air, taking headers out of the water and headers into it again; splash, splash, went this marvellous shoal of animals, till they went splash through the surf on to the black stony beach, and there struggled and jumped up amongst the boulders and revealed themselves as wet and dripping penguins, for such they were.

Much as I had read about the habits of penguins, I never could have believed that the creatures I saw thus progressing through the water, were birds, unless I had seen them to my astonishment thus make on shore. I had subsequently much opportunity of watching their habits.

We landed on the beach; it was bounded along its whole stretch at this point by a dense growth of tussock. The tussock (*Spartina arundinacea*), is a stout coarse red-like grass : it grows in large clumps, which have at their base large masses of hard woody matter, formed of the bases of old stems and roots.

In penguin rookeries, the grass covers wide tracts with a dense growth like that of a field of standing corn, but denser and higher, the grass reaching high over one's head.

The Falkland Island " tussock" (*Dactylis cæspitosa*), is of a different genus, but it seems to have a similar habit. Here there is a sort of mutual-benefit-alliance between the penguins and the tussock. The millions of penguins sheltering and nesting amongst the grass, saturate the soil on which it grows, with the strongest manure, and the grass thus stimulated grows high and thick, and shelters the birds from wind and rain, and enemies, such as the predatory gulls.

On the beach were to be seen various groups of penguins,

either coming from or going to the sea. There is only one
species of penguin in the Tristan group; this is, *Eudyptes saltator*,

GROUP OF "ROCK HOPPERS" AT INACCESSIBLE ISLAND.
(From a photograph.)

or the "well diving jumper." The birds stand about a foot
and a half high; they are covered, as are all penguins, with a
thick coating of close set feathers, like the grebe's feathers that
muffs are made of. They are slate grey on the back and head,
snow white on the whole front, and from the sides of the head
projects backwards on each side a tuft of sulphur yellow plumes.
The tufts lie close to the head when the bird is swimming or
diving, but they are erected when it is on shore, and seem then
almost by their varied posture, to be used in the expression of
emotions, such as inquisitiveness and anger.

The bill of the penguin is bright red, and very strong and
sharp at the point, as our legs testified before the day was over;
the iris is also red. The penguin's iris is remarkably sensitive
to light. When one of the birds was standing in our "work
room" on board the ship with one side of its head turned
towards the port, and the other away from the light, the pupil

on the one side was contracted almost to a speck, whilst widely dilated on the other; Captain Carmichael observed the same fact.* The birds are subject to great variations in the amount of light they use for vision, since they feed at sea at night as well as in the day time.

It seems remarkable that there should be only one species of penguin at the Tristan da Cunha group, since in most localities several species occur together. It would have seemed probable that a species of "jackass" penguin (*Spheniscus*), should occur on the islands, since one species (*S. Magellanicus*), occurs at the Falkland Islands and Fuegia, and another (*S. demersus*), at the Cape of Good Hope, intermediate between which two points Tristan da Cunha lies. The connection between these two widely separated *Sphenisci* is wanting; it perhaps once existed at Tristan, and has perished.

Most of the droves of penguins made for one landing-place, where the beach surface was covered with a coating of dirt from their feet, forming a broad tract, leading to a lane in the tall grass about a yard wide at the bottom, and quite bare, with a smoothly beaten black roadway; this was the entrance to the main street of this part of the "rookery," for so these penguin establishments are called.

Other smaller roads led at intervals into the rookery to the nests near its border, but the main street was used by the majority of birds. The birds took little notice of us, allowing us to stand close by, and even to form ourselves into a group for the photographer, in which they were included.

This kind of penguin is called by the whalers and sealers "rock-hopper," from its curious mode of progression. The birds hop from rock to rock with both feet placed together, scarcely ever missing their footing. When chased, they blunder and fall amongst the stones, struggling their best to make off.

With one of the Germans as guide, I entered the main street. As soon as one was in it, the grass being above one's head, one was as if in a maze, and could not see in the least where one

* In the "Supplement to the British Museum Catalogue of Seals and Whales," p. 7, reference is made to a like peculiarity of the iris in the case of *Otaria Jubata*.

was going to. Various lateral streets lead off on each side from the main road, and are often at their mouths as big as it, moreover, the road sometimes divides for a little and joins again : hence it is the easiest thing in the world to lose one's way, and one is quite certain to do so when inexperienced in penguin rookeries. The German, however, who was our guide on our first visit, accustomed to pass through the place constantly for two years, was perfectly well at home in the rookery and knew every street and turning.

It is impossible to conceive the discomfort of making one's way through a big rookery, hap-hazard, or "across country" as one may say. I crossed the large one here twice afterwards with the seamen carrying my basket and vasculum, and afterwards went through a larger rookery still, at Nightingale Island.

You plunge into one of the lanes in the tall grass which at once shuts out the surroundings from your view. You tread on a slimy black damp soil composed of the birds' dung. The stench is overpowering, the yelling of the birds perfectly terrifying ; I can call it nothing else. You lose the path, or perhaps are bent from the first in making direct for some spot on the other side of the rookery.

In the path only a few droves of penguins, on their way to and from the water, are encountered, and these stampede out of your way into the side alleys. Now you are, the instant you leave the road, on the actual breeding ground. The nests are placed so thickly that you cannot help treading on eggs and young birds at almost every step.

A parent bird sits on each nest, with its sharp beak erect and open ready to bite, yelling savagely " caa, caa, urr, urr," its red eye gleaming and its plumes at half-cock, and quivering with rage. No sooner are your legs within reach than they are furiously bitten, often by two or three birds at once: that is, if you have not got on strong leather gaiters, as on the first occasion of visiting a rookery you probably have not.

At first you try to avoid the nests, but soon find that impossible ; then maddened almost, by the pain, stench and noise, you have recourse to brutality. Thump, thump, goes your stick, and at each blow down goes a bird. Thud, thud, you hear from the men behind as they kick the birds right and left off the nests, and so you go on for a bit, thump and smash, whack, thud, " caa, caa, urr, urr," and the path behind you is strewed with the dead and dying and bleeding.

But you make miserably slow progress, and, worried to death, at last resort to the expedient of stampeding as far as your breath will carry you. You put down your head and make a rush through the grass, treading on old and young hap-hazard, and rushing on before they have time to bite.

The air is close in the rookery, and the sun hot above, and out of breath, and running with perspiration, you come across a mass of rock fallen from the cliff above, and sticking up in the rookery; this you hail as " a city of refuge." You hammer off it hurriedly half a dozen penguins who are sunning themselves there, and are on the look-out, and mounting on the top take out your handkerchief to wipe away the perspiration and rest a while, and see in what direction you have been going, how far you have got, and in which direction you are to make the next plunge. Then when you are refreshed, you make another rush, and so on.

If you stand quite still, so long as your foot is not actually on the top of a nest of eggs or young, the penguins soon cease biting at you and yelling. I always adopted the stampede method in rookeries, but the men usually preferred to have their revenge and fought their way every foot.

Of course it is horribly cruel thus to kill whole families of innocent birds, but it is absolutely necessary. One must cross the rookeries in order to explore the island at all, and collect the plants, or survey the coast from the heights.

These penguins make a nest which is simply a shallow depression in the black dirt scantily lined with a few bits of grass, or not lined at all. They lay two greenish white eggs about as big as duck eggs, and both male and female incubate.

After passing through the rookery, we entered one of the small coppices I have already described. Hopping and flutter-ing about amongst the trees and herbage, were abundance of a small finch and a thrush; no other land birds were seen. The finch (*Neospiza Acuhnæ*) looks very like a green-finch, and is about the same size.

The thrush (*Nesocichla eremita*) looks like a very dark-coloured song thrush, but it is peculiar for its remarkably strong acutely ridged bill. It is peculiar to the Tristan group. It feeds especially on the berries of the little Nertera; but also is fond of picking the bones of the victims of the predatory gull (*Stercorarius antarcticus*). The finch eats the fruit of the Phylica.

It was here that we first encountered that remarkable tameness, and ignorance of danger in birds which has been so constantly noticed by voyagers landing on little frequented islands, and notably by Darwin, who dilates on the fact in his account of the Galapagos Archipelago.

The thrush and finch hopped unconcernedly within a yard or two of us, whilst stone after stone was hurled at them, and till they were knocked over, and often sat still on a bough to be felled with a walking stick. By whistling a little as one approached them, numbers could be thus killed, and yet the Germans, with their house close by, had been constantly thus killing the thrushes for eating for two years. The birds are, however, not quite so tame in Tristan Island.

The finch seems to have become extinct in Tristan da Cunha itself. Von Willemoes Suhm was told that the Tristan da Cunha people had tried to introduce the bird into their island.*

We were in search of another land bird, a kind of Water-Hen (*Galinula nesiotis*), which is found on the higher plateau at Tristan, and is described by the inhabitants as scarcely able to fly. We could not meet with a specimen. Only very few inhabit the low land under the cliffs, and we were not able to land at the only place from which the higher main plateau of the island is to be reached.

The Germans said that the Inaccessible Island bird is much

---

* I presume that the *Neospiza Acuhnæ* of Cabanis, described from old specimens from Bullock's collection, is the *Emberiza Braziliensis* of Carmichael. No second species of finch was seen or heard of by us as existing now in the islands. The genus *Neospiza* is peculiar to the Tristan group, but of South American affinity. *Crithagra insularis*, the other finch described by Cabanis as found in the group, is a peculiar species allied to African forms. A list of the Tristan land birds collected by the "Challenger" has not yet appeared.

smaller than *G. nesiotis*, and differs from it in having finer legs
and a longer beak. This is, however, hardly probable, since
the Tristan species occurs at Gough Island.

The family of *Gallinulidæ* is remarkably widely spread, and
one of these birds is in several instances the inhabitant of some
isolated island group; several occur thus in the Pacific. This is
curious, since one would at first perhaps think these birds bad
flyers, but they are not, and are not uncommonly met with on
the wing at sea far from land, just as we met with Water-rails
between Bermuda and Halifax.

Sitting on the tree-tops with the thrushes were numerous
"noddies" of the same two species as those of St. Paul's Rocks.
It was strange to see birds which one had met with on the
equator living in common with boobies, here mingling with
Antarctic forms. The noddy however ranges far north also, even
occasionally to Ireland.

The whole of the peaty ground underneath the trees in the
Phylica woods is bored in all directions with the holes of
smaller sea birds, called by the Germans "night birds," a Prion
and a Puffinus.

The burrows that these birds make are of about the size of
large rats' holes. They traverse the ground everywhere, twisting
and turning, and undermining the ground, so that it gives way
at almost every step. A further account of these birds and their
habits, will be found in the account of Kerguelen's Land, in
which island they abound.

I went along the beach, and through a second wood towards
the waterfall, where was the hut of the Germans, and their
potato ground. A flock of thirty or forty predatory gulls
(*Stercorarius Antarcticus*), were quarrelling and fighting over
the bodies of penguins, the skins of which had been taken in
considerable numbers by our various parties on shore. The
Skua is a gull which has acquired a sharp curved beak, and
sharp claws at the tips of its webbed toes. The birds are
thoroughly predaceous in their habits, quartering their ground on
the look-out for carrion, and assembling in numbers where there
is anything killed, in the same curious way as vultures.

They steal eggs and young birds from the penguins when

they get a chance, but their principal food here appears to be the night birds, especially the Prions, which they drag from their holes, or pounce on as they come out of them.   The place was strewed with the skeletons of Prions, with the meat torn off them by these gulls, which leave behind the bones and feathers.

The Antarctic Skua is very similar in appearance to the large northern Skua, of which a figure is given here in default of better.   The two species were at first considered by naturalists to be identical; they differ however, especially in the structure of the bill.   The Skua is of a dark brown colour, not unlike that of most of the typical birds of prey.   We met with the

BRITISH SKUA, STERCORARIUS CATARRACTES.

bird constantly afterwards on our southern voyage, as far down even as the Arctic Circle; and a specimen was noticed by Ross further south still, in Possession Island.

The hut of the Germans was a comfortable one of stone, thatched with tussock and with a good frame window and door, and comfortable bunks to sleep on.   There used to be wild goats on the top of Inaccessible Island, and there are still plenty of wild pigs.   The ferine pigs were, as the Germans told me, of various colouring, and showed no tendency to uniformity; but the goats were almost invariably black, only one or two had a few white markings about head, neck, and chest.   The sows used to be seen with litters of seven or eight young, but in a few days the number dwindled to one or two; the sows probably eating their young.   The young suffered often from a sort of scrofula, in which the glands about the neck became much enlarged.

The pigs now remaining are mostly boars: they are very hairy and have long tusks. The hogs are fierce, and one of the Germans told me that one once regularly hunted him, as if to attempt to kill him for food. The pigs feed mainly on birds and their eggs, but eat also the roots of the tussock and wild celery; they have nearly exterminated a penguin rookery on the south side of the island, but a few penguins remain, who have learnt to build in holes under stones, where the pigs cannot reach them.

This fact is curious, as showing how easily circumstances may arise, such, that in an islánd even so small as Inaccesible, one colony of birds may develop a totally new habit, whilst other colonies of the same species preserve their original customs. And yet how strong is the tendency in birds to preserve their habits! I know of no more striking instance of this than the fact that the Apteryx of New Zealand (*A. Australis*) considers it necessary to put as much of its head as it can under its rudiment of a wing, when it goes to sleep.*

The pigs cannot get down the cliffs to the rookeries on the north side of the island.

One penguin at the Falkland Islands (*Spheniscus Magellanicus*) regularly nests in burrows, sometimes twenty feet long. Another species of the same genus (*Spheniscus minor*), breeds in neat holes burrowed in sandbanks, at New Zealand.†

On the beach are large banks of seaweed, but as at Tristan the heavy surf so batters the weeds, that it is difficult to find a serviceable specimen. An Octopus is very common amongst the stones, about the edge of the surf. I caught several attracted by the washing of the penguins' flesh and skins in the water. A *Chiton, Patella* and *Buccinum* are also common about the shore, as at Tristan.

All night long the penguins on shore in the rookery kept up an incessant screaming, no doubt lamenting the terrible invasion to which they had been subjected. The sound at a distance was not unlike that which one hears from tree-frogs in the south of

* T. H. Potts, "On the Birds of New Zealand," Vol. II, 1869, p. 75. Trans. N. Z. Institute.

† T. H. Potts, Ibid., Vol. V, 1872, p. 186.

Europe, "Cāā Quărk, Cāā Quărk, Cā Cāā Cā Cāā." In the morning we moved to Nightingale Island, taking the Germans with us.

Nightingale Island, Oct. 17th, 1873.—Nightingale Island, the smallest of the Tristan group, lies 20½ miles S.W. of Tristan Island, and about 22 miles N.W. by W. of Inaccessible Island. The island is about $1\frac{1}{20}$th miles long, by less than one mile broad; its area is thus not more than one square mile. We steamed up to the north-west side in the morning.

In the north-east is a rocky peak, from which an elevated ridge runs down to the sea on the east side, whence the Peak is accessible. On the north side it is impracticable, being too precipitous. A lower ridge stretches N.E. and S.W. on the south side of the island, and a broad valley separates the western termination of this ridge from the high ground and peaks on the N.E.; the highest peak is 1,100 feet in height, and the highest point of the lower ridge, 960 feet.

The whole of the lower land, and all but the steepest slopes of the high land and its actual summits, are covered with a dense growth of tussock, which occupies also even the ledges and short slopes between the bare perpendicular rocks of the Peak. The lower ridge is covered with the grass on all except its very summit, where amongst huge irregularly piled boulders of basalt, grow the same ferns as are found in Inaccessible Island, and *Phylica arborea* trees. The summit of the higher ridge appears to have a similar vegetation, the tussock ceasing there.

In the sea of tall grass, clothing the wide main valley of the island on its south side, are patches of Phylica trees, growing in many places thickly together, as at Inaccessible Island, with a similar vegetation devoid of tussock, beneath them. The appearance of the tall grass, when seen from a distance, is most deceptive; as we viewed the island from the deck of the ship, about a quarter of a mile off, we saw a green coating of grass, coming everywhere down to the verge of the wave-wash on the rocks, and stretching up comparatively easy looking slopes towards the summit of the Peak.

The grass gave no impression of its height and impenetrability, and one of the surveyors started off jauntily to go to the top of the Peak and make a surveying station. On closer inspection, however, the real state of the case might be inferred, for there was plainly visible a dark sinuous line leading from the sea, right inland through the thickest of the tussock. This was a great penguin road, and the whole place was one vast penguin rookery, and the grass that looked like turf to walk on, was higher than a man's head.

I made out with my glass a great drove of penguins on the rocks under the termination of the road, and I went below at once to put on my thickest gaiters.

We pulled on shore through beds of kelp, and landed on shelving rocks leading up to caves, the haunt of the Fur Seals in the proper season. We met the surveyors coming back, well pecked and dead beat, having given up the Peak in despair.

The shelving rock is composed of volcanic conglomerate, full of irregular fragments and rounded lumps of hard basalt, and various scoriaceous forms ; in places also of a similarly derivative rock of a reddish colour, but devoid of larger embedded fragments. In a cliff about forty feet in height, adjoining and rising from the shelves, are beds of fine-grained volcanic sandstone rock, banded yellow and black, and horizontally bedded, probably of submarine formation.

These beds constitute the whole mass of two or three small outlying rocks or islands lying to the N.E., and are there also horizontal. These beds appear about twenty feet thick in the cliff, and above them is a layer of basalt of about the same thickness, which extends east and west, capping the softer beds and conglomerates. This layer is evidently a lava flow of comparatively late date, as it seems to have run down the valley between the two ridges, and to have come from the south ; its upper surface is a little rounded, higher in the centre, and thinning off at the edges, as may be seen in the section exposed in the cliff.

It is on the almost level upper surface of this flow, that the great penguin rookery lies. The island has evidently, like Inaccessible Island, undergone immense denudation, and there is no trace of any centres of action remaining. In the low cliffs of the coast, numerous caves are formed by the eating out by waves of the softer strata underlying the hard cap of basalt.

The caves are so numerous as to form a striking feature in the appearance of the island as it is approached from seawards; such caves are not apparent at Inaccessible or Tristan da Cunha Islands.

The caves with the sloping ledges leading up to them, are frequented as was said by fur seals. Four years before 1,400 seals had been killed on the island by one ship's crew; they are much scarcer now, but the island is visited regularly once a year by the Tristan people, as is also Inaccessible Island. The Germans only killed seven seals at Inaccessible Island, but the Tristan people killed forty there in December, 1872. Two seals were seen by us in the water about the rocks, but none on land.

The sloping rock ledges are covered with a thin coating of dark green ulva, which, when dry, has a peculiar almost metallic glance. A short scramble up the rocks brought us at once face to face with the tall grass and penguins.

The party broke up into small groups, each choosing what it thought the best route for penetrating the enemy's country. I made along the rocks to the point where, as I had seen from the ship, the main street ended: here were hundreds of penguins coming from and going to the sea in droves, or hurrying along singly to catch up some drove, or lolling about on the rocks, basking; the moving ones going along hop, hop, hop, just like men in a sack race.

The hard rock was actually polished, and had its irregularities smoothed off where the feet of the birds had worn it down at the entrance to the street. No doubt the Diatom skeletons present in the food and dung of the penguins, and always in abundance in the mud of their rookeries, adhering to their dirty feet, acts as polishing powder and assists the wearing process.

The street did not open by a single definite mouth towards the sea, but split up into numerous channels leading down to a number of easy tracks through the rocks. A little way in there was a clear open track six feet wide, and in places as much as eight or ten feet in width.

On each side narrow alleys led at nearly right angles to the rows of nests with which the whole space on either side of the main street was taken up.

Amongst the penguins here were numerous nests of the yellow-billed Albatross (*Diomedea culminata*) called by the Tristan people "Mollymauk," variously spelt in books, Molly Hawk, Mollymoy, Mollymoc, Mallymoke. It is, as are most of the sealers' names in the South, a name originally given to one of the Arctic birds, the Fulmar, and then transferred to the Antarctic from some supposed or real resemblance.

In the same manner the name given by northern whalers to the Little Auk is given in the South to the Diving Petrel of Kerguelen's Land. And the term "clap match" given to the female southern fur seals by the sealers is the name originally given by the Dutch to the hooded seal or "bladdernose" of Greenland (*Cystocephalus*), and is a corruption of the word "Klapmuts," a bonnet, "the seal with a bonnet." It is curious that in this case the term should have been thus transferred to so very different a seal, which has nothing resembling a hood, but the word is so peculiar that there can be no doubt about its origin.

Various similar corruptions are in use as terms for southern animals. The name Albatross itself is the Spanish word "alcatraz" a "gannet." The Spanish no doubt called the albatrosses they met with "gannets," their familiar sea bird, just as common sailors will call every sea bird a gull, and a foreigner's corruption of the word became adopted as a special name for the bird.

The name Penguin is another instance in point. The word was not coined, as often supposed, by the early Dutch navigators, from the Latin word "pinguis," but is, as has been shown by M. Roulin, and others, a Breton or Welsh word, "pen gwenn," "white head," the name originally given to European sea birds with white heads, probably to the Puffin (*Mormon fratercula*). The name Pingouin is applied in modern French to the Great and Little Auk. In early voyages the name is applied to various exotic sea birds. In early Dutch travels the true meaning of the word is given, and it is stated to be English.*

* Sy worden Pinguijns ghenaemt niet van wegen haer vettigheyd, so de

K

The Mollymauk is an albatross about the size of a goose, head, throat, and under part pure white, the wings grey, and the bill black with a yellow streak on the top and with a bright yellow edge to the gape, which extends right back under the eye. The yellow shows out conspicuously on the side of the head. It is not thus shown in Gould's coloured figures. The bird is extremely handsome. They take up their abode in separate pairs anywhere about in the rookery, or under the trees, where there are no penguins, which latter situation they seem to prefer.

They make a cylindrical nest of tufts of grass, clay, and sedge, which stands up from the ground. The nest is neat and round. There is a shallow concavity on the top for the bird to sit on, and the edge overhangs somewhat, the old bird undermining it, as the Germans said, during incubation, by pecking away the turf of which it is made.

I measured one nest, which was 14 inches in diameter and

NEST OF THE MOLLYMAUK.

10 inches in height. The nests when deserted and grass-grown make most convenient seats. The birds lay a single egg, about the size of a goose's, or somewhat larger, but elongate, with one end larger than the other, as are all albatross eggs.

The egg is held in a sort of pouch whilst the bird is incubating. The bird has thus to be driven right off the nest before the egg is dropped out of the pouch and it can be ascertained whether there is one there or no.

The birds when approached sit quietly on their nests or stand by them, and never attempt to fly; indeed they seem,

schryver van dit Journael verkeerdelijck meent, maer om dat sy witte hoofden hebben, want dat betekent Pinguijns in't Engelsch, gelijck in Sir Thomas Candish voyage te sien is. " Begin ende Voortgang vande vereenigde Neederlandtsche Geoctroyeer de Ost-Indische Compagnie." I[te] deel. Long folio, Pub. 1646. " Schip.-vaerd der Hollanders nae de Straet Magaljanes," p. 28.

when thus bent on nesting, to have forgotten almost the use of their wings.

Captain Carmichael, in his account of Tristan da Cunha, relates how he threw one of the birds over a cliff and saw it fall like a stone without attempting to flap, and yet these birds will soar after a ship over the sea as cleverly as any other albatross ; indeed, the same peculiarity occurs in the case of the large albatross when nesting.

When bullied with a stick or handled on the nests, the birds snap their bills rapidly together with a defiant air, but they may be pushed or poked off with great ease. Usually a pair is to be seen at each nest, and then by standing near a short time one may see a curious courtship going on.

The male stretches his neck out, erects his wings and feathers a bit, and utters a series of high-pitched rapidly repeated sounds, not unlike a shrill laugh. As he does this he puts his head close up against that of the female.

Then the female stretches her neck straight up, and turning up her beak utters a similar sound, and rubs bills with the male again. The same manœuvre is constantly repeated.

The albatrosses make their nests sometimes right in the middle of a penguin road, but the two kinds of birds live perfectly happily together. I saw no fighting, though, small as the penguins are, I think they could easily drive out the Mollymauks if they wished it.

The ground of the rookery is bored in all directions by the holes of Prions and petrels, which thus live under the penguins. Their holes were not so numerous in the rookery at Inaccessible Island as here. The holes add immensely to the difficulties of traversing a rookery, since as one is making a rush, the ground is apt to give way, and give one a fall into the black filthy mud amongst a host of furious birds, which have then full chance at one's eyes and face.

Besides the mollymauks and petrels, one or two pairs of Skuas had nests on a few mounds of earth in the rookery. How these mounds came there I could not understand.

The Skuas' eggs are closely like those of the lesser black-backed gull, and two in number. The birds swooped about our heads as we robbed the nests, but were not nearly so fierce as those we encountered further south. All round their nests were scattered skeletons of Prions.

I, with three sailors carrying my botanical cases, attempted to scale the Peak ; we had a desperate struggle through long grass and penguins, and at last had to come back beaten, and made for the Phylica patches, where the ground was clear. Thence I fought my way through the grass up to the top of the lower ridge of the island, but though there were no penguins on this slope, I never had harder work in my life.

I had to stop every ten yards or so for breath, the growth of the grass was so dense. My men lost me and never reached the top. On the summit I found the rest of the party which had come on shore, full of the hardships they had suffered in getting through the rookery, and looking forward with no pleasure to the prospect of going back again through it.

Two spaniels had been brought on shore and were taken through the rookery, partly by being carried, partly dragged. One of these was lost on the way back ; he would not face the penguins and could not be carried all the way, so got left behind, and I fear must have died and been eaten by Skuas.

Poor old " Boss," Lieutenant Channer's pet, though one-eyed and too old to be much good for shooting, was a favourite, and we were all very sorry for him. Three volunteers charged back into the rookery in search, but it was of no use. He was frightened to death and would not answer to a call.

The dogs brought to Inaccessible Island by the two Germans ran wild in the penguin rookery, notwithstanding their exertions to keep them at home, and finally the dogs had to be shot. They fed themselves on the eggs and young.

After getting through the rookery on to the rocks, it was amusing to see the party arrive singly and in twos at all sorts of points of the edge of the rookery and on the verge of the cliff, having lost their direction, and often to their disgust having to turn back through the edge of the rookery again to reach some spot where they could get down to the sea.

The penguins were having their evening bath and pluming themselves on our arrival. The number of birds here must be enormous. At least one-fourth of the surface of the island and small outlyers, for these also are rookeries, must be covered by them; taking thus a space a quarter of a mile square, and allowing two only to a square yard, there would be nearly 400,000 penguins.

The rookery has evidently once been larger than at present, since a good part of the tall grass, now not occupied by birds, had old deserted nests amongst it. Probably the number of birds varies considerably each season.

One of the most remarkable facts about the penguins is that they are migratory; they leave Inaccessible Island, as the Germans told us, in the middle of April after moulting, and return, the males in the last week of July, the females about August 12th; and I do not think it possible that the Germans could have been mistaken. Whither can they go, and by what means can they find their way back? The question with regard to birds that fly is difficult enough, but it may always be supposed that they steer their course by landmarks seen at great distances from great heights, or that they follow definite lines of land. In the present case the birds can have absolutely no landmarks, since from sea level Tristan da Cunha is not visible from any great distance; the birds cannot move through the water with anything approaching the velocity of birds of flight; they have however, the advantage of a constant presence of food. The question of the aquatic migration of penguins and seals seems a special one, and presents quite different difficulties to that of the migration of birds of flight. The penguins certainly do not go to the Cape of Good Hope nor St. Helena, and they cannot live at sea altogether.

The migration of the turtles at Ascension Islands, seems to be possibly a parallel case. The young turtles on leaving the egg go down to the sea and disappear, returning only when full grown to breed; this is the account given by residents. If they do really leave the neighbourhood of the island, there seems no possible means by which they can find their way back.

There is little fresh water on Nightingale Island. I saw one pond in the rookery, but the water was undrinkable. In a cave, however, where we landed, there was a scanty trickling spring of excellent water filling a small basin; water enough to keep three or four persons alive might be got here.

We left Nightingale Island in the evening, and made for the Cape of Good Hope.

Besides the birds I have mentioned, the great Albatross

(*D. exulans*) breeds at Tristan da Cunha, and on the top of Inaccessible Island. At Tristan da Cunha it nests actually within the crater of the terminal cone around the lake, 7,000 feet or more above the sea.

The Mollymauk is common in Tristan da Cunha, and its eggs were brought off to us by the islanders for sale; they are not bad eating. Cape pigeons (*Daption capensis*) and the Giant-petrel (*Ossifraga gigantea*), nest in Tristan da Cunha, and one specimen

GREAT ALBATROSS, DIOMEDEA EXULANS.

of *Procellaria glacialoides*, was obtained on shore by Von Willemoes Suhm.

There are two land shells of the genus *Balea* allied to pupa; an *Oniscus*, three small Curculios, four *Geometræ*, a *Hippobosca*, *Musca*, and *Tipula*, mentioned by Captain Carmichael as found in Tristan da Cunha; we found them also, and besides an Iulus was very common, and several spiders.

From what the Germans told him, Von Willemoes Suhm concluded that there were two butterflies, a *Vanessa* and an *Argynnis* in the island; if so, these may no doubt be attracted by the scarlet blossom of the *Pelargonium*, so abundant in the island, and fertilize it, and act as a stimulus to the preservation of its colour, and to some extent account for this.

Otherwise one must regard this case as an instance of the survival in an island, where it is now without function, of a brightly coloured flower developed originally in the progenitors of the plant on a continent amongst numerous insects.

Though some of the plants in the Tristan da Cunha group

appear to flower all the year round, others have their regular blooming season. This is the case with the Pelargonium and the Tea plant. The Pelargonium blossoms, according to the Germans, in the middle of summer. Large numbers of the plants come into blossom at the same time, so that the beach is thickly strewn with the coloured petals fallen from the cliffs.

The Tea plant was nowhere found in blossom in October, though it was abundant. The Phylica trees were all in the same stage of development, bearing fully formed but green fruit.

The existence of the Cape Horn current sweeping up to the islands, may account for the presence of many South American plants in them. The part of the Brazilian current which turns from the coast of South America, and runs across to the Tristan group, brings with it many seeds to the islands, but these, being tropical, do not germinate. The seeds are cast upon the beach at Tristan, and are familiarly known amongst the islanders as sea beans, from a belief that they grow at the bottom of the neighbouring sea.

Two of these seeds were shown to me; one of them was a bean of a tropical American tree, the other was the seed of a *Guilandina*,* also tropical, which seed, singularly enough, is also cast up sometimes at Bermuda, and is there called a sea bean, and worn on watch chains as a curiosity, and I believe as an antidote to drowning.

Sir Joseph Hooker, in his lately published account of the Botany of Kerguelen's Land,† writes : " The flora of Tristan da Cunha, Nightingale and Inaccessible Islands, is essentially Fuegian, with an admixture of Cape Genera, but with none of those characteristic of Kerguelen's Island. Of Cape types it contains a Pelargonium and an abundance of both the Phylica and Spartina of Amsterdam Island, together with species of *Oxalis* and *Hydrocotyle*. The Fuegian and Falkland Island plants of Tristan da Cunha and its islets, which have not hitherto been found in the islands south and east of them, are however, more numerous than the Cape genera even, and include

* See page 17.
† Transit of Venus Expedition, Botany. "Observations on the Botany of Kerguelen's Land," p. 8. By Sir J. D. Hooker, P.R.S.

Cardamine hirsuta, Nertera depressa, Empetrun nigrum var. *rubrum, Lagenophora Commersoniana,* and *Apium australe;* and the flora contains besides the strictly American genus *Chevreulia.*"

The close similarity of the flora of the three islands of the group points to a former connection between them. Their high cliffs, composed of successive layers of lava, and the absence, except in Tristan da Cunha, of well marked centres of eruption, as well as their general features, show that they have undergone great denudation. A sounding between Tristan da Cunha and Inaccessible Island gave a depth of 1,000 fathoms; between Inaccessible and Nightingale Islands, the depth was 460 fathoms.

It is obvious, from the relative position of the three islands, that the prevalent winds blow directly from Inaccessible Island towards Nightingale Island and Tristan da Cunha.

With regard to the Cryptogamous vegetation of the group, nearly all the seaweeds, as appears from Prof. Dickie's report on the specimens collected by me, are Cape of Good Hope species, or occur at the Cape as well as at numerous other localities : two only are new and apparently endemic. Of Fungi, an *Agaricus,* which grows on the Phylica stems, is described by Mr. Berkeley as new, as *A. phylicigena.* Of the mosses and *Hepaticæ,* Mr. Mitten describes ten species as new, out of thirty-six collected by me ; of eleven lichens collected, two were new ; one, *Lecanora acunhana,* is noted by Nylander as "*bene distincta.*"

An Islander told me that the flowering plants on Gough Island were the same as those of Tristan da Cunha, but he thought there were different ferns ; he had lived there some time sealing.

Scientific Notices of the Tristan da Cunha Group.

Du Petit Thouars, flora of the island, in his "Melanges."

Captain Carmichael's account of the island in "Linn. Trans.," Vol XII, p. 496.

For descriptions of the collections of plants made by me in the Islands, see list of papers relating to the "Challenger" Expedition, at the end of this Book.

For a description of Gallinula nesiotis, by P. L. Sclater, F.R.S., &c., see "Proc. Zool. Soc., 1855," p. 146.

For notes on the Zoology of the Islands, see Von Willemoes Suhm, "Proc. R. Soc.," No. 170, 1876, p. 583.

For notes on the Geology, see J. Y. Buchanan, Ibid., p. 614.

For Birds, see "Cabanis über zwei neue Finken-Arten." Journal für Ornithologie, 1873, s. 153, 154.

The ship took ten days to reach the Cape of Good Hope; the only interesting feature of the voyage was the appearance of the various southern Oceanic birds which constantly were to be seen flying at the stern. The great albatross or Cape sheep, the Mollymauk, which however was not seen far from land; the Giant-petrel (*Ossifraga gigantea*), the Cape hen (*Procellaria æquinoctialis*), the Cape pigeon (*Daption capensis*), a Prion and a Stormy-petrel.

CHAPTER VI.

CAPE OF GOOD HOPE.

Aspect and Formation of the Country. Simons Bay. Appearance of the Vegetation. The Road to Cape Town. The Silver Tree. Habits of Baboons. The Rock Rabbit. Habits of Rodent Moles. Kitchen Middens. Burial Places of Natives. Antelopes. An Ostrich Farm. Tracks of Animals in the Sand. Great Variety of Flowering Plants. Clawless Otter. Land Planarians. Chameleon. Jackass Penguins. *Bdellostoma.* Rare Whale with Long Tusks. *Peripatus capensis,* the Ancestor of Insects. The Turacou.

Simons Bay, October 28th to December 17th, 1873.—We anchored at Simons Bay on October 28th, but found ourselves in quarantine because we had had yellow fever on board at Bahia.

The Cape of Good Hope lies at the end of a long narrow promontory running nearly north and south, and forming between itself and Cape Hangklip on the east, a large bay known as False Bay, whilst at its point of origin from the mainland and on its east side, is Table Bay with Cape Town at its head.

The promontory has a sort of backbone of mountains, which in some places come right down steep into the sea, in others are flanked by more or less extensive sand-flats. The mountains are highest towards the northern extremity of the ridge which terminates in the far-famed Table Mountain, 3,550 feet in height. Constantia Berg, about one-quarter of the distance from this point to the Cape, is 3,200 feet high; the remaining mountains range from about 2,000 to 1,500 feet.

The sandy flats are towards the southern part of the promontory almost confined to its Western side, the steep slopes of the mountains on the False Bay side, being for the most part washed directly by the sea, but at the head of False Bay a wide extent of flat sandy plain extends right across the head of the bay and round the foot of Table Mountain northwards. This plain is known as the " Cape Flats."

The Cape of Good Hope is at the tip of the promontory, and is not, as I used to think, the southernmost point of Africa. Cape Agulhas to the eastward is far south of it.

The mountains are entirely composed of a hard metamorphic sandstone, passing in many places into a white quartzite which is disposed in perfectly horizontal strata. This perfect and remarkably uniform horizontality of the rock-beds is the cause of the peculiar form of the Cape land surface and forms the chief feature in the landscape.

Everywhere the mountains rise by a series of steps with flat intervening surfaces. Table Mountain itself derives its name from its horizontal flat top, bounded by perpendicular cliffs rising straight up from the flats ; and the same formation being continued for hundreds of miles inland, the country continually rises in steps forming successive table lands, known as the Karroo Plains, about 2,000 feet above sea level, and beyond these the Roggefeld, 3,500 feet in elevation.

We steamed into False Bay past the Cape Point lighthouse up to Simons Bay, where is the dockyard. The long range of mountains extending from Hangklip along the eastern shore of False Bay in the district known as Hottentots' Holland, seen in the distance was strikingly beautiful, with soft and delicate outlines, and lighted up with beautiful pink and violet tints as in an Italian landscape. I was astonished at the beauty of the scenery, as I had been led from the accounts of Simons Bay to expect nothing but a desert of sand.

Simons Bay lies on the east side of the Cape promontory, and about half way up the west side of Faloc Bay. There is a dockyard, houses for the dockyard officials and workmen, a small barrack, a naval hospital, a small town of one street stretching along the shore, and a few houses scattered on either side of the road which leads in one direction towards Cape Town, in the other towards Cape Point. The town stands on a narrow tract of land composed of talus from the hills which rise in steep slopes behind it, buried more or less in different places in glistening white sand.

The hills about the Cape district have all an exactly similar appearance as far as their clothing with vegetation is concerned. They look not unlike Scotch moorland, being covered every-where with low bushes without trees. The vegetation has a general brownish or greyish tint; there are no bright greens in the landscape. This arises from the fact that the plants are nearly all evergreen, and have, as a rule, either narrow needle-like leaves, like the pines, or leaves covered with grey downy hairs; in fact, all sorts of contrivances for resisting their great enemy, the drought.

The most characteristic feature, however, in the landscape is the showing through in all directions of the red soil between the bushes and clumps of vegetation; the interspaces not being filled in with grasses, and no continuous covering of vegetation being formed.

In the flowering season, from June to August, which depends here on the rainy season, and falls thus in mid-winter, the aspect of the landscape is entirely changed, and whole tracts of country are coloured of most brilliant hues. We were too late for this, but nevertheless could form an idea of what it must be like, because, though the greater numbers of plants of each of the various species blossom all together at the regular season of the species, there are always to be found stragglers blossom-ing at other seasons, and nearly every plant can be collected in flower by search at almost any period of the year.

Simons Bay is 84 miles from Cape Town by road, but a railway runs from a village called Wynberg, about 14 miles distant from Simons Bay, to the town. There is practically only one road at Simons Bay, for though two others start with great promise, the one along the shore towards Cape Point, and the other up the steep hill at the back of the town (Red Hill), they soon lose their character and dwindle to the condition of mere tracks over the moorland, very difficult for a stranger to follow, as I more than once found. Hence "going up the road" or "down the road," is the term at Simons Bay for visits to and fro Cape Town.

The road follows the shore, being cut out on the side of the steep coast, and crosses at several places sandy sea beaches, where the driver keeps the horses with their feet at the very verge of the surf, because the sand is harder here, as everyone knows who has had to walk along a sandy shore.

The conveyances are two-wheeled carts with a hood cover, open in front and with two parallel seats placed transversely. There is a pole to them, and a pair of horses are always driven, great care being taken as to balancing. I never saw a pair of horses thus driven in a two-wheeled vehicle before.

The drivers are mostly Malays, of whom there are large numbers in Cape Town and Simons Town, emancipated slaves of the Dutch, or progeny of these. Those who disregard expense take four horses to one of these traps, and the mail always has four. It is a shabby cart, like the rest. The Malays drive well, and manage a very long whip to a nicety. The travelling is not dear; a cart and pair to Wynberg, i.e., 14 or 15 miles, costs 15s.

Half-way to Wynberg is a noted wayside inn, called "Farmer Peck's," with a long rigmarole about the Gentle Shepherd of Salisbury Plain, over the door, and some Latin verse, and inside some quaint old prints illustrating coarsely the Life of the Prodigal Son. Here it is the custom to stop and take stimulants, and a peculiar drink of milk, eggs, and brandy is made, and is highly recommended for anyone coming down with a bad head after dissipation at Cape Town.

The road after this leaves the head of the bay behind and stretches over part of the flats, and passing at a distance High and Low Constantia, where the celebrated wine is made, reaches Wynberg. Wynberg is by far the most beautiful spot about Cape Town, and almost as beautiful as any village I have seen; but then nearly all its beauties are derivative, not indigenous, and arise from the fact that it is situate in the midst of thick pine groves and plantations of other trees. Here one sees growing together the European pines, the oak, poplars, and the gnarled and contorted South American Cactus (*Cereus*), and numerous Australian gum-trees and acacias.

The road at Wynberg leads through a grove of pines for a mile or more, the pines meeting overhead and forming a delicious shade, and shutting in the road on either hand with their closely set stems.   No doubt the very trying heat and glare of the open sand-flat over which one drives before reaching the Wynberg grove, makes one exaggerate the beauty of its refreshing shade.   Even amongst the grove the brick-red dusty soil stains the trunks of the trees, and after long absence of rain turns the very foliage brick-red.   At Wynberg is the cricket ground where the Army plays the Navy, the Army the Cape Town Club, and so on, and also a most excellent hotel, known as "Cogill's," after the proprietor.

Above Wynberg are the talus slopes and débris mounds of Table Mountain, covered with the wonderful Silver-tree, whose leaves shine like burnished metal, and which is found nowhere else in the world but about the slopes of this mountain and its immediate neighbourhood.   It does not even grow at Simons Bay.   Nowhere on the earth but just round this one mountain.

The Silver-tree (*Leucadendron argenteum*) is one of the *Proteaceæ*, which natural order is characteristic of the flora of the Cape and South Australia, the genera being nearly equally divided between the two regions, and found scarcely anywhere else.   A few only are found in tropical Australia, in New Zealand, South America, and equatorial Asia.   Another group of plants, the *Restiaceæ*, serve further to connect the Cape with Australia, and there are other marked alliances.

The wide difference between the West and East Australian flora has been treated of by Sir Joseph Hooker, and the greater resemblance of the Western Australian flora to that of South Africa.   Sir Joseph Hooker thinks it probable, from botanical grounds, that Western Australia was connected with the Cape district by land at a time when it was severed from Eastern Australia.

How is it that Marsupials are not found at the Cape, being nevertheless found in the Great Oolite in England ?   It would seem necessary almost that they must have been present at the Cape and have died out, unless it is possible that *Proteaceæ* and *Restiaceæ* are very much older than Marsupials, in which case they would be very old indeed.

Table Mountain is most easily accessible from this side, and it was from hence that I ascended it with Dr. Mansell, F.L.S., as my guide, who gave me most useful information about the Botany.

From Wynberg the rail takes one in about half an hour to Cape Town, the train stopping at about half a dozen villages or suburbs, where many of the business men of the city live. Cape Town is not very interesting in itself. There are few fine buildings. The best is that containing the library and museum.

The officers of the ship liked Cape Town for its gaiety and dancing. I enjoyed Simons Bay most thoroughly, because it is a place where one can get at once amongst wild nature, and over the hills and moors, amongst the rocks, or along the coast, and come into immediate relation with examples of nearly all the characteristic South African animals in their wild condition. I constantly crossed the high ridge of the Cape promontory, just above Simons Bay, and made across to the shore on the other side. The whole promontory is one tract of open moorland, with only a few farms and houses of boers with small holdings, scattered at long distances from one another.

On one of my first expeditions I came across a troop of baboons, *Cynocephalus porcarius*. They are as big as a New-foundland dog when full grown. They live especially about the sea-cliffs and steep talus slopes leading down from these to the sea; but they are to be met with also on the open moor-land above. They live in droves or clans, of 30, 40, or even up to 70, and there were three such bodies of them in the country immediately about Simons Bay, and in the tract stretching down to Cape Point.

When on the feed, two or three keep watch, and one usually hears them before one sees them. The warning cry is like the German "hoch" much prolonged. As soon as they see one, three or four of them mount on the scattered rocks so as to have a clear view over the bushes and heaths, and watch every move-ment of the enemy, so that it is extremely difficult to get within shot of them. If one stands still, or does not go any nearer, merely passing by, they employ themselves, as they sit un-concernedly, in scratching in the usual monkey fashion; but still never losing sight of their object of suspicion.

Once I came across a troop on a sudden, on looking over a low cliff. They dashed off at a tremendous pace, galloping on all fours, till far out of shot, when they climbed up on to a rocky eminence, and calmly sat down to watch me. The baboons live on roots, which they dig up, and on fruits, and they turn over the stones to search for insects and such food underneath. It is striking thus to see monkeys roaming about on open moorland, where there are no trees. I had never properly realized the fact before.

The track of the baboons in the sand is unmistakable. The foot makes a mark where the animal has been galloping, just like that of a child's foot; the fore-limb makes a mark not half so deeply indented, the hand being used merely to touch on, as it were, to prepare for a fresh spring with the feet. I found the skeleton of one of the baboons in a cave at Cape Point. The animal had evidently crawled into the cave to die.

Everywhere amongst the rocks lives the Rock-Rabbit (*Hyrax capensis*). The Rabbits live in large crevices in the cliffs or under huge masses of rock, which have fallen and lodged on some ledge. In the places frequented by them the rock ledges are covered with bushels of their dung. They come out to feed in the mornings and evenings, but also bask sometimes in the hot sun at mid-day.

They are very inquisitive, and sit up on a rock, and look at one, and then suddenly dash into their hiding-place. After a time, if one remains quiet, they come out for another look, and afford a good chance for a shot. Their cry of alarm is a sort of short hissing noise, not a whistle like that of the marmots, of which animal they immediately remind one, though so widely different in structure, their nearest living ally being the rhinoceros.

They had young at the time of our visit, and I met with two litters, each of three young, which were about the size of very large rats, with soft chocolate-brown downy hair. The young play about on the rocks together like kittens, chasing one another, and darting in and out amongst the clefts. I shot two at one shot. One of these, when dying, made a regular squeal very like that of a rabbit. The old ones are hard to kill, carrying off a considerable charge of shot, and they bite very fiercely.

Amongst the heath are partridges and a few quails, at some seasons plenty of the latter; but just now, only a few were to be found, and they were breeding. I saw two nests. In the thicker bushes are so-called " pheasants " (*Francolinus*). There are introduced true pheasants about the foot of Table Mountain in considerable numbers, preserved for shooting.

A large shrike, with a yellowish breast, is the commonest and most conspicuous of the smaller birds; but the most beautiful are the little *Nectarinidæ* or Honey-birds, which here take the place of the Humming-birds of South America, and in their splendid gold and green colouring are almost equal to them. Above Simons Town is a sort of small gorge or chasm in the mountain-side, where there is a waterfall with beautiful ferns growing about it, and where above, on the cliffs, nest hundreds of swallows. I used as a boy to wonder how chimney swallows and house martins managed to nest before there were any houses.

The sandy flats and fields about the sea-shore are covered with mole-hills, and bored in all directions with tunnels, large enough to admit the hand and arm easily, by the huge Sand-mole (*Bathyergus suilus*). *Bathyergus* is a Rodent, with an excessively long pair of projecting lower gnawing teeth. It is a foot long, and covered with a light grey-brown silky fur.

There is another similar Rodent mole of about half the size (*Georychus capensis*), which rather affects higher land, but occurs also sometimes with *Bathyergus*.

The two together are in such abundance as to cover the country in all directions with mole-hills, and in galloping over the sand one is very apt to be thrown headlong by one of their galleries giving way under the horse's feet. I had two such falls in one day. A clever horse, brought up in the country, learns however, whilst turned out on the run, to lift his foot out of a hole without stumbling.

It is the custom to call the moles, such as we have in Europe, the *true* moles, and to regard these Rodent moles as animals which in some extraordinary way have adopted habits not proper to Rodents, but natural and what is to be expected in a certain group of Insectivora. But in reality, there seems to be no

L

reason why the one set should be the *true* moles rather than the other, excepting merely as a matter of home nomenclature and prejudice. The South American Rodent mole, the "*tucutuco*" (*Ctenomys*), is familiar as described by Darwin in his Journal. And besides this, there are all the *Spalacini*, or Blind-moles, of which there are nine genera, including *Bathyergus* and *Georychus*, forming steps towards the ground squirrels, *Geomys*.

Of the *true* moles, or Insectivora, with the habits and outward shape of *Bathyergus* and *Georychus*, there are only five or six genera in all. Why should not *Talpa* be looked upon as the plagiarist? There is still another very different animal, with mole-like habits, the little armadillo (*Chlamyphorus*) of the Argentine Republic. It seems remarkable that no Marsupial in Australia has become modified to suit mole-like habits. All other Mammalian habits almost have been adopted by Marsupials. *Bathyergus* has, like our Talpa, a bare snout, and strong digging hands and feet. It burrows of course in search of roots and vegetable food only, not for worms like Talpa.

The people about Simons Town have an idea that the animals work the earth at certain stated hours, and have regular periods of rest; but I was always able, by going over a good deal of ground, to find one working at any time of the day. The heaps thrown up are huge, a foot high, five or six times as big as those of our little mole. A fresh heap is betrayed at once by its darker colour, *i.e.*, its dampness; in a few hours the dry heat of the Cape reduces it to a glistening white.

One has not long to watch, standing a few yards off, before the fresh heap is seen to heave up, three or four times in succession, as the mole forces freshly scooped-out earth up into it from below. I tried at first shooting into the heap as it was thus heaving, in the hopes of getting the mole, but never with any success. In order to shoot the worker, the earth should be quickly thrown back from the fresh heap, and the hole laid open to the air.

One then has only to retire about ten paces and wait patiently. The mole does not like the fresh air, and in the course of five minutes or so, comes back to fill it up, but usually puts its head out for a moment first, to find out what's up, though it certainly cannot see far with its minute eyes, which are not bigger than the heads of carpet pins, the whole eye-ball when extracted being not bigger than a tenth of an inch in diameter.

Of course, a charge of shot at the moment the animal shows its head is effective. But the easiest method of getting specimens is on scraping away the earth from the fresh mound to insert in the hole a common rabbit gin, well secured with peg and string. I trapped a good many *Bathyergi* in this way, and one *Georychus*. *Bathyergus* is very fierce when dragged out of its hole, fast by one leg in a gin. The animal bites the air savagely with its enormous teeth, which project an inch and a half from the lower jaw, and makes an angry half-snarling, half-grunting noise.

I took several of the moles on board the ship alive in a sack. I let the sack swing by accident against one of my legs, and one of the moles gave me a very unpleasant nip, biting through the sack and my clothes.

When put in a strong wire cage the mole first tried to burrow, but finding that absolutely impossible, tried to bite the wires all round, and that failing, became sullen and quiet. The animal can evidently see for short distances.

Besides these moles, which are a great pest in gardens, there is a little Insectivorous mole (*Chrysochloris inauratus*), the Golden-mole, which is not more than half the size of our English mole, and has a dark silky fur shot with most brilliant metallic golden tints. This mole makes quite superficial runs in the ground, so near the surface that the earth is raised all along the run, and hence the track can be followed everywhere above ground. When one of these is seen at work, it can be thrown out with a stick or spade at once.

I several times went over the hills to the coast on the other side of the promontory. At White Sands, nearly opposite, are a series of shell mounds, or " kitchen middens," which occur also at Cape Point and many places along the coast. There are huge mounds of large *Patellas*, *Haliotis*, and other shells; the limpets are so large as to make convenient drinking cups.

All about the mounds are to be found various stone implements used by the people, either Bushmen or Hottentots, who made the mounds (probably Bushmen). There are flat stones, each with a long shallow groove worn on them, and small cylindrical stones lying about which fit the hand, and have evidently been used for rubbing up and down the grooves, and have indeed thus worn them. The use of these grooved stones is uncertain. The usual idea is that various•bulbs and roots used by the midden people were ground in them. Perhaps they used them partly for pounding or rubbing tender the hard muscular foot of *Haliotis, Patella,* and other Gasteropods, to prepare them for eating.

Haliotis (the large Ear-shell) is prepared now at the Cape for eating by pounding, as also at the Channel Islands. The *Haliotis,* as cooked at the Cape, is excellent, quite a luxury. No iron is allowed to touch it in preparation; it must be got out of the shell with horn or wood implements, then pounded with stone or wood and finally stewed. It is considered that if iron touches the animal it becomes rigidly contracted and hopelessly tough. It is quite possible that the popular opinion may be correct, and that contact with iron may produce a rigid tetanus of the muscles.

Some of the grooved stones have grooves on both sides, one groove having been evidently worn out. Some of the grooves are as much as a foot long and two inches, or a little more, in width.

Besides these stones there are the well-known digging stones; circular disc-shaped stones, perforated in the centre. The stone is passed over a stick, the lower end of which is hardened in the fire or thrust into an antelope's horn, and the stick thus weighted, is used by the Bushmen and Hottentots to dig roots. A Bushman whom the late Dr. Bleek, the distinguished South African linguist, had under his charge, called the apparatus a squaw's stick, because, of course, the squaws have to do the digging. He showed us how it is used.

Well-made spear and arrow-heads and scrapers are found with these things, but are comparatively scarce, and far more abundant on the Cape Flats.

Very broken pieces of a coarse pottery are common about the refuse heaps. The pottery is black, and seen on fracture to be full of fragments of quartzite. I found two pieces with handles, evidently the side handles of pots. In the Cape Museum are plenty of similar pieces, and also a drawing on a small slab of stone, from a neighbouring cave which was probably a home of the midden people.

The middens lie in places where there are banks of shifting sand. As the sand shifts, there are exposed, all about on the slopes, heaps of stones, evidently put together for some purpose. A considerable number of human bones were lying about. I turned over several of the stone heaps which had evidently been hitherto undisturbed, and excavated for a short depth beneath them without finding any interments; but in one case a complete skeleton lay around one of the heaps, and at Cape Point I saw a second one lying beneath a similar heap, having been evidently buried in a crouching position with the body unstraightened after death. The majority of the stone heaps have, however, certainly not been graves, but are very possibly the remains of places where fires have been lighted.

The sand at White Sands is calcareous. As it shifts before the wind it in many places buries bushes growing near the shore. These die, and their stems, buried in the sand, decay, and in doing so set free a certain amount of acid which brings about a solution and redeposit of calcareous matter in the sand. The sand immediately surrounding the stems is thus cemented into a solid mass which encrusts the remains of the bark. The wood decays away, and a pipe with a wall of cemented calcareous sand is the result. The sand shifting again, these pipes, which are often branched, are left exposed on the beach.*

In my excursions to White Sands I often stopped at the cottage of an old-fashioned "boer." He was a boer in a very small way: an old man who, at the age of nearly sixty, had married a young wife. He was partly of French parentage, many French having come to the Cape at the time of the Revolution. These people were wonderfully hospitable, and gave me milk, coffee,

---

* Darwin observed similar structures in Australia, but in this case the cavities left by the decaying branches had been filled in by hard calcareous matter. "Journal of Researches," p. 540.

and Cape brandy, and were delighted to hear about the "Challenger's" voyage.

The old man had a huge old Dutch bible, 150 years old, with pictures, maps and commentary. He prided himself very much on his knowledge of it, and got it down, put on his spectacles and showed me the map of the Garden of Eden, with Adam and Eve and the rivers. He knew it by heart, and evidently considered it as of perfect geographical accuracy. But the commentary was his delight. It was the true old gospel that he loved. He terribly disliked modern innovations.

I was led to cultivate his acquaintance, because he let slip at our first interview the information that he knew where, close by, there was the skeleton of a Hottentot lying under a rock. Directly he had said so I saw that he repented, and at first he would not hear of showing me the place. He said he was afraid the ghost of the skeleton would haunt him.

It was a long time before his wife could laugh him out of this notion. Eventually he showed me the place, but unfortunately the bones were rotten and the skull was battered in, the man having apparently been murdered, whether Hottentot or no, and half covered up in a hurry with a few stones.

I had naturally a desire to see wild antelopes at the Cape. I did not, however, in the least expect to see one without going into the interior, and was surprised to find that antelopes still exist in the Cape peninsula, and I had a shot at three of them on the very Cape of Good Hope itself. I had an erroneous notion concerning antelopes, that they all lived in much the same way, forming vast herds that roamed over flat plains, and performed migrations in bodies from one place to another as scarcity of food necessitated.

Now, however, I found that the various species are mostly totally different in their habits. Some are nocturnal, some diurnal; some live on the mountains, some on the plains, some amongst the bushes, some in forests; some are gregarious, others solitary.

The antelopes are all called "Bok" (goat), pronounced in the country "Buck" by the Cape people. The two antelopes about Simons Town are what the Dutch named, from its resemblance to that animal the roebuck, "Rheebök" (*Pelea capreola*) and the "Grysbök" (grey goat) (*Calotragus melanotis*).

The Rheebök lives about on the stony hills and rocks in small herds of from six to a dozen, or so. There are now forty or fifty of these antelopes on the estate of a Mr. McKellar at Cape Point, and there are plenty of Grysbök there also. I twice went over to Cape Point Farm from Simons Town to hunt these antelopes.

The Rheebök are shot either by being stalked, or more easily by being driven, they using regular passes in the hills where guns can be posted. The Rheebök is as large as a small fallow deer, and of a light-grey colour; it is extremely difficult to see it at any distance, it being so like in colour to the bush and rocks. It is only as it moves its tail and shows the white underneath it, that the hunter catches sight of it at first; the white patch under the tail is certainly a very material dis advantage and source of danger to the animal. It is very wary and difficult to stalk; it feeds in the day-time.

The Grysbök on the other hand, lies hid in the thickest bushes or beds of reed, during the day, and only comes out to feed at night time. It is very small, less than half the size of the Rheebök. When rain has fallen, it is easily tracked to its lair, and turned out and killed with shot, but in dry weather the only chance for the sportsman is to drive it up by riding through the bushes and shooting from horseback, or to turn it out with dogs. I saw one only dash for a moment through the bush, spring lightly over a mass of thick low scrub, and disappear instantly in the bush again, before I could get my gun to bear. The animal is of a dark-red colour. Mr. McKellar used to hunt the Grysbök with beagles with great success.

Mr. McKellar, who was most kindly hospitable, has an ostrich farm, but his flock of birds was not very large at the time of our visit, he having had bad luck at first in breeding. He owns the actual Cape of Good Hope and a long stretch of the moorland adjoining, and has thrown a wire fence right across the peninsula, so as to give his ostriches the run of a large tract, stretching right down to the Cape itself. One old hen ostrich was a pet about the house, but used to do sad damage in the farm-yard, eating the young goslings, swallowing them like oysters. It was amusing to go with Mr. McKellar into one of his breeding paddocks ; here a pair of ostriches were brooding on a nest of eggs, dividing, as usual, the labour between them. The cock was very savage and attacked all intruders, so his master had a long pole with a fork at the end of it, and when the ostrich ran at the party, he caught its neck in the fork. The ostrich was excessively enraged, but soon had to give in.

A kick from an ostrich is well known as very dangerous. The only thing to do when attacked without means of defence, Mr. McKellar said, is to lie flat down and let the bird walk on you till he is tired. I was astonished at the brightness of the red colouring developed on the front of the legs of the cock bird during the breeding season. The ornamental appearance of the bird is greatly enhanced by it.

A narrow but strong and high pen was provided for plucking the birds in. They are driven into it and held fast. It is found better to pluck the feathers out than to cut them off. The stumps, if left in, are apt to cause trouble.

Young ostriches, when first from the egg, have curious horny plates at the tips of their feathers, like those in the feathers of one of the Indian jungle fowls, and some other birds not in the least related to one another.

The Cape Peninsula becomes very narrow towards its termination, and ends in two capes, Cape Point, on which is the lighthouse, and the Cape of Good Hope. The Cape of Good Hope itself is a mass of rock terminating in perpendicular cliffs towards the sea, but with ledges here and there on which numbers of cormorants (*Phalacocorix capensis*) nest.

Behind the terminal rocky mass is a waste of white sand, horribly dazzling to the eyes in bright sunshine. Similar sand, loose and deep, so that one's foot sinks into it at every step, lies all around the farm-house, but is more or less covered with bushes. This sand is terribly tiring to walk on, but after a little rain the various animals can be tracked on it as easily as on fresh snow, and it is thus that they are best hunted.

The boys thus find numbers of small tortoises (*Testudo*

goemetrica), which are here in great numbers, extremely pretty ones with embossed shells. These shells are often made to do duty as ornamental paper weights, being filled with lead. Besides these there are the tracks of the various snakes. A broad groove with a much narrower groove in its centre, marked by the tip of the tail, is made by the terrible Puff-adder (*Clotho arietans*), on which one always stands a chance of treading when walking about. Then there are Cobra tracks, and tracks of numerous other snakes. Both Cobras (*naja haje?*) and Puff-adders are sufficiently abundant about Simons Town. I had four or five adders and two Cobras brought me to preserve. The Cobra was caught swimming in the sea, just off the dockyard.

Again, there are tracks of the Ichneumon (*Herpestes*), called by some name sounding like "moose haunt," and those of the Musk-Cat (*Genetta felina*), both extremely destructive, and trapped and hunted with all energy by the farmers. There are tracks of porcupines leading to their holes, which are often in the caves about the sea cliffs, and have stray quills lying about their mouths, sufficient evidence of the nature of the inhabitant. There are Rock-rabbit tracks, and there are the tracks of the Rheebök and Grysbök, all to be readily distinguished by an educated eye.

The great variety of the flowers at the Cape is a source of constant interest to the naturalist. It is also pleasant to see in their wild condition, large numbers of beautiful flowers, with which one has long been familar as the chief decoration of greenhouses at home. All over the hills grow "Everlastings" (*Helichrysum*), some with large snow-white flowers, others of various bright tints. There is an endless variety of handsome heaths, and numerous familiar Pelargoniums. Amongst bulbs, there are various showy Gladiolus and various species of Iris, and the tall white-flowered Aroid (*Richardia æthiopica*), commonly called "*Arum*," without the white spathe and golden spadix of which no English conservatory is complete ; all these are very common.

I had not before I saw the Cape flora, realized the wonderful power of change-ringing, as it were, in plants. Here may be seen a plant with a yellow flower, very like a dandelion, but with leaves dark on the upper surface, and downy beneath, yet in shape like those of our familiar plant. Close by, one meets with a similar flower with needle-like leaves, like those of a heath; close by again, is another growing on a low bush with leaves, something in the style of those of the holly : then again, another with extremely sharp stout thorny spines for leaves, then another heath-like, but with the leaves reduced to small tubercles. These are all forms with this one sort of flower (I speak only as to outward appearance). One easily finds a white-flowered daisy as it were, ringing similar changes, and so on. Lobelias, again, are to be seen with exactly similar looking blue flowers ringing all the changes of heath forms, spiny forms, &c.

Amongst the animals which live on the Cape Peninsula, the Clawless otter (*Lutra inunguis*), is, worthy of mention : it is a very large otter, twice or three times as large when full grown as the European one. It lives about the salt marshes and lakes, and is tolerably common; it hunts like the South American marine otter, in companies, but only of three or four. It has no claws on the fore feet, having lost them by natural selection in some way or other, and on the hinder feet the claws are wanting on the outer toes, and only rudiments of them remain on the middle ones. There are, however, pits marking the places where the claws used to exist. The webbing between the toes is also in this otter rudimentary ; the beast altogether is very heavily built, with the head very broad and powerful. It appears to be an otter bent on returning to land habits.

I found two species of Land Planarian worms on some American Agaves, in the grounds of the Observatory. At first I thought these Planarians might have been introduced from South America with the Agaves, but they correspond in structure exactly with the genus *Rhynchodemus* of Ceylon, and seem certainly indigenous, although Land Planarians were not hitherto known to exist in Africa.*

A small Chameleon is very abundant everywhere on the hedges near Cape Town. We had one alive in the ward-room ;

---

* For a description of these Planarians, and an account of the Land Planarians obtained during the voyage elsewhere, see H. N. Moseley, "Notes on the Structure of several forms of Land Planarians." Quar. Journ. Micro. Sci., Vol. XVII, New Ser., p. 273.

it was quite tame and rested quietly on a bunch of twigs, hung up to the lamp rail, and would whip flies out of one's fingers from a distance of at least four inches with its tongue. It gave birth to three young ones one night: they twisted their tails round the twigs on which the mother was reposing at once, and at once began catching flies; but our house-flies were too big for their mouths to swallow, and they had to chew away at them for a long time before they could get any juice out of them.

About the sea-shore at Simons Bay, are quantities of cormorants, or shags, as they are called (*Phalacrocorax capensis*); they sit in groups on all the rocks about the town, and bask in the sun, and at times appear in vast flights darkening the air. Gannets (*Sula capensis*) are constantly in sight, and gulls (*Larus dominicanus*) ever flying over the water.

I paid a visit to an island in False Bay, called Seal Island. It is a mere shelving rock on which it is only possible to land on very favourable occasions. The whole place is a rookery of the Jackass penguin (*Spheniscus demersa*). It is an ugly bird as compared with the crested penguin of Tristan da Cunha; the bill is blunter, but the birds can nevertheless bite hard with it: [all the penguins seem to bite rather than peck]. The birds here nested on the open rock, which was fully exposed to the burning sun and occasional rain. It must not be supposed that either penguins or albatrosses are necessarily inhabitants of cold climates, a species of penguin and an albatross breed at the Galapagos Archipelago, almost exactly on the equator.

There was not a blade of grass on the rock, but it was covered with guano, with little pools of filthy green water. The birds nested under big stones, wherever there was place for them; most of the nests were, however, quite in the open. The nests were formed of small stones and shells of a *Balanus*, of which there were heaps washed up by the surf, and of old bits of wood, nails, and bits of rope, picked up about the ruins of a hut which were rotting on the island, together with an old sail, some boat's spars, and bags of guano, evidently left behind by guano-seekers. The object of thus making the nest is no doubt to some extent to secure drainage in case of rain, and to keep the eggs out of water washing over the rocks; but the birds evidently have a sort of magpie-like delight in curiosities: *Spheniscus magellanicus* at the Falkland Islands, similarly collects variously coloured pebbles at the mouth of its burrow. Two pairs of the birds had built inside the ruins of the hut.

All the birds fought furiously, and were very hard to kill. They make a noise very like the braying of donkeys, hence their name; they do not hop, but run or waddle. They do not leap out of the water like the crested penguins when swimming, but merely come to the surface and sit there like ducks for a while, and dive again. We dragged off a number in the boat for stuffing, and took young and eggs; the old ones fought hard in the boat and tried to bite one another's eyes out.

There was a large flock of terns on the rock, rendering it quite white on one part, but they were not nesting. There were plenty of shags' nests, some few with young ones, but most of them were already relinquished: they were built on a higher standing-piece of the rock, and were large round deep nests made of dried seaweed.

There is a great fishery at the Cape of a fish called "Snook," a sort of Barracuda, which is salted and dried, and sent mainly to Mauritius for sale. The Snook boats were always to be seen about in the bay. The fish are caught with a hook and line, whilst the boat is in motion. The fishermen are especially careful not to get bitten by the fish as they haul them in; wounds caused by the bite of the fish are said to fester in a violent manner as if specially poisoned.

The fish, however, which is most interesting from a scientific point of view, which is caught at the Cape, is a large Myxinoid (*Bdellostoma*) allied to the lamprey. Two or three of these were caught with a hand line and fish bait from our ship whilst at anchor at Simons Bay, and they are not at all uncommon, though so very rare in European museums. The specimens caught were nearly three feet in length. They swallowed the bait far down, and astonished the sailors by the immense quantity of gelatinous slime which they discharged from the surfaces of their bodies when drawn on board. The slime forms masses of a jelly-like substance.

The villages between Simons Bay and Wynberg have fences made of various bones of whales. A whale fishery was formerly carried on here, but no longer pays. An extremely interesting and very rare whale is occasionally procured at the Cape. It is a Ziphioid, *Mesoplodon layardii*. The Ziphioids are a group of the toothed whales and allied to the sperm whale. They have the bones of the face and upper jaw drawn out and compressed into a long beak-like snout which is composed of solid bone, hard and compact like ivory. The upper jaw is devoid of teeth, having lost them in the process of evolution, and the lower jaw, which is lengthened and pointed to correspond with the upper, retains but a single pair of teeth.

In the species in question, *Mesoplodon layardii*, these two teeth in the adult animal become lengthened by continuous growth of the fangs into long curved tusks. These arch over the upper jaw or beak, and crossing one another above it at their tips, form a ring round it and lock the lower jaw, so that the animal can only open its mouth for a very small distance

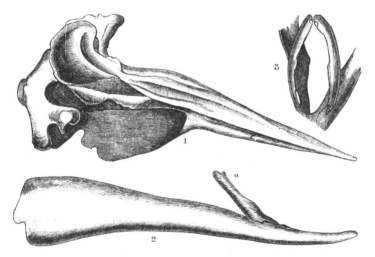

1 Skull of Mesoplodon layardii. 2 Lower jaw ; *a* small cap of dentine on the tooth. 3 Top of lower jaw seen from the front, showing the ring formed by the teeth. Copied from the British Museum Catalogue of Seals and Whales.

indeed. The tusks are seen always to be worn away in front by the grating of the confined upper jaw against them. How the animal manages to feed itself under these conditions is a mystery.

It is remarkable that the main mass of each tusk is made up of what appears as an abnormal growth of the fang.* The actual conical tooth, that is the original small cap of dentine of the tooth of the young animal, which corresponds to the part of the tooth showing above the gum in other whales, does not increase at all in size, but is carried up by the growth of the fangs, and remains at the tips of the tusks as a sort of wart-like rudimentary excrescence.

Specimens of *Mesoplodon layardii* are excessively rare, and I sought diligently for such during the whole of my stay at the Cape, and was rewarded by procuring parts of two skulls. One of these, a skull without the lower jaw, I found near Mr. McKellar's, at Cape Point. The skull was exposed on the beach, being stuck up with its beak thrust into the sand to be used as a rifle target.

The animal, as Mr. McKellar told me, had come on shore about eight years before. It yielded oil of a very superior quality, which sold for more than twice the price of ordinary whale oil. It was about 10 feet in length, and was, as far as he remembered, coloured black on the back and white on the belly, with a conspicuous line of demarcation of the colours on the side. The beast had the usual tusks.

The other specimen consisted of the snout and lower jaw, with the tusks of another example of the species. It was given me by Mr. A. M. Black, of Simons Town. The animal came on shore at Walwick Bay in 1869. It yielded 80 gallons of oil, and was from 16 to 18 feet in length. It is remarkable that these whales seem never to be met with or caught at sea. They always are procured by their running on shore. The Ziphioids are especially interesting, because many species were abundant

---

* Prof. Owen, with the single original specimen only before him, considered that the tusks had acquired "an abnormal direction and state of growth" in that particular specimen. "Palæontographical Soc.," Vol. XXIII, 1869, p. 26. Prof. Flower, though knowing of a second specimen, still seems doubtful. "Trans. Zool. Soc.," Vol. VIII, 1874, p. 211. Now that more specimens are known, there can be no longer doubt as to the normal occurrence of the condition described.

in Tertiary times, and their beaks being so dense in structure as to be readily preserved as fossils, are common in such deposits as the Red Crag of Suffolk. I had the good luck to procure another Ziphioid at the Falkland Islands during the voyage, near Port Darwin.

I stayed at the hotel at Wynberg for a fortnight, whilst working at the anatomy and development of *Peripatus capensis*. *Peripatus* is an animal of the very highest importance and antiquity, and I believe it to be a nearly related representative of the ancestor of all air-breathing Arthropoda, *i.e.*, of all insects, spiders, and Myriapods.

The animal has the appearance of a black caterpillar, the largest specimens being more than three inches in length, but the majority smaller. A pair of simple horn-like antennæ project from the head, which is provided with a single pair of small simple eyes. Beneath the head is the mouth provided with tumid lips and within with a double pair of horny jaws. The animal has seventeen pairs of short conical feet, provided

PERIPATUS CAPENSIS. (Natural size.)

each with a pair of hooked claws. The skin of the animal is soft and flexible, and not provided with any chitinous rings.

The animal breathes air by means of tracheal tubes like those of insects. These, instead of opening to the exterior by a small number of apertures (*stigmata*) arranged at the sides of the body in a regular manner as in all other animals provided with tracheæ, are much less highly specialized. The openings of the short tracheæ are scattered irregularly over the whole surface of the animal's skin.

It appears probable that we have existing in *Peripatus* almost the earliest stage in the evolution of tracheæ, and that these air tubes were developed in the first tracheate animal out of skin glands scattered all over the body. In higher tracheate animals the tracheal openings have become restricted to certain definite positions by the action of natural selection.

The sexes are distinct in *Peripatus.* The males are much smaller and fewer in numbers than the females. The females are viviparous, and the process of development of the young shows that the horny jaws of the animal are the slightly

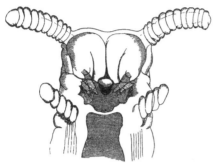

modified claws of a pair of limbs turned inwards over the mouth as development proceeds; in fact, "foot-jaws," as in other Arthropods.

Before I studied *Peripatus* at the Cape, nothing was known of its manner of development, nor of the fact that it breathed air by

HEAD OF EMBRYO OF PERIPATUS CAPENSIS, SHOWING THE DEVELOPMENT OF THE JAWS.

means of tracheæ. It was generally placed with the Annelids, though its alliance with the Myriapods had been suspected by Quatrefages.

That *Peripatus* is a very ancient form is proved by its wide and peculiar distribution. Species of the genus occur at the Cape of Good Hope, in Australia, in New Zealand, in Chili, in the Isthmus of Panama and its neighbourhood, and in the West Indies. If its horny jaws were only larger they would no doubt be found fossil in strata as old as the Old Red Sandstone at least.

The animal is provided with large glands, which secrete a clear viscid fluid, which it has the power of ejecting from two papillæ, placed one on either side of the mouth. When the animal is touched or irritated, it discharges this fluid, with great force and rapidity, in fine thread-like jets. These jets form a sort of net-work in front of the animal, which looks like a spider's web with the dew upon it, and appears as if by magic, so instantaneously is it emitted.

The viscid substance, which is not irritant when placed on the tongue, is excessively tenacious, like bird-lime, and when I put some on a slip of glass, some flies approaching it were at once caught and held fast. It appears from the observations of

Captain Hutton on the New Zealand species,* that the jet of slime is used by the animal not only as a means of offence, but to catch insects, on which the animal feeds.

I found only vegetable matter in the stomachs of the Cape species, and concluded that the animals were vegetable feeders. The animals live at the Cape in or under dead wood, and I found nearly all my specimens at Wynberg, in Mr. Maynard's garden, in decayed fallen willow logs, which were in the condition of touchwood. I tore the logs to pieces, and found the animals curled up inside.

The animals are very local, and not by any means abundant, so that an offer of half-a-crown for a specimen to boys did not produce a single example. My colleague, the late Von Willemoes Suhm, and I both searched hard for *Peripatus*. He was unsuccessful; but I was lucky enough to find a fine specimen first under an old cart-wheel at Wynberg. Immediately that I opened this one I saw its tracheæ and the fully-formed young within it. Had my colleague lighted on the specimen he would no doubt have made the discovery instead.

Peripatus capensis is nocturnal in its habits. Its gait is exactly like that of a caterpillar, the feet moving in pairs, and the body being entirely supported upon them. The animals can move with considerable rapidity. They have a remarkable power of extension of the body, and when walking stretch to nearly twice the length they have when at rest.†

Had I not been engaged for so long a time in working at *Peripatus*, I should have certainly paid a visit to the Knysna Forest, accessible by steamer from Cape Town, which contains wild elephants preserved by Government, and numerous antelopes, and other large animals. My principal object in going, however, would have been to see the curious bird, the Touracou (*Turacus albocristatus*), one of the Plantain-eaters. This bird has bright red feathers in its wings, the red colouring matter of which

* Capt. F. W. Hutton, " On Peripatus Novæ Zealandiæ." Ann. and Mag. Nat. Hist., 1876, p. 362.

† For a detailed account of the anatomy, and development of Peripatus Capensis, see H. N. Moseley, "On the Anatomy and Development of Peripatus Capensis." Phil. Trans. R. Soc., 1874, p. 757.

M

is soluble in water, so that the birds are apt to wash their red feathers white when in confinement.

The colouring matter, "Turacin," as was discovered by Prof. A. H. Church,* is distinguished by yielding a remarkable absorption spectrum, and contains a considerable quantity of copper. The bird is very common in the Knysna. and I was told by sportsmen who had shot it, that in rainy weather it will hardly fly, but crouches down under the bushes, and may sometimes be knocked down with a stick.

A most extraordinary statement concerning these birds, to the effect that the red colour, when washed out of the feathers, becomes restored, is made by M. Jules Verreaux.† It seems impossible to understand how this can happen, since there seems no means by which the colouring matter can be conducted from the body of the bird to the web of the feather. Such a result seems only possible in Horn-bills, some of which, as is well known, paint their feathers yellow by rubbing in a yellow secretion discharged from glands under the wing.

M. Verreaux states that in rainy weather, just as I was informed, the Turacous get their feathers wet through, and are, in consequence, unable to fly, but crouch on the ground, instead of resting on the tree-tops as usual. He caught several with the hand, the colour came out on his hands from the wet feathers. He washed the colour out of their wings with soap and water till the feathers were almost white. The bright red colour however, returned directly the feathers were dry, and this occurred even when the same bird was washed twice in the same day.

The red colouring matter is scarcely at all soluble in pure water, but the addition of the slightest trace of alkali to the water enables it to extract the pigment from the feathers, and yield a blood-red solution.

For notes on P. N. Zealandiæ, see H. N. M. Ann. and Mag. Nat. Hist., 1877, p. 85.

* "Researches on Turacin," Phil. Trans., 1870, p. 627.
† M. Jules Verreaux, "Proc. Zool. Soc.," 1871, p. 40.

CHAPTER VII.

PRINCE EDWARD ISLANDS. THE CROZET ISLANDS.

Appearance and Formation of Marion Island. Vegetation of the Island.
Azorella selago. Limit of Vegetation in Altitude. Relations of the
Flora. Former Extension of Land in this Region. Nesting of the
Great Albatross. Mode of Courtship. Skuas. "Johnny" Pen-
guins. Rock Hoppers. Rookeries of King Penguins. Absurd
appearance of the Young Birds. Singular Mode of Incubation.
Habits of Sheath-bills. Appearance of the Crozet Islands. Tree-trunks
found in the Island by former Voyagers.

Marion Island, December 26th, 1873.—Marion Island, which
with the smaller island of Prince Edward makes up the Prince
Edward Group, was sighted on the evening of December 25th.
The centre of Marion Island is in lat. 46° 52′ S., long. 37° 45′ E.,
that of Prince Edward Island in lat. 46° 36′ S., long. 37° 57′ E.,
the City of Lyons being in a nearly corresponding latitude
in the northern hemisphere.

The islands are distant from the Crozets (which lie a little
to the north or west of them, and are the nearest land) 450
miles. The African continent is distant from them about 960
miles, the nearest point being about Cape Recife at Algoa Bay.
From Kerguelen's Land the Marion Islands are distant about
1,200 miles, from Lindsay and Bouvet Islands about 1,400
miles, from Tristan da Cunha and Gough Islands about 2,150
miles; and, lastly, from the Falkland Islands and Fuegia
(to which, in common with all the other Antarctic islands
hitherto examined, except the Campbell and Auckland group,
they are in their flora most nearly related) they are distant
about 4,500 geographical miles.

The islands lie, as do the Crozets and Kerguelen's Land, well
within the course of the Antarctic drift, which, fusing with the
Cape Horn current, sweeps in an easterly direction across the

Antarctic sea and further within the broad belt of prevalent westerly winds. The combined action of the winds and the current have, no doubt, brought about in greater part the diffusion of the Fuegian and Falkland Island plants to the islands lying eastward of them; but it is possible that the multitude of sea-birds inhabiting the islands, and nesting, as they do, amongst the herbage, may have been of influence in the matter by transporting seeds attached to their feathers or feet. Most of the birds are of widely wandering habits.

The island of Marion, the larger of the two forming the group, and on which alone of the two an opportunity of landing was afforded, is about 11 miles in length, 8 in extreme breadth, and about 80 square miles in area. The highest point is about 4,250 feet above the sea-level. The island is entirely volcanic, and presents the usual features of volcanic islands which are of considerable age. The highest land is in the centre; and irregular slopes lead down to the sea on all sides. These slopes are of very moderate inclination, and are broken in numerous places by shallow valleys bounded by cliffs where the more ancient flows of lava have suffered denudation. These valleys are occupied by more recent lava-flows, which still retain their rough pinnacled upper surface. Further, all over the slopes and summits of the island are scattered irregularly, numerous small cones, formed mostly of conspicuously red scoriæ. The lava is basaltic, presenting in many places in the cliffs a columnar structure. Some sand gathered on the shores of a small fresh-water lake near the sea was full of augite and olivine crystals.

The island was sighted, together with Prince Edward Island, on December 25th, but was not approached closely till the morning of December 26th. The upper part of the island was covered with snow, commencing, as usual, on the slopes as patches lying unmelted in sheltered hollows, succeeded by a general thin coating or powdering over, through which the black rock showed out in all directions, and above this, again, on the highest cones and peaks, forming a continuous sheet of glistening white. The summits were enveloped in clouds, which lifted or dispersed in a partial manner from time to time. Below the snow and up amongst the patchy region, the slopes of the island were covered with a coating of green, which formed a contrast to the dark cliffs and red lower cones, which were almost destitute of verdure and had very little snow upon them. Here and there large patches of yellow showed out amidst the green, and were conspicuous even at some distance from the shore. It was found that these patches were formed of mosses. The mosses, indeed, occurring thus in patches, some dark, some nearly white, and others yellow, form the principal features in the vegetation as seen from a distance, showing out, as they do, amongst the very uniform mixture of phanerogamic plants. The small rocky projections on the rough surfaces of the modern lava-flows, standing out dark above the verdure, have at a distance exactly the appearance of low bushes with dark foliage, and were at first believed to be such. Landing was effected on the north-east side of the island. The day was remarkably fine and sunshiny.

The rocks, about high-tide mark, are covered with a dense growth of the large brown seaweed, *D'Urvillœa utilis*, which is of great assistance in breaking the surf. Beyond the ordinary reach of the sea, but still within the beach-line, the rocks are covered with a crassulaceous plant (*Tillœa moschata*, D.C.), occurring also in Kerguelen's Land Succeeding the beach is a thick growth of herbage investing a swampy black peaty soil, which covers the underlying rock more or less thickly everywhere on the lower ground and extends up with the herbage almost to the snow. The principal plants forming the thick growth are an *Acœna* (*Acœna ascendens*), *Azorella selago*, and a grass (*Poa cookii*, Hk. f.). The *Acœna* is by far the most abundant plant on the island.

The *Azorella* forms low, convex, bright green patches in intervals between the *Acœna* or cake-like masses at its roots.

Azorella selago is a characteristic plant of the southern islands, and will be frequently referred to in the sequel. It belongs to the Umbelliferæ. It forms large convex masses often several feet in diameter, which are compact and firm, and when on solid ground yield little to the tread. The masses are made up of the stems and shoots of the plants closely packed together side by side, with their flowering tips and small stiff and tough leaves forming an even rounded surface at the exterior, being all of the same length.   The interior of the masses is full of dead leaves and stems.   The whole where growing in abundance

VIEW IN KERGUELEN'S, NEAR ROYAL SOUND.

In the foreground, the rounded masses formed by Azorella selago.   From a photograph.   In the distance, Mt. Wyville Thomson.

forms sheets and hummocks which invest the soil sometimes for acres in extent at Kerguelen's Land, with a continuous elastic green coating.   An allied plant, *Bolax glebaria*, forms similar masses at the Falkland Islands, and there is a tendency in many Antarctic plants to assume a similar habit, as in the case, *e.g.*, of *Lyallia kerguelensis*.

The grass is abundant everywhere, mingled with the *Acœna* and *Azorella*. The plants are, no doubt, rendered especially luxuriant by the dung of the numerous sea-birds ; but no mutual benefit arrangement has sprung up between the *Poa* and the penguins, as it has at the Tristan da Cunha group between the penguins and *Spartina arundinacea*. The *Poa cookii* nowhere forms a tussock. The rookeries of King Penguins are entirely bare, and the grass is not more luxuriant around the nests of the Golden-crested Penguins than elsewhere. The *Poa* was the only grass found in flower in the island. Different-looking forms were observed, especially around the numerous pools of water on the hill slopes ; but they are possibly mere modifications of the same grass due to alteration of conditions ; none of them were in flower. *Pringlea antiscorbutica*, the Kerguelen cabbage,* is at least in the part of the island explored, by no means so abundant as at Kerguelen's Land. It was some time before a plant was found ; but subsequently a good many were met with, but not growing in groups of more than four or five plants. Some were found on the very verge of the shore, within reach of the spray, and the rest on the banks of a small rivulet. The cabbage was mostly in full flower and bud, with sepals and anthers complete. No plants were found with seed at all ripe. The last year's seeds were decayed. This plant at least would appear to have a regular summer flowering-season, since Sir Joseph Hooker found only the fruit at Kerguelen's Land in the winter.

Of the ferns the *Lomaria alpina* is the most conspicuous, forming thick and wide patches amongst the *Acœna* and grass, and occurring abundantly everywhere. *Aspidium mohrioides* was found growing under sheltered banks beside the small stream together with the other three ferns.

Hymenophyllum tunbridgense, the British species, and *Polypodium australe* grow abundantly on the sheltered sides of the

* For an account of this plant and figure, see under Kerguelen's Land, p. 184.

projecting rock-masses already mentioned, but are dwarfed and almost hidden amongst the mosses. They grow in greatest luxuriance on the damp banks of the stream.

The mosses are in most striking abundance,* and, in some very wet places, form continuous sheets over the ground many square yards in extent. Lichens are not in very great quantity, except the incrusting forms, which are tolerably abundant on the rocks.

An attempt was made to reach the actual upper limit of vegetation, but failed from being commenced too late in the day. The ascent was up the bed of the small stream already mentioned, which lay at the verge of one of the modern lava-flows, where it abutted on a low cliff exposing a more ancient flow in section. The more recent flow had a very gradual inclination of not more than 8°. The first scattered patches of snow were encountered at about an elevation of 800 feet. A patch of the cabbage was met with at 1,000 feet.

The highest point reached was at about 1,500 feet elevation. Here *Ranunculus biternatus* had disappeared, and where growing a little lower down was very much dwarfed. The *Azorella*, with a few mosses, formed the principal vegetation; but the green was merely dotted over the bare rock and stones. The patches of snow were here frequent. The *Azorella* appeared from this point to be continued on for about 300 feet more, becoming scantier and scantier. The absolute limit of vegetation may probably be placed at about 2,000 feet. The part explored was somewhat sheltered. A red cone of scoriæ more exposed was quite bare of green from about 1,000 feet elevation upwards.

At about 1,400 feet elevation, the water in a shallow pool exposed to the sun was found to have a temperature of 65° F., the temperature of the air in the shade being 44°. At 900 feet a similar pool, but one which had a small stream of colder water running into it from the cliff, had a temperature of 55°, the air here being at 45°. The thermometer here, when plunged into the midst of a rounded mass of *Azorella*, rose to 50°. It is

---

* Thirty-one species were collected, five of which are described by Mr. Mitten as new.

evident that these mounds retain and store up a considerable quantity of the sun's heat; and this fact probably yields a partial explanation of their peculiar form, which is that of so many otherwise widely different Antarctic plants, and of some New Zealand Alpine plants (*Raoulia, Hastia*). No doubt power gained of resistance to wind is one of the chief causes of assumption of this form.

The island being of such considerable area, and so short a time having been available for the examination of its flora, no conclusions can be drawn from the absence of certain plants, such as *Lyallia*, which might have been expected to occur there, since they occur in Kerguelen's Land associated with nearly all those found. Although the few plants on such islands as these are, as a rule, widely spread, yet some appear to be local and somewhat scarce, as, for example, the *Aspidium*, which was only found at the last moment, under the banks of one of the streams. It is thus highly probable that several plants have been overlooked, and amongst them possibly *Lyallia*. The nine flowering plants collected in the island are all identical with the species growing in Kerguelen's Land ; and the same is the case with the Club-mosses. Of the ferns, two occur in Kerguelen's Land, which has also two others not occurring here.

Fifteen vascular plants in all were found in the Island of Marion.

Mr. Darwin suggests that Kerguelen's Land has been mainly stocked by seeds brought with ice and stones on icebergs.[*] The occurrence of *Pringlea* on Marion Island, as also on the Crozets and Kerguelen's Land, probably points, however, to an ancient land connection between these islands, which the antiquity and extent of denudation of the lavas would seem to bear out. It is difficult to see how such seeds as those of *Pringlea* could have been transported from one island to another by birds ; and these seeds seem to be remarkably perishable ; besides, the distinctness of the genus points to a former wide extent of land on which its progenitors became developed. The existence of fossil tree-trunks in Kerguelen's Land points to similar conditions. Sir J. D. Hooker, in the " Flora Antarctica," p. 220, expressed

---

[*] " Origin of Species," 6th Edition, p. 354.

the above conclusion after his voyage with Capt. Ross, 35 years ago, and with singular foresight suggested that there has taken place "the destruction of a large body of land, of which St. Paul's and Amsterdam Island may be the only remains ; or the sub-sidence of a chain of mountains running east and west, of which Prince Edward Island, Marion and the Crozets are the exposed peaks." This view is directly confirmed by the discovery by the " Challenger's " soundings of the Kerguelen Plateau, which " rises in many parts to within 1,500 fathoms of the sea surface, and forms the common foundation of all the islands situated in this part of the world, viz., Prince Edward's Islands, the Crozet Islands, the Kerguelen Group, the Heard Islands, and the islands of St. Paul and Amsterdam," " as proved by the soundings of both the ' Challenger ' and the ' Gazelle.' "* The occurrence with the cabbage on Heard Island, of the helpless wingless fly, seems a further proof that the plant was not conveyed to the various islands by birds. It is hardly possible that both could have been transported. The fly could probably not exist without the cabbage. The existence of the same species of fresh water fish in New Zealand, Tasmania, the Falkland Islands and South America, points also to the former existence of more intervening land between these points.†

* " Thalassa," an Essay on the Depth, Temperature and Currents of the Ocean, by J. J. Wild, of the Civilian Scientific Staff of H.M.S. " Challenger," pp. 19 and 23. London, Marcus Ward, 1877.
† A. R. Wallace, " The Geographical Distribution of Animals," Vol. I, p. 401, 403. London, Macmillan, 1876.

The species of Phanerogamia and vascular cryptogams found in Kerguelen's, Marion, and Heard islands, are enumerated in Prof. Oliver's report upon my collection, " Journ. Linn. Soc.," XIV, p. 389, from which report the specific names above cited are taken. For the Cryptogamia of Marion Island, vide list of papers at the end of this book.

The following are the temperature-observations taken on board the " Challenger " by Staff Commander Tizard, R.N.:—

On December 26th, when the ship was off Marion Island, the ther-mometer, read at six in the evening, showed for the preceding twelve hours, maximum 45°·5 F., minimum 36°·2.

December 27th. The ship was occupied dredging off both islands ; 6 A.M. maximum 43° F., minimum 40°·5 ; sea-surface 40° to 41°.

On December 26 the temperature at 10 A.M. was 37°·8 F. ; midday, 43° ; midnight, 42°.

The tracts of lower, nearly flat, land of Marion Island skirting the sea, and the lower hills and slopes along the shore, presented a curious spectacle as viewed from the ship as it steamed in towards a likely-looking sheltered spot for landing. The whole place was everywhere dotted over with albatrosses, the large white albatross or Goney (*D. exulans*). The birds were scattered irregularly all over the green in pairs, looking in the distance not unlike geese on a common.

A boat-load of explorers went on shore, everyone having a heavy stick, as it was expected that we might meet with Fur Seals. As the boat pulled on shore cormorants flew about over our heads in numbers. A gull also was common, probably the same as at Kerguelen's Land, and I saw a small bird fly by, close to the water, which was probably *Pelacanoides urinatrix*, also of Kerguelen.

As we approached the shore we saw a pair of terns sitting on the rocks, probably *Sterna virgata*, which occurs at Kerguelen's Land; beautiful birds of a light soft grey and white plumage with coral red beaks and feet. The Giant Petrel or "Break-bones" was also wheeling about over the water, and a few large albatrosses.

As we neared the beach we saw a bird like a small white hen, eyeing us inquisitively from the black rocks, against which a considerable swell was washing. This bird was the "Sheath-bill" (*Chionis minor*), of which we afterwards saw so much.

The surf is subdued a great deal by the thick growth of *D'Urvillœa utilis* upon the rocks. The plant is a huge brown seaweed with stout stems, as thick as one's wrist, attached to the rock by large conical boss-like suckers, and with large spreading leaves on the stalks, provided with floats composed of a series of honeycomb-like air-cells within a thickened frond. With some little difficulty we scrambled out on to the rocks, which were extremely slippery.

The first to get on shore fell in immediately with a female Sea-Elephant lying on a little patch of damp grass-land at the mouth of a miniature gully, opposite to which we landed. They thought they had got a Fur-Seal, and killed the animal at once by striking it on the head with a stone.

I made my way up a steep bank and over a low hill to reach the plain where were most albatrosses. The walking was extremely tiring. The bank was steep and the soil saturated with moisture, and consisting of a black slimy mud, with holes full of water everywhere. The thick rank herbage concealed these treacherous places, and the ground being covered with Azorella tufts, these gave way under one's feet and rendered progression excessively wearying. Further, the sun coming out bright and hot every now and then, made us, who had gone on shore thickly clad, perspire very freely.

The albatrosses were all around, raised from the ground. Their nests are in the style of those of the Mollymauks, but much larger, a foot and a-half at least in diameter at the top. They are made up of tufts of grass and moss, with plenty of adhering earth beaten and packed together, and are not so straight in the sides as those of the Mollymauks, but more conical, with broad bases.

GREAT ALBATROSS ON ITS NEST, MARION ISLAND.
(From a photograph.)

The female albatross is sprinkled with grey on the back, and is thus darker than the male, which is of a splendid snow-white with the least possible grey speckling, and which was now, of course, seen in his full glory and best breeding plumage ; the tails and the wings of both birds are, of course, dark. The albatrosses one meets with at sea are most frequently birds in young plumage or bad condition, and have a rather dirty draggled look.

The brooding birds are very striking objects, sitting raised up on the nest, commonly with the male bird beside it. They sit fast on the nest when approached, but snap their bills savagely together, making thus a loudish noise. They will bite hold of a stick when it is pushed up against their bills. They need a good deal of bullying with the stick before they stand up in the nest and let one see whether they have got an egg there or no.

Then the egg is seen to appear slowly out of the pouch in which it is held during incubation. It is nearly five inches long, or about as big as a swan's, and is white with specks of red at the large end. Only one egg is laid. In most of the nests there were fresh eggs ; in some, however, nearly full grown young birds.

At Campbell Island, of the Campbell and Auckland group, the young of *Diomedea exulans* were found just breaking the shell in February by an exploring party.* Charles Goodridge, who was one of a sealing party on the Prince Edward Islands in 1820, and spent two years on the Crozets, says, that the albatrosses there lay at about Christmas, and that the period of incubation is about three months. (?) The young, he says, were wing-feathered, and good to eat about May, and did not fly off till December.†

The young albatrosses are dark grey in plumage. They snap their bills, like the old ones, to try and frighten away enemies.

The old birds never attempt to fly, though persistently ill-treated or driven heavily waddling over the ground. Very many were killed by the sailors that their wing-bones might be taken out for pipe stems, and their feet skinned to make tobacco pouches. The old males tried to run away when frightened, but never even raised their wings.

* "Notes on the Geology of the Outlying Islands of New Zealand. Reported by Dr. Hector, F.R.S." Trans. N. Zealand Inst., Vol. II, 1869, p. 75.

† "Narrative of a Voyage to the South Seas, and eight years' residence in Van Diemen's Land," p. 35, by C. M. Goodridge. London, Hamilton and Adams, 1833.

It is amusing to watch the process of courtship. The male standing by the female on the nest raises his wings, spreads his tail and elevates it, throws up his head with the bill in the air, or stretches it straight out forwards as far as he can, and then utters a curious cry, like the Mollymauks, but in a much lower key, as would be expected from his larger larynx. Whilst uttering the cry, the bird sways his neck up and down. The female responds with a similar note, and they bring the tips of their bills lovingly together. This sort of thing goes on for half an hour or so at a time. No doubt the birds consider that they are singing. Occasionally an albatross flies round and alights upon the grass, but I saw none take wing.

There were numerous nests of the Skua about amongst the herbage in dry places. Two nests of these birds are never built near together. The birds always have a wide range of hunting ground round their nest. The Skuas in Marion Island were extremely bold and savage, as they were also in Kerguelen's Land. When one approaches the nest they swoop down, passing with a rush close down to one's head, whizzing past one's ears in a most unpleasant manner.

The two birds take turns at towering above, and thus swooping. They have sharp claws and beaks, and no doubt would injure one's face or eyes severely if they touched them as they passed. One has to beat them off with a stick or gun barrel. They are very clever in avoiding the stick as they rush past, but several were knocked down. Sometimes I have had to waste a charge on them to get rid of them. Some pairs are much more savage than others. They have a harsh cry. Of course, when their young is handled they are most furious, and one has to keep a stick going as one carries it off. The birds are very like the Northern Skuas in their habits. One of them swooped down on a duck which I had shot one day at Kerguelen's Land which fell in the water. The bird picked it up when I was not more than half a dozen yards off, and was making off with it in its beak, carrying it easily, when I brought it down with a second shot, the duck thus costing me two barrels.

I searched the sea-shore along for a considerable distance in the hope of finding Fur-Seals, but saw none. Three sorts of penguins were abundant. One was a penguin called by the sealers the " Johnny " (*Pygosceles tœniata*), the "Gentoo" of the Falklands. This penguin is a great deal larger than the crested Penguins, in fact nearly as big as the King Penguin. The beak is bright red, long and sharp-pointed, the back dark blackish, the breast white. The colour of the back is continued on to the head, but a white patch on the top of the head in contrast with the dark colouring is the marked feature about the bird. These penguins we nowhere meet with nesting. They are often associated with the King Penguins. They were usually to be met with here and in Kerguelen's Land in parties of a dozen or twenty or thirty on the grass, close to the shore, and were apparently moulting at the time of our visit. At Christmas Harbour, Kerguelen's Land, some lots of them camped at 100 feet, at least, up the steep but green hill-side at the end of the harbour.

These penguins do not hop, but run, and when closely pursued throw themselves on their bellies on the ground, and struggle along, rowing themselves with violent blows of their wings on the sand or mud, dashing the mud into one's eyes, as one chases them. When in the water, as they come to the surface, they make a sort of very feeble imitation of the leap of the crested penguins, never throwing the whole of the body out of the water, but only the back. They are also to be seen swimming about when undisturbed, with their head and back out of the water, and body horizontal.

Another penguin, the "Rock Hopper" (*Eudyptes saltator*), the same species that occurs at Tristan da Cunha, but a little smaller, as far as I could judge, was nesting about the low cliffs on the shore. The ground on which the nests were made was very wet and filthy, and the nests were, like those of the Jackass Penguins at the Cape of Good Hope, made of small stones, raising the egg about an inch from the mud. These penguins were exactly like the Tristan ones in their cry, and were quite as savage, but then they were in full sight, and not amongst grass; for though there was plenty of grass just over them, nearly a foot in height, they prefer to build where the ground is quite bare. The birds therefore for some reason have adopted slightly different habits from those of the representatives of the species at Tristan da Cunha.

Most interesting, however, by far, amongst all rookeries of penguins which I have seen, was one of King Penguins (*Aptenodytes longirostris*), which I met with a little further along the shore. The rookery was on a space of perfectly flat ground of about an acre in extent. It was divided into two irregular portions, a larger and smaller, by some grassy mounds. The flat space itself had a filthy black slimy surface ; but the soil was trodden hard and flat. About two-thirds of the space of one of the portions of the rookery, the larger one, was occupied by King Penguins, standing bolt upright, with their beaks upturned, side by side, as thick as they could pack, and jostling one

another as one disturbed them. In the figure the birds' heads are drawn as if held horizontally. This is unnatural, the head and neck should be stretched out vertically, quite straight, with the tip of the beak pointed directly upwards.

The King Penguins stand as high as a man's middle, they are distinguished at once not only by their size, but by two narrow streaks of bright orange yellow, one on each side of the glistening white throat.

Penguins were to be seen coming from and going to the sea from the

KING PENGUIN. APTENODYTES LONGIROSTRIS.

rookery, but singly, and not in companies like the Crested Penguins. The King Penguins, when disturbed, made a loud sound like "*urr-urr-urr.*" They run with their bodies held perfectly upright, getting over the ground pretty fast, and do not hop at all. A good many were in bad plumage, moulting, but there were plenty also in the finest plumage.

On the small area of the rookery, which consisted of a flat space sheltered all round by grass slopes, and which formed a sort of bay amongst these, communicating with the larger area by two comparatively narrow passages, was the breeding establishment.

These penguins are said by some observers to set apart regular separate spaces in their rookeries for moulting, for birds in clean plumage not breeding, and again for breeding birds. Here the breeding ground was quite separate and the young and breeding pairs were confined to this smaller sheltered area. This was the only King Penguin rookery which I saw in full action. At Kerguelen's Land, the King Penguins were only met with in scattered groups of a dozen and twenty or so, and they were then not breeding, but only moulting.

On this breeding ground, at its lower portion, numbers of penguins were reclining on their bellies, and I thought at first they might be covering eggs, but on driving them up, I saw they were only resting. There was a drove of about a hundred penguins with young birds amongst them. The young were most absurd objects. They were as tall as their parents, and moved about bolt upright with their beaks in the air in the same manner; but they were covered with a thick coating of a light chocolate down, looking like very fine brown fur.

The down is at least two inches deep on the birds' bodies, and gives them a curious inflated appearance. They have a most comical look, as they run off to jostle their way in amongst the old ones. They seemed to run rather better than the adults, but perhaps that was fancy.

Absurd in appearance as these young are, those that are just dropping the down and assuming the white plumage of the adults, are far more so. Some are to be seen with the brown down in large irregular patches, and the white feathers showing out between these. In others the down remains only about neck and head, and in the last stage a sort of ruff or collar of

N

brown remains sticking out round the bird's neck, and then, when it cocks up its head, it looks like a small boy in stick-up collars. The manner in which these young ones cock up their heads gives them a peculiar expression of vanity, and as they ran off on their short stumpy legs, I could not resist laughing outright.

At the farthest corner of the breeding space, in the most sheltered spot, was a clump of birds of a hundred or more. The birds were most of them in a slightly stooping posture, and with the lower part of their bodies bulged out in a fold in front. As I came up and bullied these birds with my stick a little they shifted their ground a bit, with an awkward sort of hopping motion, with the feet held close together. It immediately struck me that they were carrying eggs with them, as I had read that King Penguins do. Their gait was quite peculiar, and different from the ordinary one, and evidently laboured and difficult.

I struck one of them with my stick, and after some little provocation she let drop her egg from her pouch, and then at once assumed the running motion. These birds carry their egg in a complete pouch between their legs, and hold it in by keeping their broad web feet tucked close together under it. They make absolutely no nest, nor even mark from habitually sitting in one place; but simply stand on the rookery floor in the described stooping position, and shift ground a bit from time to time, as occasion requires. I suppose the egg is not dropped till the young one begins to break the shell. Charles Goodridge says that the period of incubation is seven weeks, and that the birds commenced laying in the Crozets in November, and continued to lay, if deprived of their eggs, till March.

The birds with eggs were sitting close together. When, on my frightening them, some were driven against others, savage fights ensued, and blood was drawn freely; the birds whose ground was invaded striking out furiously with their beaks.

Round about the brooding birds were others, I think males, in considerable numbers. These males probably feed the females with which they are paired. There were also some young downy birds. If one of these latter was driven in amongst the brooders it was at once pecked almost to death.

The young ones utter a curious whistling cry, of a high pitch and running through several notes, quite different from the simple bass note of the adults.

The rookery was only inhabited in about a quarter of its extent, but it was strewed everywhere with the bones of the penguins in heaps, and on the verge of the rookery was a small ruined hut, with the roof tumbled in, and overgrown with weeds, and containing an old iron pot and several old casks, and some hoop iron; evidently an old sealer's hut. The sealers had probably employed their spare time in making penguin oil, and taking perhaps skins, which are made up into rugs and mats at the Cape of Good Hope, often only the yellow streaked part about the neck being used. Hence the many bones and emptiness of the rookery. The egg of the King Penguin is more than ordinarily pointed at the small end. It is greenish-white, like other penguin eggs.

Living also about the rookery was a flock of about thirty Sheath-bills (*Chionis minor*). The instant they saw us approaching they came running in a body over the floor of the rookery in the utmost excitement of curiosity, and came right up within reach of our sticks, uttering a " Cluck Cluck," which with them is a sort of half-inquisitive, half-defiant note. We knocked over several with big stones and our sticks ; but the remainder did not in the least become alarmed. They just fluttered up off the ground to avoid a stone as it was sent dashing through the thick of them ; but immediately pitched again, and ran up, as if to see how the stone was thrown. I only on one other occasion saw the Chionis thus living gregariously in flocks ; at Kerguelen's Land we found them already paired, except one flock which I saw near the entrance of Royal Sound, and at Marion Island many were already paired. That they should thus form flocks, when not breeding, is what might be expected from their near alliance to the Plovers.

At the rookery these birds were living on all sorts of filth dropped by the penguins, and were the scavengers of the place, and when I drove some of the brooders off their eggs, and an egg or two got broken, the Sheath-bills, who had followed us up closely, notwithstanding the slaughter we had done amongst them, came and pecked at the eggs almost between our legs.

The Skuas of course were close at hand, and swooped down at once on the body of a penguin that we skinned. Beyond the penguin rookery was a large tract of nearly flat land, very swampy, and covered with grass. On the drier parts were numerous troops of from twenty to thirty King Penguins, and in one place a smaller rookery, but as far as I saw without brooders.

There was here a shallow freshwater lake, on which some young albatrosses were swimming. I ascended the slope inland towards the snow, going up the gentle slope of the modern-looking lava flow already referred to. The ground was very boggy, and let one sink in sometimes almost up to the middle. There were numerous Great Albatross's nests scattered about, but they did not extend more than 100 feet above sea level, and hardly anywhere as high up as that.

Far above the level of these, I found a young bird, I think the young of the Giant Petrel, in a nest scarcely raised from the ground; the young bird vomited up the contents of its stomach and gush after gush of red oily fluid at me as I stirred it up with a stick. All the petrels vomit oil in this way, and the white ones thus are apt to spoil themselves for stuffing in a most provoking way, before one can get their mouths and nostrils stuffed with cotton wool.

The valley in which the lava flow up which I was going, lay, was bounded to the south by a cliff about 200 feet high, composed of a series of more ancient lava flows. The lowermost of these showed a more perfect columnar structure than the uppermost, and the columns of the lower layers were much smaller than those of the upper. A small stream ran down in the narrow depression, between the border of the lava stream and the talus slopes of the cliff. In the bed of this were at intervals small beds of a compact red earth, forming almost a rock, deposited by the stream, and subsequently in places cut through by it and exposed in section.

High up, at about 500 feet elevation, were some four or five Sooty Albatrosses (*Diomedea fuliginosa*, the Piew or Pio of sealers), soaring about the tops of the cliffs and probably nesting there. This bird is continually to be seen about cliffs and higher mountain slopes, and seems never to nest low down like the Mollymauk and Gony.

In holes in the banks at this elevation, a Prion was extremely abundant, but it was also pretty abundant down about sea level. Its peculiar angry cry, somewhat like the snarling of a puppy, uttered as it hears footsteps about its hole, is very puzzling at first as one listens to it, coming up from the ground at one's feet, but is unmistakable and quite unlike the cry of any other of the *Procellaridœ*, which we met with; I see however that Mr. Eaton in his notes, as cited by Mr. R. B. Sharpe, says: " that the cry of the petrel *Halabœna cœrulea* is exactly similar to that of the Prion." We dug out a bird with its egg.

I saw a hole with ears of grass dragged into it, and like a mouse's. It is not unlikely that there is a mouse in the island, as at Kerguelen; in Goodridge's time mice were so abundant on St. Paul's Island, that he speaks of feeding hogs, which he kept in confinement, on them. They were found lying in heaps in a dormant state in the early mornings (l. c., p. 65). A *Curculio* and two *Staphylinidœ* were found by Von Suhm on the island, and also a small land shell which was common. A fly with rudimentary wings was also found by him, apparently the same as one of those at Kerguelen's Land (*Amalopteryx maritima*). No land bird was met with, and no duck was seen, though one species of duck is so abundant at Kerguelen's Land.

Crozet Islands, Jan. 2nd, 1874.—We ran on towards the Crozet Islands, before the westerly winds, and after lying about close to this group in a dense fog, which prevented our sighting it and landing on Hog Island as we had intended, the fog at last lifted slightly on the evening of Jan. 2nd.

We ran in between Possession Island and East Island, as Ross had done thirty years before. As we steamed towards the land, the coast of Possession Island could just be discerned under a dense fog bank, the white breakers being plainly visible. The fog lifting a little more, a long range of cliffs could be seen; the tops of these, however, were still hid, together with all the higher portion of the island, in the densest fog. The fog seemed to lie some little way off the land, for the cliffs were lighted up by sunlight. Down these cliffs in several places, waterfalls poured into the sea.

As we neared the island and entered the passage between Possession Island and East Island, and came opposite the sealers' anchorage at Navire Bay, we had a clear view of this end of the island. It here presented a series of gentle slopes, bounded by low littoral cliffs. Further off, towards America Bay, the cliffs were seen to be much higher. Navire Bay is a very slight indentation of the coast line, affording hardly any shelter : it has a beach of large pebbles, and from it extends up inland a sinuous valley, appearing to my eye as rather a space left between two lava flows than the result of denudation. On one side of the beach was seen a hut and a store of oil barrels.

A shot was fired, but no one showed himself. The place was evidently deserted. There was too much surf on the beach to allow of landing. It was late in the evening, and a bank of fog appeared to be drifting up to envelope us ; so after sounding we made for Kerguelen's Land, greatly of course to my disappointment, for the flora of the Crozets was then quite unexplored. The slopes, however, appeared from the ship as if covered with a similar vegetation to that of Marion Island, which however, did not extend so high up the mountains.

The slopes were covered with albatrosses, nesting as at Marion, and the birds seen about the ship were the same as at that island, but in addition a Mollymauk was seen.

East Island presents towards Possession Island, very high sheer precipices, with most remarkable jagged summits. Only these summits, with their bold outline showing out against the sky, lit up by the light of the sunset, were to be seen ; the base of the cliffs was hidden in impenetrable fog. The Crozets are in about the same latitude as the Prince Edward Islands.

Crews of vessels have several times been cast away on the Crozet Islands. I have already referred to the account given by Charles Goodridge of his stay of two years in the islands in 1821–23.* Goodridge describes the discovery by his party at

---

* "Narrative of a Voyage to the South Seas, &c.," pp. 42, 43, by C. M. Goodridge. London, Hamilton and Adams, 1833.

above a mile from the reach of the tides, of several trunks of trees about 14 feet long, and measuring from 14 to 18 inches through, which were found lying on the ground as if thrown up by the sea. The wood was close, heavy, and hard, but being split up with wedges made very good clubs. Hence it was not fossil wood. Goodridge concluded that it was drift-wood thrown up so far during some volcanic convulsion.

We were told by the sealers that the rabbits, which are abundant on the Crozets, were not good to eat, because of their food. The wild hogs were, in Goodridge's time, very fierce and dangerous to approach single handed. The hogs have large tusks. Sealers told us that it would not be well to introduce pigs into the other southern islands, as they would destroy the birds, the main support of chance castaway mariners. The last account of a visit to the Crozets is that of Captain Lindesay Brine, R.N , who saw an iceberg 300 feet in height within sight of the group.[*]

The mean temperature of the air whilst the ship was off the islands, from December 30th to January 2nd, was about 44° or 45°. The highest reading was 50°, which occurred twice, the lowest 39·6°.

January 6th.—We sighted Bligh's Cap in the evening. It appeared as a hazy rounded cone on the horizon. Numerous birds surrounded the ship, and as on our approach to the other islands, penguins were to be seen in every direction. The birds were, *Dromedea exulans, D. fuliginosa, D. culminata,* a *Prion, Daption Capensis, Ossifraga gigantea,* and an *Oceanitis.* A Skua also was seen, though the land was eight miles distant A squall in the morning brought a slight fall of snow. The water assumed a peculiar dark colour, probably from its shallowness. Bligh's Cap is a small outlying rocky island to the north of Kerguelen's Land.

January 7th.—After lying off for the night we reached Christmas Harbour and anchored at 8.30 P.M.

For a list of Plants collected in the Crozet group by the U.S. Transit of Venus Expedition, see J. H. Kidder, M.D., " Bull. U.S. Nat. Mus.," No. 3, II, p. 31.

[*] Capt. Lindesay Brine, R.N., "Geogr. Mag.," Oct., 1877.

CHAPTER VIII.

KERGUELEN'S LAND.

Position of the Island. Its Mountains and Fjords. Active Volcano. Christmas Harbour. Sea Elephants and Fur Seals. Shooting Teal. The Kerguelen Cabbage. Wingless Flies and Gnats. Vegetation at Successive Heights. Fossil Wood. Rookeries of Rock Hopper and Macaroni Penguins. Penguins Inhabiting a Cave. Betsy Cove. Glaciation of the Land Surface. Iceborne Rocks. Excavation of the Fjords. Beds of Burnt Coal. The Sea Leopard. Killing Sea Elephants. Nature of the Trunk of the Sea Elephant. Carrion Birds. The Giant Petrel. Habits of several Burrowing Petrels. The Diving Petrel. Habits of Sheath Bills. Struggle for Existence amongst the Birds. Mode of Whaling amongst the Kelp.

Kerguelen's Land, January 7th to January 30th, 1874.— Kerguelen's Land extends from about lat. 48° 39′ S., to lat. 49° 44′ S.* Its southernmost point is therefore in about corresponding latitude to the Lizard in Cornwall, which is in a little less than 50° N. In longitude, very roughly speaking, Kerguelen's Land corresponds with the island of Rodriguez, the Maldive Islands, Bombay, Tobolsk, and the mouth of the River Obi.

The extreme length of the island is about 85 miles, and the extreme breadth 79 miles; but the coast is so much indented by sounds or fjords that the area of the island is not more than, very roughly, 2,050 square miles, or about three times as great as that of Oxfordshire.

The island lies within the belt of rain at all seasons of the

* Lat. of Cape Francis, the northernmost point, 48°·39 S., long. 69°·02 E.

Lat. of Cape Challenger, the southernmost point, 49°·44 S., long. 70°·05 E.

Extreme breadth between long. 70°·35 E. and long. 68°·42 E.

The Lizard is in lat. 49°·57 0′ 41″ N.

year, and being reached by no drying winds, and its temperature being kept down by the surrounding vast expanse of sea, has hence its soil and vegetable covering permanently saturated with moisture. Further, with this fact of constant precipitation of moisture is connected the form of the island itself, since fjord formation is accomplished only by glaciation on a large scale, and this can only occur where there is a constant supply of snow. The island further lies within the line of the Antarctic drift, as do also the Crozets and Prince Edward Group ; and this cold current must reduce the temperature considerably.

The island is in the region of prevailing westerly winds, the course of which is in the Southern Ocean, untrammelled and undisturbed by barriers of land. Since the line of greatest length of the island lies in a north-west and south-east direction, and the coast line, though much broken, trends on either side in the same direction, the north-east side is the sheltered one, and that, consequently, where are the safest anchorages, whilst the south-west side is the weather one.

The island is throughout mountainous, made up of a series of steep-sided valleys separated by ridges and mountain masses, which rise to very considerable heights. Mount Ross, the highest, is 6,120 feet in altitude, Mount Richards 4,000 feet, Mount Crozier 3,250, Mount Wyville Thomson 3,160, Mount Hooker 2,600, Mount Moseley 2,400.

The island thus, when viewed from the sea at a distance, presents a remarkable jagged outline of sharp peaks, which is most striking when the island is observed from the south. The valleys run down everywhere to the sea, broadening out as they approach it. The coast is broken up everywhere by deep sounds or fjords, which resemble closely in form the fjords of Norway, and of all other parts of the world were fjords exist. They are long channel-like excavations of the coast-line, occupied by arms of the sea, often shallower at the mouths* than at the upper extremities, and bounded on either hand by perpendicular cliffs.

The island is of volcanic formation as far as it has yet been

* The shallowness of the mouths of the fjords is well marked in the case of Royal Sound and Rhodes Bay.

investigated, and there is no doubt that it is entirely so formed, the beds of coal alone excepted, and certain beds of red earth, which are of the same origin as the coal, but merely different in that they have undergone a more intense heating.

The island has undergone immense denudation, and on its whole north-eastern and southern regions there is no trace of any volcanic cone or signs of comparatively modern volcanic action, as at Marion Island. Every appearance bespeaks considerable antiquity.

Nevertheless it seems to be certain that there exists towards the south-west of the island, a still active volcano with hot springs in its neighbourhood. We fell in with an American whaling captain, Captain Fuller, who has been often on the weather shore, and is well acquainted with the position of the volcano, and though he had not been actually at it himself, some of his men had; and in Tristan da Cunha we received independent testimony in the matter from old sealers.

The appearance of the island in the region of the volcano must thus be very different from that of the north-eastern and south-eastern portions.

As necessarily follows from the presence of fjords, the whole of the lower rock surface of the island shows most marked evidence of glaciation.

Christmas Harbour, almost on the extreme north of the island, is a small example of one of the fjords. It is a deep inlet with dark frowning cliffs on either hand at its entrance. The land on either side runs out into long narrow promontories, which separate the harbour from another similar fjord on the south and from a bay on the north. The promontories thus formed are high and bounded throughout almost their entire stretch by sheer precipices on either hand. On the northern side only of Christmas Harbour, somewhat above its mouth, does the land rise in a steep broken slope, which can be ascended directly from the sea.

At the termination seawards of the southern promontory, is the well-known arched rock of Christmas Harbour, a roughly rectangular oblong mass, evidently formerly continuous directly with the rest of the promontory, but now, separated from it, except at its very base, by a chasm, and perforated so as to form an arch. Above the high cliffs on the south side of the harbour, towers up a huge and imposing mass of black-looking rock with perpendicular faces; this overhanging somewhat towards the harbour from the weathering out of soft strata beneath it, looks as if it might fall some day and fill the upper part of the harbour. On the north side rises a flat-topped rocky mass 1,215 feet in height, called Table Mountain.

At the head of the harbour is a sandy beach and small stretch of flat land, as exists at the heads of all the fjords, and beyond this the land rises in a series of steps, separated by short cliffs towards the bases of Table Mountain and the great rock on the south.

The appearance of the whole is extremely grand, and the marked contrast between the blackness of the rocks and the bright yellow green of the rank vegetation clothing all the lower region of the land, so characteristic of the appearance of all these so-called Antarctic Islands, renders the general effect in fine weather, most beautiful. I landed on the morning of the 7th of January at the head of the harbour, with a large party, all eager to kill a Fur Seal; as the boat grounded on the black volcanic sand, some greyish brown forms were made out, lying amongst the grass just above the beach. A rush was made to the spot, but they were found to be only four Sea Elephants, reclining beside a small stream which runs down here from a little lake on a small plateau above, into the sea.

The Elephants, when stirred up, raised their heads and put on their usual savage expression which they exhibit when disturbed, which is effected by contracting the facial muscles about the nose, so as to throw it into a series of very prominent transverse folds. They opened their mouths, showed their teeth and uttered a roar, which consists of a series of quickly succeeding deep guttural explosions. They bit savagely at a stick, and twisted it out of our hands, but made no attempt to go to sea, making on the contrary into the stream, and up it inland, moving by the regular flop flop motion of the body, like that of the common British seal, but more clumsily performed.

Whilst everyone was either looking at these Elephants, or

beating the ground for ducks, I looked round for other seals, and on a shot being fired, I saw the head of an animal raised high above the grass on the flat close to the beach, and about a hundred yards off. I knew at first glance that it was a Fur Seal, and made for it in all haste. The seal, or Sea Bear, was lying in a sort of form in the grass. It contrasted most strongly in its appearance and gait with the Sea Elephants we had just left.

The *Otariadæ*, or seals with external ears, differ from all other seals in that, in progression on land, they turn their hinder limbs or flippers forwards, and rest on the backs of them, and raising the body from the ground with the fore limbs, shuffle along with a sort of awkward walking gait, by the alternate use of the hind limbs. All other seals keep their hind limbs stretched straight out behind when on land as when in the water, and these limbs are therefore of no aid in moving on land, which is accomplished entirely by undulating movements of the body. The *Otariadæ* are in fact connecting links between the true seals and such beasts as the Sea Otter; their limbs still retain some of their old land functions.

The Sea Bear has besides a thick coating of long hair, the familiar thicker layer of silky fur beneath, which renders its skin so valuable. The Sea Bears are nimble on land as compared with the helpless Sea Elephants, and can climb up on to rocky ledges, and even spring some little distance.

The seal I had found was an old male, covered with greyish-brown shaggy hair, and with a short greyish mane about the neck. He moved his head up and down uneasily when disturbed, as one sees a bear sway his head. One of the party came up as we were watching him, and running up close to the beast, as if it had been a helpless Sea Elephant, was forced to retreat in a hurry, for the beast made a savage dash at him, open-mouthed.

The seal was very difficult to kill outright. Fur Seals are easily knocked over with a blow on the nose, but are very tenacious of life, and require to have their throats cut directly they are stunned, or they escape after all.

There are still a considerable number of Fur Seals about Kerguelen's Land. I killed two; two others were killed by our party at Howes Foreland, and two others were seen there. Two of the whaling schooners killed over 70 Fur Seals on one day, and upwards of 20 on another, at some small islands off Howes Foreland to the north. It is a pity that some discretion is not exercised in killing the animals, as is done in St. Paul's Island in Behring's Sea, in the case of the northern Fur Seal. By killing the young males, and selecting certain animals only for killing, the number of seals may even be increased.* The sealers in Kerguelen's Land kill all they can find.

The sealers told us that the southern Fur Seals sometimes eat penguins, and that they had found the remains of them in their stomachs. Seals feed to a very large extent on Crustacea. Thus *Otaria jubata* is said to feed more on Crustacea and smaller fish, than on large fish, and in the Campbell and Auckland Islands to eat also birds,† and Mr. Brown, in his account of the habits of Arctic seals and whales, says that the food of the northern seals consists mostly of Crustacea, species of *Gammarus*, called "seals' food" by the whalers.‡ In summer the Northern Seals eat fish. They sometimes take down birds, but not often. Dr. Buckholtz found only Crustacea in the stomachs of *Phoca Greenlandica* in the Arctic regions, mainly *Gammarus Arcticus*, and *G. Themisto*.§

The sealers told me, that sometimes, but very rarely, they found another kind of seal, like the Fur Seal somewhat, which they called the " Sea Dog." A second species of eared seal probably thus occurs as a rarity at Kerguelen's Land.

The whole beach of Christmas Harbour was covered with droves of the Johnny Penguin (*Pygosceles tæniata*) and King Penguins, and establishments of these penguins were to be seen on small level grassy spaces far up the hill slope.

* "The Eared Seals." J. A. Allen. Bull. Mus. Comp. Zool., Harvard Univ., Cambridge, Mass., Vol. II, No. 1.

† For an account of the habits of the Southern Sea Lion, see " Twenty Months in the Campbell and Auckland Islands." Peterm. Mitt. 1866, s. 103.

‡ R. Brown, " On the Mammalia of Greenland," with succeeding papers on the Seals and Whales. "Proc. Zool. Soc.," 1864.

§ " Die zweite Deutsche Nord-Polarfahrt in den Jahren 1869 und 1870," 2. Bd. Wissenschaftliche Ergebnisse. Leipzig, F. A. Brockhaus, 1874. W. Peters, " Zeugethiere und Fische."

Teal were shot in great numbers by our party. The teal of Kerguelen's Land (*Querquedula Eatoni*) is peculiar to the island and the Crozets. It is somewhat larger than our common teal, and of a brown colour, with a metallic blue streak, and some little white on the wing. It is enormously abundant all about Kerguelen's Land, near the coast. I killed in one day, twenty-seven teal, and similar bags were frequent. Four or five guns used to bring back usually over 100 birds.

The teal feed mainly on the fruit of the Kerguelen cabbage, and are extremely good eating. They were the greatest treat possible to us, when living, as we necessarily were, almost entirely on preserved meat.

The teal are to be found mostly in flocks, or when breeding in pairs. They are, where they have not been shot at by sealers, remarkably tame, and require to be kicked up almost to afford a shot. At one valley near Three Island Harbour in Royal Sound, which had probably not been visited by man for thirty or forty years, perhaps hardly ever, after tramping some distance after teal without success, I saw a flock get up from the bed of a river which ran down the valley, about 150 yards off. I thought the birds must be wild and had been recently shot at ; but no, they got up merely to come and look at me. They pitched about 40 yards off, and then set off running towards me in line, like farm-yard ducks, seven of them in a row, headed by a drake. As a sportsman, I hesitate to describe the termination of the scene. Only those who have been long at sea know what an intense craving for fresh meat is developed by a constant diet of preserved and salt food. The teal were most excellent eating, and there were many mouths to feed. My rule was always to shoot them on the ground if I could, and as many at a shot as possible. When I could not do this I took them flying, and with tolerable success.

Some of the teal were breeding at the time of our visit; some with young full-fledged and already away from the nest; others with eggs. The nest is a neat one, placed under a tuft of grass, and lined with down torn from the breast of the parent bird. There were five eggs in one nest that I found.

The duck, when put up off the nest, to effect which the nest requires almost to be trodden upon, or when found with her young away from the nest, flutters a few yards only, as if maimed, and pitches again, and cannot be frightened into a long flight. It is curious that the bird should have retained this instinct where there are no four-footed or human enemies; possibly she finds it a successful *ruse* when the brood is attacked by the Skuas.

The young must fall constantly a prey to these ever-watchful Skuas, for in most cases I found only a single young one following the mother. There were no young met with in the condition of flappers, and the general breeding season was probably only about to begin, as it was with many birds of the island. The greater part of the birds were yet in flocks.

The flat stretch of land at the head of Christmas Harbour is covered with a thick rank growth of grass (*Festuca Cookii*), and a Composite herb with feathery leaves and yellow flower (*Cotula plumosa*), also with *Azorella* as at Marion Island, with *Acæna Montia fontana* and *Callitriche verna* about the dampest places. The soil is black and peaty and saturated with water. It is almost impossible to find anything to burn; the *Azorella* is the only thing that will burn, and sometimes pieces of this may be found that are dry enough, in places where the *Azorella* bunches overhang small precipices, and the water can thus drip away.

The feature which distinguishes the general appearance of the vegetation of Christmas Harbour from that of Marion Island is the presence of the Kerguelen Cabbage in large quantities. The plant grows on the slopes and bases of the cliffs in thick beds. The cabbage is in appearance like a small garden cabbage, but often with a long trailing stalk. It is, however, not annual, but perennial, and the flowering stalks, instead of coming out from the centre of the head, come out laterally from the sides of the stalks between the leaves.

The old flower stalks die and wither, but do not drop off. I counted on one cabbage at Betsy Cove 28 flowering stalks, of different ages; three of them only being of the current year's growth and fresh. They appeared to belong to eight successive years. The cabbage about Christmas Harbour was either in flower or green fruit, mostly the latter. It was only to the south of the island, about Royal Sound, that ripe seed was met with; but there, especially at Mutton Cove, it was abundant. The cabbage (*Pringlea antiscorbutica*), which like the familiar vegetable is a cruciferous plant, is peculiar to the Prince Edward, Crozets, Kerguelen and Heard Islands, and belongs to a genus with no near ally.

KERGUELEN CABBAGE, PRINGLEA ANTISCORBUTICA.
(From a photograph.)

Crawling about the heart of the cabbages, and sheltering there, are to be found swarms of the curious wingless fly, likewise peculiar to Kerguelen's Land, and islands where the cabbage is found. The fly (*Calycopterix Moseleyi, Eaton*) is simply a long-legged brown fly, with very minute rudimentary wings. It crawls about lazily on the cabbage, and lays its eggs in the moisture between the leaves, about the heart of the plant.

Another fly (*Amalopteryx maritima*), with wings rudimentary but larger in proportion to the body than in the other, is found about the rocks, on the sea shore, where it jumps about when hunted, as if it were a small grasshopper. It is the same as found at Marion Island, where it was discovered by Von Willemoes Suhm. Probably the fly frequenting the cabbage exists also at Marion Island; but we did not know where to look for it when there, and cabbages were not very abundant; but it is possible, also, that this fly does not extend there, for we saw no teal on Marion Island, though they exist in abundance on the Crozets, and especially on Possession Island, where, as we were told by the sealers, there is a lake full of them. However, we examined but a very small tract of Marion Island, and similar tracts are to be found in Kerguelen's Land, with very few cabbages, and consequently without teal. Both animals may abound in parts of Marion Island not visited by us.

A wingless Gnat (*Halyritus amphibius*) also inhabits the sea-shore, living amongst the sea-weed constantly wetted by the tide. I discovered at the Falkland Islands, a similar wingless gnat, and a fly which I believe to be closely allied to the Kerguelen *Amalopteryx*, and which thus adds to those already known,[*] a further interesting link between the forms of life inhabiting these widely separated islands.

I mounted up the slope towards Table Mountain. The climb is up a succession of steps, the successive flat ledges presenting glaciated surfaces scattered over with stones fallen from above. The thick rank vegetation ceases at about 300 feet altitude, and then becomes more sparse. *Colobanthus Kerguelensis*, a Caryophyllaceous plant, peculiar to Kerguelen's Land and Heard Island, affects the more barren stony ground at this elevation, and I did not meet with it anywhere about the lower slopes, or amongst the peaty soil. At Heard Island it grows at sea-level.

At about 500 feet elevation, a very handsome lichen (*Neuopogon Taylori*) commences rather abruptly. It is a very con-

[*] See Rev. E. H. Eaton. "Breves Dipterarum uniusque Lepidopterarum insulæ Kerguelensis indigenarum diagnoses." The Entomologists' Monthly Magazine, August, 1875, p. 58.

C. O. Waterhouse, "On the Coleoptera of Kerguelen's Land." Ibid., p. 50. There are five genera of Diptera in the island (four of Muscidæ, one of Tipulidæ, all cited as endemic in the southern islands. Possibly, however, two of these occur in the Falkland Islands. The beetles are all apterous, one having the elytra united. Two genera and all the species are endemic.

spicuous plant, being of a mingled bright sulphur-yellow and black colour, and of large size. It is abundant on the higher rocks everywhere. Azorella and the cabbage grow up to about 1,000 feet, the height of the ridge from which the rocky mass forming the top of Table Mountain rises. Here the cabbage ceases, but Azorella is continued in very small quantities to the top of the mountain, growing on its very summit, but only in very sheltered corners between rocks and much dwarfed.

Azorella, the cabbage, and a grass (*Agrostis Antarctica*), were the only flowering plants growing at 1,000 feet, and these only very sparsely. The land at this height presented a series of ridges of barren rock and piles of stones. At Mutton Cove and about Royal Sound, a very marked line, at about 1,000 feet, separates the green lower slopes from the barren stony ridges and peaks above. It is probably the line above which snow lies for the greater part of the year unmelted, though the hills just above it, at Mutton Cove, were quite free from snow at the time of our visit.

In a pool of water, on the summit of Table Mountain, I found a quantity of specimens of a small *Lumbriculus*, or allied form of Annelid.

The phonolith of which Table Mountain is composed, is full of olivine crystals, occurring in large rounded masses as in the Ardeche valley, and many other volcanic districts.

A comparatively low ridge separates the head of Christmas Harbour from the sea directly beyond. On a flat expanse of this ridge are two small freshwater lakes, in which grow two water plants, *Limosella aquatica* and *Nitella Antarctica*, both widely spread plants, the first occurring, amongst other places, in England; and the second being very closely allied to a common English species.

I found *Limosella aquatica* only in these particular lakes, and then only after a very long search, since it resembles extremely closely, in its general appearance, when growing in masses, a *Ranunculus* (*R. Moseleyi, Hk. f.*), which grows with it in the water.

Above the lakes the ridge rises somewhat, and then terminates in an inaccessible precipice fronting the sea, with short talus slopes below, on which are rookeries of crested penguins.

Under the peculiar overhanging rock, on the south of the harbour, are beds of fossil wood, and the excavation beneath its base is hence called Fossil-wood Cave. The wood occurs in beds lying nearly horizontal, and a few feet only in thickness.

The beds are of a soft whitish clay-like matter, which is full of black vegetable remains, all apparently so charred and decomposed, as to give little or no hope of any structure being made out in them.

The wood is in large trunk-like masses; the largest which I saw was about 1½ feet in diameter; in some the bark is preserved. The wood is in various states of fossilization, some of it being comparatively soft, other specimens extremely hard, passing even in the centre into actual basalt, containing small amygdaloidal masses of zeolites. Analcite and other zeolites are abundant in the Kerguelen lavas, as are also agates.*

On the talus slopes beneath the cliffs, along the whole south side of Christmas Harbour, are vast Penguin rookeries; the Penguins here nesting amongst the stones where vegetation is entirely wanting: and to the north of the harbour at its entrance are other similar rookeries. Towards the upper part of the harbour, the rookeries are those of the smaller crested penguin called "Rock-hopper" by the sealers (*Eudyptes saltator*), the same as that at Marion Island, but nesting scattered amongst these is another kind of penguin, *Eudyptes chrysolophus*, the Macaroni of sealers.

This bird has a most beautiful golden crest, showing conspicuously on the middle of the upper part of the head, commencing just behind the beak, and with a plume on each side as in the bi-crested species. The bird is larger than the "Rock-hoppers," and is further distinguished from them by the presence of a naked, somewhat tumid space, at the base of the beak, which is of a light pink colour. In other colouring the bird resembles the Rock-hoppers. This penguin occurs at the Falkland Islands, where it nests as at Kerguelen's Land, in small quantities amongst the Rock-hoppers.†

The birds however, only thus nest amongst the other pen-

* See J. Y. Buchanan, " Proc. R. Soc.," No. 170, 1876, p. 617.
† " Proc. Zool. Soc., 1865," p. 527.

guins where they are few in number: towards the head of the harbour, and under the natural arch, they have enormous rookeries of their own, where, singularly enough, a few of the Rock-hoppers nest as guests amongst them; they have large rookeries also in Heard Island, where their eggs are gathered in large quantities by the sealers for eating. The sheath-bills are as abundant here as at Marion Island, but they are larger and heavier than are the birds of that island, and seem to form a sub-species. They will be again referred to.

During our stay at Kerguelen's Land, we put into several harbours on the coast. At Aldrich Sound I found a cave in the sea-cliff fronting Ship's Channel and under Mount Bromley. The cave had been formed by the excavation by the waves of the volcanic rock, which had been altered, and rendered more yielding at this spot by the intrusion of a dyke which had destroyed the tenacity of the rock by its heat. The dyke which was a narrow one, and almost vertical in direction, was inclined a little, at one part of its course, so as to form the roof of the cave on one side.

The cave was long and tunnel-like. The "Rock-hopper" penguins breed in this cave. I went into it about forty yards until it was quite dark; the penguins retreated still before me. I had no means of getting a light to explore the cave further. The small penguin of New Zealand (*Spheniscus minor*) has been observed breeding in like manner in the inner chamber of a dark cave,* and this mode of nesting is in keeping with the usual habit of this species and others of breeding in deep burrows, which are of course quite dark.

About Betsy Cove and Royal Sound, to the southward the valleys are broader, and there is more open flat land than there is around Christmas Harbour, and there are thus here large expanses covered with vegetation.

At Betsy Cove we stayed about ten days surveying the surrounding district. The Cove is also called Pot Harbour, from there being an old broken iron pot on the beach, a whaler's try pot, used for boiling down blubber. As we came into the harbour and anchored, though not more than a quarter of a

* "Trans. N. Zealand Inst.," Vol. II, 1868, p. 75.

mile from the beach, from some peculiar condition of the atmosphere, the pot looked of immense size, even when viewed with a glass, and two King Penguins (*Aptenodytes longirostris*), standing beside it, looked like men in white and black clothes. I went on shore with a boat at once at the desire of Sir Wyville Thomson, to get the penguins, for we thought they must be stray specimens of the huge antarctic penguin *Aptenodytes Fosteri*. I cannot understand how the delusion came about, it was certainly complete.   The pot has been for forty years on the beach.

There are two skulls of the southern Whalebone whale (*Eubalæna Australis, Gray*) lying here in the surf: such skulls are common all along the coast, remaining with other bones where whales have been towed on shore to be boiled down.

At Three Island Harbour in Royal Sound, there is a long row of them on the shore.

The neighbourhood of Betsy Cove is very interesting from a geological point of view, for it is here that the glaciation of the surface is most marked, and the glaciated surfaces most easy of access.   Close to the harbour, on the north, are a series of *roches moutonnes*, but the best examples are on the road from Betsy Cove to the head of a fjord adjoining, called Cascade Reach, because there is a waterfall on a stream which falls into its upper extremity.

Betsy Cove and Cascade Reach are both indentations in a larger bay called Accessible Bay, which lies at the end of a wide valley stretching far inland, and bounded on either hand by long elevated ridges.   In this broad valley, the bottom of which forms one of the flat expanses already referred to, project up a number of flat topped rocky hills, with smooth ground upper surfaces bounded all round by vertical cliffs; some of the most characteristic of these hills are to be met with on the way up the south side of Cascade Reach from Betsy Cove.

The tops of these hills show everywhere rounded surfaces, most obviously ground smooth by ice action, but the rock is not sufficiently hard to retain striation marks, and since the whole surface of the land has evidently undergone immense denudation subsequently to its glaciation, these are nowhere to be made out, and moraines have also disappeared.

The ridges north and south of the broad valley look at first glance as if they might be moraines, but their main structure is rock, in its original position, though covered mostly by talus.   A similar ridge to the south of the great fjord, Royal Sound, has likewise very much the appearance of a moraine ; but here also the main constituent is volcanic rock *in situ*.   There is nowhere to be seen a free-standing ridge composed entirely of moraine matter; but about the flat-topped hills, just described, there are beds of sand and stones that may represent broken-down remains of moraines.

Resting on the rounded surfaces of the flat-topped hills, and scattered over them in all directions, are immense quantities of stones of all sizes.   The stones have all their angles sharp and

ICEBORNE ROCKS RESTING ON GLACIATED SURFACES, NEAR
BETSY COVE, KERGUELEN'S LAND.

unweathered, and they rest in all sorts of positions on the smoothed rock, and they have most evidently been dropped into their present position by ice floating over the glaciated surfaces when these were in a submerged condition.

The summits of the flat-topped hills are formed of caps of basalt, showing usually columnar structure in their cliff faces. These caps of basalt of the several hills appear, undoubtedly, to have formed at one time a continuous sheet.

Exactly similar flat-topped hills occur everywhere about in Kerguelen's Land, and notably in Royal Sound, which is a deep and grand fjord studded all over with numerous rocky islets, probably 100 or more in number.   These islets are all flat-topped with erratics on their upper surfaces, and they appear to increase gradually in height towards the head of the Sound. The hills are of the same constitution as those about Betsy

Cove, and if the great valley at Betsy Cove were submerged, we should have on its northern side the hills projecting as islands, and giving a miniature representation of those in Royal Sound.

There can be but little doubt that the whole of these islands in Royal Sound were once connected, and that there was thus a broad sheet of lava rock with a gentle inclination from inland towards the sea. This slope was covered with a huge glacier, which was bordered by the mountain ridges now bounding the Sound to the north and south, and, perhaps, deposited some of the talus at present forming part of the ridge above Mutton Cove. After grinding the whole surface of its bed, the glacier shrunk and cut deeper channels between masses of rock, which were left standing, and thus formed the present islands.

Either during this period, or after glaciation had ceased, the whole was submerged till the upper surfaces of all the islands were under the sea, and then ice drifting seawards from the remnants of the shrunken glaciers at the heads of the fjords, dropped upon the rock surfaces the erratics which at present lie upon them. At this time all the moraines were washed away.

At the base of the hills about Betsy Cove, the bottoms of the secondary valleys are as distinctly glaciated as the main valleys themselves, and the slopes of the smoothed surfaces seem to lead towards the cavity and mouth of the present Cascade Harbour.

About Betsy Cove, thin beds of a red earthy matter a foot or two in thickness are very common, underlying beds of basalt and weathering out in the cliffs so as to leave ledges and low-roofed caverns. They occur in exactly the same manner as the beds of coal at Christmas Harbour; and when this coal is burnt in the fire it bakes to a compact mass of red earthy matter, exactly resembling that above referred to. There seems no doubt that these red beds, as well as the coal beds, represent old land surfaces. The soil consisting of black peaty matter as now, not many feet thick, has been overflowed by lava streams, which in the case of the coal have been only hot enough to char all the vegetable matter, in the other case have burnt it to an ash.

The coal at Christmas Harbour consists of abundant earthy

matter, full of charred remnants of vegetable tissue, but I could find no recognizable leaves or definite forms, except something which resembled a *Chara*. Even microscopic structure seems entirely destroyed. From the glaciated condition of the beds overlying the coal and red earth, the great antiquity of the Kerguelen vegetation is evident. It has been dwelt upon by Sir J. D. Hooker.

At Betsy Cove are the graves of some whalers, none of very old date. They have small white painted wooden monuments. It was at Betsy Cove that the best teal shooting was enjoyed, there being several small rivers in the neighbourhood, and plenty of small ponds and marshy ground with abundance of cabbages. On one of my teal shooting excursions I met with a Sea-Leopard (*Stenorynchus Leptonyx, Gray*). The beast is very like the common British seal in appearance. It is spotted yellowish white and dark grey on the back, the under surface being of a general yellowish colour.

The one in question was small, not more than five feet long. It was asleep, lying almost on its back on the grass in a little bay. The poor beast showed no fight at all, and never snarled or showed its teeth. I killed it with a stone and my hunting knife, and sent it on board to be made into a skeleton.

The Sea-Leopard seems still pretty abundant on the coasts. I saw one much larger in Royal Sound, and Von Willemoes Suhm killed another. The sealers said they intended to visit Swain's Island, a small outlier, to kill a herd of 400 of these seals reported to be in a rookery there.

Farther along the coast, on the same day, I encountered a small herd of Sea-Elephants consisting of four females and two males. One male was much larger than the other, and the four cows were reclining beside him, the younger and less powerful male lying apart from the rest. All were resting on a thick bed of seaweed cast up by the tide on a beach of large pebbles.

The male was 12 feet long and enormously heavy and fat. The females were about eight feet in length. All were of a light fawn colour except one female which was shedding her coat, and was covered over with patches of reddish hair. Though I fired my gun at some teal close by, the Elephants were little disturbed. The males just raised their heads and then went to sleep again; the females took no notice.

I went up close to the older male and excited him in the hopes of seeing him raise his trunk-like snout, and he was roused again later on, but this had not the effect of making him move from his ground or frightening him at all; but on one of the ship's cutters, for which I had sent a petition to the ship, coming into the bay full of men in order to kill specimens of the Elephants and take them on board, the Elephants became immediately alarmed as if accustomed only to expect danger from boat parties.

I had forgotten that the Tristan da Cunha people had told me that they always shot the male Sea-Elephant and lanced the cows, and I thought the beast could be stunned by blows on the snout like Fur-Seals, so Lieutenant Channer, who had been out shooting with me, went up to the big male and began hammering him on the snout with a stick heavily loaded with lead, but without any effect beyond enraging the beast to the utmost. The animal was not stunned by the blows, because the skull of the Sea-Elephant is protected above by a high inter-muscular ridge or crest, and the bones around the nostrils are very strong. In these point s the Elephant is very different from the Fur-Seal. The beast raised itself on its fore-flippers and at the same time twisted up its tail into the air, just as represented in "Anson's Voyages," where the Sea-Elephant was figured for the first time as the Sea-Lion of Juan Fernandez.

OLD MALE SEA-ELEPHANT OF JUAN FERNANDEZ.
(Copied from Anson's Voyages.)

The beast raised its head and opened its huge mouth to the widest, showing formidable teeth and a capacious pinkish gullet, from which proceeded loud and angry roars.

The animal was too young to have a largely developed

trunk.   There was merely an arched projection thrown up for some little distance above the nostrils, partly by inflation, partly by strong contraction of muscles on each side of the nose.   If the beast had got hold of Channer he would have bitten a limb to pieces at one crunch.   The head of the stick came off, and so I ran up and put a bullet into the animal's heart.

This male Sea-Elephant when enraged had its snout much in the condition as that shown in Leseur's plate * in that one of

SEA-ELEPHANTS.
(Copied from Leseur's Plate.)

the animals of the group represented, which is just going to land from the sea on the left-hand side of the landscape.   The old male elephants were described by the sealers of Heard Island as having a trunk 10 inches in length.   These old males were called " Beach-masters."   Anson's sailors called the largest male at Juan Fernandez the " Bashaw."

I obtained from a harponeer on board one of the whaling schooners which we fell in with at Kerguelen's Land, a very well executed carving in a soft volcanic stone from Heard Island, which represented two men skinning a dead Beach-master.   Unfortunately, this was lost with other curiosities in transit from the ship, after we reached home.   In this, the trunk of the old male Elephant was shown hanging like a short flaccid tube from the snout.   It is shown somewhat thus in Leseur's figure, drawn for Peron, in the case of the animal represented as lying on beach in the foreground; but the trunk there is probably shown

* " Voyage de Découvertes aux terres Australes."   Peron et Leseur. Paris, 1807.   Atlas Pl. XXXII.

much too prominent and solid looking. The old sealers used to eat the trunks as a tit-bit, calling them " snotters." Goodridge speaks of it as " a sort of fleshy skin, which hangs over the nose." In Anson's Voyage it is described as hanging down five or six inches below the end of the upper jaw. Peron says very little in his account of the Sea-Elephant about the trunk.*

I give here a woodcut, from a rough drawing made for me by the harponeer above referred to, of a " Beach-master," with its trunk in the inflated condition.

The trunk, when the animal is enraged, is inflated and erected, being blown full of air. From the drawing it appears that Anson's figure is probably nearly correct in the matter of the trunk, as it certainly is in the manner in which the tail is curled up into the air in the enraged beast.

DRAWING OF OLD MALE SEA-ELEPHANT.
(By a Harponeer.)

The trunk is produced by inflation of a loose tubular sac of skin placed above the nostrils, just as is the " cap " in the northern Bladder-nose seal (*Cystophora proboscidea*). The trunk is evidently, as appears from both the drawings, sacculated, and hence irregular in form when inflated. In the Bladder-nose the nasal cap develops, only at advanced age, just as in the case of the trunk of the Sea-Elephant.

I bought the stone carving from the harponeer for a sovereign

* For Peron's " Histoire de l'Eléphant Marin," see l.c. T. II, p. 32. A translation of it is given in Brewster's " Edinburgh Journal of Science," 1827, Vol. II, p. 73.

and a bottle of whisky.   He would not have taken five pounds less the whisky, as it was a matter of honour with him that he should get a drink for his shipmates out of the proceeds.

Whilst we were killing the male Elephant, two of the cows had been killed by the sailors; one of them got away for a time to our extreme regret, badly wounded, into the sea, and the unfortunate animal had to be shot several times before it was killed.   Being wounded, it made back for the shore.   I was astonished at this, since it is directly contrary to the ordinary habits of seals.   I presume the animal sought safety with the rest of the herd.

The Sea-Elephants have a most enormous quantity of blood in them.   This wounded female stained all the water of the head of the little bay, red.   The blood, so black as it is in the body of the seal, and dark like the muscles, became of a bright arterial red as it mingled with the sea water.   Mr. R. Brown (in his account of the Arctic Seals and Whales inhabiting the Coasts of Greenland, " Proc. Zoolog. Soc.," 1864), refers to the remarkably dark colour of the flesh of seals, due to the gorging of the muscles with venous blood; and states further, that in the young seals, which have never been in the water, the muscles are red, and that the blood of the seal, dark when shed, turns thus red, when exposed to sea water or the air.

These Sea-Elephants, which were prepared as skeletons on board the ship, were found to have only a greenish slime in their stomachs.   Neither the *Otariadœ* nor the Sea-Elephants feed during the breeding season, but live upon their fat, becoming gradually thinner and thinner.   The Sea-Elephants have a regular layer of blubber on their bodies like that of whales and porpoises.   So perfect a protection is this non-conductor against loss of heat, that a dead walrus, which like most seals has the same covering, has been found to retain its internal temperature after having lain 12 hours in ice-cold water.*   In the Fur-Seals (*Arctocephalus*), there is no such thick layer of blubber developed, but only a small quantity of fat attached to the skin.

* "Die zweite Deutsche Nord-Polarfarht in den Jahren 1869 und 1870." 2. Bd.   Wissenschafttliche Ergebnisse, Leipzig, F. A. Brockhaus, 1874. W. Peters, Zeugethiere und Fische.

The muscles also are redder than in other seals, more like beef, or muscles of land animals generally, not black, and the meat was found very good to eat by some of our crew. Mr. Brown (loc. cit.) speaks of a green slime found by him in the stomachs of the northern Bladder-nose (the northern representative of the Sea-Elephant). He ascribes it to seaweed adhering to Mollusca (*Mya truncata*) eaten by the seal. It is, however, probably only bile pigment. Peron found cuttle-fish beaks and Fucus in the Sea-Elephants' stomachs. The walrus, like the Bladder-nose, feeds on Mollusca. In a walrus, dissected by the second German North Polar Expedition, the bodies of from 500 to 600 (*Mya truncata*) were found in the stomach, with only one single small piece of shell, the animal evidently rejecting the shells with great care. Stones are found in all seals' stomachs, apparently just as in those of penguins.

There seems little fear of the Sea-Elephant dying out, notwithstanding that everyone that can be got at is killed and boiled down by the sealers. I saw myself, at Kerguelen's Land, eighteen Elephants, and one at Marion Island. On the weather-side of the island is a beach, where are thousands of Sea-Elephants. These can be got at from land, but shallow water and a heavy surf prevents the approach of a boat. Hence, if the animals be killed and their blubber boiled down, the casks cannot be got off to a ship, nor can they be transported over land.

The beach is called Bonfire Beach, because some English sealers made a lot of oil here, headed it up in casks, and then found they could make no use of it. So they piled the casks up and set fire to them, in the hopes of driving some of the Elephants to more convenient quarters. The numbers of seals at Kerguelen in ancient times must have been enormous. Their vast old empty rookeries are still marked by trough-like hollows in the ground, where the seals used to lie.

We rolled the dead Sea-Elephant down to the water, and got him afloat with some difficulty, then towed the three animals off to the ship with great labour, by rowing against the wind, through the thick beds of kelp (*Macrocystis pirifera*). Whilst we were at work on the beach, crowds of birds began to assemble, especially the Giant Petrel or "Breakbones" (*Ossifraga gigantea*).

the "Nelly" or "Stinker" of sealers. This bird in its habits is most remarkably like the vulture.

It soars all day along the coast on the look-out for food. No sooner is an animal killed, than numbers appear as if by magic, and the birds are evidently well acquainted with the usual proceedings of sealers—who kill the Sea-Elephant, take off the skin and blubber, and leave the carcass. They settled down here all round in groups, at a short distance, a dozen or so together, to wait, and began fighting amongst themselves, as if to settle which was to have first bite.

The birds gorge themselves with food, just like the vultures, and are then unable to fly. I came across half a dozen together at Christmas Harbour in this condition. We landed just opposite them; they began to run to get out of the way. The men chased them, they ran off, spreading their wings, but unable to rise; some struggled into the water and swam away, but two went running on, gradually disgorging their food, in the utmost hurry, until they were able to rise, when they made off to sea.

The northern Fulmar (*Fulmarus glacialis*) seems to resemble the "Breakbones" very closely in habits. Like it, it does not nest in holes like most *Procellaridæ*. It feeds in the high north on carrion, and becomes so gorged with meat from a whale's carcass as to be unable to fly without disgorging.*

I was astonished at the comparatively small quantity of food, that is, the smallness of the extra weight, which made all the difference between the bird's not being able to rise at all, and its being able to soar away with almost its usual power. It would be interesting to test various birds with weights and compare their power in this respect. A *Procellaria* is evidently very much below an *Accipitine* in strength in this matter though so perfect a flyer.

But the "Breakbones" were not the only birds which assembled to feast on the remains of the Sea-Elephants. With them came the Skuas, but not in great numbers, and multitudes of gulls and Sheath-bills, which latter were the most impudent, and the first to dare approach a dead cow Elephant which we left on the rocks. The whole of the birds must have been dis-

* MacGillivray, "British Water Birds," Vol. II, p. 436.

appointed, when they found we were not sealers, for they apparently could not penetrate the skin of the dead cow, and a day or two afterwards only the eyes were pecked out : but the Breakbones were then still hanging about the carcass, waiting, though not in such numbers as before.

On another day, beneath the cliffs, north of Betsy Cove, I found a young Fur-Seal lying amongst some boulders at the foot of the cliff. There was a broad flat shelf of rock here, nearly level with the sea, and forming an excellent landing-place for seals, so I was especially hunting for them, but should have missed this one amongst the rocks, had it not attracted my attention by a sort of half-hiss, half-snarl. I killed it, and carried the whole beast with great labour to the ship, half a mile or more, on my back, in order that a skeleton should be made of it.

On several occasions I superintended parties of stokers, who volunteered to dig up birds and eggs for our collection. This is the method in which very many of the birds of Kerguelen are most readily procured. The beaten ground beneath the Azorella is perforated everywhere with holes of various petrels. Those of the Prion (*Prion desolatus*) are most numerous. They are about big enough to admit the hand, but the nest and egg are nearly always far out of reach, the holes going in a yard and a-half sometimes.

Prion is a small grey bird, a petrel from the form of the nostrils, but with a broad boat-shaped bill, with extremely fine horny *lamellæ*, projecting on either margin of the bill inside. The bird flies like a swallow, and was nearly always to be seen in flocks about the ship, or cruising over the sea, or attendant on a whale to pick up the droppings from its mouth. Hence it is termed by sealers the "Whale-bird." Its food, as that of all the petrels except the carrion ones, seems to consist of the very abundant surface animals of the south seas, especially of small Crustacea. These form also, apparently, the only food of the penguins; for the stomachs of all the penguins which we examined were crammed with them only. The Prion lays a single white egg.

Besides the Prion there is the "Mutton-bird" of the whalers

(*Œstrelata Lessoni*), a large Procellarid, as big as a pigeon, white and brown and grey in colour. It makes a much larger hole than the Prion, six inches in diameter, and long in proportion. At the end is a round chamber with a slight elevation in the centre, where the nest is somewhat raised, with a deeper passage all round; at least, I saw this in two nests. The old bird is very savage when pulled out. It makes a shrill cry, and bites hard, the sharp decurved tip of the upper mandible being driven right through a man's finger if he is not careful in handling the bird. The egg is white, and about the size of a hen's.

Another petrel, *Majaquens æquinoctialis*, which also is often to be seen cruising after the ship, but then always solitary, is called the "Cape Hen" by ordinary sailors, and "Black Night Hawk" by the whalers. It makes a hole, larger a good deal than that of the Mutton-bird, and nearly always with its mouth opening on a small pool of water, or in a very damp place. The hole is deep under the ground and very long, two yards or more. The birds seem to make their holes in certain places in company. At one place, on the shores of Greenland Harbour, I found a number of such holes, all within a small area. The bird utters a peculiar prolonged and high pitched cry, either when dug into on the nest and handled, or on going into the hole and finding its mate there.

I saw once about a dozen of these birds swimming together at Royal Sound, but usually they hawk over the sea singly, with a long sweeping flight like that of the albatross. The young are like round balls of grey down, and, as might be expected, have the nostrils much more widely open than the adults.

Further we found a Stormy Petrel (*Oceanitis sp.*). It makes a short small hole in the turf at the verge of the cliffs, and lays a white egg, with slight red speckles at one end, large in size in proportion to the bird.

A more interesting petrel is the diving Procellarid (*Pelecanoides urinatrix*), which is a petrel that has given up the active aerial habits of its allies, and has taken to diving, and has become specially modified by natural selection to suit it for this changed habit, though still a petrel in essential structure. The habits of the bird, which occurs in the Straits of Magellan, are described by Darwin in his Journal.* This bird is to be seen on the surface of the water in Royal Sound when the water is calm, in flocks of very large numbers. On two days in which excursions were made in the steam pinnace, the water was seen to be covered with these birds in flocks, extending over acres, which were black with them. The habits of the northern Little Auk are said to be closely similar to those of this bird; so close is the resemblance, that the whalers have transferred one of their familiar names for the Little Auk to the Diving Petrel. The diving petrels dive with extreme rapidity, and when frightened, get up and flutter along close to the water, and drop and dive again. It is a curious sight to see a whole flock thus taking flight. The birds make holes in the ground like the Prions, and lay an egg white with a few red specks at one end. They breed in enormous quantities on the islands in Royal Sound. They are readily attracted by a light, and some were caught on board through coming to the ship's lights.

On one of the digging excursions I found a nest of the Sheath-bill (*Chionis minor*), and subsequently found several others. The bird has a wide range, corresponding to that of the Kerguelen cabbage, occurring like it in the Prince Edward Islands, the Crozets and Heard Islands. Another species of the genus occurs in Patagonia. It resembles the Kerguelen species closely in general appearance, though differing in many essential points. A figure of it is here given in default of one

SHEATH-BILL OF FUEGIA.   CHIONIS ALBA.

of the Kerguelen bird. It might however almost stand for this latter. The birds (the " Paddy " of the sealers) are present everywhere on the coast, and from their extreme tameness and

* " Journal of Researches," p. 290.

P

inquisitive habits, are always attracting one's attention.   A pair
or two of them always forms part of any view on the coast.
The birds are pure white, about the size of a very large pigeon,
but with the appearance rather of a fowl.   They have light pink-
coloured legs, with partial webbing of the toes, small spurs on the
inner side of the wings, like the spur-winged flower, and a black
bill with a most curious curved lamina of horny matter projecting
over the nostrils.   Round the eye is a tumid pink ring bare of
feathers ; about the head are wattle-like warts.

The birds have been examined anatomically by De Blain-
ville,* who concluded that they are nearly related to the Oyster-
catchers.   The birds nest under fallen rocks along the cliffs, often
in places where the nest is difficult of access.   The nest is made
of grass and bents, and the eggs are usually two in number, and
of the shape of those of the Plovers, and of a somewhat similar
colouring, spotted dark red and brown.   They have been described and figured by Gould, and he considers the eggs to show
further alliance of the Sheath-bills to the Plovers.   I found two
nests with three eggs, but two is the most usual number.

The young are black on coming from the egg, following the
usual law with white birds, the white colouring being a lately
acquired peculiarity.   The young one has the nostrils wide
open and merely a tumidity about the posterior margin of the
nostrils and across the beak where the sheath is commencing to
grow out.

On sitting down on the rocks where there are pairs of Sheath-
bills about, one soon has them round him, uttering a harsh, half
warning, half inquisitive cry on first seeing one, and venturing
gradually nearer and nearer, standing and gazing up at the
intruder with their heads turned on one side.   The birds come
frequently within reach of a stick, and can often be knocked over
in that way, or bowled over with a big stone, as they will sit
quietly and allow half a dozen stones, as big as themselves
almost, to be thrown at them.

At length, only after being narrowly missed several times,

* " Voyage de la Bonïte," Zoologie, Tom. I, p. 107 ; Pl. Oiss. IX.
The anatomy of the Sheath-bills has been further lately made the
subject of a memoir by Dr. Kidder.   " Bull. U.S. Nat. Mus.," No. 3.

they take flight, and make off, uttering their harsh note a succession of times. If a bird be knocked over with a stick, it is usually only stunned, the Sheath-bills are very tenacious of life. If the one thus caught be tied by the leg with a string and allowed to flutter on the rocks, in front of one as one sits, the neighbouring sheath-bills will come at once to fight with it and peck it, and can be knocked over one after another. When courting one another, the birds show all the attitudes of pigeons, the male bowing his head up and down and strutting, making a sort of cooing noise.

The birds eat seaweed and shell-fish, mussels and limpets, besides acting as scavengers, as already mentioned. They carry quantities of the limpet and mussel shells up to the clefts or holes under the rocks which they frequent. They readily feed in confinement, and we had several on board the ship, running about quite at home. One of them established itself in one of the cutters for a short time, and used to take a fly round during the voyage to Heard Island and return again to the ship.

The birds, though usually to be seen running on the rocks, can fly remarkably well, and their flight is like that of a pigeon. I have seen them flying at a great height about the cliffs of Christmas Harbour.

A Tern (*Sterna virgata* ?), the "Mackerel-bird," "King-bird," or " Kinger " of sealers, nests on the ground amongst the grass, laying a single egg, just like that of other terns. When a nest is approached the old birds are very bold, and fly round the head of the intruder, uttering a sharp cry. Their young are brown and remarkably like a thrush at first glance were it not for the web feet. When I saw one for the first time I thought a Land-bird had been found in Kerguelen, but such certainly does not exist except the Sheath-bill, if it can be considered as such. It is, however, worthy of note here, that in Antipodes Island, which lies south-east of New Zealand and a little nearer the South Pole than Kerguelen's Land, parroquets are abundant, although the island is covered with tussock,* and without trees.

* "Notes on the Geology of the Outlying Islands of New Zealand. Reported by Dr. Hector, F.R.S." Trans. N. Zealand Institute, Vol. II, 1869, p. 176.

The Gull (*Larus Dominicanus*) nests also on the open ground amongst grass tufts, and the birds breed in considerable flocks together, choosing often some dry place on the lower slopes of a hill-side. I saw two such places where there were a few nests with young and remains of many more. No regular nest is made. The young are brown-coloured. The old birds make a great deal of noise when the young are carried off, but make no attempt to protect them. The brown colour of the young is closely like that of the dead grass in which they lie, and under which they hide on approach of danger. The colour is protective to them; they are, certainly, very difficult to see amongst the grass.

A species of Cormorant (*Phalacrocorax verrucosus*), which occurs at the Falkland Islands and at New Zealand, and which is almost certainly the same bird which we saw at Marion Island, is very abundant about Kerguelen. The birds are very handsome, especially the male. The chest is white, the back dark brown and black with green metallic tints upon it. At the base of the bill are large orange warty protuberances.

The birds build on ledges of the cliffs, or on the higher part of steep declivities leading directly down into the sea. They are especially fond of the horizontal grooves and ledges in the cliffs formed where the red earth bands weather out beneath the harder overlying basalt. They are gregarious in their nesting, and in places small islands or projecting headlands, are stained yellow-white with their droppings, so as to be conspicuous from a distance at sea.

The birds make a compact neat round nest, raised about a foot from the ground, and composed of mud and lined with grass.

They lay either two or three eggs, pale blue in colour, and covered with a chalky substance, as are all cormorants' eggs. The young are ugly beasts, covered with intensely black down. When there are three in the nest nearly full-fledged they form an absurd sight, since the nest is then not big enough to hold more than one properly, so the greater part of the bodies of the three young projects out, and then, to crown the absurdity, the mother comes and sits on the top of these three young as big as herself.

An idea of the relations of the various birds to one another in the struggle for existence will be gained from the following incident: I saw a cormorant rise to the surface of the water, and lifting its head, make desperate efforts to gorge a small fish which it had caught, evidently knowing its danger, and in a fearful hurry to get it down. Before it could swallow its prey, down came a gull, snatched the fish after a slight struggle and carried it off to the rocks on the shore. Here a lot of other gulls immediately began to assert their right to a share, when down swooped a Skua from aloft, right on to the heap of gulls, seized the fish and swallowed it at once.

The shag ought to learn to swallow under water, and the gull to devour its prey at once in the air. The Skua is merely a gull which has developed itself by fighting for morsels.

We fell in with three American whaling schooners at Kerguelen. They work Heard Island for Sea Elephants and Kerguelen for whales more especially. They get their principal hands at Fogo in the Cape Verdes on the way out; the Portuguese there being very willing to embark, even for a South Sea whaling cruise, in order to escape the military conscription. The schooners, which belong to two different owners, are tended by a barque, which brings out provisions and takes home oil and skins.

A difficulty would arise from a whale when struck running through the thick beds of kelp (*Macrocystis*) which everywhere form tangled barriers at a certain distance from shore. This is got over by having large very sharp knives ready, which are held close beside the line as the boat scuds through the water, dragged by the whale, and cut a clean passage in the weed.

The whales are killed by means of a bomb, a cylindrical iron tube full of powder provided with a fuse and pointed at one end; at the other, provided with feathers like an arrow. The whole is not unlike a large crossbow bolt. The feathers are made of vulcanized indiarubber, and when the bolt is rammed into the gun from which it is fired, are wrapped round the end of the shaft. As soon as the bolt leaves the muzzle they expand, and prevent the bombs wobbling or capsizing.

The invention is extremely ingenious. The bomb is fired

from a heavy gun from the shoulder, and is good up to about fifteen paces. It is fired into the whale just behind the flipper. It goes in, and after a while makes a loud explosion, often killing the beast almost at once. Four kinds of whales are common about Kerguelen's Island, but only one, the Southern Whalebone Whale, is regularly hunted. A bomb is fired into the other kinds, if there is a chance of doing so from the ship, and if the beast hit appears maimed, it is then tackled on to with the harpoons. Similar bombs are now regularly used in the North.

I was sorry to leave Kerguelen's Land, for I enjoyed the place thoroughly. We had wonderfully good weather, and sometimes the sun was extremely hot. The sunrises and sunsets were often most gorgeous, and the view in evening or early morning up Royal Sound, with its wide expanse of sea dotted all over with rocky islands, like some large inland lake, and with Mount Ross towering blue in the distance, and capped with snow and glaciers, is most grand and beautiful.

The climate of Kerguelen's Land is, as is that of all the neighbouring islands, remarkably equable. It is never very warm, never very cold. In the middle of winter, during Ross's stay there, the thermometer rarely fell below freezing point, and the snow never lay on the lower land more than two or three days. The whalers told us that it was very rarely that ice formed which would bear; and Sir J. D. Hooker speaks of breaking ice on the Christmas Harbour Lake only two inches thick, and taking from under it *Limosella* in full flower.

During our stay, the highest reading of the thermometer was 59° F., and the lowest 39°·5 F.: the mean about 43° or 44°: this in the middle of summer, or rather slightly past the middle. The bane of the place consists in the constantly occurring sudden storms of wind, one of which made us drag our anchor at Betsy Cove, and might easily have sent the ship against the rocks, and two of which kept us tediously beating about off the land on two occasions, when we were making from one point to another.

For a complete list of the birds of Kerguelen's Land, see R. Bowdler Sharpe, F.L.S., F.R.S. "Trans. of Venus Expedition, Zoology of Kerguelen's Land. Birds." From this paper the names of birds given above are taken.

For the Crustacea, see E. J. Meirs, F.L.S., F.Z.S. Trans. Venus Expedition. Ibid.

For the Terrestrial Annelida, see E. Ray Lankester, F.R.S. Ibid.

See "Further contributions to the Natural History of Kerguelen's Island," by J. H. Kidder, M.D. "Bull. U.S. National Mus.," No. 3, 1876, II.

See also, for an account of the island, "Narrative of the Wreck of the 'Favourite' on the Island of Desolation; detailing the adventures, sufferings and privations of John Munn; an Historical Account of the Island and its Whale and Sea fisheries." Edited by W. B. Clarke, M.D. London, 1850.

CHAPTER IX.

HEARD ISLAND.

Diatoms on the Sea Surface. Macdonald Island. Whisky Bay, Heard Island. Coast-line composed of Glaciers. Structure of the Glaciers. Terminal and Lateral Moraines. Glacier Stream. Rocks Cut by Natural Sand Blast. Lava Flow and Denuded Crater. Scanty Vegetation. Range in Elevation of Arctic and Southern Plants Compared. Mode of Hunting Sea Elephants. Habits of these Animals. Sealers Inhabiting Heard Island. Birds of the Island.

February 2nd, 1874.—We sailed from Christmas Harbour, whither we had gone at the termination of our survey to erect a cairn with instructions for the Transit of Venus Expedition, on February 2nd, and made for the Macdonald Group, which lies about 240 miles to the south-east of Kerguelen's Land. The channel between the two groups is extremely variable in depth, bottom being found at times in less than 100 fathoms, and at others no bottom being obtained in from 220 to 425 fathoms.

The sea surface was full of Diatoms, which filled the towing net in large masses. These masses were found by Mr. O'Meara to be composed mainly of various species of *Chætoceros*, with spines of extraordinary length, aggregated in small masses of a jelly-like substance. Occurring with these species of *Chætoceros*, were representatives of five other genera of Diatoms, three of which were of new species.*

Heard Island, February 6th, 1874.—On February 6th, after beating about for several days in fog, and lying becalmed during one day, we sighted the northernmost island of the Macdonald Group. It was alternately brightened up by sunshine, and

---

* Rev. E. O'Meara, M.A., "On the Diatomaceous gatherings made at Kerguelen's Land, by H. N. Moseley." Linn. Journ., Botany, Vol. XV, pp. 56, 57.

hidden in the drifting scud and mist. It consists of a small main rocky mass, and two outliers with a very irregular outline and weather-beaten appearance.

The main mass is Macdonald Island, and gives the name to the group. It is bounded on all sides by cliffs, which are high towards the eastward, but lower towards the westward. There was no snow on the island; on one stretch of sloping flat land, a covering of vegetation could be made out no doubt similar to that of Heard Island. One of the outliers is in the form of a pinnacle, projecting straight up from the sea.

We anchored at Heard Island, in Corinthian or Whisky Bay, as it is named by the sealers, in the afternoon; I landed at once with Captain Nares and Mr. Buchanan. Heard Island is in about lat. 53° 10′ S., long. 73° 30′ E. It is thus in about the same latitude as the eastern entrance of the Straits of Magellan, and in a corresponding latitude in the southern hemisphere, to our city of Lincoln in the northern; it is in nearly the same longitude as Bombay. It is about twenty-five miles in extreme length, and six in extreme breadth, and has an area of about 80 square miles. The island is elongate in form, stretching in a direction about N.W. by W., and S.E. by E. The southernmost extremity turns eastward, and runs out into a long narrow promontory.

Whisky Bay is near the northernmost extremity of the island. To the south-east of the ship, as she lay in the small bay, were seen a succession of glaciers descending right down to the beach, and separated by lateral moraines from one another; six of these glaciers were visible from the anchorage, forming by their terminations the coast-line eastwards. They rose with a gentle slope, with the usual rounded undulating surface upwards towards the interior of the island, but their origin was hid in the mist and cloud; and Big Ben, the great mountain of the island, said to be 7,000 feet in height, was not seen by us at all.

One of the glaciers, that nearest to the ship, instead of abutting on the sea-shore directly with its end, as did the others, presented, towards its lower extremity its side to the action of the waves, and ending somewhat inland, formed a well-marked but scanty terminal moraine.

WHISKY BAY, HEARD ISLAND : ON THE LEFT, GLACIERS COMING DOWN TO THE SEA.

To the sea-shore this glacier presented a vertical wall of ice, resting directly upon the black volcanic sand composing the beach. In this wall was exposed a very instructive longitudinal section of the glacier mass, in which the series of curved bands produced by differential motion were most plainly marked, and visible from the distance of the anchorage.

The ice composing the wall or cliff was evidently being constantly bulged outwards by internal pressure, and masses were thus being split off to fall on the beach, and be melted, or floated off by the tide. The ice splits off along the lines of the longitudinal crevasses, and falls in slabs of the whole height of the cliff; a freshly fallen slab, a longitudinal slice of the glacier, was lying on the beach.

The fallen ice floats off with the tide. Some stones, which were dredged in 150 fathoms between Kerguelen's Land and Heard Island, were believed by Mr. Buchanan to have been recently dropped by floating ice from Heard Island. The stones in question were as yet not penetrated by the water.*

The other glaciers in sight cut the shore line at right angles, and thus had no terminal moraines, the stones brought down by them being washed away by the sea.

The glaciers showed all the familiar phenomena of those of Europe with exact similarity. There are here the same systems of crevasses, more marked in some regions than others, and dying out towards the termination of the glacier, where the surface is smooth and generally rounded. The crevasses were of the usual deep blue colour, and the ridges separating them of the usual fantastic shapes.

Above, the glaciers were covered with snow, which, as one looked higher and higher, was seen to gradually obliterate the crevasses, and assume the appearance of a névé. The extent of glacier free from snow was very small ; the region in which thawing can take place to any considerable extent being confined to a range not far above sea level.

Here and there were to be seen, on the surface of the glacier, the usual deep vertical pipe-like holes full of water. These were lined by concentric layers of ice, composed of prisms

* J. Y. Buchanan, " Proc. R. Soc." No. 170, 1876, p. 609.

disposed radially to the centres of the holes and produced by successive night frosts.

Cones of ice covered with sand, and appearing as if composed of sand alone, but astonishing one by their hard and resistant nature when struck with a stick, were also to be seen on the glacier. I have seen closely similar cones in Tyrol; and, when a tyro at alpine climbing, have jarred my hand in attempting to thrust my alpenstock into them. Here the sand was black and volcanic. Small table-stones were not uncommon upon the glacier, and, in fact, all the phenomena caused by thawing from the action of direct radiant heat were present.

The usual narrow longitudinal lines or cracks caused by the shearing of the ice in its differential motion were present, and gave evidence of the grinding together of the closely opposed surfaces forming them.

The dirt and stones on the surface of the ice were as usual more abundant towards the termination of the glacier and the moraine, but they were not so abundant as usual, and there were no large stones amongst them, nor were such to be seen in the moraine.

The harponeer of the " Emma Jane," the whaling schooner with which we fell in at Kerguelen's Land, told me that he had always wondered where the stones on the ice came from at all, and no wonder, for Big Ben is usually hidden from view, and the glaciers seem to have nothing above from which the stones might come. Most of the stones, no doubt, reach the surface and see the light only when they are approaching the bottom of the glacier.

The terminal moraine showed the usual irregular conical heaping, and marks of recent motion of the stones and earth composing it from the thawing of the ice supporting them, and a small stream running from the glacier-bed cut its way to the sea through a short arched tunnel in the ice, as so commonly occurs elsewhere. A small cascade poured out of the ice-cliff on to the seashore from an aperture about half-way up it. All the moraines showed evidence of the present shrinking of the glaciers.

The view along the shore of the successive terminations of the glaciers was very fine. I had never before seen a coast-line

composed of cliffs and headlands of ice. None of the glaciers
came actually down into the sea. The bases of their cliffs
rested on the sandy beach and were only just washed by the
waves at high water or during gales of wind.

The lateral moraines were of the usual form, with sharp
ridged crests and natural slopes on either side. They formed
lines of separation between the contiguous glaciers. They were
somewhat serpentine in course, and two of them were seen
to occur immediately above points where the glaciers on either
hand were separated by masses of rock *in situ*, which masses
showed out between the ice-cliffs on the shore and had the
ends of the moraines resting on them.

A stretch of perfectly level black sand about half a mile in
width forms the head of the bay and intervenes between the
glaciers and a promontory of rocky rising land stretching out
northwards and westwards, and forming the other side of the bay.
It was on the smooth sandy beach bounding this plain that we
landed. The surf was not heavy, but we had to drag the boat
up at once.

In this we were helped by six wild-looking sealers, who had
made their appearance on the rocks as soon as the ship entered
the bay, with their rifles in their hands, and had gazed on
us with astonishment. The boss said, as we landed, he "guessed
we were out of our reckoning." They evidently thought no one
could have come to Heard Island on purpose who was not
in the sealing business.

The sandy plain stretches back from the bay as a dreary
waste to another small curved beach at the head of another
inlet of the sea. Behind this inlet is an irregular rocky mountain mass forming the end of the island, on which are two large
glaciers very steeply inclined, and one of them terminating in a
sheer ice-fall. At its back this mountain mass is bounded
by precipices with their bases washed by the sea.

The plain is traversed by several streams of glacier water
coming from the southern glaciers. These streams are constantly changing their course, as the beach and plain are washed
about by the surf in heavy weather. At the time of our visit
the main stream stretched across the entire width of the plain

and entered the sea at the extreme western verge of the beach. We had therefore to ford it.

The stream was about 20 yards across, and knee-deep. It was intensely cold, and pained my legs worse than any glacier water I have ever waded in. The water of the stream was brown, opaque, and muddy, charged with the grindings of the glaciers. Running into the sea it formed a conspicuous brown tract, sharply defined from the blue-green water of the sea, and extending almost to the mouth of the bay.

The sandy plain seemed entirely of glacial origin; it was in places covered with glacial mud, and was yielding, and heavy to walk upon.

Mr. Buchanan observed that the isolated rocks which had been rolled down upon this plain from the heights above were cut by the natural sand-blast into forms resembling trees on a coast exposed to trade winds. The effect of every prevalent wind was shown by the facets cut by the blown sand upon the surfaces of the rocks, the largest facet in each case being that turned towards the west.*

The plain was strewed with bones of the Sea-Elephant and Sea-Leopard, those of the former being most abundant. There were remains of thousands of skeletons, and I gathered a good many tusks of old males. The bones lay in curved lines, looking like tide lines, on either side of the plain above the beaches, marking the rookeries of old times and tracks of slaughter of the sealers. Some bones occurred far up on the plain, the Elephants having in times of security made their lairs far from the water's edge. A few whales' *vertebræ* were also seen lying about.

On the opposite side of the plain from that bounded by the glacier is a stretch of low bare rock, with a peculiar smooth and rounded but irregular surface. This rock surface appears from a distance as if glaciated, but on closer examination it is seen to show very distinct ripple marks and lines of flow, and the rock-mass is evidently a comparatively recent lava flow from a small broken-down crater which stands on the shore close by.

The remains of the crater are now in the form of three

* J. Y. Buchanan, M.A., Report, "Proc. R. Soc." No. 170, 1876, p. 622.

fantastic irregularly conical masses, composed of very numerous thin layers of scoriæ, conspicuous because of their varying and strongly contrasted colours and very irregular bedding. The

BROKEN-DOWN CRATER, WHISKY BAY, WITH SNOW UPON IT.

lava flow is seen in section in the low cliffs forming the coast-line of the harbour.

The present condition of Heard Island is evidently that which obtained in Kerguelen's Land formerly. Glaciers once covered Kerguelen's Land almost entirely and dipped down into the sea. It is, however, an extraordinary fact that Heard Island, only 300 miles south of Kerguelen's Land, should thus still be in a glacial epoch, whilst in Kerguelen's Land, a very much larger tract, the glaciers should have shrunk back into the interior, and have left so much of the land surface entirely free of ice, the ice epoch being there already a thing of the past.

The great height of Big Ben, and consequent largeness of the area where snow constantly accumulates and cannot be melted, no doubt accounts to a considerable extent for the peculiar conditions in Heard Island. A similar rapid descent of the snow-line within a few degrees of latitude occurs in the Chilian Andes,[*] so great is the chilling influence of the vast southern sea.

* Grisebach, "Die Vegetation der Erde." Leipzig, 1872. 2. Bd. s. 467. Ibique citato.

Heard Island is in a corresponding latitude to Lincoln. No doubt, when England was in its last glacial epoch, Heard Island enjoyed a much milder climate, and it was possibly then that the large trees grew, the trunks of which are now fossil in Kerguelen's Land, and that the ancestors of *Lyallia* and *Pringlea* flourished.

A stretch of land on the north-west side of the plain was covered pretty thickly with green, which was on closer view seen to be composed of patches of Azorella,* growing on the summits of mud or sand hummocks, which were separated from one another by ditches or cavities, of usually bare brown mud.

Some of these Azorella patches were of considerable extent, and the plant was evidently flourishing and in full fruit. On some hummocks grew tufts of the grass *Poa Cookii*, in full flower and with the anthers fully developed; and on the sheltered banks of the hummocks the Kerguelen cabbage (*Pringlea anti-scorbutica*), grew in considerable quantity, but dwarfed in comparison with Kerguelen specimens, both in foliage and in the length of the fruiting stems. Most of it was in fruit, but some still in flower, as at Kerguelen's Land.

Around pools of water in the hollows grew a variety of a British plant, *Callitriche Verna* (sub *sp. obtusangulata*), in quantity, and it occurred also in abundance submerged; in company with a Conferva. In the same sheltered spots grew *Colobanthus Kerguelensis*, in greater abundance even than at Kerguelen's Land.

These five flowering plants,† all occurring also in Kerguelen's Land, were the only ones found in the island, and it is improbable that any others grow there. Heard Island has thus a miserably poor flora, even for the higher latitudes of the southern hemisphere. The Falkland Islands, in lat. 51° to 52° S., have 119 phanerogamic plants, and Hermit Island, far to the south of Heard Island, in lat. 56° S., has 84 phanerogams, and amongst them trees of which this island is the southern limit.

An Antarctic flora can in reality hardly be said to exist, since there are absolutely no phanerogamic plants within the Antarctic circle, and on Possession Island, lying off the coast of

---

* See p. 166.

† Prof. Oliver, F.R.S., "Journal of Linn. Soc.," Vol. XIV, p. 389.

Victoria Land, in about lat. 72° S., within the Circle, Sir Joseph Hooker found* only 18 cryptogams, mosses, lichens, and algæ, no trace of phanerogams. Yet in Saltdalen, in Norway, north of the Arctic Circle, there are fine timber forests and thriving farms, yielding abundant crops of hay and barley. Melville Island, in lat. 74° 75′ N., 500 miles north of the Arctic Circle, has a vegetation of 67 flowering plants.

Sir J. D. Hooker, in his latest memoir on the botany of Kerguelen's Land, says: "The three small archipelagos of Kerguelen Island (including the Heard Islands), Marion and Prince Edward's Islands, and the Crozets, are individually and collectively the most barren tracts on the globe, whether in their own latitude or in a higher one, except such as lie within the Antarctic Circle itself; for no land, even within the North Polar area, presents so impoverished a vegetation."†

About the sides of the hummocks already described grew scantily four species of mosses, one of which proved to be new and peculiar to the island.

The majority of the land surface of Heard Island, free from ice, besides the green tract described, is entirely devoid of vegetation. Only on the talus slopes of the hills on their sheltered sides, are seen scattered in a very few places scanty patches of green. These composed lower down mainly of Azorella stretch up the slopes, and terminate at an elevation of a few hundred feet in bright yellow patches, consisting entirely of mosses, just as at Marion Island, on the higher slopes. I searched in vain for lichens of any kind.

There seems to be a very great difference with regard to the vertical range of plants in these southern islands, and in the Arctic regions. In Marion Island, I estimated the absolute limit of vegetation at an altitude of about 2,000 feet; in Kerguelen's Land, the limit seems to lie at about 1,500 feet or lower; plants of any kind are there already scarce at 1,000 feet above sea level. In Heard Island vegetation seems to cease at 300 or 400 feet altitude. Yet in East Greenland, the same

* " Flora Antarctica," p. 216.
† " Observations on the Botany of Kerguelen Island by Sir J. D. Hooker, P.R.S.," &c. Transit of Venus Expedition, Botany, pp. 2, 3.

plants are found to range from sea level up to 3,000 feet, and there is no real limit of altitude; even at 7,000 feet elevation a thick cushion of moss, several inches in length, was found by the German North Polar Expedition covering the ground.*

This remarkable condition in the Arctic regions is mainly accounted for by Dr. Pansch, by the fact that, with the sun always near the horizon in high latitudes, the hill-slopes receive its rays nearly vertically on their surfaces, and thus receive more radiant heat, even than the flat land below them. There is little cooling at night, the clouds and mist preventing radiation.

In Kerguelen's Land, of course, in its low latitude, the inclined surfaces do not profit so much by their inclination. There, as in the high north, the mosses and lichens are the highest plants in range. In the successive groups of islands, Marion, Kerguelen, Heard, they come lower and lower down the mountain-slopes, and in Possession Island, south of the Antarctic Circle, the few flowering plants remaining below them at Heard Island have disappeared, and they are left growing alone.

In all the southern islands the density of the phanerogamic vegetation, the extent of development of the individual plants, and the number of species present, decrease directly with the height. The facts show how much more the constant absence of warmth, and a continuous moderately low temperature, is inimical to plant development, than is periodical cold of the severest kind.

The condition of the vegetation in various localities in East Greenland depends more on the distance of these from the ice barrier, than on their position more or less north or south. The vegetation becomes more abundant as progress is made inland, away from the ice-bound coast. Exactly the opposite seems to hold in Kerguelen's Land, where the chief source of warmth, though at the same time the constant cause of the equalization

---

* " Die zweite Deutsche Nord-Polarfahrt in den Jahren 1869 und 1870." 2. Bd. Wissenschaftliche Ergebnisse, Leipzig. F. A. Brockhaus. " Klima und Pflanzenleben auf Ostgrönland," von Adolf Pansch in Kiel.

of temperature, is the sea: and where the accumulated snow inland, and its attendant mists, render the soil there barren.

In East Greenland all phanerogamic water plants are absent, because of the long freezing of the water in winter; in the southern islands there is a Limosella, and a large number of the other Phanerogams seem to take on a special aquatic habit.

To return to Heard Island. At Corinthian Bay large masses of seaweeds were banked up on the sandy shore. I collected eight species, which have been described by Prof. Dickie.* Amongst them were two new species, two which occur at Kerguelen's Land, whilst the remainder occur in Fuegia. The main mass appeared considerably different from the masses of algæ found on the Kerguelen shore. *Durvillœa utilis* grew attached to the rocks under the cliffs, but the kelp (*Macrocystis pirifera*) does not grow at all about this group of islands, according to the sealers, which is a remarkable fact, in consideration of its great abundance at Kerguelen's Land.

The sealers said that the climate of Heard Island was far more rigorous than that of Kerguelen's Land. In winter the whole of the ground is frozen, and the streams are stopped, so that snow has to be melted in order to obtain water. In December, at Midsummer, there is plenty of sunshiny weather, and Big Ben is often to be seen. It is possible to land in whale boats, on the average of the whole year only once in three days, so surf-beaten is the shore, so stormy the weather.

We saw six sealers; two were Americans, and two Portuguese from the Cape Verde Islands. They were left on the island by the whaling vessels which we met with at Kerguelen's Land, their duty being to hunt Sea-Elephants. The men engage to remain three years on the island, and see the whale ships only for a short time in the spring of each year.

On the more exposed side of the island there is an extensive beach, called Long Beach. This is covered over with thousands of Sea-Elephants in the breeding season, but it is only accessible by land, and then only by crossing two glaciers or " ice-bergs " as the sealers call them. No boat can live to land on this shore, consequently men are stationed on the beach, and live there in

* " Journal of the Linn. Soc.," Vol. XV, p. 73.

huts; and their duty is constantly to drive the Elephants from this beach into the sea, which they do with whips made of the hide of the Elephants themselves. The beasts thus ousted swim off, and often "haul up," as the term is, upon the accessible beaches elsewhere, and there they are killed and their blubber is taken to be boiled down.

In very stormy weather, when they are driven into the sea, they are forced to betake themselves to the sheltered side of the island; hence the men find that stormy weather pays them best. Two or three old males, termed "beach-masters," hold a beach to themselves and cover it with cows, but allow no other males to haul up. The males fight furiously, and one man told me that he had seen an old male take up a younger one in his teeth and throw him over, lifting him in the air. The males show fight when whipped, and are with great difficulty driven into the sea. They are sometimes treated with horrible brutality.

The females give birth to their young soon after their arrival. The new-born young are almost black, unlike the adults, which are of a light slate brown, and the young of the northern Bladdernose, which are white. They are suckled by the female for some time, and then left to themselves lying on the beach, where they seem to grow fat without further feeding. They are always allowed by the sealers thus to lie, in order to make more oil.

This account was corroborated by all the sealers I met with. I do not understand it; probably the cows visit their young from time to time unobserved. I believe similar stories are told of the fattening on nothing of the young of northern seals.

Peron says that both parent Elephant seals stay with the young without feeding at all, until the young are six or seven weeks old, and that then the old ones conduct the young to the water and keep them carefully in their company. The rapid increase in weight is in accordance with Peron's account.

Charles Goodridge gives a somewhat different account, namely, that after the females leave the young, the old males and young proceed inland, as far as two miles sometimes, and stop without food for more than a month, and during this time lose fat. The male elephants come on shore on the Crozets for the breeding season at about the middle of August, the females a little later.

There were said to be forty men in all upon Heard Island. Men occasionally get lost upon the glaciers. Sometimes a man gets desperate from being in so miserable a place, and one of the crew of a whaler that we met at Kerguelen's Land said, after he had had some rum, that occasionally men had to be shot; a statement which may be true or false, but which expresses at all events the feelings of the men on the matter.

The men that we saw seemed contented with their lot. The "boss" said, in answer to our inquiries, that he had only one Fur-Seal skin, which he would sell if he was paid for it, but he guessed he'd sell it anyhow when he got back to the States. He had been engaged in sealing about the island since 1854, having landed with the first sealing party which visited the island. For his present engagement his time was up next year, but he guessed he'd stay two years more. He'd make 500 dollars or so before he went home, but would probably spend half of that when he touched at the Cape of Good Hope on the way.

The men had good clothing, and did not look particularly dirty. They lived in wooden huts, or rather under roofs built over holes in the ground, thus reverting to the condition of the ancient British. Around their huts were oil casks and tanks, and a hand-barrow for wheeling blubber about. There were also casks marked Molasses, Flour, and Coal.

The men said they had as much biscuit as they wanted, and also beans and pork, and a little molasses and flour. Their principal food was penguins (*Eudyptes chrysolophus*), and they used penguin skins with the fat on for fuel. Captain Sir G. S. Nares saw five such skins piled on the fire one after the other in one of the huts.

The bay in which we anchored was thronged with Cape Pigeons (*Daption Capensis*) and Prions in astonishing numbers. The Prions were on the wing in the usual manner, in dense flocks; the Pigeons, called sometimes by the sealers "Egli Bird," were mostly feeding on the water at the mouth of the glacier stream. They were breeding in holes in the low basaltic cliffs.

On the same cliffs was a rookery of Shags. They appeared

much whiter than the Kerguelen birds, a broad band of white passing round the body, under the wings and across the back. They were probably of the same species (*Phalacrocorax verrucosus*) which is described as developing in New Zealand a broad white band at the close of the breeding season.* The sealers had remarked that the Heard Island Shag was whiter than the Kerguelen one. The season at Heard Island may have been more advanced, or a change of plumage may take place earlier; or from the sealers' remark it would appear that the Heard Island birds differ in their amount of development of white from the Kerguelen ones.†

On a steep talus slope leading down from the broken-down crater already described, to the sea, was a large penguin rookery, from which the sealers drew their supplies. A tern, the same as one of the Kerguelen ones, was nesting on the terminal moraine of the glacier at the head of the harbour. The sealers call it "King-bird" or "Kinger." I saw brooded eggs. The gull of Kerguelen's Land (*Larus Dominicanus*) was very abundant. It was curious for the first time to see gulls perched upon a glacier. The only other birds which I saw were the Skua (*Stercorarius Antarcticus*) and the Giant Petrel (*Ossifraga gigantea*), and a Stormy Petrel (*Oceanitis sp.*), which was very abundant. The "Sheath-bill" (*Chionis*) was said by the sealers to be common in the island; I saw one only.

The only insects which I saw were the large apterous fly of Kerguelen's Land, which shelters itself, as there, in the heart of the wild cabbage, and a single dead specimen of a small beetle, found amongst the Azorella, which unfortunately I lost.

I had only three hours' time on shore. I was busy hunting for insects when I saw the Captain signalling for a return, and picking up the biggest Sea-Elephant skull which I could find, and knocking a few tusks out of some others, to keep as mementos of this dismal spot, I made the best of my way across

* "Trans. N. Zealand Inst.," Vol. V., p. 224.

† Messrs. Sclater and Salvin separate Phalacrocorax imperialis from P. verrucosus, because of the development in it, and not in the latter, of white on the back. It is unfortunate that no specimens could be got in Heard Island. "Proc. Zool. Soc.," 1878, p. 650.

the muddy and yielding plain, and through the glacier stream, although the skull was almost more than I could carry, in addition to rock specimens and a big vasculum. We got off only just in time, for a considerable sea was running by the time that we reached the ship.

We were to have landed again on the following morning; but the wind shifted, and there was a thick fall of snow, covering the deck to the depth of two inches, and rendering the shore of an uniform white, excepting where a few black precipitous rocks showed out here and there in relief. The moraines were scarcely visible, and we realized how fortunate we had been in having hit upon so fine a day for landing on the island.

We got under way at about 5.30 A.M. As we left the bay we saw, even at this early hour, one of the wretched Portuguese starting off to walk the beaches in search of his prey, the miserable Elephants.

CHAPTER X.

AMONGST THE SOUTHERN ICE.

First Iceberg Sighted. Typical Forms of Southern Bergs. Preservation of Equilibrium. Wash Lines. Caverns. Bi-tabular Bergs, How Formed. Weathering of Bergs. Stratification of Ice in Bergs. Cleavage. Scarcity of Rocks on Bergs. Discoloured Bands in the Ice. Rev. Canon Moseley on the Motion of Glaciers. Colouring of Bergs. Blue Berg. Surf on the Coasts of Bergs. Scenic effects of Icebergs. Appearance of the Pack Ice. Discolouration of Ice by Diatoms. Gales of Wind amongst the Icebergs. Snowbow. Whales Blowing. Grampuses. Birds amongst the Ice. Antarctic Climate in Summer.

Amongst the Southern Ice, February 8th to March 4th, 1874.—From Heard Island we ran nearly due south for six days,

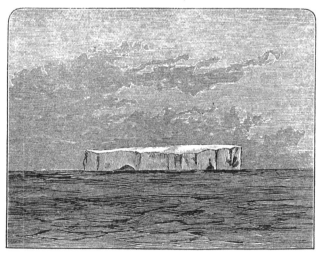

FIRST ICEBERG, SIGHTED FEBRUARY 10TH.
(From a sketch by Lieut. H. Swire, R.N.)

approaching the Antarctic Circle at an average rate of about 115 miles a day. The first iceberg was sighted on February 10th, in a latitude nearly corresponding to that of the Shetland Islands and Christiania in Norway, in the northern hemisphere.

The temperature gradually fell as we went southwards, and on February 9th went down for the first time to just below freezing point in a snow squall.

At first, all the icebergs seen were numbered each day, and their positions noted down; but when we came to have 40 in sight at once this plan was abandoned, and we subsequently had more than a hundred in sight on several occasions.

The typical form of the Antarctic iceberg, as seen above water, and apparently the form which it always has when first set free on its wanderings, is very simple. The top is a nearly flat expanse of snow, and this is bounded all around by perpendicular cliffs. The boundary lines of the expanse are no doubt always in the first instance nearly straight lines, since they must be produced by the splitting off of the berg from the parent mass, and the previous splitting of similar bergs from its own outer border when still attached.

A considerable number of the undecayed bergs seen by us were almost rectangular in outline. Some few were irregularly oval, and the weathered ones of course of all possible irregular outlines.

Since ice requires about nine times its volume to be immersed in order to float it above sea water, the portion of an iceberg which shows above water is a very small proportionate part of the mass. Mr. Buchanan made an accurate estimate

DIAGRAM TO SHOW THE PROPORTION OF AN ICEBERG IMMERSED, AND ABOVE WATER.

of the specific gravity of samples of the berg ice, and calculation of amount of immersion of icebergs. The proportionate depth of a berg below water will of course depend on the form and on the relative density of the upper and lower strata of the mass. Usually, no doubt, the mass below water is far less than nine times the vertical depth of the height of the part above water, from two considerations. Firstly, the sides of the berg are not perpendicular, but long ledges run out from the base of the cliffs below water, the immersed part being thus much larger in figure than the exposed ; and, secondly, the exposed part is of lighter, less compact ice, and often further lightened by excavation of caves, and presence of crevasses.

So large a proportion of the bergs being required to be immersed in order that the bergs broken off from the parent ice masses should float in stable equilibrium, with their surfaces originally uppermost maintained still in that position, it is necessary that the pieces thus breaking off, supposing their upper surfaces to be square, should be at least as wide as they are thick. If this were not the case, if the density of the ice masses were uniform, the bergs would necessarily topple immediately they broke free, and this fact would be shown by their strati- fication being vertical to their plane of flotation. This, however, seems never, as far as I could judge from the bergs I saw, to occur. Tilting only takes place after bergs have been long weathered. The bergs seem nearly always to be of large area in proportion to their thickness, and to maintain their original balance for very long periods. No doubt the much greater density of the ice composing the lower portions of the bergs tends to keep them in their original position.

The waves, partly no doubt because of the water at the very surface being warmed by the sun, and partly no doubt by heat resulting from their motion, cut a wash-line all round the bergs, which appears as a concave groove-like channel with a polished inner surface, just at the water-level.

When bergs rise to a higher level, or tilt, these wash-lines remain marked on the bergs, as straight polished streaks, visible from a great distance (coloured plate, fig. 5), giving evidence of the former lines of flotation of the bergs. Sometimes, several ancient wash-lines are visible on one berg, and where the cliff surfaces on which they are scored are protected at their base from the waves by secondary cliffs or projections, they may, remain intact for very long periods.

The wash-lines being hollowed out at the bases of the cliffs, these latter soon overhang, and large masses split off along the lines of joint and cleavage, and fall. The masses evidently split off tolerably evenly from the whole height of the cliffs, for these are nearly always, when thus still water-worn at their bases, perpendicular, and on our firing a shot at a berg cliff, the ice split off in this manner from the whole height of the cliff.

When there are crevasses in the ice at the level of the wash-line leading into the ice from it, the wash of the waves hollows out caverns which resemble in general form caves cut in the same manner by waves on coast-lines, and have their mouths wider at the levels of the wash-lines.

The presence of caves is a proof that a berg has floated at the level of the wash-line, along which they lie for a long period. The remains of the upper part of the crevasse which has assisted in the development of a cave, is often to be seen stretching up from its roof. Often by change of line of flotation of a berg, a line of caves is carried up far above sea level, and three or four caves disposed along an old wash-line are thus often to be seen on the surface of a berg, the line being sometimes horizontal, sometimes tilted. In a berg which has undergone extreme denudation, or on a narrow spur of a young berg, a cave may be excavated right through the berg and give rise to a natural arch. A further degeneration of the arch gives rise to an isolated pinnacle.

The base of the berg under water beneath the wash-line being supported by the water, does not split off at once like the cliff above when cut into. Hence the waves constantly deepening the wash-line as the cliffs fall, and eating their way into the berg at the water-line, a platform of ice is left behind under water, projecting at the base of the cliffs above it. After a time the part of the berg above water losing weight, the berg rises, and this platform is raised above water, and the berg thus becomes two-storied or bi-tabular.

A fresh wash-line is cut below the margin of the platform now raised, and low perpendicular cliffs are formed round it. A third platform may be formed in the same manner and raised, and the berg may become three-storied.

At the base of the older cliffs in each case, the old wash-line is usually to be seen where the cliffs are joined by the platform succeeding them, but in some instances it is obscured by the subsequent formation of a débris slope from the falling of the cliff; for the cliff, as on land, when no longer cut into by the waves at its base, tends to degenerate into a slope of natural inclination.

The resemblance in the weathering of a berg by the action of waves to that undergone by a rocky coast under the same circumstances is complete. Caves, cliffs, pinnacle-like outliers, and a shore platform at the base of the cliffs, are formed in a closely similar manner in each case.

In order that a horizontal platform of any wide extent should be formed beneath the water, it is necessary that the berg should float at almost exactly the same level for a very long period. I do not properly understand how this occurs. Each time that a

BI-TABULAR ICEBERG.
At the base of the upper cliff is seen the old wash line.   (From a sketch by the Author.)

mass of ice falls from the undermined cliff in order that the equilibrium should be maintained, it is necessary that nine times that bulk of ice should be removed from the base.

No doubt portions of the platforms below water are constantly being split off by the upward pressure and floating to the surface as "calves." The formation of a large platform under water must, however, depend on such a "calving" not taking place, unless on sides of the berg other than that on which the platform is formed. Nevertheless, by some means or other, either by melting or calving, a very uniform wasting of the berg below water must take place in order to form a platform. It cannot be supposed that the amount of snow which falls on the berg when set free can be sufficient to balance the loss by the action of the sea.

There must be a reason why the bergs which thus become

two-storied have their lower story commonly, as in the berg figured here in the text, only at one of their ends. Probably a certain amount of lower platform existed all round this berg when it first rose, but this was cut away on all the sides where it was narrow, by being undermined by the waves. The line of the main upper cliff was thus soon reached on these sides, and this cliff was then itself further undermined, so that, as shown in the sketch, the old wash-line was obliterated, and remained only at the base of that cliff which was protected by the still remaining secondary platform.

The greater undermining of bergs at one side may, no doubt, be due to their taking up, from the shape of their parts exposed above water and the relation of these parts in position to the form of the parts below water, a particular direction with regard to the wind, and maintaining this so that one particular side is usually the windward one, and therefore most battered by the waves.

It seems far more difficult to explain how it occurs that bergs suddenly rise to a considerable height further out of water than that at which they have floated before. Such a sudden rise must necessarily be supposed in order to account for the two-storied form.

In order that, in the case of the berg figured for instance, a rise should occur from the height of the old wash-line to the present water-line, a mass of the berg above water must have been suddenly removed, equal in volume to the whole part of the berg above water lying below the level of the top of the lower story.

It seems almost incredible that such a mass should break off and fall away suddenly. A splitting of a berg in two can be readily understood, but the mass in this case must come entirely from the part of the berg above water. It cannot have split off at an angle, for the walls of the berg in question were perpendicular cliffs. The berg certainly had never toppled.

A different explanation possible is, that nine times the volume of ice above referred to, was suddenly added to the part of the berg below water by its passing into cold water or a change of season. It may be that the raised story represents the effects of growth of the base of a berg during one winter when it probably still lay far south. The surface water would be colder then, and the cliffs not being so much, or hardly at all undermined, time would be allowed for the rising without destruction of the platform, and thus the process need not be so sudden.

At first sight it seemed to me easy enough that the berg should rise suddenly by the falling of part of its mass, but on considering the matter with a plan showing the vast proportion of its bulk required to be thus removed, I found the question more difficult.

The height of the main cliffs of the bi-tabular berg figured was estimated by Captain Tizard at about 200 feet, and that of the lower cliffs at 60 feet. We saw some distant bergs which were possibly 300 or 400 feet in height and three or four miles in length. A berg 200 feet in height would have a base extending to a depth of 300 fathoms or so, according to its form, and this base will be thawed at different rates at successive depths, according to the distribution of temperature in the water at the various depths. The shapes of the bergs below water must thus follow curves corresponding to those used by physicists to express successive deep-sea temperatures graphically.

A very large proportion of the bergs seen by us were as thus described, flat topped and maintained their original balance. Very many were bounded by a single range of cliffs washed by the waves all round. In some these ranges were evidently old and very much indented. These are simple bergs (see the coloured plate, fig. 4.)

Many were highly complex, combining two stories, lines of caves, talus slopes, and evidences of having tilted to a certain angle from the original line of flotation once or twice; some were excessively worn and weathered, having apparently been long in warmer regions, and were pinnacled and broken up by deep gullies or channels bounded often by rounded ridges projecting at their mouths on either side.

One much weathered pinnacled berg was passed which had its entire surface shining and polished as if it had recently toppled, and no fresh snow had fallen since this had occurred.

We saw several with the parts which had been below water partially exposed by tilting. The surfaces of these were always polished and smooth. We saw no berg tilt or turn over during our voyage. One we saw was divided into three separate columnar masses as far as the part above water was concerned. No connection of the columns was visible.

The platforms under water at the bases of the bergs often run out into spurs and irregular projections, and these may be dangerous to ships going too near. Soundings were taken on one of these platforms and gave seven fathoms at some distance from the berg and three and a half nearer in.

Nearly all the flat-topped bergs showed numerous crevasses in their cliffs near their summits, and these were always widest towards the summits, and were irregularly perpendicular in general direction.

The flat tops of the bergs had usually rather uneven surfaces, being covered with small hillocks, apparently formed by drifting of snow, or showing irregularities where they covered over the mouths of crevasses. The surfaces in fact, looked just like those of the "firn" or "névé," the cracked snow-fields at the heads of European glaciers, and appeared as if they would be equally dangerous to traverse, except with a party roped together. The second stories of bergs were always covered with snow, which had fallen on them after their emergence.

The stratified structure of the bergs is best seen in the case of flat-topped rectangular bergs, where an opportunity is afforded of examining at a corner two vertical cliff faces meeting one another at a right angle; we had several such opportunities. The entire mass shows a well-marked stratification, being com-

RECTANGULAR BERG.
Viewed at one of its corners.

posed of alternate layers of white opaque-looking, and blue, more compact and transparent ice. Staff-Surgeon E. L. Moss,

R.N., M.D., of the late Arctic Expedition, describes a similar stratification as occurring in Arctic ice. He had opportunities of examining the ice closely at leisure, and describes each stratum as consisting of an upper white part merging into a lower blue part, the colour depending on the greater or less number and size of the air-cells in the ice.*

Towards the lower part of the cliffs, the strata are seen to be extremely fine and closely pressed, whilst they are thicker with the blue lines wider apart, in proportion as they are traced towards the summits of the cliffs. In the lower regions of the cliffs, the strata are remarkably even and horizontal, whilst towards the summit, where not subjected to pressure, slight curvings are to be seen in them corresponding with the inequalities of the surface and drifting of the snow.

In one berg there was in the strata at one spot, somewhat the appearance of complex bedding, like that shown in Æolian calcareous sand formations, such as those of Bermuda.† The strata were often curved in places, but always in their main line of run, horizontal, i.e., parallel to the original flat top of the berg.

The strata in the cliff at the level of the wash-line of a rectangular berg 80 feet in height, were so thin and closely packed, that they looked almost like the leaves of a huge book at a distance, for by the lap of the waves the softer layers had been to some extent dissolved out from between the harder.

In one berg where the face of the cliff was very flat and seen quite closely with a powerful glass, the fine blue bands were seen to be grouped, the groups being separated by bands in which no lines were visible, or where these were obscured by the ice fracturing with a rougher surface, not with a perfectly even and polished one, as existed where the blue bands showed out.

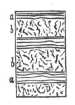

STRUCTURE OF ICE.
a a Blue bands, b b
Layers without striæ.

The cliff surfaces, where freshly fractured, show an irregular jointing and cleavage of the entire mass, very like that shown in a cliff of compact limestone. In

---

* "Observations on Arctic Sea Water and Ice." Proc. Roy. Soc., No. 189, 1878, p. 547.

† See p. 20.

one or two bergs I noticed a fine cleavage lamination like that
of slate or shale, the laminæ being pa-
rallel to the face of the cliff, and breaking
up at their edges with zigzag fracture,
almost as in diamond cleavage of slate;
this condition may have been produced
by peculiar exertion of pressure in this
particular berg.

FRACTURE OF ICE CLIFF.

When the lower cliff of the two storied
berg, described and figured in the text, had a shot fired into it,
large masses of ice fell, raising a considerable swell in the sea.
The pieces of the cliff split off in flat masses parallel with the
face of the cliff, just as I noticed to be the case in the splitting
of the glacier cliffs at Heard Island, and did not tumble forward
but slid down the face of the cliff, keeping their upper edges,
parts of the old plateau surface, horizontal.

The ice floated round the ship in some quantity; it was
opaque and white-looking, somewhat like white porcelain, and
the shattered fragments had remarkably sharp angular edges,
showing that the ice was very hard and compact, far more so
than its appearance in mass would lead one to suppose, since it
looks at a distance as if it were hardly consolidated, but merely
closely pressed snow. Its manner of cleavage only gives
evidence at a distance of its very compact nature.

Many of the floating fragments were traversed by parallel
veins of transparent ice, which were those which, when seen on
a cliff surface, look blue. A shot fired at the top of the higher
cliff produced no effect, the ball apparently going in without
splitting off any ice at all.

The greater approximation of the strata towards the base of
the bergs is no doubt due to the increasingly greater pressure
sustained by them. The blue lines seem to represent successive
slight surface thawings of superimposed falls of snow. In these
lines of clear transparent ice, a complete fusion of the snow
particles has taken place. The opaque white ice between them
though, as appears from its fracture, very compact, is less so than
these bands, as shown by its being melted sooner.*

* See preceding page.

R

There can hardly be a doubt that the ice must be of increasing density from its summit downwards.

Several small bergs were passed, which showed hardly any blue stratification in their cliffs; the top surfaces of these showed rounded conical hillocks, and a general appearance of formation by wind drifting of the snow. What few bands were present, were conformable in curve with the irregular surface. It appeared as if the denser mass were here all below water, and not large enough to float more than the lighter, more friable and recent top deposit above the water.*

Antarctic icebergs have been met with by merchant vessels in higher latitudes, varying in length from one to seven, or even ten miles in length. In 1854, a vast body of ice was passed and reported by twenty-one merchant ships in lat. 44° to 40° S., long. 28° to 20° W., a latitude corresponding to that of the northern coast of Portugal. The ice mass, which was probably a group of icebergs locked together, was in the form of a hook, 60 miles long by 40 broad, enclosing a bay 40 miles in breadth; none of the ice masses composing it exceeded 300 feet in height.†

During the short time that we were amongst the icebergs we met with none that bore upon them any moraines or rocks which could with certainty be determined as such. The scarcity of such appearances has been remarked by former voyagers. Nevertheless, there are numerous instances of rocks having been seen on southern bergs.

Several observers have met with rocks on bergs. Wilkes saw many such; Ross also, and the latter, on one occasion landed a party on a berg on which there was a volcanic rock weighing many tons, and which was covered with mud and

* For a magnificent series of large photographic views of Arctic icebergs and ice scenery, see "The Arctic Regions," by William Bradford. London, Sampson Low and Marston, 1873.

† "South Atlantic Directory," p. 94. W. H. Rosser, and J. F. Murray. London, 1870. Here will be found a general account of icebergs in the South Atlantic. On same subject see J. T. Towson, "On Icebergs in the Southern Ocean." Liverpool, 1859.

stones.* Mr. Darwin has published a note on a rock seen on an Antarctic iceberg in lat. 61° S.†

Dr. Wallich‡ remarks on the similar scarcity of the appearance of stones or gravel on northern bergs. Not one in a thousand shows dirt, &c. He attributes this to the very small disturbance of their centres of gravity which icebergs undergo when floating freely. Stones and gravel may be present in most cases, but remain most frequently invisible under water in the lower parts of the bergs. We dredged up in deep water on two occasions, near the pack-ice, fragments of gneiss and slate which were certainly transported thither by ice.

On three occasions we saw discolourations of bergs. In one case there was a light yellow band on one surface of a cliff high up, possibly the result of birds' dung which had fallen on the snow when the layer was formed; it was too high up to be due to Diatoms.

On another occasion two bergs were passed at a distance, which showed conspicuous black-looking bands, apparently dirt bands. In one of the bergs there were two or three such bands, very broad, parallel to the blue bands, and separated by considerable intervals, in which the berg showed the usual stratification. In another (coloured plate, fig. 8) two black bands existed at one end of the berg and one at the other. Both were parallel in direction to the blue bands, but the stratification at the end where the two black bands were, was inclined at an angle to that of the remainder of the berg, as if a dislocation of a part of the berg had taken place. These bergs were too far distant to allow of the exact nature of the black bands being determined.

In none of the numerous bergs did I see any bending or curved vertical bands, giving evidence of a former differential motion in the mass, such as are to be seen on every land glacier. How far the absence of these characteristic lines of motion may

* Ross, "Antarctic Voyage," Vol. I, p. 173. London, J. Murray, 1847.

† C. Darwin, "Notes on a Rock seen on an Iceberg in lat. 61° S." Geog. Soc. Journ. IX, 1839, p. 528, 529. "The Voyage of the 'Eliza Scott,' Commander John Balleny." Journal of Researches, p. 251.

‡ G. C. Wallich, M.D., F.L.S., &c., "The North Atlantic Sea Bed," Pt. 1, p. 56. London, Van Voorst, 1826.

be explained by the fact, that only about the uppermost tenth of the entire height of the bergs is seen, I do not know. A berg 200 feet in height above water, when floating, must, if it were of symmetrical form and equal density throughout, have an actual height of about 2,000 feet.

A mass detached from the edge of the barrier, and then showing lines of motion might, whilst floating, receive a sufficient addition of weight by successive falls of snow to sink it entirely below water in supporting the new structure.

Moraines and large rock masses would become hidden by such snow accumulations, both towards the free margins of the continuous glaciers, and also after the bergs containing them were detached; and a berg laden with rock need not expose it to view until after long thawing or capsizing.

The accumulation of rocks and stones in the form of definite moraines is, of course, a phenomenon which can only be produced by the accompaniment of thawing or evaporation of ice in combination with its motion. If both these processes occur to very small extent in the ice of the glaciers, whose free edge forms the Great Barrier, the rocks and stones received from the overhanging cliffs inland, or supporting beds, will be distributed evenly throughout the mass, and never be concentrated at all. The crevasses seen in the upper parts of the bergs might be produced after a berg is set free by the greater expansion, through increase of temperature, of the denser ice at the base of the mass.

I may be allowed here to make a remark with regard to the movements of glaciers, a subject to which my late father, the Rev. Canon Moseley, devoted much time and research. The theory propounded by him to account for the descent of glaciers, which, as he proved most conclusively, cannot take place by means of their weight alone, was that the motion was due to the expansion and contraction of the mass. A heavy body lying on a slope, inclined ever so little, and subject to expansion and contraction, must necessarily crawl down the slope, every change of dimensions tending to push the mass in the direction of least resistance.* This theory has been considered inadequate, and

* Rev. H. Moseley, F.R.S., "On the Descent of Glaciers," Proc. Roy. Soc., April 19, 1855. "On the Mechanical Impossibility of the Descent of

very little weight has been given to it, because, although ice expands more under the influence of heat than any other known solid, it is a bad conductor of heat, and the temperature of Swiss glaciers is said not to vary. Now, whatever may be the case with the tiny moribund glaciers of Switzerland, it seems to me that in the case of the vast continental ice of the Antarctic regions, and of the North in Greenland and elsewhere, a very important cause of motion must be expansion and contraction, due to changes of temperature. In the Arctic regions there is a considerable range of temperature below freezing point, and it is impossible but that the ice, however bad a conductor it may be, should not change its temperature very greatly, and constantly when in an atmosphere which ranges during the day, for example, between −10° F. and +19° F., a range of 29°. It is admitted on all hands that a certain amount of motion of all glaciers is due to expansion and contraction, produced by variation of temperature; but it is contended that the proportion so contributed to the general motion is insignificant in amount.

The colouring of the southern bergs is magnificent. The general mass has a sugar-loaf-like appearance, with a slight blueish tint, excepting where fresh snow resting on the tops and ledges, is absolutely white. On this ground-colour there are parallel streaks of cobalt blue, of various intensities, and more or less marked effect, according to the distance at which the berg is viewed. Some bergs with the blue streaks very definitely marked have, when seen from quite close, exactly the appearance of the common marbled blue soap, (coloured plate, fig. 6).

The colouring of the crevasses, caves, and hollows is of the deepest and purest possible azure blue. None of our artists on board were able to approach a representation of its intensity. It seemed to me a much more powerful colour than that which is to be seen in the ice of Swiss glaciers. In the case of the bergs with all their sides exposed, no doubt a greater amount of light is able to penetrate than in glaciers where the light can usually only enter at the top. A large berg full of caves and crevasses, seen on a bright day, is a most beautiful and striking object.

Glaciers by their weight only." Proc. Roy. Soc., 1869, p. 202. Also " Phil. Mag.," May, 1869. Further papers in " Phil. Mag.," 1869, 1870.

One small berg was passed at a distance which was of remarkable colour. It looked just like a huge crystal of sulphate of copper, being all intensely blue, but it seemed as if attached to, and forming part of, another berg of normal colour (coloured plate, fig. 7). Possibly it was part of the formerly submerged base, and of more than ordinary density. Only one other such berg was seen. The intensity of the blue light received from the bergs ordinarily is such that the grey sky behind them appears distinctly reddened, assuming the complementary tint, and the reddening appears most intense close to the berg.

At night bergs appear as if they had a very slight luminous glow, almost as if they were to very small extent phosphorescent.

The sea at the foot of the bergs usually looks of a dark indigo colour, partly, no doubt, out of contrast to the brighter blue of the ice. Where spurs and platforms run out under water from the bases of the berg cliffs, the shallow water is seen to be lighted up by reflection of the light from these.

The surf beats on the coast of an iceberg as on a rocky shore, and washes and dashes in and out of the gullies and caverns, and up against the cliffs. Washing in and out of the caves, it makes a resounding roar, which, when many bergs surround the ship, is very loud. So heavy is the surf on the bergs, and so steep are they as a rule, that we did not see one on which we could well have landed from a boat.

As the waves wash up into the wash-lines of the bergs they form icicles, which are to be seen hanging in rows from the upper border of these grooves.

A line of fragments is always to be seen drifting away from a large berg. These are termed wash-pieces. They are very instructive as showing the vast relative extent of submerged ice required to float a small portion above water ; the parts of the fragments below water being visible from a ship's deck.

The scenic effects produced by large numbers of icebergs, some in the foreground, others scattered at all distances to the horizon and beyond it, are very varied and remarkable, depending on the varying effects of light and atmosphere.

On one occasion, as we were approaching the pack ice, some distant bergs were seen to assume a most intense black colour.

This was due to their being thrown in shade by clouds passing between them and the sun, and the heightening of this effect by the contrast with brilliantly lighted up bergs around them. They looked like rocks of basalt.

On February 15th, a remarkable twilight effect was seen to the southward at about 10 P.M. A narrow band or line of dazzling bright yellow light shone out through a long narrow gap intervening between the lower edge of a densely dark cloud bank and the equally dark, almost black, horizon line. The horizon line was uneven, showing minute black projections or jags, due to hummocky pack ice.

The distant flat-topped icebergs showed out black and sharp, with rectangular outlines against the bright band, and some of them joined with their dark bodies, the dark cloud line to the dark horizon line, bridging over the band of light. The whole effect was very curious, and drew all on deck to gaze at it.

We frequently enjoyed the sight of brilliant red sunsets. Then the bergs directly between the observer and the illuminated sky show a hard, almost black outline. Bergs lying on the horizon, right and left of the setting sun, reflect the light from their entire faces, or from those parts of their faces which lie at the necessary angle. Hence, bright red bergs, and also fantastic red forms, due to reflection from very uneven surfaces, appear on the horizon. Bergs that are nearer take a salmon tint.

In one remarkably brilliant sunset, just before the lower limb of the sun reached the horizon, it was of a brilliant golden-yellow, which lit up the spars and shrouds of the ship with a dazzling light. Later on, the horizon became excessively dark. Above it was a streak of golden light, succeeded by a band of green sky, the two colours being separated by a narrow horizontal violet cloud. Above the green were dark clouds lighted up with bright crimson at the edges. The bergs reflected the crimson and yellow light, and assumed the brightest hues.

Bergs in the far distance, in ordinary daylight, when lighted up often have a pinkish tinge, and then look remarkably like land. The deception is very complete. No doubt Commodore Wilkes was deceived by it. Bergs often also, from the presence

of deep shadows, have the appearance of having rocks upon
them when they have not.

We entered the ice rather unexpectedly, on February 13th.
I was on deck at 11.30 P.M. Two icebergs were then in sight
aheas, only just visible in the dim foggy haze. They became
gradually more plain, and then a berg was reported right ahead.
Sail was shortened, and we glided slowly on. A line of mist,
contrasting strongly with the dark water, seemed in the uncertain light, to be creeping over the surface of the sea towards
us; in reality we were approaching it. Its edge was most sharply
defined. We passed it, and immediately the dark water showed
a sprinkling over of white dots, which looked as if they had been
snow-flakes, which for some reason had fallen on the water without
melting. These white specks became larger and larger, and closer
together, and all at once I realized that we were amongst the ice.
The thin layer of mist was hanging over its edge.

The pieces increased rapidly in size and thickness, as we
went farther and farther ahead, until, in a very few minutes, we
were forcing our way through a sort of soup-like looking fluid,
full of large pieces of ice. The pieces were as much as six feet
long, and three or four broad, all flat slabs, and standing six
inches or so out of the water. The pieces bumped and grated
against the ship's side, and the water line being near the level of
the officers' heads, as they lay in their berths asleep, several
came up on deck to see what had happened. We soon steered
out of the edge of the pack again.

Next morning I viewed the ice from the foretop, and
made a sketch of its appearance (see the coloured plate opposite). All along the horizon, southwards, was a white line of ice,
broken here and there by the outlines of bergs fast in the pack at
various distances from the ship; some partly beyond the horizon,
and with only their tops showing; others at the outer edge of the
vast expanse of ice; others at all intermediate positions.

The field of ice appeared continuous, except just near its
edge, where meandering openings, like rivers, led into it, sometimes for a mile or so. The edge of the pack was very irregular,
projecting as it were in capes and promontories, with bays
between, as on a broken coast-line. The fields of ice were made

C.F.Kell Lith. Castle St.Holborn.E.C

VIEW OF PACK ICE FROM FORE TOP

From Sketches by the Author

up of large fragments closely packed together.   The pieces were not, however, much tilted or heaped up upon one another, as commonly occurs in packs.

Off the edge of the pack, extended serpentine bands of floating ice which drifted before the wind ; they are termed, " stream ice." We dredged within one of the streams.   All the packs which we saw were similar to the one described.

Sometimes, the smaller floating masses of ice at the edge of the pack were covered with fresh snow.   The parts of them projecting above water were sometimes of very fantastic shapes. Some were like the antlers of deer, others like two pairs of antlers with three or four upstanding and branching horns, all borne aloft by irregularly shaped submerged floats.   The soft upper masses of loose or but slightly congealed snow often split off and fell away as the masses floated past.

The ice was frequently stained of the yellow ochreous tint described by Sir J. D. Hooker, and found by him to be caused by Diatoms washed up on to the ice by the waves, and hanging on its rough surface.*   The colouring was always most marked about the honeycombed wash-lines of the ice blocks.   Pancake ice is similarly discoloured by Diatoms in the Arctic regions.†

On February 25th we entered the edge of the pack, sailing amongst some loosened outliers of it.   The sea was covered with masses of ice up to 10 feet in length.   These consisted mostly of light snow ice, and did not project more than from two to four feet out of water.   The upper parts of the masses were composed of white fresh snow, or honeycombed wet frozen snow, which had been partly melted by the waves.   Very many of these ice masses were stained of an ochre tint, by Diatoms and other surface organisms.

The lower submerged ice was transparent, but extremely full of large air vesicles.   The ice below the water line, and under

* Sir J. D. Hooker's collections were described by Ehrenberg. See Capt. Ross's "Antarctic Voyage," Vol. I, p. 339, 341.   London, J. Murray, 1847.   Ehrenberg's " Report on Deposit from Pancake Ice," collected by Dr. Hooker.

† Robert Brown, " On the Discolouration of the Arctic Seas." Quart. Jour. Micro. Sci., 1865, p. 240.

the overhanging edges at that level looked blue. The upper masses were quite opaque.

I went in a boat to collect discoloured ice. The discolouration appears far less marked when the ice is seen at close quarters. It becomes almost invisible when the porous snow-ice drains dry. When however a small piece of the ice is seen floating nearly submerged, it looks almost of a chocolate brown colour.

Mr. Buchanan made experiments on the melting point, and amount of salt contained in salt-water ice. He came to the conclusion from analyses of successive meltings and the varying of the melting point, that in salt-water ice "the salt is not contained in the form of mechanically enclosed brine only, but exists in the solid form, either as a single crystalline substance, or as a mixture of ice and salt crystals."

He thinks that by fractional melting, salt water ice might be made to yield water fit to drink, although when a lump is melted as a whole, the resulting water is undrinkable.*

We crossed the Antarctic Circle on February 16th, passing about six miles to the south of it. There was open water ahead, but the "Challenger" was not strengthened for ice work, and we were not ordered to proceed further south, so we turned back. There seemed to be a deep opening in the pack here, nearly due south of Heard Island. We subsequently passed within six miles of what is marked on maps as Wilkes' Termination Land, and found that this did not exist. Wilkes no doubt was deceived by the land-like appearance of distant icebergs. It is to be noted that he merely says that he saw appearance of land here, sixty miles distant, but high and mountainous. Others have named it for him and placed it on the charts.

On two occasions, whilst amongst the southern ice, our ship was in some little danger, having to ride through heavy gales of wind amongst numerous icebergs in thick weather.

On the morning of February 24th there was a fresh breeze, in which we sounded in 1,300 fathoms, and attempted to dredge,

* J. Y. Buchanan, M.A., "Observations on Sea-Water Ice," Proc. R. Soc., No. 170, 1876, p. 609.

but the ship drifted so fast before the wind, that the dredge did not reach the bottom. The wind became fresher and fresher, and the barometer sunk to 28°·50. The atmospheric pressure is however, for some reason, normally low in the Antarctic regions, and Ross once observed it as low as 28°·35.

Before long it blew a gale, with dry powdery drifting snow, obscuring the view and rendering it impossible to see for a greater distance than 200 or 300 yards. The thermometer sank to 21° F., the lowest reading which occurred during the cruize. Before the weather became very bad we steamed up under the lee of a small sloping berg, with the intention of making fast to it if possible by means of ice anchors.

This was found impracticable, the slope of the berg being too steep to allow of men dropping on to it from the end of the jibboom, as had been intended. The ship was then placed under the lee of the berg, with the view of facilitating the reefing of top sails, as a preparation for the coming gale. Either a back current set the ship on to the berg, or the berg itself was drifting towards us with the wind more rapidly than was expected. A collision ensued, and the jibboom was forced against the side of the berg and broken, together with some parts of the rigging in connection with it.

The end of the jibboom left a star-like mark on the sloping wall of the berg, but had no other effect on the mass. The men who were aloft reefing the topsails, came down the back stays helter-skelter, expecting the top-gallant masts to fall, but no further damage ensued.

As the weather became worse we were in rather a critical position. We were surrounded by bergs, with the weather so thick with snow that we could not see much more than a ship's length, and a heavy gale was blowing. The full power of steam available was employed. Once we had a narrow escape of running into a large berg, passing only just about 100 yards to leeward of it by making a stern board, with all the sails aback, and screwing full speed astern at the same time. The deck was covered with frozen powdery snow, and forward was coated with ice from the shipping of seas.

On February 28th again there were 40 icebergs in sight at

noon. It came on to snow thickly at about 4 P.M., and another gale came on. The plan adopted by Captain Sir G. Nares, was to lay down the bearings of the adjacent bergs before the weather became too thick for them to be seen, and then steaming with all the power of the ship against the gale, to hang on as long as possible under the lee of a large iceberg, and when driven away from that, to steam rapidly across to the lee of another, the position of which was known by the bearings taken. So we went on steaming backwards and forwards through the whole of a thick dark night.

When it was at all foggy in calm weather, we hove to amongst the bergs during the night.

One evening, when there was a very slight fall of snow at the time that there was a brilliant sunset, a snow bow was seen arching high up in the sky. It did not show regularly arranged prismatic colours, but only a uniform bright pinkish yellow hazy light. It was brighter at its lower extremities, like a rainbow.

With regard to animals, we saw not a single seal, on the ice or in the water, during our Southern trip. No doubt we did not go far enough south, or sufficiently amongst the pack ice to meet with them. When we were off the pack ice, and especially when we neared the Antarctic Circle, whales were extremely abundant, apparently all of one species, a " Finback," probably the southern " Finner " (*Physalus Australis*). I saw no Right Whale amongst them at all.

As these whales moved under water close to the ship, the light reflected from their bodies lighted up the water around, and enabled one to follow their movements. I several times went away in a small boat from the ship, to shoot birds for our collection.

On these occasions the whales sometimes blew quite close to the boat. The appearance of a whale's spout as seen from the level of the sea, is very different from that which it has when seen from the deck of a ship ; it appears so much higher and shoots up into the air like a fountain discharged from a very fine rose. The whale of course in reality, does not discharge water, but only its breath ; this however, in rushing up into the air hot from the animal's body, has its moisture condensed to form a sort of rain, and the colder the air, just as in the case of our own breath, the more marked the result.

When the spout is made with the blowhole clear above the surface of the water, it appears like a sudden jet of steam from a boiler. When effected, as it sometimes is, before the blowhole reaches the surface, a low fountain as from a street fire-plug is formed, and when the hole is close to the surface, at the moment a little water is sent up with the tall jet of steam. The cloud blown up does not disappear at once, but hangs a little while, and is often seen to drift a short distance with the wind.

The expiratory sound is very loud when heard close by, and is a sort of deep bass snort, extremely loud, and somewhat pro-longed; it might even be compared to the sound produced by the rushing of steam at high pressure from a large pipe.

Smaller Cetaceans, probably of a kind of Grampus (*Orca*), were very common near the Circle; these had a high dorsal fin placed at about the middle of the length of their bodies. Im-mediately behind the fin there was a large white saddle-shaped patch, extending across their back, and they had further a con-spicuous white blotch on each side just behind the head, and in front of the flippers. The white patches contrasted strongly with the dark general colour of the body. These Grampuses swam about in small shoals with their high dorsal fins projecting far out of the water, like those of sharks do sometimes, and also those of Sword-fish. The Grampuses seemed habitually to swim thus, and the group of pointed sickle-shaped black objects moving through the water, had a curious appearance at a dis-tance. I cannot identify this Grampus with a described species.

As soon as we neared the edge of the pack ice, a petrel which we had not seen at the islands we had left, became common (*Thalassœca glacialoides*), and as soon as we reached the ice we fell in with the beautiful snow-white Petrel (*Pago-droma nivea*), which is never to be found far from the antarctic ice. The bird flies very much like the Whale Bird (*Prion*): it settles on the water to feed; it remains on the wing late at night when the other birds have disappeared. I have seen the birds flying about the ship as late as 11 o'clock at night, when it was quite dusk.

Besides these two petrels we saw when at the edge of the pack, the Sooty albatross (*Diomedea fuliginosa*), the Giant petrel (*Ossifraga gigantea*), *Majaquens æquinoctialis* and the Cape pigeon. These birds all left us when we entered the edge of the pack-ice, they appear to remain at its very margin; but in the ice we met with a Skua (*Stercorarius antarcticus*), which bird ranges very far south, and was seen in Possession Island within the Antarctic Circle by Ross.

Penguins were common at the edge of the ice. They progressed through the water like Rock-hoppers, and probably were the *Eudyptes Adeliæ* of Ross's Expedition, since they had black heads; we could not catch any, though we tried to get some which were on an ice-block; they seemed shy.

We seldom saw birds on the icebergs, but a flock of Cape pigeons was sometimes seen roosting on the top of one. The Great White Albatross (*Diomedea exulans*) accompanied the ship only about 500 miles south of Heard Island, stopping at more than 200 miles from the edge of the pack.

The Cape pigeon left us when we were in about the latitude of Kerguelen's Land, on our return from the ice northwards to Australia, and in exchange for it we fell in with a petrel like the Mutton-bird, which bird had not accompanied us south. We also met at the same time with a second species of albatross (*D. melanophrys* ?).

The last iceberg was seen by us on March 4th, in about the latitude of Heard Island. On March 9th, the South Australian current began to make itself felt, and the air became warm and pleasant. We gave up fires, and the sea being calm, were able for the first time since leaving Kerguelen's Land to take out our scuttles and air our cabins. On March 12th, we were within the westerly winds, and we had more albatrosses round the ship than we had ever had before; the Gony and *D. melanophrys*.

Appended are the summaries of the temperatures of the air during the months of January and February, observed in the Antarctic regions on board H.M. ships "Erebus" and "Terror."

Means of Temperatures observed on board H.M. ships "Erebus" and "Terror," in January, 1841, 1842, 1843, on 93 days.   Between lat. 64° and 78° S., long. 53° to 58° W. and 155° to 168° E.

4 A.M.	28°·795	8 A.M.	30°·065	Noon	31°·540
4 P.M.	31°·594	8 P.M.	29°·956	Midnight	28°·982

Hence general mean for the month, 30°·155.

Means for February observed on the same ships on 84 days.   Between lat. 60° to 78° S., long. 6° to 56° W. and 158° W. to 165° E.

4 A.M.	26°·76	8 A.M.	27°·34	Noon	28°·20
4 P.M.	28°·09	8 P.M.	27°·32	Midnight	26°·59

Hence general mean for the month, 27°·384.

From "Contributions to our Knowledge of the Meteorology of the Antarctic Regions." Published by the Authority of the Meteorological Committee.   Stanford, Charing Cross, 1873.

CHAPTER XI.

VICTORIA. NEW SOUTH WALES.

Excursions into the Bush near Melbourne. Opossum Snare. Tracks of the Aborigines on Tree trunks. Town of Sandhurst. The Highest Tree in the World. Aborigines on a Government Reserve. Ornithorynchus paradoxus. Leaves of Australian Trees, why Vertically Disposed. Fur-Seal in the Open Sea. Sydney Harbour. The Blue Mountains. Excavations in the Ground caused by Rain. Shooting Opossums by Moonlight. Fruit-eating Bats. Hunting Bandicoots. Browera Creek. Intimate Relation of Land and Sea Animals. Geological Import of this. Medusæ in Fresh Water. Kitchen Middens. Drawings by Aborigines. Handmarks. Trigonia and Cestracion.

Melbourne, March 17th to April 1st, 1874.—We sighted Port Otway in a glassy calm, and steamed past Hobson's Bay Heads into Port Philip on March 17th, and anchored off Sandridge, the seaport suburb of Melbourne.

The English house sparrow may be seen quite at home on the beach at Sandridge in flocks, picking up the refuse from the ships, and also about Melbourne generally. The bird is beginning to be a pest to the Acclimatization Society which introduced it, and finding good food in the cages of the animals in the Society's Gardens, refuses to leave them, but consorts with the parrots in the trees and bushes, and steals the food on every opportunity.

I made three excursions from Melbourne. The first was with Mr. Stephenson, the chief of the railway department, to a piece of wild bush-land belonging to him, about 25 miles distant from the city. We started with our host in a light bush waggon, with materials for camping out. We were not seven miles away from the city before the road became a sort of slough, through which the horses could hardly drag the waggon, although we all got out; and before we reached a camping ground it was pitch dark, and one of the springs was broken.

We had some difficulty in finding our way in the bit of bush to the best camping place, and then in finding the water hole and leading the horses to it. We set fire to a great fallen log, made tea in a "billy," a simple tin pot with wire handle, the universal Australian camp teapot, and had hardly lain down to sleep under our tent before it came on to rain heavily. It continued to rain all the next day.

Waking in the night I heard Opossums (*Phalangister vulpina*) caterwauling in the gum trees close by, and in the early morning the Laughing-jackasses and Piping Crows kept up a curiously contrasted concert; the loud harsh laugh of the former mingling with the flute-like musical notes of the latter.

Notwithstanding the rain, I shot a beautiful paroquet, of which and other birds numerous flocks were flying about. With the help of a neighbouring farmer, who rented the bush for grazing, an Opossum was driven out of its hole in a dead-branch or " pipe " of a gum tree and secured.

The scratches of the claws of the Opossum on the bark of the tree, show at once whether a tree is inhabited or not. All the bigger trees were scored deeply and marked with a regular track right up to the various pipes in the dead-branches far overhead. The timber of many of the gum trees decays away in the heart with great rapidity. Hence, whenever a branch is broken off, a pipe is soon formed, and it is especially these holes with abrupt entrances which the opossum affects.

The tracks are always on the side of the tree trunk on which the slope renders ascent most easy. The opossum economizes his force, or is lazy, and this fact is turned to advantage by trappers, who snare the opossums in order to make the opossum rugs, of which so many are used in Australia and exported.

A short piece of a stout branch with a fork at the end, is placed leaning against the butt of a tree meeting the opossum path, the jaws of the fork embracing the round of the trunk a little, so as to keep all steady. About a foot or so from the fork a noose is placed on the lean-to, being kept in place by a notch.

The Opossum always comes down head foremost, and finding an almost horizontal path to the ground ready made for it,

OPOSSUM SNARE.

takes it at once, gets its head in the noose, falls off and is hung. The only precaution necessary, is to allow the animal room enough to swing free so that it cannot catch hold of the trunk. A trapper had lately been camping on this bit of bush, and nearly all the large trees had their lean-to's remaining.

To ascend to a hole in a tree to drive opossums out in the daytime, a light sapling with convenient lateral branches is cut down and placed against the tree, and forms a ready ladder.

One of the most curious sights in the bush was that of the ancient tracks of the Aborigines up the trees, which had been climbed by them to obtain opossums or wild honey. These tracks are the series of small notches made each by three blows of the tomahawk, to admit the great toes, and thus act as a ladder to the Black man. The tracks, which are to be seen everywhere in Australia, lead to the most astonishing heights, up bare perpendicular smooth-barked gum-trees. Knowing bushmen can distinguish the ancient ones made by the stone tomahawk before the Blacks obtained iron from the English. Many are to be seen on old dead barkless tree-trunks, and now that the Blacks are gone they remind one of fossil foot-prints of extinct animals.

Marvellous as this power of climbing with so little support is, it can be done by Whites, and I was assured in New South Wales, when on the Hawkesbury river, that there was a White man in the neighbourhood who could beat any Black at this sort of climbing, doing it in exactly the same way, and being often employed by my informant in collecting wild honey for him at so much a nest. In the same way there are said to be Whites who can throw the boomerang better than any Blacks. In fact, a White man, when he brings his superior faculties to bear on the matter, can always beat a savage in his own field, except perhaps at tracking.

We looked up into all the trees for a native bear (*Phascolarctos cinereus*), and saw tracks of Kangaroos, but not the animals themselves. We stayed out only one night, and got back as we arrived only at nightfall, after a protracted struggle with the mud. The roads were mostly short cuts, and were what are called "made, but not metalled." Making a road is simply clearing of trees a line of ground of a certain breadth and marking the bounds with a plough. In using such a road, constant divergencies have to be successively made in order to avoid deep mud and swampy bits, or occasionally fallen trees, and the track gradually widens and straggles in the adjoining bush.

My next excursion was to Sandhurst, a rapidly grown mining town, which has arisen since 1851 at the site of the most paying Victorian diggings. The railway for a long distance, as it nears Sandhurst, passes through the midst of various sites of old diggings. The surface of the ground on each side of the line for miles at a stretch has been turned over, scooped out and heaped up, and presents the appearance of an endless succession of deserted gravel pits. Here and there a few solitary diggers, mostly Chinamen, were rewashing the dirt, but nearly all was waste and bare. The vast extent of the fields, and amount of work done, astonished me.

Sandhurst, or Bendigo, is a large town with a newly run-up appearance, built amongst the openings of the shafts of the numerous mines. The surface gold was long ago worked out, and the rich quartz reefs below are now being mined by means of shafts and drives. A new shaft was being sunk in the very centre of the town, in front of the principal banks and the verandah-covered pavements, which were crowded with share-brokers, doing business in the open streets. The great winding wheel and its supports looked out of place in the middle of the principal square and public garden of the city.

I went down two of the mines, and saw specks of gold in the richest quartz reef. Some of the very richest quartz, however, hardly shows the gold to the eye, for the metal lies hid in black dirty-looking streaks in the white rock, and is only brought to light after the process of crushing and amalgamation. I saw also the crushing establishments, where the din of the heavy iron stampers falling with a crash upon the quartz was absolutely deafening. Although the men employed in feeding the stampers are from habit able to converse, notwithstanding the noise, I could not hear in the least when my companion shouted into my very ear. I saw the pasty amalgam and the gold fresh from the retort, known as " cake," and finally I handled heavy masses of melted cake fuzed into solid ingots worth many thousand pounds. The mining people were most hospitable.

My last excursion was up the valley of the Yarra, to the beginning of the " ranges," the Australian word for mountains, at a place called Healesville. I went with one of the assistants of Baron Von Müller, the celebrated botanist, who kindly offered me his assistant as a guide. My object was to see some of the enormous Eucalyptus trees which grow in the " ranges," and which, as discovered by Baron von Müller, are the highest trees in the world, exceeding in height the *Sequoia gigantea* of California. One of these trees, measured when fallen, was found by Baron Müller to be 478 feet in length.*

We travelled about 50 or 60 miles by coach. The coaches are very like Californian coaches, and are rough but very strong, the bodies being slung by thick leather straps to wheels as stout as cart wheels. The road is scarcely anywhere better than is

---

* The highest estimate ever made of the height of a Sequoia gigantea is that of Bigelow, who put the height of one at from 420 to 470 feet. Bigelow, in " Whipple's Expedition," p. 23 (Pacific Railroad Explorations) ; cit. by Grisebach, " Veg. der Erde."

Sir Joseph Hooker, in a lecture delivered at the Royal Institution of Great Britain, April 12th, 1878, and published in separate form, p. 12, cites Prof. Whitney's careful measurements of the heights of Californian Big Trees as the best available estimate up to date. Average height 275 feet ; maximum height a little over 320 feet.

an English green lane in a clay soil district. In wet weather deep ruts are cut in it; then these are baked dry and hard, and at the next shower form watercourses and get scooped out deeper than ever. The road at last comes to consist of a series of sharp ridges separated by intervening troughs, often two feet deep. The consequence is that as the coach rattles and leaps bumping over these, the suspended body of the coach heaves and sways, and this to such an extent that my companion and a lady in the coach were sea-sick all the way.

We travelled over some of the roughest of the road at night, which, of course, made matters worse, since the "driver" could not see the pitfalls; but, like a Californian "stage driver," he well knew all the dangerous ones, even in the dark, and in one or two places made *détours* through the bush for a little way.

The ranges are covered with a dense forest of gum trees, in many places of enormous height, standing with their smooth trunks close together, and running up often for a height of 200 feet without giving off a branch. The light-coloured stems are hung with ragged strips of separated bark.

The great slenderness of the trunks of these giant gum-trees in proportion to their height is striking, and in this respect they contrast most favourably with the Californian "big trees," which, in the shape of their trunks, remind one of a carrot upside down, so disproportionately broad are they at their bases. The large species of gum tree, the tallest tree in the world, is *Eucalyptus amygdalina.* As Baron von Müller says, "the largest specimens might overshadow the pyramid of Cheops."

Beneath, in the gullies, is a thick growth of tree-ferns and underwood on the banks of a mountain stream. The undergrowth is the haunt of Bush Wallabies (*Halmaturus ualabatus*). I put one of them suddenly to flight as I was creeping through the tangled, almost impenetrable, vegetation in the hopes of getting a shot at the Lyre-birds, which were to be heard calling in all directions. The animal gave a tremendous bound and seemed more to fly than leap.

Not far from Healesville is a Government reserve, where a number of Aborigines are maintained at Government expense under a missionary. The reserve is called Coranderrk. There were about 120 Blacks there. They live in a small village of rough wooden or bark houses, in the midst of which is the house of Mr. Green, the superintendent.

The Blacks have lately been employed in cultivating hops, and with tolerably good success, but they are incorrigibly lazy. They are delighted when the plough breaks down, and immediately take a holiday with glee. They had just finished picking the crop, so were playing cricket at about half a-mile from the village, and whilst they were amusing themselves, three Whites employed about the place were hard at work. In fact, the Whites do most of the work. The Black women might make much money by plaiting baskets for sale, and the men by catching fish and hunting, but they never will work till hard pressed.

We found the cricket party in high spirits, shouting with laughter, rows of spectators being seated on logs and chaffing the players with all the old English sallies ; " Well hit ; " " Run it out ; " " Butter fingers," &c. I was astonished at the extreme prominence of the supraciliary ridges of the men's foreheads. It was much greater in some of the Blacks than I had expected to see it, and looks far more marked in the recent state than in the skull. It is the striking feature of the face.

The men were all dressed as Europeans ; they knew all about Mr. W. G. Grace and the All-England Eleven. One of them tried to impress on me the heaviness of the work they had just gone through in hop-picking, and that now it was a holiday, and he wished to know how much a bushel was paid in England for such work, evidently wanting to be able to be even with Mr. Green in the matter. The great difficulty at these reserves is to manage the distribution of payment for labour. At present, or until lately, all the proceeds went to a common stock. Of course, this makes all lazy.

Close by the reserve flowed the River Yarra, in which the *Platypus* abounds, the " Water mole," as it is called here, or the " Duck-bill " (*Ornithorynchus paradoxus*). I offered the men three half-crowns for one recently shot. Some of the Blacks thought they might try and get one ; but although one half-crown is the usual price, no one thought of leaving cricket or his looking on at the game : nor, though I offered a good price for a boomerang, did any one care to fetch one from the village.

ORNITHORYNCHUS PARADOXUS.

Down by the river bank I found a Black camped by a fire, with three women, and a lot of mongrel curs. He was just going to fish. He had a gun, and was much excited at the notion of "three half-a-crown" for a *Platypus*. We crept along the bank of the river, the Black first, then I, then my companion. The Black went stealthily along, with his head stretched forward, and every muscle tense, stepping with the utmost care, so as not to rustle a twig or break a stick under foot, and assuming a peculiarly wild animal appearance, such somewhat as I had noticed in a Tamil guide of mine in Ceylon when we were hunting for peacocks and deer. Once he started back, as a snake made off through the bushes.

It was all to no purpose. I was doomed not to see a living *Platypus* or even a Kangaroo in Australia. I saw only the footprints of the *Platypus* (like those of a duck), which the Black pointed out to me, in a regularly beaten track, made by the animals from one pond to another. The Black said that he was certain the *Platypus* did not lay eggs, and that he had several times seen the young ones, and his description of them agreed with what I knew from Dr. Bennett's researches on the subject.

Next day, as I was going down in the coach, I received two specimens of the *Platypus*, shot by this man. Unfortunately, the jolting and heat of the coach, on the journey down to the coast, rather spoilt them for microscopical examination, for which I had wished to procure them. I wished especially to examine the eyes, to see if the retina contains brightly pigmented bodies, as in the case of reptiles and birds. I could not find any trace of them; but possibly, if the tissues had been fresher, I should have met with them, for Hoffman has discovered their existence in marsupials.

Whilst we were hunting along the river bank, numerous bright parroquets were flying about amongst the trees, and a great flock of white cockatoos flew overhead, harshly screaming at the danger. They settled in some trees near, but were far too wary to let me get a shot, though I did my best to creep on them. The smaller bright parroquets are not at all wary as a rule, but are most easily shot.

Grisebach, in his account of the Vegetation of Australia,* dwells on the close relation of interdependence which exists between the tree vegetation and the coating of grass which covers the ground beneath it; and remarks, that the amount of light allowed by the trees to reach the ground beneath them is rendered more than usually great by the vertical position in which their leaves grow. Hence the growth of the grass beneath is aided.

It may be that this permitting of the growth of other plants beneath them, and consequent protection of the soil from losing its moisture, besides other advantages to be derived, is the principal reason why, as is familiarly known, two widely different groups of Australian trees, the *Eucalypti* and *Acacias*, have arrived at a vertical instead of a horizontal disposition of their leaves by two different methods.

The Acacias have accomplished this by suppressing the true horizontal leaves, and flattening the leaf-stalks into vertical pseudo-leaves or "phyllodes." The gum-trees, on the other hand, have simply twisted their leaf-stalks, and have thus rendered their true leaves vertical in position. There must exist some material advantage, which these different trees derive in common, from this peculiar arrangement, and the benefit derived from relation to other plants by this means may be greater and more important than that arising from the fact that the vertical leaves have a like relation to the light on both sides, and are provided with stomata on both faces.

In support of this conclusion I was told, when at Melbourne,

* A. Grisebach, "Vegetation der Erde," p. 216. Leipzig, W. Engelman, 1872.

that when the native vegetation was cleared away from under gum-trees they ceased to thrive, and in time perished. I was shown a number of gum-trees, not far from the city, scattered over some public land, covered with only short turf, which seemed to be mostly in a dying condition.

April 2nd, 1874.—On the voyage to Sydney, two Fur Seals were seen about the ship. They were of a smaller species than that occurring at Kerguelen's Land. They swam alongside with remarkable ease and rapidity, having in the water just the appearance of porpoises. The hind limbs were stretched out straight behind, as the animals swam, and the motion mostly maintained by rapid strokes of the fore limbs. The tail, however, i.e., the fin-like expanse formed by the closely applied and outstretched flat hind flippers, was used with an undulating movement, just as is the tail fin in porpoises.

The seals swam with ease and rapidity from the stern to the bows of the vessel, though it was going 4½ knots at the time, thus going 9 knots at least. In fact they swam with all the ease of a porpoise, and as once or twice they threw their heads and backs out of the water in a forward leap, I should certainly have mistaken them for these animals, had I not seen them almost at rest several times, and with their heads well out of water.

I never before realized the close connection between the seals and whales, and how easily a whale might be developed out of a seal. The fur seal is one which on land still bends its hind limbs forwards, as do land mammals. The seals without external ears, like the sea elephants, carry them habitually stretched out behind, as this one does in swimming. Little modification would be necessary in order to turn the otherwise useless hind limbs of the earless seals into the whale's broad tail fin, which probably represents the remains of the seal's webbed hind flippers. We afterwards, in the Straits of Magellan, became familiar with the motions of Fur Seals in the water, and frequently saw them there in shoals, progressing through the water by a series of leaps exactly like porpoises or Rock-hopper penguins.

A bird followed the ship in some numbers, which is apparently intermediate in its habits between the gulls and terns, a delicate beautiful little sea-bird (*Larus Novæ Hollandiæ*). The bird was abundant about the ship in Hobson's Bay, and in Port Jackson. At Wellington, in New Zealand, a species very closely allied, but a little smaller in size (*Larus scopulinus*) * hovered round the ship in the harbour.

Sydney, April 7th to June 9th, 1874.—The ship arrived at Sydney on April 5th. Port Jackson is famed for its beauty. It is a broad stretch of water, opening to the sea by a narrow passage, between "heads" as they are called, and running far inland, into branches and bays, in great number. Towards the upper part of the harbour, the vegetation extends down the water, and the little cliffs of sandstone rock with their covering of green are extremely picturesque. Port Jackson is one of the many harbours said to be the best in the world; but it lacks shelter, and the passage at the heads is not deep enough for a large ironclad to pass through.

I made various excursions from Sydney, during our stay. One of these was to Botany Bay; a sixpenny omnibus journey. The country here is flat and open, and the vegetation would be very like that of the Cape of Good Hope, in general appearance, were it not for the Grass-trees and Banksias. The far-famed bay is a quiet sandy inlet, resorted to for excursions and the enjoyment of sea air by the Sydney people, and now inhabited principally by keepers of tea gardens. Not far off, across the Bay, the curious Monotreme, the Porcupine Ant-eater (*Echidna*), is abundant, and can readily be found by means of terriers. Some men procured one living for Von Willemoes Suhm.

Another excursion was to the Blue Mountains. A trip to the Mountains was given as an act of hospitality by the ministers of the New South Wales Government to the officers of the "Ancona," a German war-vessel, which was at Sydney, and to those of the "Challenger." It is the custom for the Ministers thus to give picnics to parties of men, ladies not being invited.

The Blue Mountains are piles of horizontally stratified sandstone, rising behind Sydney to about 2,500 feet, with remarkably abrupt terminations on either side, and cut into extraordinary

* Howard Saunders, "On the Larinæ," Proc. Zool. Soc. 1878, p. 187.

deep gullies and chasms, with perpendicular walls, which bound
projecting headlands.

Prof. Dana treats at great length of the question of the mode
of formation of these extraordinary gullies and precipices, in his
" Geology of the U. S. Exploring Expedition," and gives various
reasons for showing that the whole has been due to aqueous
erosion; as have also the exactly similarly formed harbours of
the coast, with their very numerous branches. These, however,
have been subjected to lowering of level, and thus filled by the
sea.

These multi-ramified inlets of the sea resemble fjords in
many points, most curiously, but are very different in origin,
being in fact cañons, which by the sinking of the land have
been invaded by the sea.

The rains, both at Melbourne and Sydney, are extremely
violent, and in the friable and easily decomposed soil, have a
marvellously excavating effect. At Camden Park, 40 miles
from Sydney, I was shown by Captain Onslow, R.N., a deep
chasm in a perfectly level expanse of grass-covered land, which
was at least 20 feet deep and 20 yards across. All this had
been scooped out in a dozen years or so by the rain. In its
precipitous walls and isolated pinnacles of undisturbed soil, it
curiously represented the Blue Mountain configuration on a
small scale. It is only necessary to plough a furrow anywhere
in the soil about Camden to lead to the formation in a short
time of such a chasm.

I twice enjoyed the kind hospitality of Sir William McArthur,
at Camden Park. The park is 10,000 acres in extent. Here I
went out on several occasions to shoot opossums by moonlight.
The opossums are out feeding on the trees at night or are out
on the ground, and rush up the trees on the approach of danger.
They are very difficult to see by one not accustomed to the
work, but those who habitually shoot them discover them with
astonishing ease.

In order to find the animals, one places himself so as to get
successive portions of the tree between his eye and the moonlight, and thus searching the tree over, at last he catches sight
of a dark mass crouching on a branch, and usually sees the ears

pricked up as the animal watches the danger. This is called "mooning" the opossums. Then with a gun in one's hand one fully realizes for the first time the meaning of the saying "'possum up a gum-tree."

The unfortunate beast has the toughness of his skin alone to trust to; "bang" and down it comes with a heavy thud on the ground, falling head first, tail outstretched, or it clings with claws or tail, or both, to the branches, swaying about wounded, and requires a second shot. It must come down at last, unless indeed the tree be so high that it is out of shot or it manages to nip a small branch with its prehensile tail, in which case it sometimes contrives to hang up even when dead and remain out of reach.

Nearly all the female opossums which I shot had a single young one in the pouch. The young seemed to be attached with equal frequency to the right or left teat. I shot the animals in the hopes of obtaining young in the earlier stage, but found none such. Amongst stockmen, and even some well-educated people in Australia, there is conviction that the young kangaroo grows out as a sort of bud on the teat of the mother within the pouch. We killed about 20 opossums in a couple of hours on each occasion on which I went out.

Sometimes we got a Native Cat, *Dasyurus viverrinus*. It is not easily seen in the trees unless there are dogs to pick out the tree. On one occasion we came upon a small animal allied to the Native Cat, but much rarer, *Phascogale penicillata*.

Once I visited a great "camp" of fruit-eating bats, "Flying Foxes" as they are here called (*Pteropus poliocephalus*). In a dense piece of bush, consisting principally of young trees, the trees were hung all over with these bats, looking like great black fruits. As we approached the bats showed signs of uneasiness, and after the first shot were rather difficult to approach, moving on from before us and pitching in a fresh tree some way ahead.

The bats uttered a curious cackling cry when disturbed. They were in enormous numbers, and although thousands had been shot not long before by a large party got together for the purpose, their numbers were not perceptibly reduced. They do great harm to the fruit orchards about Paramatta, and the fruit growers there organise parties to shoot them. They have the cunning to choose a set of trees where the undergrowth is exceedingly dense, and where it is therefore difficult to get at them. I shot seven or eight, but they are very apt to hang up by their hooked claws when shot, and I lost several. I could find no *Nycteribia* living on these bats, although these insects are usually so common on the various species of Pteropus.

At Pennant Hills, near Paramatta, there is plenty of bush-land and a fine large "common" as it is called, *i.e.*, a tract of wild uncleared land of several thousand acres, in which all the neighbouring landowners have the right to cut timber and firewood. It is a fine wild track, with gullies, in which run small streams amongst the sandstone rocks and steep rocky banks covered with ferns, orchids, and Grass-trees, and other plants, forming a varied and beautiful vegetation.

Here there are still plenty of Bush Wallabies (*Halmaturus ualabatus*), and three were shot for me one morning. They are wary and difficult to approach, and I rode all day in the bush without seeing one. There are nests of wild European bees also in the dead limbs of the gum-trees, and we felled a tree and got out about thirty pounds of fine honey.

Once we started a Kangaroo Rat, *Hypsiprimnus*, from its round ball-like nest, which was lying on the ground under a tuft of grass. It was like a large wren's nest. The rat is said to be wary enough never to return to the nest when once disturbed, but always to make a fresh one.

At night we went out with a pack of terriers and mongrels of all kinds, to hunt Bandicoots (*Perameles nasuta*). Only one little terrier was of much use, but he was worth a great deal for this kind of work.

He has not been long off into the fern before we hear his short sharp bark, and know he is on the scent. Off go all the curs that have been hanging at our heels, lazy and doing nothing, to join in the fun. At last a peculiar whining bark is heard, and " Snap's " master knows that the Bandicoot is run to earth; the earth in this case being the hollow pipe running down the stem of some fallen gum-tree.

A long stick is cut and thrust into one end of the pipe whilst a bag is held at the other, and the Bandicoot is soon bagged. The Bandicoot does not attempt to bite, but requires to be held exceedingly tight or else easily escapes the hands by the power of its spring. One female had three young in the pouch. Often the tree is too long for the stick, and then a hole has to be chopped to get the animal out.

I made two excursions to Browera Creek, one of the many branches of the main estuary, or rather inlet, into which the Hawkesbury river runs. The creek is a place full of interest. Suddenly, after traversing a high plateau of the horizontal sandstone, the traveller meets with a deep chasm about 1,000 feet in depth, but not more than a quarter of a mile wide.

This chasm or channel has precipitous rocky walls on either side, with more or less talus slope, and at the bottom runs the river, a small stream, over which one can easily jerk a pebble when standing at its brink. The chasm or creek takes a winding course, so that only short sweeps of it can be seen at a time, and as it widens out and turns sharply or again contracts, one seems, when in a boat on its waters, to pass through a succession of long narrow lakes.

The river, or rather stream, at the place where we approached the creek, is tidal. It is impossible to say where the river ends and the sea begins. The main part of the creek is a long tortuous arm of the sea, ten or fifteen miles in length, and is itself provided with numerous branches and bays. These frequent branchings are perfectly bewildering to a man not accustomed to row on them every day in his life. The whole is, in fact, like a maze.

The side walls of the creek are covered with a luxuriant vegetation, with hugh masses of Stagshorn Fern (*Platycerium*) and "rock lilies" (orchids), and a variety of timbers, whilst there are Tree-ferns and small palms in the lateral shady gullies.

The descent to the river is very steep, and it was a difficult matter to lead the horses down. As we descended, we heard the Lyre-birds calling all round; at the bottom, on a little patch of flat alluvium covered with grass, is a small house and barn, where a man lives with his family all alone, and shut out from the world. He is extremely industrious, and by fishing, wood-cutting, honey gathering, and the proceeds of his farm-yard, must be doing well; we stopped at his cottage for two nights, and hired his boat.

Browera Creek is of varied interest. As an example of denudation, it appears to correspond exactly to what is seen at a much higher level in the Blue Mountains. The extraordinary proximity into which animals found usually only in open sea, are here brought with those only occurring inland, is of great interest from a geological point of view; it recalls at once to the mind such mixtures of marine and terrestrial animal remains, as those occurring in geological deposits, such as the Stonesfield beds.

Here is a narrow strip of sea-water, twenty miles distant from the open sea; on a sandy shallow flat, close to its head, are to be seen basking in the sun, numbers of Sting-rays (*Trygon*), a kind of skate provided with a sharp saw-edged bony weapon (the sting), at the base of its tail. All over these flats, and throughout the whole stretch of the creek, shoals of Grey Mullets are to be met with; numerous other marine fish inhabit the creek, some growing to 150 lbs. in weight, and often caught weighing as much as 60 or 80 lbs. A Diodon or Trunk-fish, is amongst the fishes. Porpoises chase the mullet right up to the commencement of the sand-flat.

At the shores of the creek the rocks are covered with masses of excellent oysters and mussels, and other shell-bearing mollusks are abundant, whilst a small crab is to be found in numbers in every crevice.

On the other hand, the water is overhung by numerous species of forest trees, and by orchids and ferns, and other vegetation of all kinds; mangroves grow only in the shallow bays. The gum-trees lean over the water in which swim Trygons and mullets, just as willows hang over a pond full of carp. The sandy bottom is full of branches and stems of trees, and is covered in patches here and there by their leaves.

Insects constantly fall on the water, and are devoured by the mullets. Land birds of all kinds fly to and fro across the creek, and when wounded may easily get drowned in it. Wallabies swim across occasionally, and may add their bones to the *débris* at the bottom.

Hence here is being formed a sandy deposit, in which may be found Cetacean, Marsupial, bird, fish, and insect remains, together with land and sea-shells, and fragments of a vast land flora; yet how restricted is the area occupied by this deposit, and how easily might surviving fragments of such a record be missed by a future geological explorer! The area occupied by the deposit will be sinuous and ramified like that of an ancient river-bed.

The inlet being so extremely long and so narrow, although the rise of the tide is two feet or more at the head of the creek, the interchange of water with the ocean is very small; the water in the upper parts of the creek, is merely forced back to a higher level by the tide below at flood-tide, and similarly lowered again at ebb. Hence, after heavy rain, the surface water in all the upper parts of the creek is so diluted by the torrent of fresh water from the stream, that it becomes almost fresh ; indeed, at the time of our visit, it was for three or four miles down, which was, as far as we went, so little brackish as to be drinkable. At a short depth, no doubt, the water was salt.

Here are the most favourable conditions possible for turning marine animals into freshwater animals ; in fact the change of mode of life presents no difficulty. Below, no doubt, the water is always salt, but the fish find a fluid gradually less and less salt as they rise to the surface.

We caught the mullets in the almost fresh water, with a net. The oysters were flourishing in the same water, and with them the mussels and crabs ; I even saw an abundance of *Medusæ*, and a species of *Rhizophora* swimming in the creek above the sand-flats, where there was scarcely any salt at all in the water, yet evidently in most perfect health.

Occasionally, in times of long drought, the water becomes as salt as the sea. The fishermen told me that after sudden very heavy freshets of water from the river, some of the shell-fish sickened and died. He accounted for the presence of numerous dead cockle-shells (*Cardium*) in the bed of the creek, since he had never found the animals there alive, by supposing that they had all been killed off by some unusual influx of fresh water many years before.

But beyond all that has been described, and beyond the extreme beauty of its wild and rocky scenery, the Browera Creek has yet another interest; it was in old times the haunt of numerous Aborigines, who lived on its banks in order to eat the oysters and mussels, and the fish.

On every point or projection, formed where a side branch is given off by the main creek, is to be seen a vast kitchen midden or shell mound. So numerous are these heaps of refuse, and so extensive, that it has been a regular trade, at which White men have worked all their lives, to turn over these heaps and sift out the undecomposed shells, for making lime by burning them; unfortunately the numerous weapons thus found in the heaps, have mostly been thrown away.

There is now not a single Black on the creek. Many of the mounds are very ancient, and it must have taken a very long time for such heaps to accumulate. Stone hatchet blades are still to be picked up in considerable numbers, and I obtained several. The heaps are very like those at the Cape of Good Hope in appearance, but there were none of the peculiar piles of stones about them, which I noticed at the latter locality.

The softer layers weathering out from under the harder slabs of the horizontally bedded sandstones, form numerous shelters and low-roofed caves, along the creek banks. It was in these caves or "gunyas," that the blacks used to camp, and in front of all of them, a mass of shells slopes down towards the creek just as the Cape of Good Hope.

I dug into one of the heaps; places were found where fires had been made, and there were numerous bits of burnt stick and charcoal, a piece of Wallaby bone charred by the fire, and the thigh bone of a Black woman. This latter was found without any of the remaining bones of the skeleton, the woman having been perhaps eaten piecemeal. These relics were buried in a mass of cockle, oyster and mussel shells, mingled with much black powdery matter composed of decayed shells, and other *débris.*

The walls and roofs of the caves are covered all over with drawings executed by the blacks in charcoal on the rock. These

T

are interesting from their rude character, and sketches of them are given in the accompanying woodcut.

AUSTRALIAN NATIVE DRAWINGS.

1 Opossums; 2 a fish; 3 uncertain; 4 a white man—drawn with charcoal, in caves, Browera Creek. 5 figure of kangaroo, five feet in height—cut in a slab of stone—same locality.

The row of four figures (1) evidently is intended to represent four-footed animals, probably opossums (*Phalangista*), the drawing being of about the size of that animal. Two of the figures are roughly shaded. There were several similar rows of the same figures in one cave.

Figure 2 is a tolerably good representation of one of the fish of the Creek. It also is shaded.

Figures 3 I do not understand. The larger may be intended for a shark. Figure 4 is evidently intended for a white man. North American Indians are said to have distinguished white men in their drawings by putting a tall hat on them. Such a form of headdress must be astonishing to a savage at first acquaintance.

Near one of the caves, on a flat slab of stone standing naturally erect, is a figure of a Kangaroo cut out in the stone itself. The figure is five feet in height. It is marked out by means of an incised groove, which is an inch and a half in depth. The figure is shaded, or rather rendered more conspicuous by the chipping of irregular small holes all over the area representing the body, and also as in the charcoal drawings of opossums, by means of lines.

The fore-legs of the Kangaroo seem not to have been finished, or the artist has been especially unsuccessful in his attempts to represent them, and perhaps has tried to correct them, as appears possible from the number of lines. The contour line of the body is carried across the root of the tail. Similar drawings, executed by cutting grooves in stone, are common about Sydney.

In Peron and Leseur's " Voyage,"* a plate is given of similar drawings of fish and Kangaroos by Blacks, from Port Jackson, and one of the drawings shows a similar attempt at irregular shading, as seen in some of the present figures. Another plate of the same work, shows the Blacks living on the shore, about caves under cliffs, such as those here described. The plates in question are unnumbered, and I could not find reference to them in the text of the book.

Besides the drawings, in almost every cave were hand marks. These marks have been the subject of much discussion, and various speculations have been made as to some important meaning of the " Red Hand of Australia." These hand marks have been made by placing a hand against the flat stone, and then squirting a mixture of whitish clay and water from the mouth all around. The hand being removed, a tracing of it stands out in relief, and where the sandstone is red, appears red on a whitish ground.

The hand marks have evidently been made hap-hazard, just as the drawings. They are now often out of easy reach, the former floors of the caves having slipped away. They are grouped in all sorts of ways, and amongst them I saw one in which a finger was missing, the native having possibly had a finger cut off as a matter of ceremony. The figure of a whole man is said to exist thus executed, in Cowan Creek, close by.

* " Voyage des Découvertes aux terres Australes." Peron et Leseur. Paris, 1807, Atlas.

Delightful though it was at Sydney to make so many friends amongst one's countrymen, after so long a voyage from home, and to enjoy their far-famed hospitality, one could not as a naturalist, help feeling a lurking regret that matters were not still in the same condition as in the days of Captain Cook, and the colonists replaced by the race which they have ousted and destroyed, a race far more interesting and original from an anthropological point of view.

Whilst we were at Sydney, the ship's steam pinnace was constantly employed in dredging for Trigonia shells in Port Jackson. These shells, in shape very like cockles, are immediately known by their brilliant pearly lustre within, and curious complicated hinges. They vary very much in the tint of the nacre inside. Some are orange-tinted, others pink or purple, some without colour. The shells are worn very much by the ladies of Sydney, as earrings and other ornaments, being set in gold.

The shell is especially interesting to the naturalist, because it occurs fossil in secondary deposits in Europe, and was long supposed entirely a thing of the past, until discovered living in Sydney Harbour. Moreover, with it occurs in the harbour a most remarkable fish, the Port Jackson Shark (*Cestracion Philippi*) which is also closely allied to fish, remains of which are found in the deposits together with the Trigonias.

It was believed for some time that the modern Trigonias were very restricted in their distribution. A species occurs however at Cape York, and Mr. S. C. J. W. van Musschenbrook, Governor of Ternate in the Moluccas, told me that he had obtained specimens of the genus from the coast of Halmahera (Gilolo). A Port Jackson Shark is also found far away from Australia, in the Japanese seas, and at intermediate localities.

CHAPTER XII.

NEW ZEALAND. THE FRIENDLY ISLANDS. MATUKU ISLAND.

Wellington, New Zealand. The Rata Tree. Kingfisher with Littoral Habits. Peripatus. Egg Capsules of Land Planarians. The Vegetation of the Kermadec Islands. Red coloured Muscles of the Shark. Island of Eua. General appearance of the Island of Tongatabu. Tongan Natives. Mode of Hairdressing. Facial expression of the Natives. A Pea Jacket a Badge of Distinction. Town of Nukualofa. Dress of Tongan Women. Getting Fire by Friction. Deserted Plantations. Fruit-bats Feeding on Flowers. Herons, Tree-swifts, and other Birds. Parasitic Algæ in Foraminifera. Matuku Island Fiji Group. The Island an Ancient Crater. Its Vegetation. Encircling Reef. Flocks of Lories. Periophthalmus, a Fish Living on Land. Living Pearly Nautilus. Its Mode of Swimming. Account of the Nautilus, by Rumphius.

Wellington, June 28th to July 7th, 1874.—We encountered constant gales on the voyage from Sydney to Wellington in New Zealand. The voyage lasted 14 days, and we arrived at Wellington on June 28th. The ship had to be anchored for two nights under the lee of D'Urville Island, on the south side of the entrance to Cook's Straits, until the weather moderated sufficiently to allow of the ship's passing up the straits to Wellington.

We found deep water, 2,600 fathoms, between the Australian coast and New Zealand, as might have been confidently predicted from the vast difference of the New Zealand from the Australian fauna and flora. Around New Zealand itself, there is a considerable extent of shallow sea with very uneven bottom, and from this shallow a stretch of comparatively raised bottom is extended to Northern Australia, including on its surface Norfolk Island and Lord Howe Island.

The stay at Wellington was very short, and as I was not in good health I saw very little of the country. The town necessarily contrasts unfavourably in appearance with Sydney. The buildings are all of wood, even Government House. There is one long principal street following the shore, and the remainder is more or less scattered. Tattooed Maories were to be seen commonly walking about in the streets, but all in European costume, reminding one somewhat of English gipsies.

The coast hills in the general appearance and colour of their vegetation, as seen from sea, recalled Kerguelen's Land, especially the shores about D'Urville Island, but all the valleys and inland slopes are covered with a dense forest and almost impenetrable bush. The trees are covered with epiphytic ferns, and Astelias, Liliaceous epiphytes, which, perched in the forks of the branches, remind one in their habit and appearance of the Bromeliaceous epiphytes of Tropical America.

One of the most remarkable trees which was pointed out to me by Mr. T. Kirk, F.L.S., is the Rata, a *Metrosideros*, *M. Robusta*. This, though a Myrtaceous plant, has all the habits of the Indian figs,* reproducing them in the closest manner. It starts from a seed dropped in the fork of a tree, and grows downward to reach the ground; then taking root there, and gaining strength, chokes the supporting tree and entirely destroys it, forming a large trunk by fusion of its many stems. Nevertheless, it occasionally grows originally directly from the soil, and then forms a trunk more regular in form. Another *Metrosideros*, *M. florida*, is a regular climber.

I did not see many birds. The gull of Kerguelen's Land (*Larus Dominicanus*) was common in the harbour. On the telegraph wires along the shore sat a Kingfisher (*Halcyon sanctus*) in abundance, and dashed down from thence on its prey into the shallow water of the harbour. It interested me because it was the first Kingfisher that I had thus seen leading a littoral existence and feeding on sea fish. I afterwards became familiar with Kingfishers thus inhabiting the seashores in the Straits of Magellan and the coast of Oregon in North-west America. In the poulterers' shops the curious parrot, or Kaka, *Nestor Meridionalis*, is hung up for sale. Mr. Potts describes this

---

* T. Kirk, F.L.S., "On the Habit of the Rata, Metrosideros robusta." Trans. New Zealand Inst., Vol. IV., 1871, p. 267.

bird as tearing away the dead wood of trees in search of insects, and appearing to replace to some extent in its habits in New Zealand, the totally absent Woodpecker.

The New Zealand Peripatus (*P. Novæ Zealandiæ*) is abundant near Wellington amongst dead wood, and I had 40 or 50 specimens brought to me as the result of a day's search in the Hutt Valley. As in the case of the Cape of Good Hope species, the males are much less abundant than the females.

In essential structure and habits the animal closely resembles the South African species. It is distinguished by having fewer pairs of feet, viz., 15 instead of 17. The females all contained young although it was mid-winter.

Land Planarian worms are also pretty common near Wellington. In their anatomical structure, these New Zealand species are more nearly allied to South American forms of the genus *Geoplana* than to the Australian Land Planarians. These latter belong to a special genus, *Cænoplana*, which has affinities with the genus *Rhynchodemus* of India and the Cape of Good Hope.[*]

Mr. W. T. Locke Travers, F.L.S., to whom I am indebted for much kindness and scientific information during my stay at Wellington, brought me specimens of *Peripatus N. Zealandiæ*, and also of Land Planarians, together with the egg capsules of the latter, which were hitherto unknown.

They are spherical in form, of about the size of sweet-pea seeds and of a dark brown colour. The capsules have a tough chitinous wall, and contain four or five young Planarians each. The production of these capsules by the Land Planarians I regard as further evidence in favour of the affinity of these worms to the leeches, on which I have dwelt elsewhere.[†]

---

[*] Captain F. W. Hutton informs me that, as far as he knows, the genus Bipalium does not exist in New Zealand. His assertion that it did exist there in his well-known and admirable paper, " On the Geographical Relations of the New Zealand Fauna," Trans. New Zealand Inst., Vol. V., 1872, p. 227, was due to imperfect determination of the genus in the case of the species of Geoplana of the locality.

[†] H. N. Moseley, " On the Anatomy and Histology of the Land Planarians of Ceylon." Phil. Trans. 1875, p. 148. Also " Notes on the Structure of Several Forms of Land Planarians." Quart. Journal, Micro. Sci., Vol. XVII, p. 275.

Off the Kermadec Islands, September 14th, 1874.—We were in the morning in sight of Raoul or Sunday Island, and Macaulay Island, of the Kermadec group. No landing was effected on any of the islands. This small group of islands forms with New Zealand, McQuarrie Island, and the Tonga group, a direct line of volcanic action, stretching about N.E., and thus at right angles nearly to the north-west lines, which are followed by most of the remaining Pacific groups, such as the Fijis, for example. The Kermadec Islands are all very small. The flora of Raoul Island was described by Sir J. D. Hooker* from collections made by Mr. MacGillivray, of H.M.S. "Herald." Forty-two vascular plants are known from the islands, of which five are endemic species. Half of the number consist of New Zealand ferns. The large proportion of ferns in the flora is most remarkable, and also their New Zealand character. There are no currents leading from New Zealand towards the Kermadecs. The group lies in the fork of the great current which, stretching westward from the region of Ducie, Pitcairn and Tubai Islands, follows the line of the Tropic of Capricorn, and branching, sends its northern half to the east coast of Australia to form the East Australian current, whilst its other half passes down S.W. to sweep past the east coast.

The group lies just at the northern limit of the zone of westerly winds, and within that of calms and changeable winds, but so close to the limit that the winds may well have transported many of the plants, and the preponderance of ferns may be due to the possible fact, that the winds have been the main agents in the colonization of the islands, and have sufficed to carry the minute fern spores, whilst heavier seeds have seldom reached the island, and by other means of transport.

If fern spores are diffused mainly by wind, it should be especially difficult for them to cross the zones of constant rains, and there ought to be a marked separation of fern forms in distribution about those lines.

There is no connection between the flora of the Kermadecs and that of Norfolk Island, although such would have been

* Sir J. D. Hooker, "Botany of Raoul Island." Jour. Linn. Soc., Bot. Vol. I., 1857, p. 125.

expected, as Sir J. D. Hooker states, on all considerations to occur. The soundings of the " Gazelle " and " Tuscarora," have proved that a channel of more than 2,000 fathoms in depth, passes up between New Zealand and the Kermadec Islands. Hence, an ancient land connection cannot be looked to as an explanation of the New Zealand affinities of the Kermadec flora.

Whilst dredging was going on off the islands, a shark (*Carcharias brachyurus*), which was attended by a pilot fish (*Naucrates sp.*), was caught ; it was, as is commonly the case, covered by a small parasitic Crustacean, a species of Pandarus. Some specimens of this parasite had, curiously enough, a Barnacle (*Lepas*) attached to them as large as themselves.

On the shark being skinned, I noticed that a layer of superficial or skin muscles extending all over the animal, and only about one-fourth of an inch in thickness, is coloured dark-red by blood-colouring matter (*Hæmoglobin*), as are all the muscles of Mammalia. The main internal muscular mass of the shark is pale, almost white.

Prof. Ray Lankester has described several similar instances of the restriction of the red colouring matter to certain muscles only in animals which possess it.[*] A closely parallel case is that of the little fish, the " Sea-Horse " (*Hippocampus*), in which the muscles of the dorsal fin only are red.

Mr. Lankester accounts for the presence of the *Hæmoglobin* in the dorsal fin muscles only in this case, by the special activity of the fin in question, but such an explanation fails in the case of the shark, the skin of which is apparently immovable ; moreover, the structure of the skin precludes the idea of the red matter beneath it having a respiratory function.

Mr. Lankester has shown that *Hæmoglobin* is entirely wanting in one fish at least, the white transparent oceanic surface fish *Leptocephalus*, and I believe that small oceanic Flat-fish, *Pleuronectids*,] will prove also to be devoid of red-blood colouring.

I was extremely vexed that no landing on the Kermadec

---

[*] E. Ray Lankester, " On the Distribution of Hœmoglobin." Proc. Royal Soc., No. 140, 1873.

Islands was arranged.   Further information concerning the flora of the islands is very much wanted, and it seemed hard to be dredging off the islands and not to be able to land.

Tongatabu, July 19th to July 22nd, 1874.—Our approach to the Friendly Islands group was heralded by the appearance of a Tropic Bird, which was seen flying behind the ship, although we were 150 miles as yet distant from Tongatabu.

We sighted the island of Eua in early morning, and passed to the north of it.   The island is elevated in its highest point 600 feet above the sea, and is volcanic, with coral rock at its base.   An ancient, now upraised sea-cliff of the coral rock, is conspicuous from the distance, forming a line above the present coast-cliff, as described by Dana.*

The island appears covered with bushes, with very few trees, and isolated palms on the summits of the high ground.   The bushes on the higher land appear to be all bent over in the direction of the trade wind.

The sky was dull, covered with grey clouds, and the air even somewhat chilly, and the islands did not look bright and sun-shiny, as I had expected these, the first South Sea Islands I had seen, to look.   At the base of the Eua, the surf in places raised jets of spray, looking from a distance like thin white smoke.

Tongatabu was seen seven miles distant from the small Eua, stretched along the horizon as a long narrow neutral tint band, with an indented upper margin : towards the northern end the band thinned out into isolated rows and groups of palm-trees, which looked like dots on the watery horizon.   As we ap-proached nearer, the forms of the cocoanut-trees became more and more distinct.   At length we shortened sail and steamed through the reefs with a long stretch of palm-covered land on the one hand, and numerous islets on the other, some bearing many cocoanut palms, others with few.

The main island is exceedingly flat and low, its highest point being only 60 feet above sea level.   It thus stretched itself before our view as a horizontal streak of green of uniform width, the width being due merely to the height of the vegetation ; here and there at the water's edge, were seen small inlets and

---

* J. D. Dana, " Coral Reefs and Islands," p. 30.

stretches of white sandy beaches, or low honeycombed and weathered clifflets of coral rock.

Above these, appeared a band of dark foliaged shrubs, and shrubby trees with shore-loving plants at their foot, growing in the sand ; and as a background behind, rose a mass of cocoanut-trees of various heights, but densely packed together, and thus forming with their crowns a tolerably even line ; no palms other than cocoanuts were to be seen in the mass.

On the small scattered islets which were near at hand, Screw-Pine trees were conspicuous, their stems surrounded with prop-like aerial roots, whilst on the main island these trees, which are numerous along the shore, were almost lost to view against the general backing of dark foliage.

As we steamed on, we could see beneath the cocoanut-trees on the shores, the villages of the islanders, composed of small houses of palm mats and grass thatch, and, as the news spread, we saw the villagers assemble on the beach in their conspicuous white or red clothing, to gaze at the ship.

In the harbour were several American whalers, waiting for the whales expected to come into the bay in a few days, and also a small German vessel of the firm of Goddefroy Brothers, the famed collectors of South Sea Island productions.

Not until we had passed the most difficult twist in the passage into the harbour, did the pilot come out, in a small English-built boat manned by four sturdy Tongans.

These Tongans were naked, except that they had a cloth round the waist, and one of them a further girdle of green Screw-Pine leaves ; they had all, however, linen shirts, which they put on as they got cool ; and the coxswain, formerly a Mataboolo, or lord, but degraded for drunkenness, wore besides a pea jacket.

The boat was a whale-boat, belonging to the King. As is always the case, the men being so little clothed, looked to us bigger than they really were. They were, however, remarkably finely made men, with all their muscles well developed, and all of them were extremely well nourished. The Tongans have large broad foreheads and faces, the lower jaws being wide at their articulation, the chins narrowing off rather abruptly from

the face.   The nose is flattened, but not very much ; the eyebrows are straight, the lips not large or protuberant.

The colour of the Tongans is of a light brownish-yellow with
a tinge of red.   Their hair forms the most remarkable feature in
their appearance ; it is worn in a sort of mop sticking straight
up from the head, and composed of a mass of small curls ; it is
black naturally, as are the eyebrows, beard, and moustache,
which latter are, however, scanty as a rule ; but it is altered to
a rust colour by the application of coral lime.

The colouring is usually only applied partially so as to give
a contrast between the black and red locks.   Sometimes the
centre of the head is left black, and a marginal zone coloured
red ; at others isolated locks all over the head are reddened so
as to show a black mop variegated with red.   Various other
fashions are adopted.   The Tongans often sit on their heels like
Indian races, but more usually sit cross-legged in the posture in
which Buddah is ordinarily represented.

Having studied Mr. Darwin's work, " On the Expressions of
the Emotions," I was immediately struck on seeing the men
conversing in the boat with one another, by the unusually
marked development of facial expression exhibited by them.
The muscles of the forehead during animated conversation,
are contracted and relaxed incessantly, and in a most varied
manner ; the brow is strongly wrinkled, and the eyebrows are
jerked up to such an extent as to remind the observer at once of
the jerking up of the eyebrows in monkeys.

I made as careful a study as time would permit of the
various expressions of the emotions ; all of them appear to coin-
cide in their intimate character with those of Europeans, and
this holds good also in the case of the expressions of children,
but the movements made use of are much more strongly
marked in the Tongans than in Europeans : thus, for example,
in the expression of astonishment I noticed the eyebrows thrown
up with a succession of strong jerks, not merely raised once as
with Europeans.   The use of the forehead muscles is very
peculiar, and it indeed seems to be the most characteristic
feature noticeable about a Tongan.   I saw no similar exaggerated
facial expression amongst Hawaians or Tahitians.   There was

nothing interesting to be noted about the means of expression of these latter islanders; probably they have copied European modes of expression to a large extent.

In some of their gestures, the Tongans differ remarkably from us; in beckoning, to call a person, they use, like the Malays and others, the hand with its back turned towards their bodies, and the palm directed towards the person called; the hand is moved downwards and inwards, instead of upwards and inwards as with us.

In affirmation the head is jerked slightly upwards, the eyebrows being raised a little at the same time. I asked one of the missionaries who visited the ship, about this matter, and to test it he pronounced the word for yes, and involuntarily threw up his head. The gestures accompanying the language are necessary to its perfect use, and to speak without them would be like speaking a European language with a false accent.

In negation, the head is sometimes moved slowly from side to side, but never shaken. In pointing out the way to a place, the lips are pouted in order to indicate direction at the same time that the hand is used to point with in the ordinary manner. The use of the arms and head in gesture language, is very remarkable, and conversations are carried on thus in an extremely animated manner, with the help of very few actual words.

The coxswain of the pilot's boat, the ex-member of the nobility, wore, as I have said, a pea-jacket; a photograph was taken of the boat's crew. I could not persuade the coxswain to take off the pea-jacket, in order to make the group uniform; he would only promise that if he were photographed with the jacket on in the group, he would allow himself to be taken with it off, separately afterwards. The jacket was a thick garment of the usual pilot cloth, fit only for an English winter, but the man evidently regarded it as a mark of distinction and decoration, and a proof that he was coxswain.

I had much difficulty in getting a lock of hair from one of the boat's crew, and only succeeded by the help of a missionary, who explained that I did not want it for purposes of witchcraft. The man also evidently was loth to part with a single lock of what was his chief pride. I often, in collecting hair of various races subsequently, for scientific purposes, had amusing difficulties to contend with, and I suspect some of the girls, from whom I got specimens, thought I was desperately in love with them.

The most prominent feature in the town of Nukualofa, as the principal place in the island is called, is the small white church which stands on the summit of a rounded hill about 40 feet in height. Conspicuous also is the King's house, a respectable-looking small one-storied wooden building with a verandah. There is, further, the Government building, a neat wooden structure with a tower in the centre and a wing on either side, each containing a single office-room. Here the revenue of the Friendly Island Group, which amounts to about £7,000 or £8,000, is dispensed, and the King's seal is attached to documents. At a small printing office close by, an almanac, a magazine, bibles, and a few books, are printed in the native language.

The remainder of the town consists almost entirely of native houses. The houses of the Tongans are small and oblong in shape, about 20 feet by 10 feet in dimension. The walls are of reed mats or plaited cocoanut leaves, and the thatch of reeds. The posts and beams, often of cocoanut stems, are lashed together with plaited cocoanut fibre. The ground within is simply covered with Pandanus mats. There are usually two doors or openings opposite one another in the middle of each side of the house, which are closed with a mat only. In most houses a sleeping chamber is partitioned off at one end by means of mats.

The only furniture to be seen within is the kaava bowl and the pillows, wooden rods supported on four legs, on which the neck is rested in sleep in order that the elaborately dressed hair may not be disarranged. Most Polynesians use similar pillows, and very various other races, such as the ancient Egyptians and the modern Japanese. Long practice is required to allow of their use. I have tried a Japanese pillow, but found it far too painful to be endured for even half an hour.

Near the houses are small sheds, underneath which a hole in the ground serves as an oven for cooking.

The houses at Nukualofa are clustered under the cocoanut

trees, with three or four open roadways between them. The people are remarkably hospitable, and delighted to get a strange visitor into their houses to sit and communicate what little can be managed in this way between persons knowing almost nothing of each other's languages. They offer kaava or cocoa-nuts as refreshment.

The women are large, they have fine figures and are, most of them, handsome. They wear a cotton cloth round the loins reaching down below the knees, or often, and especially on week-days, a "tappa" or native cloth, made from the Paper Mulberry. The missionaries have compelled them to cover their breasts, which is done with a flap of cloth thrown up in front, and a fine is imposed on any woman seen abroad without this additional covering. The women, however, evidently have little idea of shame in the matter; and often the cloth is put on so loosely that it affords no cover at all.

The hair of the women was formerly cut short as amongst so many savages where the men keep to themselves the right of cultivating and decorating the hair, but now it is often allowed to grow long and fall down the back. It is oiled and powdered with sandal-wood dust as a perfume. On Sundays a few women appear in complete European dress, wearing muslin gowns, and hats profusely decorated with gaudy artificial flowers. The girls are most accomplished coquettes.

The missionaries have prohibited dancing, and also the chewing of the kaava root, which is now grated instead. The chewing method was believed to spread disease. The people are diminishing notwithstanding all the efforts of the mission-aries. There are now only about 8,000 islanders in the whole group.

The Tongans are a fine manly race, and delighted us all. We should all have liked a longer stay in their island. They are an extremely merry race, fond of practical jokes; and as I was rowed on shore by a crew of them, they kept playing all kinds of pranks on one another between the strokes of the oars, such as bending over and catching at each other's legs, and they were full of laughter the whole time.

I had some difficulty in persuading one of the natives to get fire for me by friction of wood. Matches are now so common in Tonga that the natives do not care to undergo the labour necessary for getting fire in the old method, except when driven by necessity. No doubt the younger generation will lose the knack of getting fire by friction altogether.

The method adopted in Tonga is the usual Polynesian one of the stick and groove. The wood of the *Hibiscus tiliaceus* is made use of. It is extremely light when dry. It must be extremely dry in order that it can be used for getting fire. In order to procure fire, a stick or stout splinter of the wood about a foot in length is cut at one end so that it has a sharp edge bounded by two sloping surfaces on one side of the end. The side of the tip is thus in the form of a wedge with a sharp edge.

This stick is held in a slanting position between the two thumbs crossed behind it, and the fingers of the two hands crossed in front of it. The sharp edge of the wedge is applied to the surface of a large billet or stem of the same dry wood, and the stick is rubbed backwards and forwards, a certain amount of pressure being exerted. A V-shaped groove about four or five inches in length is thus cut into the billet. If the piece of wood to be grooved is rounded and smooth, a slight score is sometimes made upon it with a knife beforehand in order to prevent the stick from slipping.

Of course everything depends on the larger billet being kept absolutely immovable during the process. Sometimes the operator holds it with his own feet, or often gets some one else to stand on it for this purpose. The stick is rubbed backwards and forwards, slowly at first. It must not be pressed on too hard or the rubbing surfaces become polished, nor too softly or no heating results. In applying the exact amount of pressure, a great deal of the knack of getting the fire readily, no doubt, depends.

If the operation is proceeding well, there should be a constant feeling of slightly grating friction to the operator as he rubs, and a fine powder should be rubbed off from the surface of the groove and pushed along by the end of the stick, so that it accumulates at the far end of the groove in a small heap.

Great care must be taken that this small heap of powder is not shaken or blown away.

The friction being kept up slowly and steadily, the sides of the groove begin to blacken and soon to smoke. Rapid strokes are now resorted to, the fine dust rubbed off becomes black like soot, and at last ignites at the end of the stroke just as it is pushed into the small accumulated heap, which acts as tinder. A tiny wreath of smoke ascending from the heap shows that the operation has been successful. A gentle blowing soon sets the whole heap aglow.

The operation is excessively tiring to the wrists, since it has to be prolonged for a considerable time, but the greater the practice the less the waste of force. I have never succeeded in getting fire myself, though Mr. Darwin succeeded at Tahiti; and I have seen several Englishmen do so after practice, and especially Dr. Goode, R.N., who frequently lighted a candle in this way to show me the process on board H.M.S. "Dido" at Fiji. It is easy enough to get smoke and char the wood a little, but very difficult to get the actual fire. The slightest halt during the friction is fatal.

The old stone implements have entirely gone out of use in Tonga, and they are not plentiful, but I bought several from natives who had them put away in their houses. They call them "toki Tonga," Tongan axe, or adze, in distinction to foreign axes, whereas the Sandwich Islanders spoke of their adzes when I was buying them as stone adzes, "pohaku koi." All the stone adzes which I saw were unmounted; no doubt the handles had been used long ago, when iron was introduced, to fasten hoop-iron blades on to in the place of the discarded stone ones. The stone adze blades I procured were all of simple form like those of Fiji, and not with complex curved surfaces and shanks like those of Tahiti and some other Polynesian Islands.

The manners and customs of the ancient Tongans are probably better understood than those of any other Polynesian Islanders, because of the existence of Mariner's well-known account of them.*

The Island of Tonga is about 27 miles in extreme length,

* "An Account of the Natives of the Tonga Islands. Compiled from

U

and 10 in extreme breadth. The island is entirely composed of coral-reef rock, without, as far as is known, any blown-sand formation. The sand on the beaches is scanty. The presence of blown sand-rock on coral islands must depend on the freedom of some part of the coast from breakwaters of coral, in order that a heavy surf may form sand in abundance. In Bermuda the sand is derived from the unsheltered side of the island.

In some rock, about 30 feet above sea level, I saw, as Dana describes, some Brain Corals imbedded in the position in which they had grown. About the reefs are to be seen curious cylindrical blocks of coral standing on end, and often hollowed out at the top. These arise from the growing of a mass of ordinarily rounded coral until the top reaches the surface of the water or an insufficient depth to allow of further growth. The top of the mass then dies, whilst growth goes on at the sides, and the dead core is hollowed out by decay.

The surface of the rock in Tonga is covered with a reddish soil, like that of Bermuda. It is so hidden with soil and vegetation that it is very difficult to observe the rock structure. The wells, round holes sunk to a depth of four or five feet close to the shore, show a mere continuation of the reef-structure of the shore covered by about a foot of soil.

I was interested to recognise amongst the littoral plants of Tonga, many forms which I had gathered on the shores of the far-distant Bermuda. They were cosmopolitan tropical plants, and· became familiar objects on nearly all the tropical shores visited subsequently. One plant grows in Tonga which is almost identical with one occurring in Kerguelen's Land, but it again is cosmopolitan, and a water weed, *Nitella flexilis*. To remind one of Australia, there are Casuarina trees in Tonga, but they are nowhere abundant.

In every direction in Tonga are large tracts of land which have been under cultivation, but are now overrun with a wild growth, affording plain evidence of the reduction of the population. These tracts are overrun with a dense low tangle of several species of convolvulus and a trailing bean. The position

Communications by Mr. W. Mariner, severa years resident in those Islands." By John Martin, M.D., London, 1817.

of the more recent clearings is marked in the distance by the projection from the main mass of dark foliage of the dead branches of trees that have had their bark ringed. These, with a species of *Acacia* (?), which at the time of our visit in winter had a yellow tint upon its foliage, formed a marked feature in a general view of the vegetation from a distance.

There are naturally no indigenous mammals in Tonga except bats. A large Fruit-bat, probably *Pteropus keraudrenii* which occurs in Fiji and Samoa and also in the Caroline Islands,* is very abundant. These Fruit-bats appear on the wing in the early afternoon in full sun-light, and at the time of our visit were feeding on the bright red flowers of one of the indigenous trees. Flowers form an important proportion of the food of Fruit-bats. In New South Wales, at Botany Bay in May, numbers of Fruit-bats were to be seen feeding on the flowers of the gum trees. The bats must probably often act as fertilizers, by carrying pollen from tree to tree, adherent to their fur.

As dusk comes on, the Fruit-bats on the wing become more and more plentiful. It is probably only those specially driven by hunger that come out before dusk. Besides these large bats, there are small Insectivorous bats in Tonga, which dart about amongst the cocoanut trees, but we obtained no specimens. The heavy flap flap of the Pteropus is as strongly contrasted with the rapid motion of the true bats, as is the flight of a goose with that of a swallow. There are plenty of horses and cattle in Tonga, and the high ground of Eua is occupied as a sheep run.

A small Heron (*Demiegretta sacra*) wades about on the coral reefs at Tonga, and catches small fish, and is also to be seen frequently inland all over the island. This bird changes its plumage from pure white to uniform grey, and all stages of parti-coloured plumage were to be seen during our visit. Contrary to the usual rule, the bird is white when young, and dark in the mature state. Hence the ancestors must have been white, and the race is assuming a darker plumage for protection.

In the groves, the most abundant bird is one about the size of

* "Journal des Museum Godeffroy, Heft II. 1873." "Die Carolinen Insel Yap oder Guap."

a sparrow; brown with yellow wattles (*Ptilotis carunculata*). The bird has a sweet and very loud song, and fills the woods with its melody. A Kingfisher (*Halcyon sacra*) is constantly to be seen sitting on dead twigs, ready to dart on its prey. Amongst the cocoanut trees a beautiful little Swift (*Collocalia spodiopygia*), of the same genus as the species by which the edible birds' nests, the well known Chinese luxury, are made, and which is a Swift, and not a Swallow, as it is commonly called, skims about with a constant twittering. These Tree-swifts are especially abundant about the villages, though they nest in the crowns of the cocoanut palms.*

In the thickest masses of foliage, a most beautiful small Fruit Pigeon, of a bright green, with a patch of the purest purple on its head (*Ptilinopus porphyraceus*), is to be heard cooing gently, and the great Fruit Pigeon (*Carpophaga pacifica*), the note of which is harsh and drawling, but still derivable from a coo, is to be shot with ease by creeping up to the trees on the berries of which it is feeding at this season.

There are two Parrots known from Tonga, but they are very scarce. One of them, *Platycercus tabuensis*, is found only in Tonga and in the neighbouring island of Eua. It is called the Pompadour Parrot, from the peculiar purple red of its head and neck. The natives procure it alive from Eua, where it is abundant. One was bought for a shilling in the port during our stay. The other is a parroquet (*Coriphilus fringillaceus*), but is also scarce in Tonga. I saw neither of the parrots in the wild condition.

Lizards are abundant in Tonga, but of only two or three species. *Otosaurus microlepis*, one of the Scincidæ, is peculiar to the group. On the reefs an Eel (*Muræna*), whitish yellow-coloured spotted with brown, occurs. It is very snake-like in its movements, and I took it, on encountering it in the water, for the true Sea Snake (*Pelamys bicolor*), which also occurs here.

A large Foraminifer (*Obitolites*) is very common on the reefs.

---

* For an account of the nesting of Collocalia, see Bernstein, " On the genus Collocalia." Acta Societatis Scientiarum Indo-Nederlandicæ, Vol. II. For the nesting of the closely allied "Tree-swift," Dendrochelidon, see Bernstein, "Habits of Javan Birds," Ibid. Vol. III.

The shells, as large as threepenny pieces and like them in form, but of a chalky white colour, were to be seen in hundreds in the shallow pools. I preserved some of these in absolute alcohol, and observed that a green colouring matter was dissolved out in the spirit. On examining the soft structure of the animals, I found they were full of minute cells with very distinct transparent walls, which had all the appearance of unicellular algæ. It is possible that the green colouring of the spirit was due to the solution of chorophyly contained in the cells. The cells are evidently identical with those described by Dr. Carpenter, as existing in *Orbitolites,* and which he regarded as animal in origin, and describes as having a crimson hue in spirit specimens.* It seems just possible that they may be algæ, existing as parasites within the *Foraminifera.* If so, their presence would, as my friend Prof. Ray Lankester has pointed out to me, give further support to the hypothesis that the well-known yellow starch-containing cells of *Radiolarians,* are likewise parasitic vegetable organisms, and not essential components of the *Radiolarians,* in the bodies of which they occur.

Matuku Island, Fiji, July 24th, 1874.—We hastened along with the trade wind, and on July 24th were off the island of Matuku, one of the Fiji group, lying about 70 miles east of Kandavu. The island is volcanic, and surrounded by a barrier reef, which is about 16 miles in circumference. The highest peak is about 1,200 feet in height. I climbed to the top of this peak. From the summit the island was seen to consist of a single crater, the edge of which had been denuded and cut into a series of fantastic peaks, with intervening steep sided gullies. The ancient crater itself now forms the harbour, the inlet to which is through an opening in the girdling reef, at a spot where the border of the crater has been broken down. The surfaces of the irregular hills showed the peculiar sharp angled ridges so characteristic of volcanic cones denuded of pluvial action.

The windward side of the main peak was precipitous, and covered with thick vegetation, whilst the leeward side was

* W. B. Carpenter, F.R.S., &c., "Introduction to the Study of the Foraminifera," Ray Society, 1862, p. 35, Pl. IV. fig. 1.

open, covered only with grass and Pandanus trees. I was uncertain whether this condition was due to clearing by the natives or to the greater access of moisture from the trade wind on the windward side. Seemann describes such a condition produced by aspect, as common to all the Fiji Islands. There are however dense patches of wood here and there on the leeward side also of the crater in Matuku, and it may be that all the grass-covered area has been cleared at some time for cultivation, the island being too small and low to vary much in atmospheric conditions.

At all events the most prominent feature in the appearance of the vegetation of Matuku, is the contrast of the light green open grass slopes with the dark patches of wood. The grass is high and reedy, and very tiring to force one's way through, as are also the wooded tracts. Through these latter a road had to be cleared with the knife. In some places the grass had been fired by the natives, as a preliminary to cultivation.

The view from the summit of the island was most interesting as well as beautiful. We stood on what is now the highest point of the edge of the weathered crater. Beneath, on the one side, a steep slope led down to a narrow tract of flat land bordering the sea. This was partly open and swampy, covered with sedges and ferns, and with Pandanus trees dotted about over it, and partly covered with groves of cocoanut trees. On the other side, a vertical precipice, terminating in a similar steep slope, led down into the crater itself.

The cliff and internal slope of the crater were covered with thick and tangled wood, amongst which grew, even close to the summit, a few cocoanut palms, and one or two trees of the palm called "Niu Sawa" by the natives (*Kentia exorhiza*).

All round the island, except for a very short interval at the entrance to the harbour, was a circling zone of white breakers, marking the position of the barrier reef. The zone was separated from the shore of the island by a band of water, which had a slightly yellowish tinge, caused by its shallowness and the colour of the coral-built bottom.

The vegetation of Matuku is very different from that of Tonga-tabu, though no doubt much like that of Eua. Ferns are numerous instead of scanty, and amongst them a beautiful climbing species (*Lygodium reticulatum*) is abundant. I saw but few Casuarinas. In the woods the trees are almost hidden by a network of convolvulus.

The most conspicuous trees, except the Screw-pines and Cocoa-nut palms, at the time of our visit were those of a species of *Erythrina*,* which was in full scarlet blossom. On the honey of the flowers of this tree a most beautiful Lory (*Domicella solitaria*) was feeding, and with it some little Honey-birds (*Myzomela jugularis*). The Lory is one of the most beautiful little parrots existing, showing a splendid contrast of the richest colours, jet black, red, and green. It is peculiar to the Fiji Islands. It flies in flocks, and hence the term "*solitarias*" might lead to an erroneous impression.

A swallow (*Hirundo tahitica*) was flying about in considerable numbers, at the summit of the peak.

Hopping about on the mud, beneath the mangroves on the shore, was the extraordinary fish, *Periophthalmus*, at which I had often been astonished in Ceylon. This little fish skips along on the surface of the water, by a series of jumps, of the distance of as much as a foot, with great rapidity, and prefers escaping in this way to swimming beneath the surface. I have chased one in Trincomali Harbour, which skipped thus before me until it reached a rock, where it sat on a ledge out of the water in the sun, and waited till I came up, when it skipped along to another rock.

The fish are very nimble on land, and difficult to catch. They use their very muscular pectoral fins to spring with, and when resting on shore the fore part of their body is raised and supported on these. There seems to be no figure of this very remarkable fish which shows it at all in the attitude which it assumes when alive. The accompanying woodcut has been drawn from a specimen kindly lent to me by Dr. Günther, and I have put the fish as nearly in the natural position which it assumes when on land, as I can from memory.

* Erythrina Indica. The "Araba" flowers in August, the time to plant yams; hence the flowering of this tree is the basis of the Fijian Calendar. Seemann, "Flora Vitiensis."

The eyes of the fish, which is one of the Gobies, are remarkably prominent, projecting directly upwards from the skull,

PERIOPHTHALMUS KOLREUTERI.
On land; in act of leaping.

The fish in mangrove swamps often sits on the lower branches and roots. From what I have seen of its habits, I should expect that it would be drowned by long immersion in water. The Fijian species is *Periophthalmus Kolreuteri*. Dr. Günther, in his description of the genus, remarks: " these fishes are able to progress out of the water, on humid places, and to hunt after their prey which consist of terrestrial insects," &c.*

The natives of Matuku were mostly regular Fijians, though there were some pure Tongans amongst them, immediately to be distinguished by their use of the forehead muscles in expression. There is no doubt also mixed blood in the island. The houses of the people were miserably dirty, and built on filthy black muddy flats close to the sea.

I saw a boy make his way over a mangrove swamp, with remarkable rapidity, by crawling over the tops of the mangrove roots, and thus avoiding the mud below. Just so, the coast natives in parts of New Guinea are said to traverse the low swampy shore.

In dredging off Matuku Island, in 320 fathoms, on a coral bottom, some *Phorus*, *Turritella*, and a few other shells were brought up, as well as numerous specimens of the blind crusta-

* Dr. A. Günther, " Brit. Mus. Cat., Fishes," Vol. III. p. 97.

cean, *Polycheles,* and other animals, showing the fauna to be a true deep water one, and with these a living specimen of the Pearly Nautilus (*Nautilus pompilius*). This was the only specimen obtained during the voyage of this animal so rarely seen in the living condition by any Naturalist.

The animal was very lively, though probably not so lively as it would have been if it had been obtained from a less depth, the sudden change of pressure having no doubt very much disarranged its economy. It, however, swam round and round a shallow tub in which it was placed, moving after the manner of all Cephalopods, backwards, that is with the shell foremost. It floated at the surface with a small portion of the top of the shell just out of the water, as observed by Rumphius. The shell was maintained with its major plane in a vertical position, and its mouth directed upwards.

The animal seemed unable to sink, and the floating of the shell, as described, no doubt was due to some expansion of gas in the interior, occasioned by diminished pressure. The animal moved backwards slowly by a succession of small jerks, the propelling spouts from the siphon being directed somewhat downwards, so that the shell was rotated a little at each stroke, upon its axis, and the slightly greater area of it raised above the surface of the water.

Occasionally, when the animal was frightened or touched, it made a sort of dash, by squirting out the water from its siphon with more than usual violence, so as to cause a strong eddy on the surface of the water.

On either side of the base of the membranous operculum-like headfold, which, when the animal is retracted, entirely closes the mouth of the shell, the fold of mantle closing the gill cavity was to be seen rising and falling, with a regular pulsating motion, as the animal in breathing took in the water, to be expelled by the siphon.

The tentacular-like arms contrast strongly with those of most other Cephalopods, because of their extreme proportional slightness, and also their shortness, though they are not shorter proportionately than those of the living Sepia. They are held by the animal whilst swimming extended radially from the head, somewhat like the tentacles in a sea anemone ; but each pair has its definite and different direction, which is constantly maintained. This direction of the many pairs of tentacles at constant but different angles from the head, is the most striking feature to be observed in the living Nautilus.

Thus, one pair of tentacles was held pointing directly downwards. Two other pairs, situate just before and behind the eyes, were held projecting obliquely outwards and forwards, and backwards respectively, as if to protect the organs of sight. In a somewhat corresponding manner, the tentacular arms of the common cuttle-fish, whilst living, are maintained in a marked and definite attitude, as may be observed in any Aquarium.

The very great abundance of the shells of the Pearly Nautilus is most strangely contrasted with the rarity of the animal belonging to them. The circumstance is no doubt due to the fact that the animal is mostly an inhabitant of deep water. The shells of *Spirula* similarly occur in countless numbers on tropical beaches, yet the animal has only been procured two or three times. We obtained one specimen during our cruize, which had evidently been vomited from the stomach of a fish.

I expect that both *Nautilus* and *Spirula* might be obtained in some numbers, if traps, constructed like lobster-pots and baited, were set in deep water off the coasts where they abound in from 100 to 200 fathoms. *Nautilus* is occasionally caught both at Fiji and in the New Hebrides, in this manner, in comparatively shallow water, and the animals were so taken in the time of Rumphius, at the end of the seventeenth century. Traps seem never to have been tried for them in deep water.

The fact that the living *Nautilus* was obtained from 320 fathoms, shows that it occurs at great depths. It is probably a mistake to suppose that it ever comes to the surface voluntarily to swim about. It is probably only washed up by storms, when injured perhaps by the waves. The living specimen obtained by us seemed crippled, and unable to dive, no doubt because it had been brought up so suddenly from the depths.

The following is a translation of the account given of the habits of the animal by Rumphius, whose figure of the animal, as seen when taken out of the shell, is probably still the best extant.* " When the living *Nautilus* floats at the surface of the water, it protrudes its head with all the tentacles out, and spreads these out in the water, keeping the hinder part of the curl of the shell all the while above water. On the bottom, however, the animal creeps with the other side uppermost, with the head and tentacles on the bottom, and makes tolerably fast progress.

" The animals remain mostly at the bottom, creeping some-times into hoop nets set for fish, and lobster-pots; but after a storm, when the weather becomes calm, they are to be seen floating in troops on the surface of the water. They are doubt-less raised up by the waves caused by the storms. It follows that they keep themselves together in troops on the bottom also. The floating, however, does not last long, for drawing in all their tentacles, the animals turn their boats over, and go down again to the bottom.

" On the other hand, the empty shells are frequently to be found floating or cast up on the shore, for the defenceless animal, having no operculum, is a prey to crabs, sharks, and crocodiles ; and therefore the shells are mostly found with the edges bitten off. Since the animal does not adhere fast to its shell, its enemies can easily drag it out, leaving the empty shell to float.

" The young of this *Nautilus*, not larger than a Dutch shilling, are of a clean mother-of-pearl colour within and with-out. The rough shell substance overgrows the mother-of-pearl only after a time, and this overgrowth commences from the foremost part of the boat.

" The *Nautilus* is found in all the Moluccan islands, and also around the Thousand Islands off Batavia in Java, yet mostly only the empty shells are met with, for the animal is seldom found unless it creeps into the lobster-pots.

" The animal is used for eating, like other ' Sea cats '; but it is somewhat harder in flesh and difficult of digestion. The shell is in much greater request, for the manufacture of the beau-tiful drinking vessels so well known in Europe."

It appears from Dr. Bennett's notes on various species of

* D'Amboinsche Rariteitkamer door, G. E. Rumphius. Amsterdam, 1705, p. 61, Taf. XVII. Fol. 62.

Nautilus, that the natives in the New Hebrides dive for *Nautilus Macromphalus,* and also take it in fish-falls baited with an *Echinus,* whilst the Fijians trap *Nautilus Pompilius,* with a boiled "Rock lobster" for a bait.*

* Dr. G. Bennett, F.R.S., &c., "Proc. Zool. Soc. 1859," p. 226–229.

CHAPTER XIII.

FIJI ISLANDS.

Position and Area of the Islands of the Group. Kandavu Island. Grind-
stones for Stone Adzes. Shooting Birds in the Woods. Terrestrial
Hermit Crabs. Visit to a Barrier Reef. Ovalau Island. Excursion
to Livoni. Fijian Convicts. Log Drum. Native Hairdressing.
Kaava Drinking. Buying Stone Adzes. Excursion to Mbau Island.
Structure of the Island. Na vatani tawaki. Relics of Cannibalism.
Interview with King Thackombau. Connection of Wooden Drums
and Bells. Excursion up the Wai Levu. Sugar Plantations at Viti.
Freshwater Sharks. Joe the Pilot. Fijian Fortifications and Tombs.
A Chief's House and his Children. A Missionary Meeting. Various
Modes of Painting the Body. Grand Dancing Performances. Primi-
tive Origin of Music, Poetry, and the Drama. Wesleyan Missionary.
Albino Native. Congregation of Races at Levuka. Fijian Modes of
Expression. Laughter. Cicatrization. The Ula. Particulars con-
cerning Cannibalism.

Fiji Islands, July 25th to August 11th, 1874.—We arrived at
Kandavu Island, Fiji, on July 25th, and stayed at Fiji till
August 10th, the ship making a short trip to Levuka in Ovalau,
and returning to Kandavu to complete a survey of the harbour.
Ngaloa Bay.

The Fiji Group is scattered over an area of about 40,000
square miles on either side of the meridian of 180° W., between
lats. 16° and 20° E. The meridian of 180°, roughly speaking,
runs northward through the western end of the Aleutian chain,
and between Behring's Straits and Kamschatka, and southward,
passes just to the east of New Zealand. In latitude, the Fijis
correspond roughly with Tahiti, Rio Janeiro and Rodriguez. In
the northern hemisphere with St. Thomas in the Virgin Islands,
St. Vincent, Cape Verde Islands, and Bombay.

The land surface of the Fijis is about 7,000 square miles in
area, or about 1,000 square miles in excess of that of Yorkshire,
and there are about 150 islands in the group, excluding the very

small ones.* Viti Levu, the largest island, is 94 statute miles long by 55 broad.

The town or village at Ngaloa Bay, in Kandavu Island, was, at the time of our visit, miserably small, consisting of a few native huts, with three or four small stores kept by Europeans, and a whisky shop.

The main bulk of the island of Kandavu, as of that of Ovalau, is made up of a coarse conglomerate, composed of rounded fragments of volcanic rock. The surface of the islands is worn by denudation in such a manner as to present, as viewed from a distance, the appearance of a series of obtuse-angled triangles, rising one above the other. These are more numerous and less distinctly defined towards sea-level, whilst above, their apices form a line of peaked mountain-summits. The lower triangles are the foreshortened secondary ridges, formed on the mountain slopes by denudation. They struck me as having a more than ordinary uniformity of slope and general features in the Fiji Islands.

The whole of these slopes and ridges in Kandavu and Ovalau are covered with a dense dark green forest growth, except where, in some places, patches of land, often of large extent, and always very conspicuous, have been cleared for cultivation. The village at Ngaloa Bay is built at the mouth of a small rocky mountain stream which affords a pleasant bath. The Fijians still make use of a bow and arrow to shoot small fish in the stream, using arrows with several jagged prongs. On the banks of the stream, the surface of the live rock is in several places covered with deeply scored grooves, having been used formerly by the natives for grinding and shaping their stone adzes. I fancy most of the grinding work was done by the women, and when I see a finely polished Celt, I always picture to myself the male savage getting a stick and hammering his wife occasionally until the stone assumed the desired form.

---

* The whole Fiji Group, *exclusive of Coral islets*, includes an area of about 5,500 square miles of dry land, while at the period when the coral commenced to grow, there were at least, as the facts show, 15,000 square miles of land, or nearly three times the present surface. J. D. Dana, "Coral Reefs and Islands," p. 94. N. York, Putman, 1853.

Thus the man procured it with the least possible expenditure of labour on his part. Similar grinding places, with grooves cut in the rock, whither natives used to come to grind their stone axes, are known in Australia.

There are no roads in the island of Kandavu, merely narrow tracks through the woods and along the shores, which it is excessively tiring to traverse. I made one shooting excursion at Kandavu. The route lay first amongst beds of reeds on a small expanse of flat land at the mouth of the valley in which the stream runs ; then skirting a mangrove swamp bordering the shallow interior lagoon part of the bay, led amongst " taro " beds, and up a steep slope into the densely tangled woods. Here the trees were matted together with creepers overhead, and climbing ferns (*Lygodium*) twined up the trunks in the shade beneath.

Two young Fijians went with me. We climbed the steep dark path for a long time without hearing any bird at all. To see a bird without having heard it first was, from the denseness of the foliage, impossible. At last we heard a curious low whistling cry of two constantly repeated notes. The natives soon made out the bird overhead, but it was long before I could get a glimpse of it amongst the leaves, and as they kept bringing me nearer and nearer, in order to show me the bird, I was so close at last that it was nearly knocked to pieces by a charge of No. 12 shot. It is a constant difficulty in collecting birds in these dense tropical woods, that the birds are only able to be distinguished at very close quarters.

The bird proved to be a new species of Pigeon, *Chrysœna viridis* (Layard), peculiar to Kandavu Island. It is small and of a yellowish-green colour with a yellow head. The pigeons of the genus *Chrysœna* have a very remarkable structure in the feathers of the breast and neck. The barbs of these feathers are devoid of barbules, but are provided instead with a series of small swellings, ranged at intervals along them. The plumage of the bird has thus, to the naked eye, a peculiar loose appearance.

The Kandavu Island birds were formerly erroneously supposed to be the young of another Fijian species, *Chrysœna luteovirens*, and we thus, considering all our specimens to be young, concluded that this circumstance explained the peculiar whistling note of the birds, which is quite unlike that of other full grown pigeons.   We obtained a specimen of a closely similar bird from Taviuni, in which the plumage is of the brightest orange (*Chrysœna Victor*).

As we crossed a small clearing, I shot a large Fruit-pigeon (*Carpophaga pacifica*) which flew across; the same bird which is so common in Tonga.   On returning to the bottom of the valley, we heard the loud screams of the brightly coloured parrot, *Platycercus splendens*.   There were a pair of the birds, but they were so wild that I could not get a shot.   They are, however, not usually wild, and a large number were shot by some of the officers of the ship.   By the bank of the stream I found a pair of the Kingfisher, which is so common in Tongatabu, *Halcyon sacra*.

A large green Lizard, which is found at Kandavu and, I believe, in the other members of the Fiji group, was brought to us alive.   The Lizard (*Chloroscartes fasciatus*) is an Agamid, of a genus peculiar to the Fiji group.   It measures more than two feet and a-half in length.   It has a pouched throat with a cross fold.   All the scales of the body are keeled, and it has a low crest of triangular scales on the neck.*

In all parts of the Fijis which I visited, I met with abundance of a land-inhabiting Hermit Crab of the genus *Cœnobita*, allied to the well-known crab *Birgus latro* of the Philippines and elsewhere, which feeds on cocoanuts.   *Birgus latro* is apparently a Hermit Crab, which has given up using a shell to protect itself, because it has grown too large to be contained by any shell.   It has therefore developed, as a substitute, a hardened covering to the hinder part of its body, which was, no doubt, soft, as in other Hermit Crabs, when it wore a shell.   The Hermit Crabs of the genus *Cœnobita* are smaller, and always wear shells like marine Hermit Crabs.

On one small coral island, off the mouth of the Wai Levu, the beds of the littoral Convolvulus (*Ipomœa*) were swarming with these air-breathing Hermit Crabs, carrying about with

For a description of this lizard, by Dr. Günther, see " Proc. Zool. Soc. 1869," p. 189, Pl. XXV.

them all kinds of shells in the hot sunshine. In Kandavu they climb the hills and go far inland, bearing their shells with them, as do the terrestrial *Paguridæ* in St. Thomas and other West Indian islands.

On the shores of Wokan Island, in the Aru group, a small species of *Cœnobita* was extremely abundant on the stones and about the dry rocks above tide-mark. When alarmed they withdraw their claws and heads suddenly into their shells, and drop off their support as if feigning death. In one place at Aru I came upon such numbers of them, that their shells made quite a distinct slight rattling noise, as a drove of them alarmed let go their hold, and their shells fell amongst the stones.

But what has impressed most deeply upon my memory the fact of the existence of these terrestrial Hermit Crabs, was a surprise which I encountered at the Admiralty Islands. When collecting plants there, I thought I saw a fine large Land Snail resting on one of the topmost twigs of a bush about four feet in height. I grasped the specimen, but instead of feeling the slimy snail's body, I got a very unpleasant bite from a large Hermit Crab, and I then saw that the shell was a marine one (*Turbo*).

The genus *Cœnobita* has one of its nippers especially stout and powerful. In the Admiralty Islands a species gnaws the roots of one of the littoral trees (*Calophyllum inophyllum*). I have seen 20 or 30 of these crabs gnawing at one long wound made by them in a root, apparently feeding on an exuding gum.

Professor Semper of Wurzburg has examined the breathing apparatus of the Cocoanut Crab (*Birgus lutro*), and finds* that a large cavity on the back, commonly called the gill cavity, has the function of a true lung. By means of blood-vessels in its walls the animal breathes air directly. This cavity has been commonly said to contain water, by which the animal was supposed to moisten its gills, in order that it might breathe through its gills alone. The breathing of the animal by the gills when on land is considered by Semper as secondary. Similarly, the gill cavity acts as a true lung in other Land crabs.

* "Ueber die Lunge von Birgus latro." Zeitschrift für Wiss. Zoologie, 1878, s. 282.

At Kandavu I had an opportunity of visiting the outer margin of a barrier reef. It was one of the reefs stretching across the mouth of Ngaloa Bay. As such a reef is approached from behind in a boat, and viewed from sea level, nothing is visible of the reef itself at a distance but a line of small detached masses of rock which appear here and there, standing out dark against the horizon. As the waves approach successively the different portions of the reef, their crests are seen rising dark above the reef-line. Then as the waves break against the margin of the reef, the isolated rock-masses show out in relief against the white background of foam.

As the reef is approached more closely, the water becomes shallower, and assumes a yellow tinge, caused by the light reflected from the growing corals. The boat now requires to be steered with care along a zigzag path between coral patches, and at last grates on the growing coral as the water shallows rapidly towards the margin of the reef, and it becomes necessary to wade in order to proceed further.

It is in the shallow sheltered water, inside the actual edge of the barrier, that the finest and best grown specimens of the corals are to be found. The tufts, bushes, and rounded masses of the various corals are to be seen growing here in abundance, but yet scattered over the area, with plenty of more or less barren interspaces in the "coral plantation," as Dana terms it. The various forms of the spongy tissued Madreporas, are the characteristic feature in these Fijian reefs, there being no less than 26 species of Madrepora known from Fiji.

The outer margin of the reef is raised above the level of the coral plantation in the still waters within, and the water on it is thus very shallow at low tide, and often the margin is laid dry. At Ngaloa Bay the barrier reef springs from the fringing reef, running out from the coast across the mouth of the bay. Its elevated margin was not more than 20 to 30 yards wide. There is an elevated strip of about this width stretching all along the reef; its surface is remarkably even, and but few stunted corals were growing upon it, but *Alcyonarians* were abundant, and the whole surface was covered with a crust of calcareous seaweeds (*Corallinaceæ*).

The water on the reef edge was usually not much more than ankle-deep, but the breakers sent from time to time so strong a current inwards across the barrier, that it was difficult to keep one's footing. On the reef were resting irregularly shaped masses of solid stony corals, portions of various *Astrœidæ, Poritidæ*, or of reef rock, thrown up upon the marginal platform of the reef by the surf, and reminding one, as they rested in all sorts of positions, of the scattered rock fragments on a glacier. Sometimes they even rest on a narrowed support like " table-stones," having become first cemented to the platform, and subsequently gradually undercut by the waves. Dana has figured such tablestones. It is these thrown-up fragments which are, as has been described, the only portions of the actual reef visible from a distance.

The chief differences between the fauna of the Fijian reefs and those of Bermuda, are the absence at Fiji of any large quantities of coral formed by *Milleporidæ* and large branching *Oculinidæ*, and the absence of the large flexible *Gorgonidæ*, which form so striking a feature at Bermuda. The great abundance of Madrepores forms the characteristic feature in the Fijian reefs. I saw, however at Fiji, no Madreporas so large and fine in growth as those of St. Thomas.

On the reef-margin, by turning over the cast-up rock frag-

ACROCLADIA MAMILLATA.

ments, I found a few cowries, some huge *Trochi*, also specimens of *Turbo operculum*, and other shells. Various Holothurians and a large bright ultramarine-coloured Starfish (*Ophidiaster*), were in countless numbers, and some splendid Sea-urchins, with huge thick spines (*Acrocladia mamillata*), were found. A Shark appeared in the shallow water showing its back fin high out of it ; the fish was chased with boarding pikes by the Blue-jackets, but was too wary to allow its pursuers to come within reach. Captain Nares set up his theodolite on the reef, and took angles whilst we collected specimens.

Whilst at Levuka (in Ovalau Island), I made a trip with Lieut. Suckling, R.N., over the steep mountain ridge which backs the town, to the native villages of Livoni and Bureta. A corporal of the Fijian army and two prisoners, natives of Livoni, were sent by Mr. Thurston with us as guides.

The track led up the bed of a rocky mountain stream, and at times up nearly perpendicular faces of rock, which were, however, easy·to climb because of the nature of the rock already alluded to the harder embedded masses in the conglomerate weathering out so as to project and form foot-rests and convenient grasping places for the hands As we ascended, the soil became moister, the wood denser, and the trees more and more covered with epiphytes.

Now and again we passed small cascades tumbling into basins amongst the black boulders. The rocks around were overgrown with ferns and mosses in great variety ; wild plantains and beautifully variegated *Dracœnas* grew in abundance, and amongst them the scarlet *Hibiscus* in full flower. The overhanging tree-stems were green with climbing ferns, or served as supports to climbing Aroids with large fenestrated leaves. The beauty of the various features of this mountain stream are, however, far beyond my powers of description.

Near the summit of the ridge, the tree stems and branches became covered with orchids, and in places were loaded with dense masses of the bird's-nest fern (*Asplenium nidus*), and large Lycopods and mosses. On the summit, a hard chase after a rat ensued, as I offered a shilling reward for the animal, which might have proved at this elevation, I thought, a Native Rat, though the black rat and Norway rat are abundant in Levuka. There was, however, so much cover for the rat under the decayed logs and undergrowth, that it soon escaped.

The ridge where we crossed it was very narrow, and we almost immediately commenced a steep descent down the bed of a stream on the other side. On the way down, a flock of Lories (*Domicella solitaria*, "Kula," Fijian), flew by, whilst the trees were full of warbling birds (*Ptilotis procerior*).

We reached Livoni, formerly a populous village, and the head-quarters of the Kaivolo or mountaineers of Ovalau, who long defied King Thackombau, murdered one of his envoys, and were the terror of the Levuka people. The place was now entirely in ruins, the inhabitants having been made prisoners, and their town burnt by Thackombau. There remain now, only the oblong mounds of earth on each of which formerly stood a house, and the ditch and bank of earth, with which the village was fortified.

The place is used now as a convict station, and here a number of prisoners, mostly Kaivolos, or "devil men," from the hill tribes of the large island "Viti-levu," were undergoing their various terms of imprisonment. Eight Tongan soldiers and an old English drill-sergeant were sufficient to keep the convicts in subjection. The men were made to work at clearing the surrounding land, and planting sweet potatoes and yams; whilst they were at work, the Tongans mounted guard over them with loaded muskets, and though the opportunities in the thick bush seemed so great, they were said never to escape; they are very much afraid of the Tongans.

I was shown amongst the convicts one of the Burns murderers, who was said to have been caught when dragging the body of a white woman by the hair through the bush, with a view to eating it. I put a few questions through an interpreter: the man protested that he had never eaten human flesh, and that he would have no desire to eat me if he had a chance. He had evidently learnt that this was the proper attitude to assume with regard to this question. I expected that he would have made no scruple in confessing to former Cannibalism.

A drum was used at Livoni for summoning the prisoners, which was new to me in its construction: three cylindrical holes were cut in the ground in a row, the central one being about twice as large as the others. They were about 1 foot and 6 inches in diameter respectively; over these holes a log of light *Hibiscus* wood was supported on two cross rests of rolled up palm-leaf mat, placed in the interspaces between the holes. The holes in the earth acted as resonators, and when

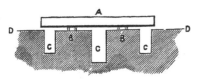

SECTION OF FIJIAN LOG DRUM.

a Log; *b b* rests; *c c c* resonating holes; *d* surface line of the ground.

the log was struck with a wooden mallet, a loud sound was produced as from the ordinary Fijian drum or " lali," which consists of a log hollowed out like a canoe; this was a rough substitute. The use of holes in the earth as resonators is remarkable.

Hearing that there was to be a " meke meke " or native dance at the next village, Bureta, we went on to this place, the path crossing and recrossing continually a stream running here through comparatively flat land, and in places as much as 20 yards across. We found numerous visitors in Bureta, many of whom had passed us on the road. All were dressed in their best, with bright new girdles of yellow and scarlet dyed Pandanus leaves, bodies and hair freshly oiled, ornaments displayed, and faces painted black or red or a mixture of both.

The various methods of dressing the hair are so numerous as to be indescribable. The thickly growing crisp mop of fine close curls is trimmed just as an old-fashioned yew hedge used to be. Sometimes a single thick tuft is left projecting from the back of the head, sometimes a diagonal ridge-like tuft, sometimes one, two, or more small plaited tails only, sometimes a curtain-like fringe shading the neck.

The hair is constantly dressed with shell or coral lime, both to kill vermin and to change the colour, and also, certainly, as a fashion. Most of the young Mbau chiefs that I saw had their hair always in this condition. These young chiefs cut their hair in front in a straight line across the forehead and square at the temples; and, in fact, trimmed it so that when whitened with lime it reminded one most forcibly of a barrister's wig. A young Mbau chief was on a visit at Bureta, and besides having his hair whitened, his face was blackened for the meke, and the contrast between black and white was most effective.

Kaava* drinking was going on in the chief's house at the time of our arrival, the young Mbau chief presiding at the ceremony. It is usual to decry kaava as a drink altogether, because, no doubt, of the nasty manner in which it is prepared, but some persons who habitually drink it praise it as extremely pleasant and cooling. Many of the resident whites at Fiji, as I was told, took kaava once or twice daily, and I knew personally of a German planter and an English settler who did so. It seems, however, to be only at Fiji, in Polynesia, that this occurs. In the Sandwich Islands and in Tahiti the Whites never think of drinking kaava, but scout the idea.

The taste is at first strange and unpleasant, and has often been compared to that of Gregory's mixture. Travellers usually never make more than one trial of the drink. The taste is, however, certainly not more unpleasant than that of London porter, for example, must be on the first occasion to Frenchmen. Great satisfaction must be derived by Polynesians from the use of kaava, or it would not have been so universally upheld as a drink amongst them, nor would its use have become associated as it is with an elaborate ceremonial.

Usually, when the party with which I travelled in the large island of Fiji entered a village, the chief of the village made a request, as an offer of hospitality, that we would drink kaava with him; and we sat on his right and left hand at the head of the circle, or rather long loop, formed by those present on such occasions. At the bottoms of the two sides of the loop were seated the servants, or a few of the lower orders of the village, who crawled in crouching and cringing, expressing their humility before the chief in the most ostentatious manner, looking indeed, sometimes, as if they were really half afraid to come at all.

The kaava is prepared at the opposite end of the loop from that at which the chief sits. Young men with good teeth are chosen to do the chewing, and they pay great regard to cleanliness, rinsing their mouths and hands carefully with water

* A solution in water of the chewed root of a Pepper (Piper methysticum). An intoxicating drink.

before they commence their task. There is a considerable amount of knack to be acquired in the chewing of the kaava root. If it is well chewed very little saliva should be mixed with it, and it should be produced from the mouth in an almost dry round mass about as large as the mouth can contain.

The masses produced by several chewers are mixed with water and the infusion is strained, as has been often described. The bowl is placed in front of the chief. It is a four-legged wooden bowl cut out of a single block. It has a string of cocoa-nut fibre fastened to it underneath to a loop cut in the wood. By this string the bowl, when not in use, is hung up against the wall in the chief's house. When the prepared bowl is placed before the chief it must always be so turned that the string is directed away from him. The chief is served first in his own private cocoanut shell. Then the others present, in order of their rank and position of their seats, receive shells full. We were always served immediately after the chief. It is the correct thing to drink the cocoanut-shell full off at a draught, and then spin the cup on its pointed end on the mat in front of one and say "amava," or a word sounding closely like this, meaning, I was told, "it is emptied;" in fact, "no heel taps." After the chief has drunk, the company all clap their hands in token of respect.

A considerable quantity of kaava, of a strength such as that of the infusion ordinarily drunk at Fiji, must be taken in order to produce intoxication; but I have known a single cocoanut-shell of strong Fijian kaava make an Englishman unaccustomed to the drink feel a little dizzy and shaky about the legs. There is a very great difference in the strength of kaava, depending very much on whether the portion of the root employed is young or old, and of course on the amount of water employed.

The infusion of the pepper-root is not allowed to stand so as to ferment, but some change probably is effected in the active principles by the action of saliva, for grated kaava, which is now used in Tonga, by order of the missionaries, as a substitute for the chewed preparation, is not so good as the latter. I have known three-quarters of an ordinary tumbler-full of Awa (the Hawaian form of the Polynesian name for the drink), specially prepared by an old woman, in Hawai, Sandwich Islands, as of extra strength, make an Englishman intoxicated within ten minutes of the time at which it was drunk.

The effects are very like those of alcohol, in that the gait becomes very unsteady, and the slightest touch sends the person affected off his balance. An elation of spirits is produced also, but apparently no drowsiness.

At Bureta I was able to buy, for sixpence each, a dozen stone Adzes, such as were used for canoe making in the Fijian group, before iron implements were imported. The adze blades are of basalt. They are bound to the handles with twisted or plaited cocoanut fibre. Many of these were still mounted on their

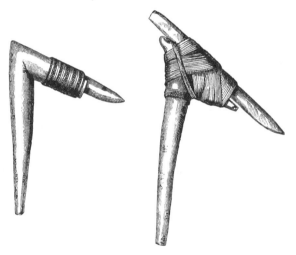

FIJIAN STONE ADZES.
Showing two methods in which the blades are mounted.

handles, and are now used by the people who have not parted with them, for cracking nuts. For an exactly similar adze I had paid six shillings in Levuka, and clubs which here were to be bought for a shilling, cost a dollar on the other side of the ridge. It is wonderful how little knowledge has penetrated as yet from Levuka to Bureta, so short a distance off. The natives could not understand a half-crown, nor could they be induced to give four sixpences for a florin. Threepenny-bits they would not take at all. " Sixpenny " and shilling they knew well. The young Mbau chief of course understood these things, and also thoroughly understood the working of my central fire breech-loading gun, he having one of his own at Mbau. Most of the chiefs have good English fowling-pieces and rifles.

After a long delay, and constant promises of a commencement, the dance was begun in a flat oblong open space in the village, which had a raised bank on two sides of it, on which the spectators assembled. As it got dark, bunches of reeds were lighted and held up around by girls to light up the dance, for the moon did not come up till late.

Only the young men, all visitors at Livoni, and belonging to the army, danced. We waited on, hour after hour, for the girls to commence, but they took so long in decorating themselves and getting ready, that after four hours' delay we were obliged to leave in a canoe which we hired for a dollar to make the journey to Levuka by sea.

We had no sooner left than the girls commenced dancing, and they probably waited for us to leave. I saw exactly the same dance as that performed by the young men executed afterwards in Viti Levu, many of the performers even being the same; I will therefore describe it further on.

We started in the canoe in the tidal part of the Livoni River at about 10 P.M., and it being low tide, and there being no wind, the canoe had to be poled the whole way down the river, and along the shore, except for short stretches, where deep water compelled the men to paddle. We had imagined that we had only five miles or so to go, but found that the river on which we were came out on the coast of Ovalau, beyond the end of the adjacent island of Moturiki, or almost at the very opposite side of Ovalau from Levuka. We stretched ourselves on the small outrigger platform of the canoe, but the motion was too irregular and the bed too unsteady to allow of much sleep. It was not till half-past 4 A.M., that we reached Lieut. Suckling's schooner.

At 6 A.M., on the same day, July 31st, I started on a cruise in one of the ship's boats, called the barge, to the island of Mbau, and the Wai Levu, with a party which was to join the ship again at Kandavu.

There being little wind all day, we failed in reaching Mbau on the first day, but arriving in its neighbourhood about dusk, we mistook a projecting headland of Viti Levu,* some miles north of Mbau, for the island of Viwa, and a small island lying off this headland for Mbau. It was impossible to distinguish in the gloom what were islands and what promontories, against the dark background of the Viti Levu coast.

All around Mbau, Viwa, and the neighbouring coast are extensive shallow coral and mud flats, the mud being brought down by one of the mouths of the River Wai Levu, which opens in the direction of Mbau. After making several attempts to reach the island which we supposed to be Mbau, and constantly grounding on the coral, we anchored in a deep channel between the coral flats for the night. In the early morning we made out Mbau, conspicuous from the white house of the missionary upon its summit, and soon reached it.

Mbau is a very small island, not more than half a mile in circumference. It consists of a central hill, of about 50 feet elevation, with a flat area at its top, and bounded by steep grass-covered slopes, surrounded by a tract of flat ground. The central mass is composed of a friable stratified rock, of a greyish or reddish colour. An exactly similar rock composes the main land immediately opposite the island, and the strata there correspond in inclination with those of Mbau. The central mass of the island is thus a small detached fragment left standing by the denuding waves. The passage between the mainland and Mbau is so shallow as to be fordable at high water, and is nearly dry at low water.

The flat lower part of Mbau which is raised only a few feet above the sea, consists of made ground, built up of blocks of coral, and mud and stones collected from the vicinity at low water, and secured all around against the action of the sea by means of large slabs of a sandstone (said to come from the main island), having been brought in canoes a distance of several miles. These stone slabs are set up on end, so as to form a parapet, and keep the earth from washing down. The slabs

* Viti Levu (pronounced Veetee la̅yvoo). Levu means " great." Settlers often clip the u, and talk of " Viti lib."

project far above the level of the land surface, and thus form at the same time a sort of fence or wall.   At intervals, openings are left in the parapet, where the water flows up short channels into the area of made ground, and allows canoes to put in at high water into small harbours as it were.

The top of the hill was formerly used as a general refuse heap by the natives, but it is now occupied by the house of the missionary.   The native houses all lie on the flat low tract close to the sea.   Mbau has been long a native fortress of great strength.   Hence the immense labour which has been spent on its formation.   It is now the residence of King Thackombau, and almost everyone in the island is a chief or of high family.

The whole surface of the island, including the hill-ground, is covered almost everywhere with a thick kitchen-midden deposit of black soil, full of large trochus-shells and cockles (*Cardium*), which abound on the mud flats all around.   Mingled with these are quantities of human bones; Mbau having been one of the places in Fiji at which cannibalism was most largely practised. There are very few trees growing on Mbau, and the food, such as taro and yams, is all brought from the main land, where there are extensive plantations.

One of the most interesting features in Mbau is perhaps the stone against which the heads of the human victims destined for the oven were dashed, in the ceremony of presenting them to the god Denge.   This stone stands close to one corner of the remains of the foundations of the ancient temple of Denge, the "Na Vatani Tawake."   The temple itself was destroyed when the Mbauans became Christians, but the mound on which it stood remains, and is of great interest.

It is a large oblong tumulus of earth, supported by two series of vertically-placed slabs of stone, exactly similar to those used for the sea parapet.   The slabs of the lower series are much larger than those of the upper, and the upper series is placed further inwards, a sort of step being thus formed in the tumulus all round.   The mound must be about 12 feet high, and some of the stones of the lower series are more than six feet in height.

Opposite the centre of one side is set up a large column of basalt, and there is another opposite the strangers' house.   These columns are said to have been taken in war, from some enemies on Viti Levu, and intended to have been used as posts for the king's house. The columns are however said by Dana* to have been brought by a Mbau chief from a small island in the harbour of Kandavu, which is composed of them, and where they were long desperately defended by the inhabitants, who held them sacred.

The whole mound most strikingly reminds one of ancient stone circles and such erections at home. Were the earth of the mound to wash away, numbers of the stone slabs might remain standing on end. I give a copy of a rough sketch which I took of the place in its present condition. Its condition before its destruction is to be seen in a book entitled "Fiji and the Fijians," by Thos. Williams (London, Hodder and Stoughton, 1870). The tumulus supported a large "Mbure" or temple, with the usual high-peaked roof and long projecting decorated ridge pole.

Now the mound is falling into decay and covered with grass, and a small pony (there are very few horses in Fiji, and of

Sacrificial Stone.
NA VATANI TAWAKI, MBAU, FIJI.

course only room for this one in Mbau) belonging to Ratu David, the king's eldest son, found the top of it a pleasant place to graze on. The pony had a quiet life, for Ratu David having

* Dana : " U. S. Expl. Ex., Geology," p. 348. The columns at Mbau are referred to by Capt. Erskine, "Islands of the Western Pacific," p. 193, London, J. Murray, 1853, who, however, did not recognise them as of unartificial formation.

been kicked off on his first attempt at riding, had not tried again.

The sacrificial stone, against which the heads of the victims were dashed, is an insignificant looking one, in no way different from the other slabs, except that it is smaller and stands by itself a little in front of them, near one corner of the mound. In front of it, in old time, bodies have been heaped up till they formed a pile ten feet high. Whilst I was sketching the mound and its stones, a very pretty daughter of one of the chiefs came and looked on, and at my request wrote her name and the Fijian name of the mound in my sketch-book, in a very good round hand.

There are several similar slab-built foundations of temples about the open space near the site of the Na Vatani Tawaki, but except in the case of one small one, they are not in such good preservation. The slabs from one of these are now being used to construct the foundations for a Wesleyan church. Conspicuous amongst the buildings close by, is the large "visitors' house," where guests were entertained, and, if of distinction, always provided with human flesh, at least once, by their hosts.

Beside the building, a slight depression in the turf is the remains of one of the ovens used for cooking the "long pig," for this is the actual name by which human flesh always went in the Fijian language. I always thought it a joke, until I was told by the interpreter. On a tree overhanging the ovens are to be seen notches, cut in the trunk from its base to its summit, an old score of the number of victims cooked beneath.

There is another stone, not far from Thackombau's house, which is smooth, and somewhat like a millstone in appearance. The ground around this is paved with slabs of coral rock, which had been perforated with holes by boring mollusks and worms before it was taken from the water. So many heads have been dashed against this stone, that it has happened that human teeth have fallen into almost all the holes in the slabs, and have becomed jammed there. The slabs were quite full of them.

This second stone was seen by Captain Wilkes' officers, and is mentioned by Brenchley. We were told by the people that a second ceremony was performed at it, the heads of bodies being a second time pounded to pieces here, in honour of the slayer, who drank kaava from some grooves which are to be seen in the slab in front. The grooves are however very irregular, and look much rather as if they had been made in sharpening stone axes. I think this second stone must have been used by a separate tribe, occupying this quarter of Mbau, for even on this small island the people were often much divided.

On going up the hill we came suddenly upon two old women bathing in a fresh water pool. They made for deep water in a hurry, but I saw that they were tattooed of a uniform indigo blue colour, from the hips to near the knees, just like the Samoan men.

King Thackombau was visited in the morning by two of our party, who took him by surprise; he was found lying on his stomach, reading his Bible. I went with a party and we were regularly announced. The king, who was dressed in a flannel shirt, and a waist cloth reaching to his knees, rose to receive us, and came forward and shook hands. He is a very fine looking man, six feet high, with his dark face set off with abundance of grey hair. His eyes are bright and intelligent, and his face full of expression, and in this respect very different from that of the ordinary Fijian of lower rank.

Three chairs were produced, but this was the whole stock in the house, and those of our party without chairs sat on the matted floor. The king reclined on his stomach as before, on his own peculiar mat, at the head of our circle, with his Bible and Prayer Book neatly piled on the right hand front corner of the mat. We said, through our interpreter, that we were glad to see His Majesty looking so well, and explained the nature of the voyage we were making in the "Challenger." I was then deputed to give an account of the wonders of the deep sea. In this subject Thackombau took the liveliest interest, inquiring about what kinds of animals existed in the deep water, evidently knowing the shallow-water ones well. He was very much interested in the fact that they are so often blind. He said he could not understand the depth in miles, but comprehended it perfectly in fathoms.

He then inquired the strength of the various navies, asking after that of England, Germany, France, Russia and America, and wanting to know even the numbers of wooden and iron ships. The information we gave him drew from him the remark that the English were a wonderful people, far greater than the Fijians.

The house was a large barn-like one of ordinary Fijian structure, with tall open roof, and a sleeping place separated off at one end with a "tappa" curtain. There was the usual square hearth, with its edging of stone. Overhead were stored the heads of canoe masts. A European chest of drawers, a table, a lamp, and two tin coffee pots, were the only visible articles of luxury. Against the door-post hung a fine club, freshly painted blue, belonging to the king's youngest son.

We asked the king for a pilot, to take us up the mouth of the Wai Levu, the great river which opens nearly opposite Mbau. He sent out at once to order one for us, and we took our leave of this knowing old Christian, who is currently reported to have partaken of 2,000 human bodies, and is certainly known to have cut out, cooked and eaten a man's tongue, in the man's sight, as a preparation to putting the rest of him in the oven, and that merely to spite the man because he begged hard not to be tortured, but to be clubbed at once.

The contrast between Thackombau and King George of Tonga was very striking, at least as far as concerns their behaviour before visitors. Thackombau took the liveliest interest in everything, and put question after question, whereas it seemed impossible to interest King George in any subject. He said nothing at all during our interview. Both are warriors of renown, and fought their way to their positions.

Ratu David the eldest son of Thackombau was very hospitable, and invited us to drink kaava with him in the evening, when he produced a bottle of brandy also. We wished to see a dance, but this was not possible, because it was Saturday evening, which is by order of the missionaries kept in a certain way sacred, as a preparation for Sunday. For the same reason Ratu David dare not allow his retinue to sing a chant used during kaava drinking, and which we were anxious to hear.

We pitched a sort of tent on a very small islet, about forty yards off Mbau, and slept there. Ratu David sent us off a young pig and a couple of fowls all alive, a most welcome present. They were killed and consumed within an hour of their arrival. The islet on which we slept is made up of blocks of coral, weathered and bored by various animals, piled up by the waves. The blocks near tide-mark are so blackened by exposure, that I took them at first for vesicular lava.

Around Mbau are extensive shallow mud flats, the mud being brought down by the Wai Levu. Across these flats we sailed next morning, with scarcely a breath of wind, though our pilot, whom we christened " Joe," kept constantly calling for a breeze, using an old Fijian pilot's chant, " Come down, come down, my friend from the mountains."

As we drifted slowly away over the glassy water, the view behind us was beautiful. Far away, blue in the distance, was a long range of the lofty peaked mountains of Viti Levu, still the abode of the Kaivolos, the long-haired mountaineers, the canni-bals. Nearer lay a streak of dark green undulating low country, bounded seawards by low cliffs, and showing near the coast the numerous cultivated clearings of the natives. Just off the cliffs of Viti Levu lay the small island of Viwa. In the foreground was the island of Mbau, with its crowded reed houses, its strange stone parapets, and its green hill topped by the missionaries' white house. From the centre of the village came the sound of what was the old cannibal death drum, beating now for morning prayers.

There were two of these drums in front of the strangers' house. They are simply logs of wood, hollowed out above into troughs, and supported horizontally on posts at about three feet above the ground, looking like horse-troughs. One was larger than another. They were beaten with two wooden billets alter-nately, and gave out different low bass booming notes. Very similar drums are used amongst the Melanesians, as at Efate in the New Hebrides,* and at the Admiralty Islands, where however they are stuck upright in the ground, and the mouths of the trough-like cavities are contracted to narrow slit-like openings, the trunks being hollowed out through these. The

* "A Year in the New Hebrides," p. 111, by F. A. Campbell. Melbourne, George Roberston, 1873.

Japanese wooden bell, or narrow-mouthed wooden drum, seems to be merely a more perfect development of these drums, and no doubt the actual bell was derived from the copying of some such wooden instrument in metal. The addition of a clapper to a bell is a late improvement. Japanese bells still have none, but are sounded by means of a beam of wood, swung against them from outside. The term "drum" should perhaps be restricted to instruments with a tense membrane.

As a musical instrument, our ordinary English Chapel Bell is much on a par with the Fijian drum, and makes an equally uncultivated and unpleasant noise.

The great river, the Rewa River, or Wai Levu (great water) opens into the sea by several mouths. We ascended by the northernmost. About the mouth of the river the land is flat and alluvial, and the river is bordered on either hand by a thick growth of mangroves. Below these trees, slimy mud slopes are left bare at low tide, on which a *Periophthalmus** hops about on the feed just as a frog might hop about. Close to the sea the mud is covered with a sea grass (*Halophila*), and hence looks greenish when left uncovered. Ducks (*Anas superciliosa*) are common on the mud at the river's brink, as is also a Heron (*Demiegretta sacra*), which pitches often in the Mangroves. The *Ptilotis* sings amongst these mangroves, and the Parrot *Platycercus splendens* screams amongst them.

After a stay at Novaloa, where there is a mission college for training native teachers, and where Fijians learn even rudimentary algebra, we drifted up with the rising tide, grounding once and having to wait an hour to float off again. We passed many villages, and several canoes full of people. We slept at Nadawa, where a small paddle steamer, the property of a trader living there, Mr. Page, and built by him there, was under repairs and waiting for new engines from Sydney. Here also was a sort of Hotel kept by two Englishmen. Mr. Page, who was extremely hospitable, gave me a bed.

In the morning we had to beat against the land breeze up the main river, which we had entered just below Nadawa. The Wai Levu is a fine large river, in some reaches 300 yards across,

* See page 296.

and in occasional flood time pouring so much fresh water into
the sea, that ships at anchor three miles off its entrance are
able to take in their store of drinking-water from the water
alongside them.* Dana calculates the volume of water poured
in Rewa Harbour at 500,000 cubic feet per minute, and that
discharged by all the mouths of the river together at 1,500,000
cubic feet. The area of the Delta is 60 square miles.

The mangrove thickets had ceased before the main river
was reached, and here above Navusa the low banks on either
hand were hidden by a dense mass of a tall grass, a species of
Saccharum, or wild sugar-cane. For the first twelve miles or so
of its lower course, the river flows through its delta, and hence
the banks are low and the country flat. Some few miles above
Navusa the banks become steeper, and low hills commence.
These gradually become more frequent as the ascent is continued,
until steep slopes, with intervening stretches of flat land, are of
constant occurrence on either hand. The view up the river now
shows a succession of ridges, one behind the other, rising gradually in the distance, and terminating in a line of distant blue
mountains.

The steep slopes are covered with a thickly interwoven vegetation, the large trees being covered with Epiphytes, Ferns,
Lycopods, and climbing Aroids, and festooned with creepers.
These creepers in places form a continuous sheet of bright green,
falling in gracefully curved steps from the top of the slopes to
the bottom, and almost entirely concealing their supports. Here
and there tall Tree-ferns rear their heads amongst the tangled
mass, and palms (two species of *Kentia*) form a conspicuous
feature amongst the foliage.

We were forced to anchor in the evening to await the turn
of the tide. As it became dusk numbers of Fruit-Bats flew overhead, whilst in the beds of reeds a constant cry was kept up by
the coots and water rails. On the tide turning we had to take
spells of an hour each at the oars as our time was short, and by
paddling on gently all night we reached before daylight a spot
about 35 miles from the mouth of the river called " Viti."

At Viti, a Mr. Storck and his wife live. Mr. Storck is a

* Dana, " Geology of United States Expl. Exp.," p. 348.

German, and was the assistant of Mr. Seemann during his investigation of the plants of Fiji. He was extremely hospitable. He had taken to growing sugar, as cotton had failed, and had a splendid crop, which he calculated to weigh 62 tons of cane to the acre. Mills were about to be erected, and there seemed every prospect of sugar paying well. There were already 20 plantations of sugar on the Rewa River. It was curious to see a man from the New Hebrides islands, so notorious for the murders of white men committed in them, acting as nurse to one of Mrs. Storck's children, and hushing the baby tenderly to sleep in his arms. He was one of the imported labourers, concerning whom so much has been written.

About Viti there are abundance of large Fruit-Pigeons, of the pigeons with purple heads, identical with those of Tongatabu (*Ptilinopus porphyraceus*); also of the "Kula" (*Domicella solitaria*), and the "Kaka" (*Platycercus splendens*). The Kaka attacks the sugar-canes, and does considerable damage. There are some huge fig-trees at Viti, with the typical plank-like roots and compound stems. Here also grow one or two cocoanut-trees, which are rarities so far up the river, for at the inland villages along the river there are no cocoanut-trees, and a regular trade is carried on by the natives in bringing the nuts up the river from the coast, in canoes, to barter them with the inland people.

The Black Rat and Norway Rat are abundant at Viti, and there is also a native Field Mouse, according to Mr. Storck, but I could not procure one in our short available time. I do not know whether a field-mouse is known from Fiji. A large fresh-water Prawn is common, and is caught for eating by the Fijian women, and in their baskets I saw also an Eel (*Murœna*).

A red stratified sandstone, with a slight inclination of its strata, is exposed in section opposite Mr. Storck's house. It is said to contain no fossils. An exactly similar rock is exposed at various spots for several miles down the river.

On the way down the river, the barge constantly grounded on shoals, our pilot, Joe, knowing nothing of the upper part of the river. We had to strip our clothes off constantly and jump overboard to shove the boat over the shallows, which at last stuck fast and had to remain in that condition till the tide came up and turned again. Joe, the pilot, cautioned us about jumping over into the water, as he said there were sharks. A shark, about three feet long, is common as far up as Mr. Storck's plantation, and large sharks are believed to be common in the lower parts of the stream, and are mentioned in Jackson's Narrative, in the appendix to Capt. Erskine's "Islands of the Western Pacific," as often taking down natives in the neighbourhood of Rewa. At Nadawa, however, Mr. Page had never seen one, and I saw women there constantly standing up to their necks in the water, collecting freshwater clams (*Unio*), evidently without fear.

The Shark of the Wai Levu is *Carcharias gangeticus*, found also in the Tigris at Bagdad, 350 miles distant in a straight line from the sea, where it attains a length of $2\frac{1}{2}$ feet. It is common in large rivers in India. It breeds in fresh water in Viti Levu, inhabiting a lake shut off from the sea by a cataract.[*]

There are sharks inhabiting fresh water in other parts of the world, as in South America, in the Lake of Nicaragua; [†] and in a freshwater lake in the Philippines there lives permanently a "Ray," a species of Saw-fish. A peculiar genus of *Mugilidæ* occurs in the Wai Levu, *Gonostomyxus* (" sa loa " Fijian). It has been described by Dr. Macdonald.[‡]

Joe, our pilot, was I suppose, about 35 years old. He had no notion of his age, but said, when asked by the interpreter.in his own language (he knew no English at all), that he was five years old. When asked if he had eaten human flesh, he said " No "; that he had killed four men, but had never been allowed a taste by the chiefs. He evidently thought himself in this respect an injured man. He had had four wives. He suffered much from cold on the river in the early morning; but, dressed up in a blanket suit by the Blue-jackets, who were very kind to him, managed to keep alive, and seemed to enjoy himself pretty well, especially at meal times.

We passed a hill, opposite which the water of the river is supposed to have the effect of making the whiskers and beard grow, and the spot is resorted to by young Fijians, in order to

---

[*] "Ann. and Mag. Nat. Hist.," Ser. 4, Vol. IV., No. 79, July 1874, p. 36.

[†] Thos. Belt, "The Naturalist in Nicaragua," p. 45.

[‡] J. D. Macdonald, R.N., M.D., F.R.S., " Proc. Zool. Soc." 1869, p. 38.

force their hair.   Joe said that he had been and bathed there when young.

We passed numerous villages on the river side and landed at some to buy clubs, spears, kaava bowls, and other implements, and the river was lively with canoes laden with yams and cocoanuts.   In most places the people crowded to the banks to stare at us, and the girls and boys shouted as we passed.   On the upper part of the river I heard a call used which reminded me somewhat of a European mountaineer's jödel; it sounded like " Hē, Hāh, hŏ, hŏ, hŏ."   Our guides to the top of the mountain in Matuku, used the same call when at the summit.   Mountaineers in all parts of the world seem to have some such cry.   The echo no doubt provokes it.

One village, Navusa, some few miles above Nadawa, interested me, as having its fortifications still perfect.   It occupies an oblong rectangular area, two sides of which are protected by a natural water defence.   On the other two a deep ditch is dug and the earth has been thrown up inwards to form a bank, on the summit of which is set a strong palisade, which is extended around the whole area.   Three narrow openings, only wide enough to admit one man at a time, give means of access.   The openings are guarded by a sort of stile, over which a slab with notches for the foot leads up on one side, a similar one leading down the other.

The whole site of the village has been levelled and raised.   Nearly all the houses rest upon raised platforms of earth, a foot or six inches in height; the chief's house being especially elevated.   Around all the houses were immense heaps of the shells of the fresh-water mussel (*Unio*), which is very common in the river.   The site of an old village on Mr. Storck's estate was made up of beds of these mussel shells.   We saw at Navusa canoe-building going on.   For an adze, a broad chisel was used, fixed into what had been the handle of an old stone adze, just as the Admiralty Islanders fix blades of iron tub hoop into the old handles of their shell adzes.   A chisel of hard wood was used for caulking, shaped just like our own caulking irons.

Near Nadawa, on the road to Nakello, is the village of Tongadrava, which has also been strongly fortified.   It is of an oval form, with two deep broad ditches encircling it, a zone of flat ground intervening between these. Narrow cross banks on opposite sides of the village lead across the ditches. Formerly all Fijian towns were fortified. Those in the Rewa district appear to have been remarkable for their strength,* especially a town called Tokotoko, where there was a perfect labyrinth of moats and ditches.

The people of Nakello, a large village, about two miles from Nadawa, according to Jackson's Narrative, were peculiar amongst the Fijians for not eating human flesh; it being forbidden "tambu" with them. In the centre of Nakello are the tombs of two chiefs. They consist of two large tumuli of earth, adjoining one another, one being older than the other. The older tumulus is oval in form, about 20 yards in diameter at the base, with sloping sides, and about 10 feet in height. At the top is a flat circular space, which is enclosed by a wall formed of slabs of coral and coral rock, set on edge; none of the slabs being very large. Another line of slabs surrounds the mound about halfway up, and here there is a sort of step on the side of the mound. Within the upper circle of stones are some slabs of Tree-fern stem set on end like the stones. The more recent mound has no circles of stones, and is oblong in form.

SECTION OF TUMULUS AT NAKELLO.
1 Lower circle of stones; 2 upper circle : 3 Tree-fern stems.

Our object in visiting Nakello was to be present at a grand dancing performance, which happens in each district only once a year, and which we were lucky enough to arrive just at the right time to see. The dance takes place on the occasion of the collection of the contributions made to the Wesleyan Missionary Society, by the natives. Such dancing performances used always to be held when the annual tribute was paid over to the chiefs, and dancing on their collection days has been encouraged by the missionaries. The policy of the Wesleyan Society pursued in Fiji is very different from that maintained by the missionaries in Tonga, where dancing is suppressed.

---

* Capt. Erskine's "Islands of Western Pacific." App. A, p. 459.

The village was full of visitors, and everyone was dressed in his best. The Dancing Green in front of the chief's house was cleared, and a white tappa flag was stuck up in the centre. We called on the chief, and found him sitting on his mat in a fine large house, about 40 feet long by 20 broad, 10 feet in height to the slope of the roof, and 25 feet to the ridge pole. The house was built of a wooden frame, the rafters and beams being secured with plaited cocoanut fibre or sennet. The walls are of reed, the roof a thatch of grass. The sleeping place at one end was on slightly raised ground, six inches above the rest of the floor, and·was divided off by a curtain of tappa suspended from a cord stretched across. The floor was merely the earth covered with mats. This description will suit any Fijian house except as to dimensions.

The chief sat on his mat near the middle of the house, whilst four or five servants and a visitor sat at the far end. The chief's small boy was being polished up by his nurse for the festivities, and another woman was making girdles of jasmine twigs for the chief's little daughter, holding one end of the garlands between her toes, as she twined the twigs into the sennet with her fingers at the other.

When the small boy was handed from one nurse to another, each nurse, after handing him, went through the usual ceremony of respect to a chief, sat still a moment and clapped her hands four times reverently, and did the same after handing the boy to his father. The clapping was not done so as to make a noise, the palms of the hands were merely brought together quietly four times. The women looked reverently on the floor whilst doing it, as if saying a prayer. It was not at all done as an act of ostentation—indeed the women's backs might be turned to the company at the time—but appeared much more like a ceremony of private devotion. The posture of the hands whilst clapped together is the same as that of Europeans and Japanese and so many races, during prayer.

The chief dressed his son's head himself. The head dressing consisted in shaving off all the boy's wool, except a vertical ridge which was left intact at the back, and looked somewhat like the crest of a Greek helmet, and in smearing the whole of the shaved part with a thick coating of a bright vermilion red.

We drank kaava and tasted Fijian puddings, which are glutinous semi-fluid masses, made of taro and cocoanut, and flavoured with molasses. The puddings are kept done up in a bag of banana leaf, and are very nasty, though specially prepared as a luxury on this occasion. The chief showed us two clubs, family heirlooms, which had killed a large number of illustrious enemies; but since, as he told us, they are always kept very carefully oiled, just as we oil our cricket bats, there was no hair or remains of blood or brains about them.

It was past noon before the people began to assemble in considerable numbers, and seat themselves on the banks and rising ground, commanding a view of the dancing-place. The dancing was begun by the body of young men which I had before seen practising the same dance for this grand occasion at Bureta, in Ovalau.

There were about 80 men in this company. A party stood together in the centre and kept up a sort of chant, one of their number beating time with two sticks upon a small bar of light wood, which was held by the hands of another. The remainder danced round to the chorus in a ring, but every now and then, changes between members of the ring and chorus took place. One of the chants I took down as " Rāihī vāl sāl sātĕ ă dūmm." The last sound was uttered with a peculiar lingering humming sound. The words chanted, usually have no meaning, corresponding to our fal la la, and similar sounds.

The chant was commenced always as a solo, the chorus joining in after the first few notes. Combined with the music, with excellent effect at various stages of the dance, was the loud clapping of hands, which was done in most perfect time, the claps of all the dancers and chorus sounding as one. Two kinds of claps were used, one with the hands hollowed, and the other with them flat. The two sounds thus produced served further to diversify the effect, and there was also added a loud shrill cry used in some of the figures just before their conclusion, and uttered by one performer only, and which came in very well.

The dancing consisted in most varied motions of the head, arms, body, and legs, the same motions exactly being gone through by every member of the circle in most perfect time. At one time the head and shoulders were bent forward, and the hands swung clapping together, at the same time as short side steps were made, carrying the performers round in the circle. Then a half-squatting position was suddenly assumed and the head was thrown first on to one shoulder, then the other. Then the performers would move on again, and stretch their arms out with a fixed gaze, as if shooting with the bow. The motions were none of them very quick, and none very fantastic.

The men wore fringes of various kinds, hanging from round their waists, mostly a combination of the yellow and red Panda-nus leaf strips and the black fibrous girdles of the fungus (*Rhizomorpha*). Most of them had also fringes of *Rhizomorpha* just below the knee, often with beads strung upon them. All had their bodies well covered with cocoanut oil, and their hair trimmed with great care.

By the time the first dance was over, there was a dense concourse of spectators round the Green. The missionary arrived, a table was set out under a tree opposite the chief's house, and three native teachers, two of them Tongan men, sat behind it to receive the money. The inhabitants of the various villages and smaller districts now advanced in separate troops, walking up in single file to the table and throwing down, each man or woman, their contributions upon it, with as loud a rattle as possible.

As each contribution fell, the three teachers and some of the members of a further large body of teachers from the college, who were squatting close by, shouted " Vinaka, vinaka " (slowly), " Vinaka, vinaka, vinaka " (quickly), which means "good, good," or "hear hear." Many bystanders joined in the applause. The money consisted of all sorts of silver coins, and a very few copper ones, and over £100 must have been collected in coin.

The people of the various villages, and the districts subject to the chiefs of these, prepare dances for this yearly occasion for many months, and they vie with one another in the splendour and perfection of the performance. As each band came up and made its contribution, a part or the whole of it at once proceeded to perform the prepared dance, and when this was over another party approached the table, and so on.

The people as they filed up to the table formed a wonderful spectacle. The girls were most of them without coverings to their breasts, but the upper parts of their bodies were literally running with cocoanut oil, and glistened in the sun. The men and boys were painted in all imaginable ways, with three colours, red, black, and blue. There were Wesleyans with face and body all red, others with them all blackened soot black, others with one half the face red, the other black. Some had the face red and the body black, and *vice versâ*. Some were spotted all over with red and black. Some had black spectacles painted round the eyes. Some had a black forehead and red chin. Some were blue spotted, or striped on the face with blue, and so on to infinite variety. How amused would John Wesley have been if he could have seen his Fijian followers in such guise !

For many of the dances the men were most elaborately dressed. They were covered with festoons of the finest gauzy white tappa, or cuticle of the shoot of the cocoanut tree. These hung in long folds from the backs of their heads, and were wrapped round their bodies as far as up to the armpits and hung from the waist down to the knees in such quantity as to stick out almost in crinoline fashion. Round the men's heads were turbans, or high cylindrical tubes or mitres of white tappa, whilst hanging on their breasts were pearl oyster shells set in whales' teeth, the most valuable ornament which a Fijian possesses, and which he is forbidden by the chiefs to sell.

Some of the men had remarkable head-dresses. One of them for instance had, sticking out from the front of his head, and secured in his hair, a pair of light thin twigs of wood, which were a yard in length. They were slightly bent over in front of his face, and at their extremities were fastened plumes of red feathers. The whole was elaborately decorated. As he danced, the red plumes swayed and shook at each jerk of his head with great effect.

The most interesting dances were a Club Dance and a Fan Dance, in each of which a large body of full-grown fighting men, some of them with grey beards, performed. In all the dances, except the first one already described, the chorus sat on the ground at a corner of the Green, and usually contained a number of small girls and boys, and used in addition to the wooden drum, a number of long bamboo joints open at the upper end, which, when held vertically and struck on the ground, give out a peculiar booming note.

In each of the dances there was a leader, who gave the word of command for the changes in the figures, and his part was especially prominent in the Club Dance, In this dance all the attitudes of advance, retreat, and the striking of the blow, were gone through with various manœuvres, such as the forming of single file and of column. Clubs are carefully decorated when used for dancing; some clubs indeed seem to be kept for dancing with, and to correspond to our Court swords in being merely decorative. There are flat spaces near the heads of the curved clubs, which on festive occasions are freshly smeared with red, blue, or white paint. Coloured strips of Screw-pine leaf are often wound round the clubs, and some clubs are decked with beads strung on Rhizomorpha fibres. Thackombau's son's club was, as I have said, freshly painted blue near the top. Thackombau on State occasions had a decorated club carried before him, just as at home the Vice-Chancellor of Oxford, and even the President of the Royal Society. No doubt at some future time, when fire-arms have been superseded, rudimentary guns, richly ornamented, will be carried in state before distinguished personages.

In the Fan Dance, all the dancers were provided with a fan of tappa stretched on a wooden frame. They divided themselves into two parties, which formed into single file in the same line with one another, but with a considerable interval between the two parties. The two bands took up the chant and danced alternately, answering each other as it were. The fans were waved in various attitudes, and at the end of each movement thrown suddenly up over the head (still held in the hands), a wild war-cry, uttered by the whole line simultaneously, accompanying the movement. The war-cry was of a single prolonged high-pitched note, and sounded intensely savage.

In another dance, performed by a large body of men, about

120 I think, the dancers formed a sort of rectangular group, arranging themselves in eight rows, the leader being in the centre of the front row. Once or twice the leader came forward to the chorus, and addressed a few words in a dramatic manner partly to them, exhorting them to do their duty well, partly to the spectators.

A club dance by boys was one of the performances. In one figure of this the boys, standing in a line with their bodies bent forwards, jerked their hips with a most astonishing facility, first to one side and then to the other. The motion, especially in cases where the boys had a large quantity of tappa projecting behind as a sort of bustle, was most ludicrous, and the audience, instead of crying the oft repeated "Vinaka, vinaka," fairly shouted with laughter.

A band of women of the district, headed by the Queen of Rewa and her daughter, who were both dressed in bright blue striped prints, marched slowly forwards across the Green to deposit their offerings, singing a chant, descriptive of various incidents from the New Testament, the descriptive part being a solo, whilst the whole band joined in a constantly repeated chorus containing the words Allelujah, Amen. This song was in lieu of a dance.

The principal interest of the performances, however, lay in the obvious fact that here were to be observed the germs of the drama, of vocal and instrumental music, and of poetry, in almost their most primitive condition in development. In these Fijian dances they are all still intimately connected together, and are seen to arise directly out of one another, having not as yet reached the stage of separation.

The dance is evidently first invented by the savage, then rhythmical vocal sounds are used by the dancers to accompany it, and simple instruments of percussion are employed to keep time. As the dance becomes gradually more varied and complex, the accompanyists are separated as an orchestra, the actual performers joining less and less in the vocal part until, as here, they merely utter a single loud cry or note occasionally during the dance.

The instrumental music of the orchestra remains long subordinate to the vocal and very simple, being represented at Fiji, as described, by the single small wooden drums and the bamboos. The orchestra, continuing its performance in short intervals in the dancing, and commencing somewhat before the first figures, in order to allow the dancers to be ready to take up the measure, as was the case at Nakello, comes at length to perform solos; and hence the origin of music apart from dancing. The gradual complication of the music and improvement and multiplication of instruments follows, until vocal and instrumental music change places in importance and become also at length separated from one another.

The dances being descriptive of victorious battles and such exploits, the chants, at first mere musical sounds and war-cries, become short descriptions of the fight, or praises of the warriors, and hence the origin of poetry. I could get no explanation of the meaning of the chants used at Nakello; as far as I could gather, they were without meaning, mere convenient sounds; but Fijian songs do exist, for Joe, our pilot, sang part of one one day and explained that it related to the superiority of the Mbau men to the Rewa men.

The origin of the drama is clearly seen in the stepping forward of the leader of the dance, as described, and dramatic enunciation by him of a short speech. A further step was to be seen in one of the other dances, when the leader, before his troop came on to the ground, rushed forward brandishing two spears in his hands, and gave a short harangue descriptive of what he was going to do.

The separation of the dancers in the Fan Dance into two parties, performing alternately and responsively, is also interesting, and brought the Greek chorus and drama into one's thoughts. It was of course not necessary to have recourse to Fiji in order to trace the origin of dancing, music, and the drama. This has been done fully long ago. But nowhere, I believe, is the primitive combination of these arts so forcibly brought before the view, as a matter of present-day occurrence, as in this group of islands.

The most extraordinary feature in the Nakello performance was the extreme order and decorum of this concourse of three or four thousand people. It seemed astounding, whilst looking on at these blue, red, and black-painted Fijians flourishing their clubs and shouting their war-cries, to reflect that this was a Wesleyan Missionary meeting. The representative of the power which has tamed these savages was a little missionary, with battered white tall hat and coat out at elbows, who stood beside us and who took no prominent part in the ceremonies, but yet had full sway over the whole, no dance having been prepared without his previous sanction.

There could be no doubt as to the amount of good which had been done to these people, and it is sincerely to be hoped that the Wesleyan Missionaries will be left unmolested to continue the work in which they have been so successful, and which they have begun and carried out often at the risk, in some instances with the loss, of their lives.

The men and children attending the meeting vied with one another in getting money to contribute, and were ready to sell anything they had almost for what we would give them. One boy pestered us to buy an old hen, and followed us about with the bird. Others sold us clubs and ornaments. The great wish was to have several pieces of silver to make a rattle on the table, and two sixpences were worth much more than a shilling, two shillings more than half-a-crown. Immediately the ceremony was over everything went up in value, and a good many articles pressed on us before, were not now to be had at any price.

Amongst the crowd was an Albino Boy. He was perfectly white, his skin having a peculiar look, almost as if covered with a white powder, in places. His eyes appeared as if the iris were of a pale-grey colour. He hid his eyes either from the light or because of shyness. His parents said he could see perfectly. I could not examine him closely as he roared at the prospect. Albinos seem unusually common amongst Melanesians, and are constantly mentioned by travellers. Hence these savages, when first seeing Whites, no doubt often took them for a race of Albinos. I saw several hunch-backed dwarfs amongst the crowd.

We sailed from the Wai Levu, or Rewa River, to Kandavu, stopping at a small island on the way, to buy a pig and some fowls. A voyage in an open boat has many discomforts, especially when the boat is crowded. The managing to sleep six together in the confined space of the stern-sheets of a ship's barge, was a difficult matter, especially as the available surface was rendered extremely irregular by the various articles necessarily stowed upon it, such as provision boxes and beer cases. We all slept with our shooting-boots on, to ensure mutual respect, as we lay packed like herrings in a barrel. On the whole the trip was pleasant enough, and the inconveniences as nothing compared with the interest of a visit to such places as Mbau and Viti Levu.

One feature of interest in the Fijis, which I have forgotten to mention, arises from the importation of labour. At Levuka are to be seen men from the New Hebrides and Solomon Islanders. Further, the curious straight-haired most characteristically featured Tokelau race, or Union Islanders, mostly girls: also Tongans and Samoans and a few Negroes from the United States. Representatives from almost all Polynesia, assemble here and may be studied by the Anthropologist.

Nothing surprised mè more than the great power of the chiefs in Fiji, and the absolute subserviency of the lower classes to them. The reality of the various grades of rank amongst such savages, and the abject condition of the slaves, were facts which I had not previously realized.

Facial expression is far less marked in the Fijians than the Tongans. Amongst the lower classes there is a remarkable want of expression; there is also, as far as I saw, entire absence of gesticulation during conversation. The methods of affirmation and beckoning are the same as in Tonga; the throwing up of the head in affirmation is common to many races, being used by the New Zealanders, Abyssinians, and Tagals of Luzon.* The forehead muscles are little used, at least by the ordinary people. Amongst the families of the chiefs there is much Tongan blood. Thackombau wrinkled his forehead constantly during his conversation with our party, and one of the mountaineers, prisoners whom I saw at Livoni in Ovalau, knit his brows frequently when I was asking him about his eating human flesh.

* C. Darwin, "The Expressions of the Emotions," p. 275.

Our interpreter, an Englishman, who had married a Fijian woman, and who knew the people well, told me that old women sometimes clap the hands twice in expressing astonishment. This habit of expression is evidently derived from the clapping of hands in expressing respect to a chief, and is interesting as showing how peculiar means of expression may thus be of entirely artificial origin. The clapping of hands is used as a ceremony of respect to superiors in Japan, as at the funeral of Okubo, the minister lately assassinated in Yedo, at which " all present saluted the deceased with three claps of the hands."*

The interpreter further said that the mountaineers in expressing astonishment, shake backward and forwards transversely once or twice, the right hand held hanging back foremost from the half-extended arm ; a similar gesture is stated by Darwin to be used by Northern Australian natives, to express negation.

A short click made with the tongue and repeated several times, is also used by the mountaineers to express astonishment, and also to express pain, as on striking the foot against a stone, or even by a man when hit by a bullet, louder exclamation being repressed through bravery. The same sound is used by us in pain, but more often to express disappointment, as on saying " what a pity ! "

The audience at Nakello, when they shouted with laughter, produced a general sound exactly like that proceeding from a European audience. No doubt the sound of laughter is one of the very earliest and oldest of human cries. It is certainly an astonishing sound, and one that it is very difficult to listen to and analyze without prejudice and a remote feeling of sympathy. The best way to study it that I know, is to seize on opportunities when one is being constantly interrupted, say at one's club, in reading a serious book, by shouts of laughter from a party of strangers ; one can then note the curious variety of spasmodic sounds produced, and marvel that men in the midst of rational conversation should be compelled by necessity to break off suddenly their use of language, and find relief and enjoyment in the utterance of perfectly inarticulate and animal howls, like those of the " Long-armed Gibbon."

* The Japan Mail, June 6, 1878, p. 306.

Z

It is a curious fact that the cries of the Gibbon are uttered in a similar manner in a series, on slight provocation. When one lately in the Zoological Gardens, Regent's Park, was in the proper mood, a very slight snatch of a whistle from the keeper would set the animal off into the utterance of a regular peal of howls, which appeared to follow one another spasmodically.

Cicatrization of the skin is practised by the Fijians, but the scars produced are not so much raised as are those of the men of Api in the New Hebrides. I saw a series of circles thus marked on one chief's arm ; he said they were done with a fire stick, and on the occasion of the death of a relation, or out of respect on the death of a chief. In the women, scars are sometimes made to enhance beauty. Young boys when troublesome, are sometimes caught by the old men, and have their flesh gashed in various places to make them sore, and keep them quiet for a time. The little finger is commonly absent on the right hand, having been cut off as a ceremony.

FIJIAN ULA.

With regard to Fijian weapons, the annexed figure represents a well-known wooden weapon, which consists of a slender handle about a foot in length, and a heavy rounded knob cut out of the same piece ; the knob is in fact the base of the tree stem, from which the weapon is made. The weapon is one of the commonest of those brought to Europe from Fiji, and exhibited in museums. It is not a club, as it is usually called and labelled, but a missile weapon, which is thrown with great force with the hand, revolving rapidly in the air as it flies, and striking a very formidable blow, often in the face. Settlers in Fiji told me it was the only native weapon which they feared when fighting with Fijians. The native name of the weapon is " Ula," The head of the ula is usually beset with a circle of large oval knobs, as shown in the figure. These knobs are the stumps of the lateral roots of the tree, from which the weapon is cut. When the ula is carved out of solid wood, a circle of knobs is often cut round the head of it, in imitation of those derived in the original weapon from the lateral root stumps. Some ulas have perfectly smooth heads.

With regard to Cannibalism, I gather many of the following details from our interpreter: When visitors of distinction paid a great chief a visit, he was expected to provide human flesh for their entertainment. If there were no prisoners, a man whose special office it was to obtain such food for the chief, went in search and often killed some girl or woman he met with alone, belonging to a village not far off.

Young woman was considered to be the best eating; Europeans were not thought so good to eat as natives, no doubt because of their very mixed diet, and much greater consumption of animal food. The bodies were prepared with care for cooking, and were usually baked in the well-known oven in the ground. A special vegetable, a species of Solanum (*S. anthropophagorum*), was eaten with the baked flesh, just as was the case in New Zealand. The vegetable was eaten with human flesh as a suitable condiment, not as an antidote. There is no reason to suppose that ill effects followed the eating of human flesh any more than from the consumption of any other kind of flesh. The sturdy health of the grey-haired Thackombau is sufficient evidence against such a supposition.

The flesh was eaten cold as well as hot, and the cold cooked flesh was often sent as a present to a distance by one chief to another. A four-pronged fork of wood was used in eating human flesh, and was held more or less sacred, but it was also used for eating other food occasionally.

The New Zealanders were, however, probably the most profusely cannibal race that has existed. As many as 1,000 New Zealand prisoners have been slaughtered at one time after a successful battle, that their bodies might be put into the ovens.

In 1828 the captain of an English merchant ship, named Stewart, made an agreement with a tribe of Maoris under a renowned chief, Te Rauparaha, to convey a war party to a distant village on the coast, for the remuneration of a cargo of New Zealand flax. The warriors were landed at night, exterminated the village, and brought off the bodies of the slain to

the ship, where they cooked them in the ship's coppers; the
captain nevertheless duly received his cargo.*

In 1832 or 1833, a large party of Maoris was landed by
another English merchant vessel on the Chatham Islands, small
outliers of New Zealand.   The islands were inhabited by a
weaker race, "Maoriori," 1,500 in number.   The Maoris simply
ate their way through the islands, killing the Maorioris as they
required them for food, and making the victims dig the ovens
they were to be cooked in, and collect wood for the purpose.†
Their object in going to the island was to feed upon the inhabitants, a Maori who had visited the islands, when engaged as
a seaman on a European vessel, having reported the islanders as
plump and well fed.

Whilst the New Zealanders considered the palms of the

FIJIAN DOUBLE CANOE.
(From a photograph.)

hands and the breast as the best eating,‡ the Fijians especially
preferred the flesh of the arm above the elbow, and that of the

* W. T. L. Travers, F.R.S., "The Life and Times of Te Ruaparaah."
Trans. New Zealand Inst. Vol. V, 1872, p. 78.

† H. H. Travers, "On the Chatham Islands," Ibid. Vol. I, 1860, p. 176.

‡ E. Dieffenbach, "Travels in New Zealand," Vol. II. p. 129. London,
J Murray, 1843.

thigh.* Not more than five-and-twenty years ago, White residents are said to have joined the natives in their cannibal feasts
at Ovalau, Fiji.†

Whilst we were at Fiji, the burning question with the
settlers was whether the group was to be annexed by Great
Britain or not. The planters and all the store-keepers were
eagerly hoping for the annexation, and many had staked their
fortunes on the event. The missionaries, on the other hand,
were praying in the best interests of the natives, as they viewed
them, that the place might remain as it was. The result is well-
known; the Fijis are now British. Thackombau and his suite
were taken to Sydney for a trip in a man-of-war, and they
returned bringing the measles with them, by which about one-
third of the native population was at once swept off.

* C. Wilkes, "Narrative of U.S. Exploring Expedition," Vol. V,
p. 101. New York, 1856.

† J. D'Ewes, "China, Australia, and the Pacific Islands," p. 151.
London, 1857.

CHAPTER XIV.

NEW HEBRIDES. CAPE YORK. TORRES STRAITS.

Api Island, New Hebrides. Fringing Reefs. Proofs of Elevation. Coral
Living Detached. Natives of Api, their Ornaments and Weapons.
Condition of Returned Labourers. Expression of the Emotions.
Raine Island. Its Geological Structure. Its Vegetation. Nesting
of Wideawakes. Gannets and Frigate Birds. Dead Turtles.
Somerset, Cape York. Nests of White Ants. Combination of Indian
and Australian Features in the Vegetation. Various Birds. Habits
of the Rifle Bird. Birds Fertilizing Plants. Camp of the Blacks.
Habits of these Natives. Curious mode of Smoking. Food of the
Blacks. They Cannot Count Higher than Three. Absolute Nudity
of the Men. Coral Flats. Collection of Savage Weapons at Cape
York. Wednesday Island, Torres Straits. Structure of Coral Flats.
Giant Clam. Native Graves. Booby Island. A Halting Place for
Birds during Migration. Many Land Birds on an Almost Bare
Rock.

Api Island, New Hebrides, August 18th, 1874.—We left
Kandavu on August 11th, and made a week's run before
the trade wind to the island of Api, in the New Hebrides,
having on board the ship some labourers, natives of that island,
who had worked out their time in Fiji, and were to be returned
to their home.

We were off the east coast of Api, on August 18th, having
passed several small adjacent islets, "Three-hill" island amongst
them, all volcanic. Api lies south of Ambrym and Malicolo, and
between these islands and Efate or Sandwich Island. It is in
about the same latitude as the northern part of the Fiji group.
The island is upwards of 20 miles long and its highest peak is
about 1,500 feet above sea-level.

The island rises in steep slopes from the sea with here and
there only a stretch of flat shore land. It consists of a series of
peaks and steep-sided valleys and ridges. The whole is entirely
covered with the densest possible vegetation, excepting on very

small spots, with difficulty discerned with a glass, where plots are
cleared by the natives for cultivation.

The ship steamed close in to the island, opposite a spot where
a valley terminated towards the sea with a widened mouth,
evidently containing a river. There was a stretch of flat land
at the bottom of the valley on which were conspicuous amongst
the other foliage some cocoanut palms and another species of
palm. As we came near natives appeared on the shore, some
hiding in the bushes, others running along at full speed, whilst
some shouted a loud " hoa." One man stood on the shore and
waved a green branch with untiring perseverance.

These natives were said to be hostile and dangerous, and
therefore the first party, the " Captain's," which landed, was
armed, but the returned labourers acted as an introduction and
made matters smooth ; still, as all the natives were armed,
either with bows and poisoned arrows, clubs, or trade muskets,
and as the inhabitants of these islands are noted for treachery,
no one was allowed to leave the beach, and our stay lasted for
only a few hours. Thus we saw very little of this island, which
had certainly never been landed upon before by any scientific
man or naval officer.

The shore is made up of a banked-up beach, composed of small
fragments of volcanic rock and volcanic sand, mingled with a
large proportion of coral fragments, and is fringed by a narrow
shore platform of coral, which, in the place where I examined
it, was not much more than 100 yards wide. The New Hebrides
have no barrier reefs but only narrow fringing reefs. The cause
of this, Dana concludes to be the fact that volcanic action has, in
this group of islands, been very recent. There are still several
active volcanoes in the group, and one was said by our returned
labourers to exist in Api. (The word Api means in Malay,
" fire "). Submarine ejections of carbonic acid and the falling of
fine dust might render the growing of reef corals round an active
volcanic island nearly impossible.

The Api shore reef is remarkable for its extreme flatness.
Almost everywhere the living corals embedded in it are growing
only laterally, the upper surfaces being dead from want of sufficient depth of water. In some small specimens of a massive

Porites the consequent flattening of the top and expansion of the lateral dimensions was most excellently shown in pieces convenient for museum purposes.

The Corals, which were few in number of species, were finer grown towards the outer verge of the reef, as is always the case on shore platforms, the very opposite condition to that which holds in case of barrier reefs. In some places were deep holes in the coral platform, reminding one of glacier crevasses on a small scale, evidently arising from the loose nature of the sloping beach on which the coral structure here rests. On the reef rest weathered remains of a more ancient shore platform which are honey-combed and wave-worn. The rock composing them is, however, undoubtedly *in situ*, and proves elevation of the islands to the extent of five feet or so. Similar fragments of raised reef were found by Mr. Murray at a short distance up the bed of the stream already mentioned. A massive porites was one of the corals on the reef. Some specimens of this species were unattached, though living, being in the form of rounded masses, entirely covered with living polyps, and I suppose from time to time rolled over by the waves. They reminded me of the similarly detached rounded masses formed by some Lichens (*Lecanora esculenta*), which are rolled about over the land by the winds as are these coral colonies by the waves.

On the reefs were comparatively few free living animals, but here I saw for the first time one of the huge *Synaptas*, which are abundant amongst the East Indian Islands and at the Philippines. The animal was a yard long and two inches in diameter, and looked like an ugly brown and black snake. The instant I touched one I knew what it was, for I felt the anchor-shaped hooks in its skin cling to my hand.

One animal on the reefs I could not understand the nature of. About six white tentacles, each nearly six inches in length, and of a uniform thickness of not more than $\frac{1}{16}$th of an inch, were expanded on the reef in a radiate manner. On irritation they were slowly but entirely retracted. I could not succeed in digging the owner of them out of the reef rock. I have never seen this animal elsewhere.

Above the shore the first land plant met with is the ubiquitous tropical Littoral plant (*Ipomœa pes capræ*). It is always the first plant above the high-water mark in these tropical shores. Above a skirting of this commenced a thick growth of largish trees, a species of Barringtonia, a Fig, and the common Pandanus of the Pacific Islands occupying the shore margin. A few paces inside the wood it was gloomy, from the thickness of the growth of trees and creepers overhead. The same climbing Aroids grew here as at Fiji, and a *Dracæna* was common, and also a beautiful climbing Asclepiad (*Hoya*) with white waxy flowers, and one or two ferns. I could not penetrate the wood

far enough to get any adequate idea of the nature of the vegetation. Five birds were shot in Api, *Artamus melaleucus* (a Shrike), a Swallow (*Hirundo Tahitica*), a Swift, a Fruit Pigeon, and the Kingfisher (*Halcyon juliæ*). I saw no sea birds.

The Api men wore as clothing nothing but a narrow bandage of dirty European fabric of various kinds. They are a small race, few, I should say, being above five feet in height. Their limbs, and especially their legs, are small and badly shaped. They are much darker in colour than Fijians; they seemed quiet enough. Several amongst those we saw were returned labourers, and were at once known by their having fastened to their waist cloth the key of the chest which every labourer brings back with him, containing the fruits of his toil. The labourers thus retain the property for which they have worked even in Api. Two men joined me on the reef. One had been in Queensland, the other in Fiji. Both spoke a good deal of English; and one said he was willing to go to Fiji again.

CHARM CUT OUT OF THE SEPTUM OF A NAUTILUS SHELL, AND EARRING MADE OF TORTOISESHELL, AND A PIG'S TAIL. API ISLAND.

Nearly all the men wore a small triangular ornament, cut out of one of the septa of the pearly Nautilus shell, and threaded by the syphon hole in it, tied round their necks.   Many had broad flat tortoiseshell bracelets, and nearly all earrings made of narrow strips of tortoiseshell moulded into a flat spiral, from which hung sometimes, as ornaments, the tips of pigs' tails.

The bows used by the natives are made of hard wood.   The arrows are without feathers, but notched for the string, and made of reeds with heavy wood ends, and tips of human bone.   The tips are all covered with poison, which is in the shape of a black incrustation.   The arrows have an elaborate and artistic coloured decoration in the binding round the part where the bone tips are inserted.   The men were unwilling to part with these arrows, which they prize highly.   They carry them rolled up in an oblong strip of plantain leaf, and showed by signs that they considered the poison deadly, and were much in awe of it.

The men have all of them cicatrization on their bodies, usually representing a human face, and placed sometimes on the shoulder, but more often upon the breast, and sometimes on both breasts.   They understood the value of the usual trade articles very well.   Knives, tobacco, and pipes were what they wanted most, but they were not eager at all to trade, and few weapons or ornaments were obtained from them.   The tortoise-shell bracelets they would not part with at any price.   It was very trying to leave a totally unknown island like Api after two hours only spent on the shore.

I had an opportunity of watching the expressions of the Api men on board during the voyage.   During their whole stay they had a peculiar dejected look, and, like the lower order of Fijians, a marked want of expression in conversing with one another. In laughing they were affected and childlike, or girlish, hiding their faces with their hands.   The hands in doing this were half-clasped, the face turned away on one side, and the clasped hands held over the shoulder in front of the face, just as in the case of a shy child.   Often the thumb was held in the mouth, the hand half-hiding the face in laughter.   I heard no loud laughter, but a steady look at one of the men nearly always called forth a grin, which expression was used invariably to show consciousness of being gazed at.   The forehead muscles were little used.   When the men were talking amongst themselves their faces showed little expression. When a little excited they ran their voices up into a sort of affected falsetto.

Amongst the men on shore I noticed a shrugging of one shoulder, the head being leant over towards the same side, constantly used to express disinclination to accept proffered barter, and a pouting of the lip, the under lip being much thrown up, was used at the same time, or alone, to express the same meaning. To signify "Farewell," the hand was held up, palm outwards, and with the fingers extended.

Raine Island, August 31st, 1874.—The ship passed Raine Island on the afternoon of August 30th, and anchoring about five miles off, under the lee of a reef, returned and landed a party on the island next day. A very full account of Raine Island is given by Jukes.* The island is at the entrance of the most usually employed passage through the Great Barrier Reef of North Eastern Australia. It is about three-quarters of a mile long, and composed of calcareous sand rock, closely similar to that of Bermuda, excepting that it is remarkably evenly bedded.

The strata dip towards the shores with a slight inclination. I measured the dip on the north-east side of the island, near the beacon, and found it 7°. I cannot say whether it is uniform all round the island. Towards the centre the strata seemed to be horizontal. Jukes observed a similar dipping of the strata in Heron Island,† but does not mention it as occurring at Raine Island. This condition would arise from the island being formed as a single low sand dune, in which consolidation subsequently took place; though why a series of smaller dunes and ridges should not here have been formed, and hence a rock like that of Bermuda, with contorted strata, have arisen, I do not see : perhaps from the constancy of the direction of the winds, or from the smallness of area, or the absence of adequately binding plants.

The shore of Raine Island was of glistening white calcareous sand, made up of fragments of shells, corals, and Foraminifera. Immediately above the beach line, where grass commenced and with it the breeding-place of the terns, the colour of the sand

* "The Voyage of the 'Fly,'" Vol. I, pp. 126 and 338.
† Ibid., p. 7.

became redder, and consolidated crusts were here common upon its surface, as at Bermuda. The sand rock is mostly redder than the beach sand from which it is formed. Perhaps this is due to the loss of a certain quantity of lime, and consequent greater proportion of iron; or perhaps to the action of the birds' dung.

On the island I found eleven flowering plants; I believe there are no more. Two of these are grasses. The grass covers tracts bordering the shores, where no other plant grows, and it is here that the terns breed. I could find no moss, fern, or lichen on the island, so that here from the action of drought and extreme heat, the conditions are just the opposite of what they are in an Antarctic island, such as Possession Island, where Cryptogams only grow. Some Fungi, and low algæ possibly, on the birds' dung, and perhaps some parasitic fungi on the plants, were probably the only Cryptogams in the island. There were even no seaweeds to be seen cast up on the beach.

There were no vestiges remaining of gardens made on the island in 1844, by the crew of the "Fly," and planted with cocoanuts, pumpkins, and other plants; all has been overwhelmed by the drift sand. I found what I hope may prove a favourable spot, and planted pumpkin, tomato, capsicum, water melon, and Cape gooseberry seeds. I think the latter plant very likely indeed to grow. There is very good black vegetable soil in places on the island.

The most striking feature at Raine Island is formed by the birds. They are in such numbers as to darken the air beneath as they fly overhead, and the noise of their various mingled screams is very trying to the ears at first, but not so painful as that of a penguin rookery. Eleven species of birds were seen on the island. A heron, seen only at a distance, the cosmopolitan "Turnstone," aud a small Gull (*Larus Novæ Hollandiæ*) appeared to be casual visitors to the island, as they were not nesting there; the Turnstones being seen in flocks on the shore.

The birds breeding on the island were as follows:—A Landrail (*Rallus pectoralis*), a widely spread species, occurring commonly in Australia, Central Polynesia, the Moluccas, and the Philippines. These birds were tame, and were knocked down with sticks and caught by the hand.   They had full-fledged young running about.

A Tern (*Sterna fuliginosa*), a widely spread species, the well-known " Wideawake " of Ascension Island, was exceedingly abundant.   The stretches of flat ground above the shore line covered with grass were absolutely full of the brown fledged young of this bird.   Eggs were already very scarce.   A Noddy (*Anous stolidus*), the same bird as that at St. Paul's Rocks and Inaccesible Island, so far off in the Atlantic, makes here a rude nest of twigs and grass amongst the low bushes, but often nests also on the ground.   There were plenty of eggs of this bird, it being not so advanced in breeding as the tern.

Two species of Gannets, *Sula leucogaster* and *Sula cyanops*, were nesting on the ground, and especially on a plot of ground quite flat and bare of vegetation ; probably the site of the dwellings of the men employed in 1844 in putting up the beacon on Raine Island.   *Sula leucogaster*, the Booby of St. Paul's Rocks, makes a slight nest of green twigs and grass on the ground. *Sula cyanops* makes a circular hole in the earth, about 1½ inches deep.   This species is nearly white, with the naked parts about the head of a dull blue, and with a bright yellow iris, which gives the bird a ferocious look as it ruffles its feathers and croaks at an intruder.   It would almost seem as if the cause of the colouring of the eye might be the savage appearance which it gives the bird, which may thus be protected from attack.   A third smaller species of Gannet (*Sula piscatrix*) has red feet, which distinguish it at once from the other two.   I saw one or two of its nests made in the bushes, like those of the noddies, raised six inches from the ground.

There remain to be mentioned the " Frigate Birds " (*Tachypetes minor*).   Their nests were nearly all confined to a small area near the cleared patch already referred to.   They are like those of *Sula piscatrix*, raised on the bushes, and are compact platform-like masses of twigs and grass matted together with dung, about eight inches in diameter.   There were no eggs of the birds in the nests, but mostly far advanced young, which were covered with frills of a rusty coloured down.   The old birds soared overhead, and could only be obtained by being shot ;

whereas the gannets were easily knocked over on the nests with sticks. It is curious to see the Frigate birds, the nesting-place of which is usually on high cliffs, as at Fernando Norhona, here, through the entire security of the locality, nesting on the ground. The main body of the Frigate birds remained during our stay soaring high up in the air, with their eagle-like flight, far above the cloud of other birds beneath.

On the island were lying about the shells of numerous turtles which had died there. In one place there was quite a heap of these at a spot where there was a sort of miniature gully, bounded by a perpendicular wall of rock about two feet in height. It appeared as if the turtles had crawled up from the sea-shore to spawn, and being stopped by this small cliff, had been unable to turn round or go backwards, and had died there. A Locust (*Acridium*) was very common amongst the grass on the island, and a large Earwig (*Forficula*) under the stones.

Cape York, Australia, Sept. 1st to Sept. 8th, 1874.—The "Challenger" reached Somerset, Cape York, the northernmost point of Australia, on the evening of September 1st. The coast leading up from the south towards Somerset, presents a succession of sandy bays, which looked glaring and hot as we passed them in the distance. Behind these sands the country rises in a succession of low hills, and is covered with a thick vegetation. Somerset lies in a narrow channel, formed between the small island of Albany and the mainland. The island, and also parts of the mainland bordering the sea, at the entrance to the channel from the south, are bare of trees, excepting "Screw pines," and covered only with a grass, in the dry season withered into hay.

These open grass-covered spaces are rendered most remarkable objects, because they are covered in all directions with the nests of Termites (White ants). These nests are great conical structures of a brick red colour, often as much as ten feet in height. Standing up all over the open country, they give the scene almost the appearance of a pottery district in miniature, beset with kiln chimneys.

The tide runs in a regular race through the channel between Albany Island and Somerset, and we drifted rapidly with it to an anchorage opposite the small bay in which Somerset lies.

On the one hand is a small strip of Mangrove swamp; in the centre, a long beach of sand; on the other hand, the commencement of a range of low cliffs.

Behind the shore of the Bay the land rises steeply, and is covered with wood, except where cleared around two conspicuous sets of wooden buildings, the one the residence of the magistrate, the other the barracks of the water police.

Three other wooden houses, one on the beach used as a store, the other two nearly in ruins, and only temporarily inhabited, make up with these the whole settlement of Somerset. There were only five or six permanent White residents. At the time of our visit there were in the place besides, others belonging to a small Mission Steamer intended for New Guinea, and also the skippers of two vessels employed in the pearl shell trade.

The country is wooded in every direction, but with constantly recurring open patches covered with scattered acacias, gum trees, and Proteaceæ with grass only growing beneath. In the dense woods, with their tall forest trees and tangled masses of creepers, one might for a moment imagine oneself back in Fiji or Api, but the characteristic opens, with scattered Eucalypti, remind one at once that one is in Australia. The principal features of Australian and Indian vegetation, are, as it were, dovetailed into one another.

In the woods, the tree trunks are covered with climbing aroids, and often with orchids. Two palms, an Areca with a tall slender stem not thicker than a man's wrist, but fifty feet high, and a most beautiful Caryota, strong evidence of Indian affinities in the flora, are abundant. The Cocoanut Palm, as is well known, is not found anywhere growing naturally in Australia, though it is abundant in islands not far from Cape York. At Cape York some trees had been planted, but they appear not to thrive. One of these, already more than eight years old, at which age it ought to have been bearing fruit, had as yet a trunk only a few feet in height. A Rattan Palm, trailing everywhere between the underwood, is a terrible opponent, as one tries to creep through the forest in search of birds.

The number and variety of birds at Cape York is astonishing. Two species of Ptilotis (*P. crysotis* and *P. filigera*), different from those at Fiji, but closely resembling them, suck the honey from, or search for insects on, the scarlet blossoms of the same Ery-thrina tree as that at Fiji. With these are to be seen a Myzo-mela, and the gorgeous little brush-tongued Parroquet (*Tricho-glossus swainsonii*), which flies screaming about in small flocks, and gathers so much honey from the flowers, that the honey fairly pours out of the bird's beak when it falls shot to the ground. Amongst the same flowers is to be seen also a true Honey-bird (*Nectarinia frenata*), with brilliant metallic blue tints on its throat.

The common white-crested Cockatoo (*Cacatua galerita*) is here wary and difficult to get near, though not so much so as in the frequented parts of Victoria. The great black Cockatoo (*Microglossum alerrimum*) is to be found at Cape York, but I did not manage to see one. The Pheasant Cuckoo (*Centropus phasianus*) rises occasionally from the long grass in the opens, and though of the cuckoo tribe, has exactly the appearance of a pheasant when on the wing.

On one of my excursions I shot a large brown Owl (*Ninox boobook*), which was sitting at daybreak in the fork of a large tree, and which my native guide espied at once, though I had passed it. The great prize at Cape York is however the Rifle-bird (*Ptilorhis Alberti*) one of the Birds of Paradise. The bird is of a velvety black, except on the top of the head and breast, where the feathers are brightly iridescent with a golden and green lustre. In the tail also are two iridescent feathers. The bird lives in the woods, where the trees and undergrowth are twined with creepers. It does not frequent the higher forest trees much, but the tops of the shorter sapling-like growths and masses of creepers binding these together.

The call of the bird consists of three loud shrill short whistling notes, followed by a similar but much lower pitched note. The third of the first three whistles is somewhat louder and shorter than the two preceding. This is the full call of the bird, sometimes only two notes are uttered before the low note, and sometimes only a single whistle.

The call is most striking and peculiar, and guided by it, one steals gradually through the wood, treading cautiously upon the dead leaves, and tries to creep within shot of the birds. The call is uttered usually only at intervals of several minutes; it is very easily imitated by whistling, and thus a call may often be elicited, and the bird's whereabouts discovered.

The bird is extremely shy, and the snapping of a dead twig is sufficient to scare it, and it requires great patience and perseverance to shoot one. It several times happened to me that I got within fifteen or twenty yards of a Rifle-bird, and stood gazing into the thick tangled mass of creepers overhead, where I knew that the bird was, without being able to get a glimpse of it, until at last it darted out without my catching sight of it.

The bird takes short rapid flights from one part of the bush to another, the rounding of the front of the wings giving it a peculiar appearance when on the wing. The Blacks pointed out the red fruit of the Areca palm as the food of the bird, and I found abundance of the seeds of this palm in the stomach of a bird which I shot. The one bird which I shot was hopping about up and down amongst a thick piece of bush, much in the way of a wren or warbler. The male in full plumage is indeed a splendid object; the female and the young birds of both sexes are of a dull brown colour, as is the case with all the Birds of Paradise.

When walking in the woods in search of birds, a slight rustling in the fallen leaves attracts one's attention, and the Black guide becomes greatly excited. It is a pair of the "Moundbirds" (*Megapodus tumulus*), which are disturbed and are seen running off like barn-door fowls, and when thus luckily hit upon are easily shot. Several "Brush Turkeys" (*Talegalla Lathami*), were shot during our stay at Somerset, and the huge mounds thrown up by them were common objects at the borders of the scrubs, but the season was not far enough advanced for them to have commenced laying eggs.

A brilliant Bee-eater (*Merops ornatus*) was common at Cape York, and to be seen seated, as is the wont of Bee-eaters, on some dead branch, and darting thence from time to time after its prey. A little Ground Pigeon (*Geopelia*), not much bigger than a sparrow, was also abundant.

A species of Swallow-shrike (*Artamus leucopygialis*) was

very common, sitting in small flocks in rows on wires stretched for drying clothes near one of the houses, just as swallows sit on telegraph wires in England. The birds made excursions after flies, flying just like swallows, and returned to their perching place. Those which I shot all had their feathers at the bases of their bills clogged with pollen from the flowers, in which no doubt they had been searching for insects; like some humming-birds, they must act as fertilizers, carrying pollen from one flower to another.

In all my excursions I was accompanied by Blacks. An encampment of natives lay at about half a mile from the shore; the camp was a small one, and composed of the remnants of three tribes. There were 21 natives in this camp when I visited it early one morning in search of a guide, before daybreak, before the Blacks were awake. Of these 21, about six were adult males, one of whom was employed at the water police station during the day time; there were four boys of from ten to four-teen years, two young girls, two old women, two middle-aged women, and the remainder were young women.

One of the old women was the mother of Longway, who acted as my guide, and who had a son about ten years old. The Blacks were mostly of the Gudang tribe, a vocabulary of the language of which is given in the Appendix to MacGillivray's "Voyage of the 'Rattlesnake.'"* The natives were in a lower con-dition than I had expected. Their camp consisted of an irregu-larly oval space concealed in the bushes, at some distance off one of the paths through the forest. In the centre were low heaps of wood ashes with fire-sticks smouldering on them. All around was a shallow groove or depression, caused partly by the constant lying and sitting of the Blacks in it, partly by the gradual accumulation of ashes inside, and the casting of these and other refuse immediately outside it. On the outer side of this groove or form, were stuck up at an angle, large leaves of a Fan Palm here and there so as form a shelter, and under the shelter of these the Blacks huddled together at night to sleep.

A camp of this shape with a slight mound inside, and a

* For a further account of Cape York, see Jukes, "Voyage of the 'Fly.'"

bank outside, formed involuntarily by primitive man, may have given the first idea of the mound, the ditch, and rampart. The large amount of wood-ashes accumulated in such a camp, accounts for their occurrence in such large quantities in kitchen-middens, where camping must have been in the same style. A good many shells brought from the shore lay here and there about the camp.

There were besides in the neighbourhood remains of shelters of the common Australian form, long huts made of bushy branches set at an angle to meet one another above, and partially covered with palm-leaves and grass ; these the Blacks used occasionally.

In the daytime the young women and the men were usually away searching for food, but two miserable old women, reduced nearly to skeletons, but with protuberant sto-machs, with sores on their bodies and no cloth-ing but a narrow bit of dirty mat, were always to be seen sitting huddled up in the camp. These hags looked up at a visitor with an apparently mean-ingless stare, but only to see if any tobacco or bis-cuit were going to be given them ; they exhi-

OLD WOMAN, CAPE YORK.
(From a rough sketch by Lieut. A. Channer, R.N.)

bited no curiosity, but only scratched themselves now and then with a pointed stick.

The younger women had all of them a piece of some European stuff round their loins. Some of the men had tattered shirts, but one, who acted as my guide, was invariably absolutely without clothing, as was his son, who always accompanied him. The only property to be seen about the camp were a few baskets of plaited grass, in the making of which the old women were some-times engaged and which were used by the gins for collecting food in. Two large Cymbium shells, with the core smashed out, had been used also to hold food or water, but were replaced for the latter purpose now by square gin bottles, of which there were plenty lying about the camp, brought from the settlement.

The most prized possession of these Blacks is, however, the bamboo pipe, of which there were several in the camp. The bamboos are procured by barter from the Murray islanders, who visit Cape York from time to time, and the tobacco is smoked in them by the blacks in nearly the same curious manner as that in vogue amongst the Dalrymple Islanders. No doubt the Australians have learnt to smoke from the Murray Islanders.*

The tobacco-pipe is a large joint of bamboo, as much as two feet in length and three inches in diameter. There is a small round hole on the side at one end and a larger hole in the extremity of the other end. A small cone of green leaf is inserted into the smaller round hole and filled with tobacco, which is

BAMBOO TOBACCO-PIPE USED BY THE NATIVES OF CAPE YORK.

lighted at the top as usual. A man, or oftener a woman, then opening her mouth wide covers the cone and lighted tobacco with it and applies her lips to the bamboo all round it, having the leaf cone and burning tobacco thus entirely within her mouth. She then blows and forces the smoke into the cavity of the bamboo, keeping her hand over the hole at the other end and closing the aperture as soon as the bamboo is full.

The leaf cone is then withdrawn and the pipe handed to the smoker, who, putting his hand over the bottom hole to keep in the smoke, sucks at the hole in which the leaf was inserted, and uses his hand as a valve meanwhile to allow the requisite air to enter at the other end. The pipe being empty the leaf is replaced and the process repeated. The smoke is thus inhaled quite cold. The pipes are ornamented by the Blacks with rude drawings.

* J. Beete. Jukes. "Narrative of the Surveying Voyage of H.M.S. 'Fly,'" Vol. I, p. 65. London, Boone, 1847.

The bamboo pipes of Dalrymple Island are described as having bowls made of smaller bamboo tubes instead of the leaf cone. There are many such in museums. Possibly the leaf is only a makeshift. The Dalrymple Islanders, however, sucked the bamboo full of smoke from the large hole at the end instead of blowing.

It is remarkable that the Southern Papuans should have invented this peculiar method of smoking for themselves, since there can be little doubt that they derived the idea of smoking from the Malays, probably through the Northern and Western Papuans. There seems no doubt that the habit of smoking, as well as the tobacco plant, were first introduced into Java by the Portuguese,* and the habit and plant no doubt spread thence to New Guinea. The Papuans at Humboldt Bay smoke their tobacco in the form of cigarettes.

No other property than that mentioned was to be seen about the camp of the Gudangs, but on our asking for them, Longway produced some small spears and a throwing stick, which were hidden in the bush close by; and a second lot of spears was produced afterwards from a similar hiding-place. The Blacks keep what property they have thus hidden away, just as a dog hides his bone, and not in the camp; hence it is impossible to find out what they really have. I saw no knife or tomahawk. No doubt the practice of thus hiding things away from the camp has arisen from constant fear of surprise from hostile tribes.

The Blacks feed on shell fish and on snails (a very large Helix), and on snakes and grubs and such things, which are hunted for by the women. The women go out into the woods in a gang every day for the purpose of collecting food, and also dig wild yam roots with a pointed stick hardened in the fire. They have not got the perforated stone to weight their digging-stick, and are thus behind the Bushmen of the Cape in this matter. A staple article of food with these Blacks is afforded by the large seeds of a Climbing Bean (*Entada scandens*), and their only stone implements are a round flat-topped stone and another long conical one, suitable to be grasped in the hands. This is used as a pestle with which to pound these beans on the

* A. de Candolle, " Géographie Botanique," T. II, p. 850.

flat stone.   Both stones are merely selected, and not shaped in
any way.

These Blacks seem never to have had any stone tomahawks,
and their spear-heads are of bone.   They seem not to hunt the
Wallabies or climb after the Opossums, as do the more southern
Blacks, but to live almost entirely on creeping things and roots,
and on fish, which they spear with four-pronged spears.   Staff-
Surgeon Crosbie of the "Challenger" saw Longway and his boy
smashing up logs of drift-wood and pulling out Teredos and eating them one by one as they reached them.

I tested Longway and also several of the Blacks together at
the camp, by putting groups of objects, such as cartridges, before
them, but could not get them to count in their language above
three—piama, labaima, damma.*   They used the word nurra†
also, apparently for all higher numbers.   It was curious to
see their procedure when I put a heap of five or six objects
before them.   They separated them into groups of two, or two
and one, and pointing to the heaps successively said, "labaima,
labaima, piama," "two," "two," "one."   Though another of my
guides had been long with the Whites he had little idea of counting.   After he had picked up two dozen birds for me and seen
them packed away, I asked him how many there were in the
tin: he said Six.   I wish I had paid more attention to the
language of these Gudangs.   No doubt amongst such people
language changes with remarkable rapidity; especially as here,
where tribes are mixed, and some of the words at least seem
to have changed since MacGillivray's time.

The Blacks are wonderfully forgetful, and seem never to carry
an idea long in their heads.   One day when Longway was out
with me he kept constantly repeating to himself "two shilling,"
a sum I had promised him if I shot a Rifle-bird, and he constantly
reminded me of it, evidently with his thoughts full of the idea.
After the day was over, and we were near home, he suddenly
left me and disappeared, having been taken with a sudden desire
to smoke his bamboo, and gone by a short cut to the camp.

* MacGillivray, "Gudang Dialect."   He gives "epiamana elahaiu
dāma."

† = unora?   MacGillivray.

When I found him there he seemed astonished and to have forgotten about his day's pay altogether.

The Blacks spend what little money they get in biscuit at the store.   And they know that for a florin they ought to get more biscuit than for a shilling, but that is all.   Food is their greatest desire.   Their use of English is most amusing, especially that of the word "fellow."   "This feller gin, this feller gin, this feller boy," said Longway, when I asked whether some young Blacks crouched by the fire were boys or girls.   They apply the term also to all kinds of inanimate objects.   There are several graves of Blacks near Somerset.   I asked Longway what became of the Black fellows when they died; he said "fly-away," and said "they became White men."

About 35 miles from Somerset is a tribe of fierce and more powerful Blacks, of which the Gudangs are in great terror. When I wanted some plants which were a little way up a tree, Longway was not at all inclined to climb, but let a sailor who was with me do it.   Longway's boy said he could not climb.

As I have said, Longway was always completely naked.   He not only had no clothing of any description, but no ornament of any kind whatsoever, and he was not even tattooed.   Further, he never carried, when he walked with me, any kind of weapon, not even a stick.   His boy, who was always with him, was in the same absolutely natural condition.   It was some time before I got quite accustomed to Longway's absolute nakedness, but after I had been about with him for a bit, the thing seemed quite familiar and natural, and I noticed it no more.

On one of our excursions, Longway begged me to shoot him some parroquets to eat.   I shot half a dozen at a shot.   I should not have done so if I had known the result.   Longway insisted on stopping and eating them there and then.   I was obliged to wait.   Longway and his boy lighted a fire of grass and sticks, tore a couple of clutches of feathers off each of the birds and threw them on the fire for the rest of the feathers to singe partly off.   Before they were well warm through, they pulled the birds out and tore them to pieces, and ate them all bleeding, devouring a good deal of the entrails.

On one occasion, when I wished to start very early on a shooting expedition, in order to come upon the birds about daybreak, which is always the best time for finding them in the tropics, I went to the camp of the Blacks to fetch Longway, just as it was beginning to dawn. The Blacks were not by any means so easily roused as I had expected; I found them all asleep, and had to shout at them, but then they all started up scared, as if expecting an attack. I had great difficulty in persuading Longway to go with me at that early hour, and he complained of the cold for some hours. I think the Blacks usually lie in camp till the sun has been up some little time, and the air has been warmed.

With regard to expression, I noticed that the Gudangs used the same gesture of refusal or dissent as the Api men, namely, the shrugging of one shoulder, with the head bent over to the same side. Their facial expressions were, as far as I saw them, normal, I mean like those of Europeans.

Altogether, these Blacks are, I suppose, nearly as low as any savages. They have no clothes (some have bits of European ones now) no canoes. no hatchets, no boomerangs, no chiefs. Their graves, described in the "Voyage of the 'Fly,'" are remarkable in their form. They are long low mounds of sand, with a wooden post set up at each of the corners. There is far more trouble taken with them than would be expected.

The beach at Somerset is composed of siliceous sand. One becomes so accustomed when amongst coral islands, to see the beaches made up of calcareous sand, that it appears quite a novel feature when one meets again with siliceous sand, to which only we are accustomed in Europe. The sandy beach slopes down, to end abruptly on a nearly horizontal mud flat, bare at low water, which is mainly calcareous, and in fact a shore platform reef, but with few living corals on it. At low water, during spring tides, blocks of dead massive corals, such as *Astrœidœ* are seen to compose the verge of these mud flats, and it is from the detritus of these that the mud is formed. Amongst these blocks are but few living corals, a species of *Euphyllia*, small *Astrœas*, and cup or mushroom-shaped *Turbinarias*.

There is a considerable variety of species of seaweeds on the flats. There are also several forms of Sea-Grasses : a species of *Halophila*, the large hairy *Enhalus*, and a *Thalassia* grow all together, and spread in abundance over the mud, which is matted with their roots in many places.

The channel between Somerset and Albany Island is shallow, being nowhere more than 14 fathoms in depth. The dredge here brought up a rare species of Trigonia, and the " Lancelet " *Amphioxus lanceolatus*, which seems to have an extremely wide range in distribution. The fauna on the whole was very like that of Port Jackson.

Cape York is a sort of emporium of savage weapons and ornaments. Pearl shell-gathering vessels (Pearl shellers as they are called) come to Somerset with crews which they have picked up at all the islands in the neighbourhood, from New Guinea, and from all over the Pacific, and they bring weapons and ornaments from all these places with them. Moreover, the Murray Islanders visit the port in their canoes, and bring bows and arrows, drums, and such things for barter.

The water police stationed at Somerset deal in these curiosities, buying them up and selling them to passengers in the passing steamers, or to other visitors. Hence all kinds of savage weapons have found their way into English collections, with the label "Cape York," and the Northern Australians have got credit for having learnt the use of the bow-and-arrow. I believe that no Australian natives use the bow at all.

Weapons from very remote places find their way to Cape York, and thus no doubt the first specimens of Admiralty Island javelins reached English museums. Accurate determination of locality is of course essential to the interest of savage weapons. Staff-Surgeon Maclean, of the " Challenger," had a large New Guinea drum of the Crocodile form thrust upon his acceptance, as a fee for visiting . a patient on board one of the " Pearl shellers "; he gave it to me.

Wednesday Island, Torres Straits, Sept. 8th, 1874.—We left Cape York on September 8th, and made for the Prince of Wales Passage through Torres Straits. I landed at Wednesday Island a distant outlier of Cape York, which, with Hammond Island, is passed close by in the track through the passage. The island is about two miles long, it is made up of quartz porphyry, forming hill masses, a couple of hundred feet or so in height, with sandy flats at their bases.

In places, the hill slopes come right down to the sea, forming small headlands, and here the beach is composed of boulders with small stretches of quartz sand derived from the rocks between them. Along a wider bay to the north, the whole beach is made up of calcareous sand and broken and dead shells. A shore platform reef extends all along this side of the island; in some places it is made up of consolidated coral rock, full of large masses of dead corals cemented together with coral mud, or seen projecting here and there between muddy pools of water.

In other places the coral rock passes gradually into regular mud flats. There were very few living corals indeed about the shore platform; it required careful searching to find them. I found only the species of *Euphyllia*, which was at Somerset, and a small *Astræa*. One large mass of *Astræa* thrown up by the waves and embedded in the mud, had a small patch on one side of it still alive, the rest was quite dead.

Though stony corals were so scarce, soft Alcyonarians were in great abundance. The rock was full everywhere of the Giant Clam (*Tridacna*), the largest bivalve shell which has ever existed, a familar adornment of fountains and oyster-shops in England. This mollusk lives sunk in a cavity of its own in the rock, with only its brilliant blue or green mantle fringes showing and betraying its retreat. These protruded mantle lobes have the appearance of huge expanded elongate sea anemones, and at first sight one takes them for such. The shells must be quarried out of the rock with a hammer and chisel if they are wanted.

The main peculiarity of these coral flats, as at Somerset, is their extreme muddiness and the small quantity of life about them. A Sargassum grows abundantly on the rock masses, with several other algæ. No doubt the decomposition of these and the soft Alcyonarians is that which renders the coral mud so dark and slimy. The occurrence of beaches of calcareous and siliceous sand close together, both rising from the same coral flat, is an interesting fact, as showing how easily beds of such very different materials may become associated or superposed. A large Chama shell is very abundant, cemented to the hard porphyry rocks, and recalled to one's mind forcibly the extinct Hippurites.

The hills of the island are covered with a scrub, nowhere very dense or high, whilst there are small mangrove swamps at the edge of the mud flats. The low sandy tracts are open, covered with scattered gum trees with long grass growing beneath them, just as at Cape York. The long grass and bushes were parched and dry, and burnt rapidly when we fired them. On the shore were an Oyster-catcher, a small Plover, and a Sandpiper, in flocks. The few Land-birds seen, were Cape York species, the common Bee-eater, little Ground Dove, *Artamus*, White Cockatoo, and a Brush Turkey.

Close to the shore were two native graves, and the remains of shelters made of branches, and of fires. The island is often visited by the natives of the Straits when on their voyages, but not permanently inhabited. There were two graves placed side by side, consisting of oblong mounds of sand, each with six wooden posts placed regularly at the corners and middles of the longer sides. The posts had many of them large shells placed on their tops as decorations; the mounds were decorated with ribs of Dugongs, placed regularly along their sides and arching over them, whilst Dugong skulls, all without the tusks, and large shells adorned their summits.

In dredging in shallow water off Wednesday Island, a monster Starfish was obtained, apparently a species of *Oreaster*; it measured 1 ft. 9 ins. from tip to tip of its arms, and 5 inches in the height of its central disc.

Booby Island, Torres Straits, Sept. 9th, 1874.—On the following day I landed on Booby Island, which acts as a sign-post to ships entering the Prince of Wales Passage from the Arafura Sea, on the other side of Torres Straits. The island is of the same coarse quartz and felspar rock as Wednesday Island; it is only about two-thirds of a mile in circumference, and 30 to 40 feet in height. The greater part of the rock is white with the dung of sea birds, the Booby and the " Wideawake," which frequent it in vast numbers. The birds were, however, not breeding here at the time of our visit: one egg of the tern only was found. These birds were hence shy, and left the rock on the approach of the boat, and remained flying round it until our departure.

Most astonishing is the number and variety of land-birds, which is to be found on this small island. It is so small that, when the boat party had landed and had spread over it, it became dangerous to shoot in almost any direction, for fear of hitting some one. Yet here I shot seven species of land-birds, and saw three others.

Most of the birds of Cape York are constantly migrating, and the resident official at Somerset told me that the constant change from month to month of the birds seen about his place was most astonishing. The Torres Straits Islands serve as resting places for the birds crossing from New Guinea; Booby Island is evidently thus used, and the number of its land-birds is thus to be accounted for.

This island corresponds thus in this respect with such an island as Heligoland in Europe, which is a well-known halting-place of birds of passage, and at certain seasons swarms with land-birds, resting on their journey, so that ornithologists visit it to procure the rarest of birds. Heligoland also, like Booby Island, is almost devoid of trees, and the birds have to pitch there in the potato-fields. Upwards of 300 species of land-birds rest on the island, which is a point in the direct lines of migratory flight.*

A small cleft runs up into Booby Island, and nearly across it, and, affording shade and shelter, allows of the growth of a small thicket of shrubs of a species of fig. Besides these shrubs the island has little vegetation, except scanty grass, and about half-a-dozen species of herbs. Amongst the branches of the figs, lives a most beautiful Fruit-Pigeon (*Ptilinopus superbus*), with head of a brilliant purple, the body green, and shoulders red. A Painted Quail (*Turnix melanonotus*), was found amongst the grass. The other birds which I saw or shot were a Land-rail, a

* J. F. Naumann, "Ueber den Vogelzug mit besonderer Hinsicht auf Helgoland," s. 18. Rhea, Leipzig, 1846.

H. Seebohm, "Supplementary Notes on the Ornithology of Heligoland," Ibid. 1877, p. 156.

Mound-bird (*Megapodius tumulus*), a Bee-eater (*Merops ornatus*), a Zosterops (*Z. luteus*), very like *Z. flaviceps* of Fiji, a Pachycephala, a Kingfisher (*Halcyon sancta*), and a thrush-like bird, of which I saw only one specimen.

The Pigeon seems to be a permanent resident in the island. The Megapodius astonished me most; I did not know that the bird possessed powers of flight sufficient to take it to such an island; it must have been migrating. The fact no doubt explains the occurrence of species of Megapodius in various Pacific Islands. The quails are present at some times in Booby Island in enormous numbers. On August 13th, 1841, the officers of the "Beagle" shot on it 145 quails, 18 pigeons, 12 rails of two species, and three pigeons.*

The Dove and the Rail were here for the first time procured by Mr. Bynoe, and named by Gould from Booby Island specimens. It is the last place in the world, as viewed from the sea, with clouds of Boobies hovering over it, from which one would expect two new land-birds to hail. Our officers laughed at the notion of there being Quails or anything to shoot upon it. The officer of the "Beagle" found a native grave on the island. There are several caves on the island, in one of which a store of provisions is kept for shipwrecked seamen. The caves are now several feet above high water-mark, and possibly they point to a slight elevation of the island.

* "Discoveries in Australia. Also An Account of Capt. Owen Stanley's Visits to the Islands of the Arafura Sea," Vol. II, p. 329. By J. Lort Stokes, Commander, R.N. London, Boone, 29, New Bond Street, 1846.

CHAPTER XV.

ARU. KE. BANDA. AMBOINA. TERNATE.

Appearance of the Aru Islands. Trees Transplanted by the Waves.
Masses of Drift Wood. Malay Language. Ballasting a Guide.
Management of Clothes during Rain. Back Country Natives. Great
Height of the Trees. Nests of the Metallic Starling. Parrots and
Cockatoos. Bird Winged Butterflies. Shooting Birds of Paradise at
Wanumbai. Deposit of Lime in Streams. Boat Crews from the Ke
Islands. Fungus Skin Disease. Ke Island Dancing. Houses at Ke
Dulan. Leaf Arrows. Bird caught in a Spider's Web. Ascent of
the Volcano of Banda. Algæ Growing in the Hot Steam Jets.
Numerous Insects at the Summit. Alteration in Sea Level, Marked
on Living Corals. Nutmeg Plantations. Transportation of Seeds by
Fruit-Pigeons. Saluting at Amboina. Danger to the Eyes in Diving
for Corals. Raised Reefs. Myrmecodia and Hydnophytum. Moluc-
can Deer. Ternate Island. Chinese and their Graves. Sale of Birds
of Paradise. Ascent of the Volcano. The Mountain Vegetation. The
Terminal Cone. View from the Summit.

The Aru Islands, September 16th to September 23rd, 1874.—
On our way to the Aru Islands we crossed the Arafura Sea,
which lies to the west of New Guinea. The sea is extremely
shallow, being only from 30 to 50 fathoms in depth. After a
voyage of six days, from Torres Straits, we sighted the southern
part of the Aru Islands, so familiar to naturalists from Mr.
Wallace's account of them, in his " Malay Archipelago," and so
full of interest to us as the home of Birds of Paradise.

We sailed along the western coast of the islands. The
southern portions are not covered with forest, but appeared in
the distance as open grassy downs, and immediately further
north similar open country occurs frequently, amongst the
forest in patches. The grass, though it appears like turf in
the distance, is probably tall and reed like. A line of cliffs
of no great height forms the coast line. The low cliffs are

broken at intervals and there the coast is wooded and shows
a white sandy beach.

The cliffs appear as if formed of a stratified ferruginous red
rock. Here and there on the rocks were conspicuous white
patches on the cliffs, the nesting-places of Boobies, of which
large flocks were seen flying to roost as evening came on.

Masses of closely-packed tree-stems with dense foliage
masses above, appeared lining the shore where it was flat.
There were no cocoanut palms to be seen amongst them.
After coasting during the whole night, Dobbo, the port of the
islands, was reached in the morning. Dobbo lies on the small
island of Wamma, which is separated opposite the town by a
narrow channel from the large island of Wokan.

The striking feature in the vegetation of Wamma, as viewed
by one who has just been amongst the Pacific Islands, is the
very small proportion of palms showing amongst the general
mass of foliage. There are only two small clumps of cocoanut
trees near the town. The leafy masses rising above the white
beach might almost be taken to be made up of elm trees, the
tree tops being rounded in the same manner. The whole has
a dull blueish tint.

As we neared Dobbo, turning up the passages between the
two islands, we passed large quantities of leaves, fruits, and
flowers, and branches of trees floated off from the shores, and
now drifting about mingled with a floating seaweed (*Sargassum*).
Off the Ke Islands we met with similar drifts of land vegetation
and also amongst the Moluccas ; and I was astonished at the
large quantities of fresh vegetable matter thus seen floating on
the sea.

The sea birds, especially terns, habitually resort to the floating logs as resting places, and it is curious to see them in the
distance, appearing as if standing on the surface of the water,
the logs themselves being often invisible. Not only are large
quantities of fruits capable of germinating thus transported from
island to island,* but entire living plants, even trees, are washed
from island to island and transplanted by the waves.

* Mr. Darwin has recorded the experiments which he made in con-
junctate with Mr. Berkeley to determine the period of time during which

On the shores of Little Ke Island I found on the beach, above the ordinary reach of the waves, a large mass of the pseudo-bulbs of an epiphytic orchid with its roots complete. It was partly buried at the foot of a tree and seemed quite lively. It had evidently been washed up in a storm. At Malanipa Island, off the coast of Mindonao Philippines, I found a young Sago Palm, which was just beginning to form a stem, washed up just above the ordinary beach line, and firmly rooted, though in an inclined position, and growing vigorously. Several authors have described the large quantities of floating vegetable matter to be met with in the Malay Archipelago and neighbourhood. Chamisso remarked on the quantity of floating seeds off Java, and the casting up of Barringtonia, *Aleurites triloba*, and Nipa Palm seeds on the shores in germinating condition.*

These large drifts from the forests have a further interest, in that they let drop their remains to the bottom of the deep sea, thereby not only serving as food to the deep-sea animals, but leaving their husks to be preserved as fossils in deep-sea deposits. I shall refer to this latter point in considering deep-sea questions in the sequel.

We anchored off the town of Dobbo, not in the least altered in the few years since Wallace's visit, with its line of Macassar trading vessels drawn up on the beach; its " prau " builders at work, and a crowd assembled to gaze at us. We were visited by Malay notables in their finest dresses of coloured silks, and by Dutch half-caste missionaries who came in tail coats and tall hats.

The sun was excessively powerful at Aru, and I felt the glare on the white sandy beach more severely than anywhere else during the voyage. In wading in search of seaweeds on the coral shore platform, I positively found the water much warmer than was pleasant to my legs. The water was very shallow, only half way up my knees, and was heated by the reflections from the white bottom.

various seeds will resist the action of sea-water, in the " Origin of Species," 6th Ed., 1876, pp. 324, 325.

* Chamisso, " Bemerkungen auf einer Entdeckungs-Reise, 1815–1818," p. 366-401. Weimar, 1821.

We encountered the Malay language for the first time at Dobbo, and since no one there, except the missionaries, who spoke Dutch, understood any European language, it was fortunate that our navigating officer, Staff-Commander Tizard, had learnt the language when engaged in surveying in the China Seas and on the coast of Borneo. He arranged for guides and started us with a small stock of the language.

It is the easiest in the world to pick up a little of. There is no grammar, and anyone who has got a Malay dictionary can talk Malay. " I go," " I shall go," " I went," are all expressed by the same word in Malay, and one is irritated on discovering how thoroughly satisfactory such a simple arrangement is, to reflect on the endless complications of verbs and their inflexions in so many other languages and on the time which one has wasted over them.

I made several excursions on shore with one or more guides. One whom I generally took with me was a very active fellow, and I soon found him too quick for me in the close hot forest. I have always found it a bad plan to let native guides suppose that one is easily tired and unable to keep up with them, so I adopted an expedient with the man which has served me in good stead on other occasions, and which can be recommended to naturalists. Soon after I got on shore I examined a large stone with care and interest, turning it over once or twice, and then gave it him to carry, and when he had this ballast in addition to my vasculum, I found that I could keep on pretty good terms with him. In the evening, when we reached the boat, I conveyed the stone on board the ship with due solemnity and threw it overboard.

I was amused at the manner in which my guides met a heavy storm of rain. They had of course no umbrellas, but did not wish to get their clothes, which consisted merely of two cloths, one worn round the shoulders and the other round the loins, wet. They simply stripped naked, rolled their clothes up tight inside a large Pandanus leaf, and so walked along with me till the rain was over, when they shook themselves dry and put their clothes on again. Meanwhile my clothes were wet through and had to dry on me.

B B

A very large species of Screw-pine (*Pandanus*), with a fruit as big as a man's head, is common along the shore. It is a common East Indian littoral plant. The stem, though large, is soft and succulent, and hence with a small axe one can enjoy all the pleasure of felling a large tree without any fatigue. The deep cut made by a single blow is most gratifying to one's feelings of power, and having cut down one tree to obtain a specimen of the fruit, I found myself felling two or three others wantonly.

On the Island of Wokan, not far from the anchorage, Sago palms abound in the swamps. Several parties of natives from the back country were living near the shore, having come from a distance in their boats, to prepare a store of sago to take home with them.

They lived in small low-roofed houses made of poles and reeds, and raised on posts about two feet above the swampy ground. These temporary houses were so low that the natives

HOUSE OF BACK-COUNTRY NATIVES. WOKAN ISLAND.

could only squat or lie in them. The men were darker than the inhabitants of Wokan in the neighbourhood, and looked to me more Papuan in appearance. They were armed with finely-made spears with iron blade-like points, six or eight inches long, and ornamented worked wooden handles. They would not part with these at any price.

They resented my looking into their house, no doubt because the women were there. The women seemed extremely shy, and huddled together out of the way, and the same was the case at Wanumbai. The men had wrist ornaments, closely similar in make to those common in New Guinea, at Humboldt Bay, and at the Admiralty Islands. These are broad band-shaped wristlets made of plaited fibres (of Pandanus ?), yellow and black worked into a pattern.

These bracelets of the Aru Islanders were ornamented with European shirt buttons in lieu of the small ground-down shells (*Neritina*) used at New Guinea and in the Admiralty Islands for the same purpose. The buttons came, no doubt, from the Chinese traders, and probably the natives thought they were intended for this purpose, as they look not so very much unlike the shells. The men had a number of leaf buckets full of sago, ready prepared, and we saw their rude kneading-trough and strainers of palm fibre, in a swamp close by.

The trees are excessively high and large in the Aru forests. To a botanical collector, with no time to spare, such a forest is a hopeless problem. Only the few low-growing plants can be gathered, and the orchids and ferns that hang on the stems low down, especially along the coast. A few palms can be cut down. The flowers and' fruits of the trees, the main features of the vegetation, and those most likely to prove of especial interest, are far out of reach.

The trees cannot be cut down. It would take a day at least to fell one. The only hope is to lie on one's back and look for blossoms or fruit with a binocular glass, and then try and shoot a branch down. Very often, however, the trees are far too high for that, and then the matter must be given up altogether.

Growing on some of the high trees in Wokan Island, I saw most enormous Stag's-horn ferns (*Platycerium*). I certainly imagined they must be at least eight feet in the height of the fronds. I could not reach any but very small specimens.

A species of Fig, a wide-spreading tree with large leaves, seemed to me remarkable, because the fruit was borne only on the pendent aerial roots. A tree of another species of fig amused me, because its pendent roots had wound spirally around the parent stem of the tree itself, and had nearly choked it. It seemed just that a fig, so accustomed to choking other trees, should thus once in a while choke itself; but no doubt the tree suffered little, the roots taking fully the place of the strangled trunk.

The Rattans are a serious obstacle in excursions in the forests. The tendrils of these trailing and climbing palms are beset with rows of recurved hooks, which as they are drawn across one's flesh, in a dash made to get a shot at a bird, cut into it as readily as knives, but with a more unpleasant wound.

An immense tree, with a tall stem free from branches, until at a great height it spread out into a wide and evenly shaped crown, was full of the nests of the Metallic Starling (*Calornis metallica*), a very beautiful small starling with dark plumage, which displays a brilliant purple metallic glance all over its surface. The birds breed thus gregariously. There must have been three or four hundred nests in the tree; every available branch was full of them. The birds were busy flying to and fro, and were quite safe, for the tree was so high that they were out of shot of my gun at least, which was not a choke bore.

On one of my excursions in the forest I met with a flock of brilliant plumaged Parrots. They were apparently feeding in company with a flock of White Cockatoos. I managed to stalk one of the parrots, and shot it. The cockatoos set up the most angry harsh screaming, and evidently made common cause with the parrots. They sat and screamed at me on a tree close by, as angrily as if one of their own flock had been shot, and flew over my head high up out of reach of the gun, looking down at the dead bird and still screaming.

Once, as I was making my way through thick undergrowth in a swampy place, my guide touched my arm and pointed and said " Cāsuārī." I was too late to see the big bird, but I saw the tracks of its feet in the mud; and now, for the first time, realized the fact that the Cassowary, a large Struthious bird, can inhabit a dense forest. I had always coupled Struthious birds in my mind with open downs or plains, or at all events with brushwood and occasional trees. I had also not before understood that " Cassowary " was the Malay name of the bird.

I searched for Land Planarians without success. There can

however, be no doubt that they exist in Aru, since they occur in Australia, Ternate, and the Philippines.

The splendid large Bird-winged Butterfly, with brilliant green and velvety black wings (*Ornithoptera poseidon*) was common in the woods, but flew high and was difficult to catch. I shot one or two with dust shot, without their being utterly damaged. I once, however, was lucky enough to find a flock of about a dozen males, fluttering round and mobbing a single female. They were then hovering slowly, quite close to the ground, and were easily caught.

The female had thus a large body of gaudy admirers from which to make her choice. Interesting results might possibly be derived from a series of experiments, in which, in the case of brightly coloured and decorated butterflies, the colours should be rubbed off the wings of a few amongst a number of males, or painted over of a black or brown colour. It might be tested whether the females would always prefer the brightly coloured ones. Dark coloured butterflies might possibly have the wings of the male touched up with a little colour.

Similar experiments might be made with more chance of success in the case of gaudy birds, the feathers of the cock being dyed dark, or enhanced in colouring in the case of a little decorated male. The hen might be kept in a cage between two males, and it might be noted to which she gave the preference, and then, whether an alteration in the colours of the plumage caused a change in her inclination. If the artificial increase of colouring succeeded as an experiment, then experiments might be made to learn what colours, or mixture of colours, is most attractive in various cases.*

A party visited Wanumbai, Mr. Wallace's old hunting-ground, in the ship's steam-pinnace. We steamed across a sort of lagoon, shut in by the islands, passing on the way a large Sea

---

* Mr. Tegetmeier stained some pigeons with magenta at Mr. Darwin's request, but the birds were not much noticed by the others. Mr. Darwin cites the case of the pied peacock, and that of the silver pheasant which had its plumage spoiled, and which was then rejected by the hens. No systematic experiments, however, seem to have been made on this subject though they could easily be carried out in the case of birds. C. Darwin, "The Descent of Man," Vol. II, pp. 118, 120, London, Murray, 1871.

Snake on the top of the water, and made our way up the remark-able canal-like channel, for the formation of which Mr. Wallace found it difficult to account. The people of Wanumbai were very much scared at the appearance of the pinnace, full of men with guns, but we had taken some Malays from Dobbo with us to act as pilots, and introduce us, and they jumped on shore and addressed the people of Wanumbai ("*Orang Wanumbai*, Ye men of Wanumbai,") and soon made matters right. They told them that we had only come to shoot "dead birds" (*Burong mate*), the trade term by which the Birds of Paradise are known..

On the margin of the narrow sea channel, was a compound house, an oblong building raised on numerous posts above the ground. Inside it had a central passage, leading from the door to the back wall, and on either side of this it was divided into small pens by low irregularly made partitions. Each of these pens held a family, and the women huddled together to hide themselves in the corners of them, just as did those in Wokau Island.

We purchased bows and arrows from the natives. The arrows are very like New Guinea arrows in the various forms of their points, but are all provided with a notch and feathers, the latter being often bright parrots' feathers. Some have a blade-like point of bamboo, and a man who was watching a native plantation, to keep wild animals off from it, told me he used these for shooting pigs. Some are tipped with Cassowary bone, some are many-pronged, and these are used for shooting birds, and are not exclusively fish arrows, as is often supposed.

Besides these, there are the arrows with a large blunt knob at the end, used for stunning the large Birds of Paradise, with-out spoiling their skins, as described by Wallace. Pointed arrows are however used more frequently for this purpose, as Mr. Wallace relates, because the birds are so strong as to escape being stunned, and the points are more certain weapons. It is curious that closely similar knobbed arrows are used in South America by certain tribes, to kill Trogons and other fine plumaged birds. One man brought for sale a large Bird of Paradise, dried in the usual manner for sale, but he wanted the full price for it asked by the Chinese dealers at Dobbo.

I procured two guides, a boy and a man, and promised them a florin for every Bird of Paradise that I shot. I had previously

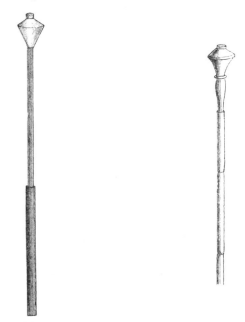

HEAD OF A SOUTH AMERICAN BIRD ARROW,
IN THE CHRISTY COLLECTION.

HEAD OF AN ARROW USED FOR SHOOTING BIRDS
OF PARADISE IN THE ARU ISLANDS.

been in pursuit of the birds at Wokan, but they were not so common there, and I believe that the native guides did not exert themselves to show us the birds, as they no doubt regard them more or less as property, and a source of wealth.

My first acquaintance with the great Bird of Paradise (*Paradisea apoda*) was at Wokan. I was making my way through the forest with a guide in the very early morning, when a flock of birds flew by in the misty light, passing right over my head. They flew like a flock of Jackdaws somewhat, and I was disgusted to realize, when too late, that they were a flock of the very birds I was in search of. I did not fire for fear of disturbing the woods. I heard them cry soon after " wauk, wauk," but could not come up with them.

At Wanumbai with my guides, I first encountered a number of Fruit-Bats, which were on the wing in the early morning, and

I killed one with a young one hanging at its breast. We soon heard the cry of the great Bird of Paradise, "wauk, wauk." I crept up within shot with my guides several times, but as usual, though they saw the bird plainly amongst the foliage, I could not make it out in time, though I saw the leaves rustle. I did not want to fire without making sure. The guides in view of the florin, were as excited as I was, and kept seizing my arm and pointing, "burong mate, burong mate," but away went the bird without showing itself to me.

The birds seemed to keep constantly on the move in the trees, hopping from branch to branch, and were very quick and silent in their flight away to a fresh spot. Several times I saw the birds amongst the branches of trees, so high that it was use-less to shoot at them, and my cartridges, specially prepared with nearly four drachms of powder, had no effect.

The birds seemed to be as often single as in companies, and were evidently on the feed in the early morning. At last a hen bird flew up off the ground close to me, with a small lizard in her beak, and pitched on a dead branch to eat it, and I shot her. But what of course I wished, was a male in full plumage. This however was not to be obtained. It is remarkable what a very large proportion of young males and females of the great Bird of Paradise there seem to be, to the comparatively small number of males in full dress. Not one of these latter was shot. I believe I saw one at the top of a high tree, but am not certain. Probably the old males are warier, being often hunted, and keep out of the way. They require four or five years to develop full dress.*

At the breeding season, when the natives kill most of them, they assemble, and are easily obtained.

The cry "wauk," is not so far removed from such cries as those of the Rook and others of the *Corvidæ*, to which the Paradise

* It is improbable that P. apoda, loses its breeding plumage as soon as the breeding season is over. P. minor, as has been observed in the case of specimens kept in confinement in the Regent's Park Gardens, certainly loses its plumes only at the moulting season, like other richly ornamented birds. P. apoda moults, according to Wallace, in January or February, and is in full plumage in May. At all events there must have remained birds with plumes in September.

birds are allied. The voices of birds need however no more necessarily be a test of the pedigrees of the birds themselves, than need language be a test of true race connection amongst mankind.

Many birds imitate one another's cries, and the Hon. Daines Barrington,* long ago showed by experiment, that nestlings learn their song from their parents, and even their call note, and if taken away very early from the nest, learn the song of any other bird with which they are associated, and then do not acquire that proper to their own species, even if opportunity be afforded.

If nestling birds were brought up apart from other birds, they would no more sing, than would men similarly reared have any idea of talking to one another.

Under these circumstances the birds would utter only what Barrington terms their chirp, a cry for food, which, peculiar to each species, is uttered by all young birds, but which is entirely lost as the bird reaches maturity. Untaught men would be as speechless as apes, far less able to communicate with one another than deaf mutes who watch the communications of others. It is a pity that it is impossible, on humanitarian grounds, to repeat now the experiment of King Psammetichus. It would be interesting to watch the result.

In the case of the other smaller species of Paradise Bird found in the Aru Islands, the King-bird (*Cicinnurus regius*) the males in full plumage seemed as common as the simple brown young males and females. The natives knew these latter well, as forms of the brilliant red bird, though so vastly different, and several times pointed them out to me, as " Gobi, gobi," their name for the " King-bird."

The King-birds were even more abundant at Wanumbai than the larger species. The males, when settled in the trees, constantly uttered a cry which is very like that of the Wryneck or Cuckoo's Mate. I saw most of them in the lower trees of the forest, at about 30 feet from the ground. One shot by

* " Experiments and Observations on the Singing of Birds," by the Hon. Daines Barrington. Phil. Trans. Vol. LXIII. 1773, p. 249. A. R. Wallace, " Contributions to the Theory of Natural Selection," p. 220. London, Macmillan, 1875.

Mr. Abbott, engineer of the "Challenger," when we were together in Wokan, hovered and hopped for some time about a mass of creepers hanging from a large tree, apparently searching for insects. As it hovered, it showed its bright scarlet back like a flash of fire.

Usually, the bird sitting on the twigs and seen from below shows none of its beauty.

The birds seem very tame, but like the Rifle-bird, and the Great Bird of Paradise, are usually in constant motion. One full-plumage bird sat on a twig, about four feet from the ground, and looked at me for a while at not more than three yards distance, and then darted away, more out of natural impulse, I imagine, than fear.

I shot five of the birds in one day. One of them had the wonderful spiral green tail feathers, only just growing out. The bright lapis-lazuli blue colour of the bird's legs and feet when fresh, greatly enhances its beauty. Luckily the skin of the Paradise Birds is tough, and I found the King-bird easy to skin. The short red feathers encroach on the base of the bill, on its upper surface in an unusual manner, the tip of the bill only being free, and this gives the head a curious appearance.

The coral rock of Wokan Island is exposed in section, on the shore not far from Dobbo, in a cliff about 11 feet in height. The strata are inclined towards the sea at an angle of about 20°. Inland, the surface is marked by a series of ridges of small elevation, and from the presence of numerous bivalve shells, seems to have been raised above sea level.

There is a fresh-water stream not far from Wanumbai, which flows over the coral rock, overhung by dense vegetation. In the bed of the stream, a constant deposit of carbonate of lime is taking place, and the bed is partitioned into a series of pools, separated by ridges and projections of stalactite-like substance, which lines also the pools themselves. Similar deposits in tropical streams have been observed elsewhere, as in Roaring River, Jamaica.*

It was elicited by Captain Tizard, from the Malays at Dobbo,

* Sir H. T. de la Beche, F.R.S. "The Geological Observer," p. 13, 2nd Ed. London, Longman, 1853.

that a deer abounds in the northernmost of the Aru Islands; no doubt it is of the same species as the deer of Amboina (*Rusa moluccensis*): I was shown the horns. It must have been introduced either by the Malays or Dutch.

The Chinese dealers in Manchester and Birmingham goods and arrack at Dobbo, used cajuput oil as a preservative for their Birds of Paradise skins, to keep off ants and other insects.

Books referring to the Aru Islands. "Discoveries in Australia," also "An Account of Capt. Owen Stanley's Visit to the Islands of the Arafura Sea," by J. Lort. Stokes, Commander, R.N., Vol. II., p. 333. London, Boone, 1846. "Voyage of the Dutch Brig 'Dourga.'" Trans. by W. Earle. Madden & Co., London, 1840. A. R. Wallace, F.R.S., &c., "The Malay Archipelago."

The Ke Islands, September 24th and 25th, 1874. We crossed over from the Aru Islands to the Ke Islands, taking a day on the passage and dredging and sounding between the two groups, finding a depth of 300 fathoms. Whilst we were off the coast of Great Ke Island several boats full of natives put off to the ship. The boats were described by Wallace. They are shaped like whale boats and are fastened together with rattans.

The crews used paddles with long blades pointed at the ends and cross handles. They paddled in time with a chanted cadence identical with one used by the Fijians in their dances, "ē ai ō tum tum." At intervals the sound rose loud from the approaching boats as it was taken up in chorus.

The chant was accompanied by a drum with a tense membrane, on which two sounds were made by striking it slightly with the tips of the fingers or more violently with the palm of the hand, the sound reminding one that one was getting, in one's travels, nearer towards India.

The men, a boat-load of whom came on board, were like the Aru Islanders, but mostly, I thought, stronger built. They wore their hair long and loose, and had no ornaments. Most of them wore only an apron of cloth. All of them were in the most horrible state of skin disease, their skins being in a rough scurfy condition in many cases all over the body. I have not seen elsewhere such bad cases of vegetable itch. The disease is due to a parasitic fungus and closely allied to or identical with

Pityriasis versicolor.   Dr. Crosbie, Staff-Surgeon of the "Challenger," made a careful microscopical examination of it.   The disease is widely spread in Melanesia and Polynesia.*

The men kept constantly scratching themselves violently, and life can be hardly worth having in Great Ke Island.   Yet the disease is one easily cured.   After all, the natives are no worse off than were Cambridge under-graduates in the middle of the seventeenth century, and they used to be nearly physicked to death into the bargain, absolutely in vain.†

The men begged for all kinds of things, and especially spirits and tobacco.   One of the boats had well-made pottery, nicely ornamented with patterns in red, for barter.   The men, as did also the Malays at Dobbo, used a slight click with the tongue, accompanied by a very slow shaking of the head, to express astonishment.

We anchored off Little Ke Island.   Several boats came off paddling to a different but very similar chant.   The men being ship-builders by profession, were delighted with the ship, and ran all over it and climbed into the rigging.

A dance was got up on the quarter-deck.   The drum was beaten by two performers and a song accompanied it, but there was no clapping of hands, as in Fiji.   The whole mode of dancing was absolutely different, and the attitudes of the dancer were sufficient alone to have told one that one was amongst Malays and not Melanesians or Polynesians.

The dance, in which only two or three performers danced at a time, consisted of a very slowly executed series of poses of the body and limbs.   There was no exact keeping of time to the accompaniment nor unison of action between the dancers.   The hands and arms during the action were slowly moved from behind to the front, the palms being held forwards and the thumbs stretched straight out from them.

In another dance a motion, as of pulling at a rope, was used. The chant to one dance was the words "uela a uela."   There

* See Tilbury Fox, M.D., "On the Tokelau Ringworm and its Fungus." The "Lancet," 1874. p. 304.

† John Strypes' "Letters to his Mother, Scholæ Academicæ," p. 293. Christopher Wordsworth, Cambridge, 1872.

was also a dance of two performers with pieces of sticks, to represent a combat with swords. The whole was closely like the dancing of the Lutaos which we saw later at Zamboangan in the Philippine Islands, but not so elaborate.

The ship moved to an anchorage off the small town of Ke Dulan. The houses were all raised on posts, except the Mahommedan Mosque, which building shows a curious development of the high-peaked Malay roof into a sort of half tower, half spire, representing no doubt an equivalent of the dome. Under the caves of the houses baskets were hung up for the fowls to nest in.

Some boys were playing near the village, and, as a toy, they had a very ingeniously made model of a spring gun, or rather spring bow, a trap by which a large arrow is shot into a wild pig, on its setting loose a catch. Our guide, a boy, wearing a turban, placed his hand on his turban and said, " Mahommed," and explained to Captain Tizard that the small boys at play, whose heads were bare, were not such as he, but heathen. He was evidently very proud of his religion.

The Ke Islanders, besides arrows like those of the Aru Islanders, use others which are peculiar. They are light thin narrow strips cut out of the long leaves of what I believe is a species of Canna. The strips are so cut that the stiff midrib of the leaf forms the shaft of the arrow, and portions of the wings of the leaf are left on at the base of the arrow to act as feathers. The point is simply sharpened with the knife.

These leaf arrows when dry are hard and stiff. They are very easily made by a few strokes of the knife, and a large bundle of them is carried by the archer. They are shot away

ARROW CUT OUT OF A CANNA LEAF.

at a bird in the bush without the trouble being taken to find them again, as in the case of other arrows. They are so small and light that they make very little show in their flight, and no noise; and I saw a youth shoot at least a dozen of them, at a large Nutmeg Pigeon, without the bird's doing more than move its head, and start a little as they flew by almost touching it.

These Nutmeg Pigeons (*Carpophaga concinna*) are very large heavy birds. Some of those shot weighed 2 lbs. I shot two at one shot as they sat on a branch on a high tree right over my head. They fell one on each side of me with a very heavy thud, and I believe would have stunned me had they not luckily just missed my head. I had never considered this danger before.

Mr. Darwin in his Journal* refers to *Epeira clavipes*, as said by Sloane to make webs so strong as to catch birds. At Little Ke Island Von Willemoes Suhm actually found a strong and healthy " Glossy Starling " (*Calornis metallica*) caught fast in a yellow spider's web, and he took the bird out alive and brought it on board the ship to be preserved.

The Banda Group, September 29th to October 2nd, 1874.
From the Ke Islands the ship proceeded to the Banda Group, famed for its nutmegs. On the voyage, which consumed three days, a small island named Bird Island was passed, from which at one spot smoke was issuing from amongst rocks covered with a white incrustation. The smoke was evidently a volcanic fumerole.

Banda Island was reached on September 29th. The ship anchored in a harbour, shut in by three surrounding islands. On one of these was the town, the old fort built by the Portuguese, and the residences of the Dutch Officials. Another island is the small active volcano of the group called Gunong Api (mountain fire); the Malay equivalent of the word volcano. On the third island (Great Banda) are the principal nutmeg plantations. I accompanied a party which ascended the volcano, which is 1,910 feet in altitude only. It appears to be very seldom climbed, either by Dutch residents or natives. The mountain is a steep simple cone. The ascent was made on the east side. The cone is covered with bushes up to within about 700 or 800 feet of the summit, and with the help of these climb-

* " Journal of Researches," p. 36.

ing is easy though arduous. Above the limit of the bushes there are steep slopes of loose stones, wearying to climb and constantly falling. Above these, again, the surface of the cone is hard, the fine ashes and lava fragments of which it is composed, being cemented together so as to form a hard crust. This is roughened by the projection of fragments, but still smooth enough to require some care in the placing of the feet to men wearing boots. The Malay guides with naked feet stood with ease upon it anywhere.

The inclination of the slope is about 33°; and to a man who easily becomes giddy no doubt would be rather formidable in descent. An American traveller, who had probably never been up any other mountain before he ascended the Banda Volcano, has written a most appalling account of the danger which he encountered in descending. To a man with an ordinarily good head there are no difficulties in the ascent or descent.

At the summit the fragments of basaltic rock were under-going slow decomposition under the action of heated vapours issuing in all directions from amongst them, and were softened and turned white, like chalk. Any of these fragments when broken showed part of their mass still black and unaltered, and the remainder white; the decomposition not having reached as yet through the whole.

Jets of hot steam issued in many places from fissures. Around the mouths of these were growing gelatinous masses formed by lowly organized algæ closely similar in appearance to those found growing around the mouths of hot springs in the Azores.* Here, however, there was no water issuing, the only moisture being supplied by the condensation of the steam. There was no accumulation of water, but drops of moisture hung on the sides of the fissures.

In some places the gelatinous algæ, and a white mineral incrustation, formed alternate layers coating the mouths of the fissures. The steam on issuing within the fissure had a tem-perature of 250° F.; and where the crust of algæ was flourishing the thermometer showed 140° F. The steam had a strongly acid and sulphurous smell.

On the summit of the mountain, where the ground is cool, a

* See page 36.

Fern, a Sedge, and a Melastomaceous Plant grow. Besides these, I found another flowering plant, growing in a crack in the midst of a strongly sulphureous smoke which issued constantly from it. The thermometer when laid on the surface of the ground where this plant was growing showed a temperature of 100° F. ; and at a depth of one and a half feet below it the soil about the fissure had a temperature of 220° F.

At the summit of the mountain were numerous flying insects of various kinds, although there was nothing for them to feed' upon, and large numbers of them lay dead in the cracks, killed by the poisonous volcanic vapours. So numerous were they that the Swallows had come up to the top of the mountain to feed on them.

I noticed similarly large numbers of insects at the summit of the volcano of Ternate, at an altitude of more than 5,000 feet. Insects are commonly to be seen being carried along before the wind in successive efforts of flight. No doubt they are blown up to the tops of these mountains, having towards the summits no vegetation to hold on to. The winds pressing against the mountains form currents up their slopes; and in the case of volcanos, which are heated at the summits, no doubt there is a constant upward draught towards their tops, caused by the ascending column of hot air.

I dwell on the accumulation of insects at the tops of these mountains, because when blown off into the free air from these great elevations by heavy winds, as no doubt they often are, the insects are likely to fly and drift before the wind to very long distances, and thus be aided in colonizing far-off islands.

I found the skull of an Opossum (the Woolly Phalanger, *Cuscus*) on the mountain. The animal is common in the Banda Group. It occurs also in the Moluccas and elsewhere. Its occurrence on the Banda Islands seems most easily accounted for on the supposition that it escaped from confinement, having been brought to the islands at some time by Malay voyagers. Malays seem fond of keeping wild animals in confinement. or taming them. There were several such pet animals about the houses at Dobbo, at the time of our visit.

At the base of the Banda Volcano, on the shores of the

island, a belt of living corals composed of a considerable variety of species is easily accessible at low tide. Of these corals the largest bulk is composed of massive *Astræids*, of which about ten different forms were collected. A massive *Porites* is also very abundant.

One species of "Brain Coral," and an *Astræa*, form huge masses, often as much as five feet in diameter, which have their bases attached to the bare basaltic rock of the shore. The tops of all of these coral masses are dead and flat and some-what decayed : but on these dead tops fresh growth is now taking place, showing that slight oscillations in the level of the shore of a foot at least have taken place recently. The tops of the corals have been certainly killed by being left exposed above water.

Such slight oscillations are to be expected at the base of an active volcano. The present re-growth is due to the corals being now again submerged. The fact that these corals are to be seen growing on the bare rock itself, and not on *débris* of older corals, shows that the coral growth is very recent.

The Brain Coral grows in convex, mostly hemispherical, masses ; the *Astræa* more in the form of vertically standing cylindrical masses, or masses which may be described as made up of a number of cylinders fused together. The masses of the *Astræa* are usually higher than those of the *Mæandrina* by about a foot, because they are able to grow in shallower water, and they thus range also higher up on the beach.

Many of the masses of this *Astræa* in the shallower water are left dry at each low-tide, and appear to suffer no more in consequence than do the common Sea-anemones of our English coasts, which are so closely allied to them. I have not seen any other species of coral thus growing where it is exposed at low tide. The "Brain Coral" apparently cannot survive ex-posure, and hence the tops of its masses have been killed during the change of depth of the water at about a foot below the height at which those of the *Astræa* have perished.

The common Mushroom Coral, so often to be seen as a chim-ney ornament in England (*Fungia sp.*), is most extraordinarily abundant on the shore, at a depth of one or two feet at low water, and with it an allied larger, similarly free-growing coral

(*Herpetolitha limax*). The Mushroom corals cover the bottom in places in such large quantities, that a cart-load of them might be picked up in a very short time; I have nowhere seen them so common.

I visited one of the Nutmeg Plantations in Great Banda. The nutmeg is the kernel of a fruit very like a peach in appearance and which makes an excellent sweetmeat when preserved in sugar. The owner of the plantation, a very wealthy Malay native of Banda, told me that about one male tree to every fifty females was planted on the estate; he had a superstition that if a nutmeg seed was planted with its flatter side uppermost, it would be more likely to produce a male seedling.

Formerly, before the Dutch Government renounced its monopoly of the growth of nutmegs in the Moluccas, the trees were strictly and most jealously confined to the Island of Great Banda. The utmost care was taken that no seeds fit for germination should be carried away from the island, for fear of rival plantations being formed elsewhere; seeds were, however, often smuggled out.

The Government destroyed the Nutmeg trees on all the other islands of the group. It was, however, found necessary to send a Commission every year to uproot the young nutmeg trees sown on these islands by the Fruit-Pigeons, called Nutcrackers by the Dutch residents (*Carpophaga concinna*).

The various Fruit-Pigeons must have played a most important part in the dissemination of plants, and especially trees, over the wide region inhabited by them. Sir Charles Lyell,[*] referring to the transportation of seeds by the agency of birds, noted especially this transportation effected by pigeons, and quotes Captain Cook's Voyages to the effect that at Tanna " Mr. Foster shot a pigeon," (obviously a *Carpophaga*), in whose craw was a wild nutmeg.[†]

At the Admiralty Islands very large numbers of a Fruit-Pigeon (*Carpophaga rhodinolœma*), were shot by the officers of the " Challenger." Their crops were full of fruits of various kinds, all of which I had failed to find, or reach in the growing

---

[*] " Principles of Geology," 10th Edition, Vol. II, p. 69.
[†] " Cook's Second Voyage," Vol. II, p. 69. London, Strachan, 1777.

condition in my botanical expeditions. Amongst these fruits were abundance of wild nutmegs, and wild coffee-berries ; many of the fruits were entirely uninjured, and the seeds quite fit for germination.

No doubt, when frightened or wounded by accident, the pigeons eject the whole fruits, and they habitually eject the hard kernels, as I saw quantities of them lying about under the trees on a small island at the Admiralty Islands, on which the birds roost in vast numbers.

As soon as ever a few littoral trees, such as *Barringtonia* and *Calophyllum inophyllum*, have established themselves by means of their drifting seeds on a freshly dry coral islet, the Fruit-Pigeons alight in the branches in their flight from place to place, and drop the seeds of all kinds of other trees with succulent fruits. I have seen the pigeons thus resting on two or three small littoral trees, which as yet form almost the only vegetation of Observatory Island, a very small islet in Nares Bay, Admiralty Islands.

Hearing the sound of music in the native district of the town of Banda one evening, I made my way towards a house from which it came, in the hopes of seeing a Malay dance. Instead of this I found Malays indeed dancing, but to my disappointment, they were dancing the European waltz.

I saw a Mahommedan's dancing-party in one of the houses ; the performers were of course all men. The room in which they danced was widely open to the street, and lighted up. About twenty men dressed in their best sat on mats placed against the wall round the room, the host occupying a place at one end ; two members of the party rose at a time and danced. The movements were very slow, and frequently the two dancers led one another by the hand and presenting themselves to different sides of the assembly in turn, bowed with great ceremony ; the whole reminded me somewhat of a quadrille.

Amboina, October 5th to 10th, 1874.—On the ship anchoring at Amboina, it was found necessary that a salute should be fired. The " Challenger " being, as a surveying ship, provided with very few guns, was usually excused this ceremony, but it was thought by the Dutch authorities that the natives would not properly understand the arrival of a foreign man-of-war, without the usual honour being paid to the Dutch flag; so two small Armstrong breech-loaders were let off alternately through the bow ports.

The old Dutch saluting guns on the fort seemed to return the unpleasant noisy compliment with some difficulty, and one of them leapt off the parapet into the ditch, in the excitement of unwonted exercise. It is to be hoped, that before long the intolerable nuisance of saluting will be done away with; it is most astonishing that civilized persons can be so much the slaves of habit, as to make a painful noise of this kind when necessity does not require it; everyone concerned dislikes the noise, and there is a great waste of material.

The custom, however, shows signs of dying out, for it has reached already to some extent a rudimentary condition. In large war-vessels, the actual fighting guns are considered too big to be played with in this manner, and a special saluting battery of small old pattern guns, useless for any other purpose, is kept mounted on the forecastle for the sole sake of making this hideous noise.

I have read of a case in which in a small out-of-the-way European colony, the governor had to send on board a foreign man-of-war which had arrived in his port to beg for powder to return the customary salute. We may, however, congratulate ourselves that matters might be worse; there are some unfortunate races, the members of which have to spend their money in powder and let it off, on all occasions of petty private domestic rejoicing.

The coral banks, though abundant, were not so easily accessible at Amboina as at Banda, being in deeper water, and specimens of most of the species could only be procured by deep wading and diving. After diving for corals in a depth of about ten to twelve feet, I found my eyes very sore for some hours afterwards. I believe that this soreness was most probably produced by the stinging organs of the corals; all corals are provided with urticating organs. The stinging produced by the Hydroid corals of the genus Millepora was long ago noted by Darwin and others.* In the West Indies the coral is sometimes called sea-ginger.

* "Journal of Researches," p. 464.

In the case of most Anthozoan corals, the stinging organs are not powerful enough to make themselves felt through the skin of the hands, but I have often felt my hands tingle after having been employed in collecting corals, other than Millepora, on the reefs.

In diving, the face and open eyes are brought close to the corals at the moment that these are grasped and irritated, and it seems possible that the eyes might become seriously inflamed and injured by the action on them of the nettle-cells. I mention the circumstance as a warning to collectors; where Mille-porids are present, great care should certainly be exercised.

On the shore of the harbour of Amboina, coral reef rock occurs raised many hundred feet above sea level, forming a steep hill-slope. At the summit of the ridges so formed the rock stands out here and there, weathered into fantastic pinnacles, with sur-faces honeycombed by the action of rain, just as at Bermuda.*

Some of the smaller trees growing on these ridges are covered with the curious epiphytes, *Myrmecodia armata* and *Hydnophytum formicaum ;* these are plants belonging to the natural order *Cinchonaceæ*. Both plants are associated in their growth with certain species of ants; as soon as the young plants develop a stem, the ants gnaw at the base of this and the irritation produced causes the stem to swell; the ants con-tinuing to irritate and excavate the swelling, it assumes a globular form, and may become larger than a man's head.

The globular mass contains within a labyrinth of chambers and passages, which are occupied by the ants as their nest. The walls of these chambers and the whole mass of the inflated stem, retain their vitality and thrive, continuing to increase in size with growth. From the surface of the rounded mass are given off small twigs, bearing the leaves and flowers.

It appears that this curious gall-like tumour on the stem has become a normal condition of the plants, which cannot thrive without the ants. In *Myrmecodia armata* the globular mass is covered with spine-like excrescences. The trees I referred to at Amboina, had these curious spine-covered masses perched in every fork, and with them also the smooth surfaced masses of a species of *Hydnophytum*.

* See pages 21, 78, and 83.

Numerous dealers brought trays of the shells for which
Amboina is famous to the ship, but the prices asked are so high,
that it would probably pay to bring some of the shells back
again from Europe to Amboina for sale to passing visitors.
Cassowaries' eggs were also offered for sale, and large quantities
of Deers'-horns (*Rusa moluccensis*).

The Deer are very abundant in Amboina.   I accompanied a
party which went in pursuit of them.   We had a letter to a
native head-man in one of the villages on the shores of the inlet
in which the harbour lies.   The head-man treated us hospitably, and collected about a dozen beaters.   The Deer were
lying down concealed on a plain of some extent close to the shore,
covered with tall grass in some places up to our middles, and
skirted by bushes.

We saw a Stag and two Hinds make off out of range, as
we made our way along the edge of the tall grass.   The men
beat the bushes at the edges of the grass, and at last drove a
Hind out of one clump to the guns, and it was shot.   The
numerous tracks in the grass showed that plenty of deer must
come there to feed.

Ternate Island, October 14th to 17th, 1874.—The island of
Ternate is an active volcanic cone rising direct out of the sea to
a height, according to "Challenger" observations, of 5,600 feet.
My small aneroid indicated the height as somewhat less, but
was no doubt in error.   The island, which belongs to the Dutch,
lies almost exactly on the equator.   Separated from it by a narrow strait is the somewhat similar cone of Tidore.   The lower
slopes are planted with nutmegs, cloves, pepper, cocoa trees,
and a profusion of fruits.

The mountain is unquiet, and there were said to occur on an
average three or four earthquakes every week; I had great
hopes that I should have an opportunity of feeling one, but was
disappointed.   The Dutch keep up a Government staff at the
island, very much to the benefit and happiness of the people,
but I believe at a considerable financial loss.

The Governor or Resident of the island at the time of the
visit of the "Challenger," was an accomplished naturalist, S. C.
J. W. van Musschenbroek; he received the Expedition with the

greatest kindness and hospitality, and even got up a ball on the shortest notice. The musicians were Malays, who were indefatigable, but knew only one tune.

The Resident presented a fine collection of Snakes and Corals to the Expedition, and gave the greatest assistance and information on all natural-history matters. There are a large number of Chinese in the population of the island, and the Captain China, or head of the Chinese under the Dutch, according to their well-known method of Government in East Indian Colonies, was one of the notables present at the ball.

The Chinese have been for hundreds of years in the island, and I was astonished to learn that some of them have, in the course of generations, entirely lost the knowledge of their own language, and now speak only Malay. I was told that it was even possible that the Captain China himself might be in this condition. I had thought this quite impossible in so strongly conservative a people, and indeed had not realized the fact that numerous generations of Chinese are born, die, and are buried in these islands under Dutch rule.

At Amboina, the large and costly tombs of the Chinese form a feature in the landscape on the hill-sides,* and there is a large Chinese graveyard at Ternate, with many tombs of great age. I had fancied that all dead Chinese were carried to China to be buried, at all events if rich. The English seem to be the only civilized migratory people who never lose their language.

Instances of such loss by all other European races are to be found in the United States.

Malay collectors are sent every year to New Guinea from Ternate, to collect Birds of Paradise and other Birds, and a regular trade with New Guinea is carried on from this port. The Malay collectors are some of them extremely expert in preparing and preserving bird-skins. They mount them with a small stick stuck into the tow stuffing, and protruding at the tail. The skin is handled by the stick, and thus the bird's feathers are prevented from being injured.

---

* Similarly at Timor, the costly Chinese tombs at which island are figured in Peron and Leseur's " Voyage," published 1807.

There are several Mahommedan dealers in bird-skins in the town of Ternate. A Papuan Bird of Paradise (*Paradisea Papuana*), well skinned, cost about eight shillings, and I gave fourteen shillings for a well-skinned Red Bird of Paradise (*P. rubra*). Skins of various Paradise Birds, prepared flat, and dried in the old native style, were common and cheap enough. Amongst these skins were a large quantity of what I believe was the very rare Black and Scarlet-coloured Parrot (*D. pequetti*). These birds could hardly have been killed and thus prepared for sale, as ornaments, like the batch they were amongst; but they were unfortunately of no good as natural-history specimens in their mangled condition.

As I wished to ascend the Peak of Ternate in search of plants, the Resident provided four Malay guides for the purpose. I started with Lieutenant Balfour. We passed a night at the house of one of the Government officials, who kindly offered us hospitality, at an altitude of about 1,000 feet. Leaving the house at 4.30 A.M. on the following morning, we commenced the climb through a field of sugar-cane. The path led nearly straight up the cone all the way, and was excessively steep, and the ground was very slippery from a heavy fall of rain the night before.

It was pitch-dark for the first hour, and we slipped and fell constantly. At an altitude of about 2,000 feet above sea level, the last cleared and cultivated land, a rice-field, was passed. On the border of the field grew several of the Saguir palms (*Arenga saccharifera*), which are abundant in the gardens at sea level. An intoxicating drink is made from the juice of this palm, and like many other palms it yields sugar.

Above the rice-fields, woods were entered at about daylight, and these extend up to an altitude of about 4,150 feet. Jack-fruit and a Wild Plantain were observed to grow up to a height of about 2,600 feet. In the woods was a small hut, used by men who come up to hunt the deer, which are abundant on the mountains. On a tree close to the hut was cut the name of Miklucho Maclay, the well-known explorer of New Guinea.

From the verge of the woods, at 4,150 feet altitude, for about 750 feet further ascent, a dense growth of tall reeds was traversed. At this height (4,800 feet above sea level), a ridge was reached from which a descent of about 100 feet was made into an outer ancient crater, corresponding to the Canadas of the Peak of Teneriffe.

There are two such outer ancient craters at the summit of the Peak of Ternate, and the ridges forming the old borders of these craters and the outer portions of the bottoms of the craters themselves are traversed in succession on the way to the terminal modern cone of eruption which stands in the inner of the two.

The outer and oldest of the craters is a wild-looking place, inhabited by numerous wild pigs and deer. It is covered with a growth of bushes and a small tree fern, and four other species of ferns,* and with these grows a Club-moss (*Lycopodium*), and a Whortleberry (*Vaccinium*). The shrubs were apparently of only two species, and the flora seemed a very poor one in number of species.

The second ridge, marking the summit of the inner extinct crater, is about 50 feet higher than the outer one. Within this inner crater there is scarcely any vegetation, a few scattered blades of grass only. Here was met with a large mass of lava, evidently recently ejected from the active crater, and hurled to this distance. The mass had a smooth reddened surface, and was deeply split all over by cracks formed evidently by contraction on cooling.

The terminal cone itself is entirely devoid of vegetation. The cavity of the inner extinct crater from which it rises is filled up, except at its margin, by the results of later eruptions. Hence the base of the terminal cone lies about 60 feet above the level of the margin of this crater, and is approached by a gentle ascent.

The cone itself rises steeply and suddenly, with a slope of 30°, and is about 350 feet in height. The guides had hesitated somewhat when we ascended the slope leading out of the first extinct crater, and had done their best to persuade us not to go any farther, telling us it that was dangerous to proceed. They lagged

* Gleichenia dichotoma, Pteris incisa, Polypodium phlebiscopum. J. G. Baker, F.R.S., "On the Polynesian Ferns of the 'Challenger' Expedition." Journ. of Linn. Soc., Bot., Vol. XII. p. 104.

behind as we approached the terminal cone, and as soon as we began to climb it, turned round and ran back as fast as they could go.

We were told afterwards that they have strong superstitious fears concerning the volcano, and believe that if anyone climbs the terminal cone, a terrible eruption and earthquake are certain to ensue. It appeared as if there might be some real risk in the ascent. The cone is not composed of ashes, but of masses of basaltic lava of various sizes; all of these on the surface appeared freshly fractured and split, as if quite recently thrown out of the crater, and broken up on cooling.

At the summit, a slope of 30°, exactly the same as that of the outside of the cone, the natural slope no doubt of the lava fragments, leads down into the crater, from a sharp ridge, along which we walked. A dense smoke rose from the interior of the crater, and hid its form and extent entirely from view.

The wind was easterly (E. by N.), and drove the smoke away from the side of the crater on which we were. The smoke is excessively suffocating, and a sudden shift in the wind might be fatal to anyone who was a short way down within the crater, or even at some places on its margin. It would not be easy to get down it in some places, at all events in a hurry. It was only possible to descend about 20 yards into the crater, and even then the vapours inhaled were very trying. Steam and acid vapours issued from cracks everywhere, decomposing the lava amongst which they passed. In most of the cracks were small quantities of sulphur.

From the margin of the crater overlooking the town of Ternate there was a magnificent view, embracing the island of Halmahera (Gilolo), which lay spread as a map beneath us, and the peak of Tidore, and many far-distant islands. Our guides rejoined us when we came down to the outer crater.

For the benefit of any future explorers of the Peak, which is very seldom ascended, I give the time required for the ascent. We left the house at 1,000 feet altitude at 4.30 A.M., reached the margin of the outer crater at 8.30 A.M., and the summit at 9.30 A.M. The temperature of the air at an altitude of 4,800 feet was 71° F. at 8.30 A.M. At the summit of the mountain it was 68°·5 F. at 9.30 A.M.

CHAPTER XVI.

THE PHILIPPINE ISLANDS.

Zamboanga, Mindonao Island. Paddy Fields and Buffaloes. The Lutaos and their Pile-Dwellings. Pile-Dwellings on Dry Land. The Ground Floor, a Late Addition to the First Story. Wide Distribution of Pile-Dwellings. Their Possible Origin. Dances Performed by the Lutaos. Bamboo Jew's Harp. Lutao Canoe and Weapons. Search for Birgus Latro. Birds' Eggs hatched in the Sea Sand. Alcyonarian Corals. Basilan Island. Cart-wheels cut from Living Planks. Galeopithecus and Flying Lizard. Cebu Island. Mode of Dredging up Euplectella. Mactan Island, Raised Reef. Large Cerianthus. Trachytic Volcano at Camiguin Island. Temperature at which Plants can Grow in Hot Mineral Water. Manila-Hemp Plantations. Manila. Shirt Worn over Trousers. Clothes Originally Ornamental only. Half-hatched Ducks' Eggs Eaten. Cock Fighting. Sale of Indulgences.

Philippine Islands, October 24th to November 12th, 1874, January 11th to February 5th, 1875.—The ship arrived on October 24th, 1874, at the town of Zamboanga, which lies at the extremity of a long promontory projecting from the west side of the large island of Mindanao, the southernmost of the Philippine group. A small area at the tip of this promontory belongs to Spain; a wide tract behind it belonging to Portugal; whilst the entire island of Mindonao is about half of it Portuguese, and half Spanish. The ship paid a second visit to Zamboanga on the return journey southwards, from January 29th to February 5th, 1875.

On landing at Zamboanga I was immediately reminded that we were nearing India, and scenes in Ceylon were recalled at once to my memory. Swampy paddy fields stretched everywhere round the town with plenty of snipe in them, and the domestic buffaloes lay about wallowing in mud pools and throwing water over their backs with their scoop-like ears. In one pool, several native women were bathing in company with the buffaloes.

Especially interesting in the Philippines are the various stages in development and modification of pile-dwellings. All the native buildings are pile-dwellings or modifications of them, and some of the better houses, built under European influence, are evidently copied directly from the same models.

Pile-dwellings are first invented as an expedient for raising houses in the water for protection; but when the race which for generations has thus dwelt surrounded by water takes to living on dry land, actuated somewhat no doubt by sanitary considerations, it follows the ancient pattern of architecture with slavish exactness, and only by gradually introduced modifications of that plan, arrives at last at a house supported directly on the ground.

At Zamboanga and at the neighbouring island of Basilan, which we also visited, are settlements of a considerable number of a race called by the Spanish "Moros" (i.e., Mahommedans), who keep themselves strictly apart from the Bisayan and other Malay races, amongst which they here dwell. The Moros at Basilan still build their pile-dwellings out in the sea, so that they can only be approached by boats. At Zamboanga, however, where the Moros seem somewhat more tamed by Spanish influence, they have so far come on shore with their houses, that these are built in a row along the beach, and at low tide are not entirely surrounded with water, whilst the shore can always be reached from them by means of a plank. The main inhabitants of the Philippines, in the course of successive generations, have taken their houses altogether on shore, except where here and there there are houses in swampy ground, which form a sort of gradation between the two conditions.

The Moros or "Lutaos" are said to have settled in Mindonao in the seventeenth century, and to have considered themselves until quite recently, as subjects of the Sultan of Ternate.* They are a fierce and warlike race, pirates by profession at all events not long ago at Basilan and Mindonao, and still so at the Sulu Islands. They seem but half subjected to the Spanish rule.† The

* Dr. Th. Waitz, "Anthropologie der Naturvölker," 5te Th. 1tes Hft. Die Malaien, Leipzig, 1865, s. 56.

† Since the above was written, the Sulu Islanders have during this year, 1878, submitted to Spanish rule on receipt of a sum of money. An

men are short and broad-shouldered, with powerful chests and thick-set bodies, and extremely active. Their features are of the Malay type, but peculiar. Their eyes are remarkably bright. Their colour is light yellowish brown. They have often a slight beard and moustache. They wear bright-coloured shirts and rather tight-fitting trousers, buttoned close round the leg at the ankle. The Moro women are short and small, and delicate-limbed, most of them very handsome when young; many of them are very light-coloured in complexion; their eyes, like the men's, being extremely bright. They are fond of bright yellows and reds in their dress, and are very fully clad. The men are armed with circular shields and spears, and also used formerly at least suits of armour made of plates of buffalo horn, linked together with wire, which are very rare objects in Ethnological Museums.

At Basilan Island, at Port Isabella, the Moros houses are

PILE-DWELLINGS OF LUTAOS AT ZAMBOANGA.

constructed on piles in a small lagoon-like offset of the channel between this island and the small outlying island of Malamaui. The houses are entirely isolated by the water. They stand together, and a wide rickety platform connects many of them with one another.*   At Zamboanga, the Moros houses are also

agreement has been signed at Manila, between the Sultan of Sulu and the Spanish Government.

* For an Account of the inhabitants of the Sulu Islands, the same race

built in a group. The main house in each case is usually supported on three rows of piles; but various additions and outbuildings are supported on irregularly added piles. There is always a platform before the entrance, and sometimes one for canoes behind. It was odd to see a horse left tied by his Moro owner to the door-post, standing up to his belly in the water, through the rising of the tide.

The houses of the other native inhabitants throughout the towns of Zamboanga and Ilo Ilo are mostly of closely similar pattern. They stand in like manner on piles, though on dry ground, and have a platform usually at one end. This is reached by a short steep ladder, with widely separated and irregular rounds, up which the house-dogs, from practice, run as nimbly and easily as the children and their mothers. The platforms are now used for drying clothes upon, and such purposes.

The first process of modification of the pile-dwelling gone on shore, is the putting up of a fence of palm leaves in the lower part of the spaces between the piles supporting the house. A pen is thus formed in which pigs or other animals are kept. Then well-made mats or reed walls are put up, entirely enclosing the space between the piles, with a regular door for entrance, and the place becomes a convenient store-house. As a further stage, boards are nailed between the piles, and a secure chamber is obtained.

A further step again, is the adoption of stone pillars for the wooden piles. Wooden houses thus supported on stone representatives of piles, may often be seen with an iron railing, passing from pillar to pillar beneath, and in this way forming an enclosure. From stone pillars the step is easy to arches, supported on pillars of masonry as a substructure, and some houses of business, although their upper structures have ceased to be wooden, and are built of more solid materials, are still to be seen amongst the rest, supported thus on the descendants of piles.

In the last stage the arches are discarded, and continuous walls of masonry substituted as a support to the wooden super-

as the Moros, with descriptions and figures of their houses, see Wilkes' "Narrative of the U.S. Exploring Expedition," Vol. V, Ch. IX. New York, 1856.

structure. Even then the ground-floor is often still used only as a store-house or piggery, but in many cases is regularly occupied.

Thus in these houses, what would seem almost an impossibility is nevertheless the fact. The ground-floor is an addition to the first story, which latter is older than it, and preceded it. The verandah is the representative of the platform originally intended for the inhabitants to land on from canoes.

I watched the building of one house, which when finished looked perfectly two-storied, the lower part being neatly boarded in, and provided with a door and windows. Nevertheless, in the construction of the house, the history of its development was exactly recapitulated, just as is the case familiarly in natural history. The roof and first story were built first complete upon the piles, and the lower structure added in afterwards.

I could not help being struck by the remarkable resemblances of many of these Malay houses to Swiss châlets. In the châlet the basement enclosed with stone walls is usually only a cattle-stall, the first story is the dwelling-house, and as in the Malay building, is constructed of wood. It seems possible that the châlet is the ancient lake-dwelling gone on shore, like the Malay pile-dwelling, and that the substructure of masonry represents the piles which formerly supported the inhabited portion of the house. There are similar balconies in the châlets representing possibly the platforms. A good deal of the carving of balconies, and some of the staircases, in the better constructed wooden houses in Ilo Ilo, reminded me very much of that of the same structures in châlets, though the resemblance in this case is accidental.

The most interesting feature about pile-dwellings seems to be their very wide geographical extension. Representatives of almost all races of man seem to have arrived at the same expedient, apparently not by any means a simple one, independently of one another. There are the well-known Pfhalbauten of Switzerland, in South America the similar houses of the Cuajiro Indians, on the Gulf of Maracaibo. In North America the Haidahs on the north-west coast construct similar habitations. Commander Cameron lately observed similar dwellings in Lake Mohrya, in Central Africa.* In New Zealand, the Lake Pas, which were mostly used as store-houses, are known from the Rev. Richard Taylor's description.† In this case, piles were driven into the bottom of the lake, and the interstices filled in with stones and mud, so as to form a platform.

There are the well-known New Guinea pile-dwellings, such as seen by us at Humboldt Bay, and there are also the pile-dwellings of all the Malay races. The Gilbert Islanders construct also houses raised on piles, and a number of these natives from the island of Arorai, who were taken to Tahiti, to serve as labourers on cotton estates, have put up houses of this kind for themselves in the latter islands, amongst the very different dwellings of the Tahitians themselves.

It seems probable that the idea of a pile dwelling has in many cases arisen from the escape of natives from enemies by getting into a canoe or raft, and putting off from shore into a lake or the sea, out of harm's way. If the attacked had to stay on such a raft or canoe for some time, they would anchor it in shallow water with one or more poles, as the Fijians do with their canoes on rivers, and hence might easily be derived the idea of a platform supported on piles.

The officers of a Spanish man-of-war in the port of Zamboanga at the time of our visit, hospitably gave us an entertainment on shore, and got the Moros to dance for our amusement. Two men danced with spears and shields, in imitation of a combat, in which the utmost rage was simulated on both sides; the teeth were clenched and exposed, the head jerked forward, and the eyes starting as they advanced to the attack. The dance of the women was like that described as performed by the Ke Islanders. The body was kept nearly rigid, and turned round slowly or moved a short distance from side to side by motion of the feet alone. The feet were kept close together, and side by side, and moved parallel to one another with a shuffling motion.

* S. L. Cameron, Comm. R.N.," Across Africa," Vol. II., p. 65. London, 1872.
† Rev. Richard Taylor, F.L.S., "On the New Zealand Lake Pas." Trans. N. Zealand Inst., Vol. V, 1872, p. 101.

The principal display in the dancing consisted in the very slow and gradual movement of the arms, wrists, and hands. One arm was maintained directed forwards and somewhat upwards, the other at about the same angle downwards, and the position of the two was at intervals gradually reversed; the hands were turned slowly round upon the wrists, and often the dancing consisted for some interval merely in the graceful pose of the body, and this movement of the hands.

The main point in the dancing seemed to be that all the motions should follow and pass one into the other with perfect gradation in time, and without any jerk or quickening. The thumbs were always maintained extended at right angles to the palms of the hands, as at the Ke Islands.

A young boy danced a somewhat similar dance to that of the girls. During his performance, he at one time put forward one leg and curved the sole of his foot so that only the toe and heel touched the floor, and turned round with the foot in that position. At another time he shuffled along slowly with the heel of one foot in the hollow of the other.

I obtained from a Moro boy a Jew's-harp made of bamboo, on which he was playing. The instrument is most ingeniously cut out of a single splinter of bamboo, the vibrating tongue being extremely delicately shaped; the tongue is cleverly weighted by means of a knob of the wood left projecting on its back. The instrument produces a tone indistinguishable from that of a metal Jew's-harp; it is quite unlike Melanesian bamboo Jew's-harps in its form.

A sharp tide runs in the channel between Zamboanga and the Island of Santa Cruz Major, which lies just opposite the town. In the tide-way, whilst

MORO JEW'S-HARP, CUT OUT OF A SINGLE PIECE OF BAMBOO.

the water was running in either direction, a most unusual abundance and variety of surface-living oceanic animals and larvæ of shore forms, was obtained with the towing net; amongst these were Tornaria, and larvæ of *Sipunculids* and *Chirodota*. The place would be a most convenient and productive one to a working zoologist.

The Brachiopod *Lingula* is so abundant in shallow water close to the town, that two boys gathered more than a hundred specimens at a single low tide at the request of Von Willemoes Suhm. Unfortunately the much prized " mariske " did not reach the " Challenger." The boy with his bottle full was met by a rival collector, who completed a bargain forthwith. There are rival collectors even at Zamboanga, and we suspected, I do not know whether rightly or not, that it was a natural-history collector from the United States who was in the neighbourhood at the time, who had thus been lucky enough to become possessed of our expected treasure.

A King Crab (*Limulus rotundicaudatus*), is not uncommon near Zamboanga, it is called "cancreio." Von Suhm thought that he had obtained a series of young larvæ of *Limulus* amongst the surface animals collected by the net, but he subsequently came to the conclusion that he had been mistaken. At low tide, by wading and turning over stones, enormous Planarians of the genus *Thysanozoon*, are to be found in plenty; they are of a dark purple colour, and measure, some of them, as much as five inches in length, and two inches in breadth.

I accompanied Von Willemoes Suhm on a visit to the Island of Santa Cruz Major. We sailed over in a Moro canoe managed by two of these natives; the boat was armed with a large number of bamboo spears, simple light bamboos cut off slanting at one end so as to form a sharp cutting point like that of a quill tooth-pick in shape. A bamboo so cut is extremely sharp, and the spears must be formidable weapons, especially against a thinly clad adversary. Two or three dozens of these spears were placed on rests on either gunwale of the boat, and there were besides two round shields of a kind of basket-work in the boat.

Our object in visiting Santa Cruz Major Island was to search

for the great Cocoanut-eating Crab (*Birgus latro*) ; it is called
" Tatos " at Zamboanga, and survives in Santa Cruz Major
because there are no Pigs in the island.   Wild Pigs destroy not
only these Crabs, but dig up Shore-crabs (*Ocypoda*), and Land-
crabs from their holes.   In Ceylon, near Trincomali, the wild
swine come down every night to the beach to dig up Crabs, and
I have seen a large tract of sandy beach which has been
ploughed up by them in the search.   The " tatos " is searched
for and eaten as a delicacy in Zamboanga.

We landed close to a Moro house built out into the sea, so
as to be surrounded at high water.   The inhabitants were lolling
about in the shade, and though we offered them good pay they
would not go a quarter of a mile to look for " tatos " for us.   At
last a boy consented to go as guide ; instead of searching for
the Crabs under the Cocoanut trees, as I had expected, we were
shown as the haunts of the animals hollows at the roots of
mangrove and other trees in swampy ground, amongst the holes
of ordinary Land Crabs, but we could not find the tatos.

Von Suhm was anxious to investigate the development of
the Birgus from the egg.   An intelligent native at Zamboanga,
who collected for us, said that the female Crab carries about large
masses of eggs with it in the month of May, and retains them
so attached until the young are developed, just like the parent ;
he said the Crabs went down to the sea occasionally to drink.

A Mound Bird (*Megapodius*), is common in the island.   The
calcareous sand amongst the bushes close to the seashore, was
scratched and turned over in many places by these birds in
burying their eggs.   Our guide dug out half-a-dozen eggs,
closely like hens' eggs in appearance, from one of these places.
The eggs were buried in the clean sand, at a depth of $3\frac{1}{2}$
or 4 feet, and with no mound over them, or vegetable rub-
bish of any kind.   The eggs are thus hatched by the simple
warmth of the sand received from the sun and retained during
the night, just in the same manner as turtles' eggs are hatched,
indeed, turtles' eggs might have been found in the same hole.
It was mid-day, and the surface sand was hot, far hotter than
the sand below, where the eggs lay, which felt as well as the
eggs distinctly cool to the touch.   I had always supposed that

these birds and their allies hatched their eggs by means of the heat derived from decayed vegetable matter.

We shot a small Cuckoo, with a beautiful greenish golden metallic lustre on its feathers (*Centrococcyx viridis*), in the bushes. On the shore were inclosures built by the Moros as fish traps, to retain fish as the tide receded. In the shallow water contained in these traps were a large number of *Medusæ* all lying on the tops of their umbrellas, with their tentacles directed upwards in full glare of the sun. They looked thus posed like a lot of Sea-Anemones, and I took them for such at first. They appeared perfectly lively, and from time to time contracted their umbrellas; It appeared almost as if they had assumed their position voluntarily, and were waiting for food in the same manner as *Actinias*.

Alcyonarians (social Polyps, distinguished by having eight tentacles), are extraordinarily abundant about the beach of Santa Cruz Major. The reef rocks are covered with the soft spongy forms of *Alcyonarians;* they form extensive beds, which are soft and boggy to tread on in wading. Amongst these grows a stony coral, which is likewise *Alcyonarian*, as I found to my astonishment on examining its minute structure. It forms thick erect plate-like masses which are of a chocolate colour when living. The coral is remarkable because its hard calcareous skeleton is of a bright blue colour instead of white, as usually the case. The coral is hence named *Heliopora cœrulea*. It is, as far as is known, the only surviving representative of a large number of extinct forms of Palæozoic age, which are familiar in the fossil condition. It is nearly allied to the well-known Red Coral of commerce.*

Again, another interesting Alcyonarian is abundant, together with those just described, namely, the red Organ-Coral (*Tubipora musica*). There were cartloads of this coral, dead and dried, lying on the beach, which was entirely composed of various coral *débris*. The "Organ-Coral" was not to be found living in shallow water on the reefs, but living specimens were dredged from a depth of ten fathoms.

* H. N. Moseley, "On the Structure and Relations of the Alcyonarian Heliopora Cœrulea, &c." Phil. Trans. R. Soc., Vol. 166, Pt. 1.

Basilan Island, Feb. 4th and 5th, 1875.—The ship went for a night to Port Isabella in Basilan Island, lying west of Zamboanga, to coal at the Spanish Government stores there. The houses of the Moros at this place have already been referred to ; the town was mostly in process of construction by families of Bisayans moved from Zamboanga, and much of it was being built on causeways and made ground constructed with coral rock on tidal mud flats ; some families newly arrived were camped on the sites of the houses they were building.

Separated from Basilan Island by a narrow strait is the very small island of Malamaui. This island is mostly covered by a dense forest of lofty trees, many of which have the curious vertically projecting plank-like roots which are so fully described by Mr. Wallace in " Tropical Nature."* The natives cut solid wheels for their Buffalo carts directly out of these natural living planks ; and the large circular window-like holes left in the roots at the bases of the trees are curious features in the forest.

I was constantly put on the alert by the rustling of what sounded like some large animal amongst the dead leaves, and expected every minute to get a shot at a deer, but at last found that the animal disturbing the silence of the forest was a huge Lizard (I believe *Hydrosaurus marmoratus*), which bolted up the trees when approached and sat in a fork. The forest was full of these reptiles.

I wished much to see the well-known aberrant flying Insectivorous mammal, *Galeopithecus Philippensis*, which, like a Flying Squirrel, has membranes of skin stretched between its legs and out on to its tail ; so that, supported on this as by a parachute, it skims through the air in its leaps from tree to tree with a partial flight. I had no interpreter, but found a Bisayan native who knew Spanish. I knew what " to-morrow morning early " was in Spanish, and also what " I want to go and shoot *Galeopithecus* " was in Malay. And to my great amusement I combined these two so widely different languages in a sentence with perfect success, " Mañana por la mañana saia mau purgi

* A. R. Wallace, " Tropical Nature and other Essays," p. 31. London, Macmillan, 1878.

passam kaguan." The man appeared accordingly next morning at daybreak and I went with him and shot the animal.

The guide led me through the forest to some clearings belonging to Moros here living inland. Their houses were raised on poles at least twelve feet above the ground. We went to one where the wife of the owner, a very handsome young woman, was sitting on the ladder with her child in her arms. Some few trees were standing isolated, not having been as yet felled in the clearing. On one of these, after much search, a Kaguan (*Galeopithecus*) was seen hanging to the shady side of the tall trunk. It was an object very easily seen, much more so than I had expected. It moved up the tree with a shambling jerky gait, hitching itself up apparently by a series of short springs. It did not seem disposed to take a flying leap, so I shot it. It was a female with a young one clinging to the breast. It was in a tree at least 40 yards distant from any other, and must have flown that length to reach it. I understood from my guide that numbers of the animals were caught when trees were cut down in clearing. They are especially abundant at the island of Bojol, north of Mindonao, and their skins were sold at Cebu, which lies near, at four dollars a dozen.

Close by on some lower trees were several Flying Lizards (*Draco volans*), which similarly have a flying membrane, but in their case supported on extensions of the ribs. I saw the little lizards spring several times from tree to tree and branch to branch; but they pass through the air so quickly that the extension of their parachute is hardly noticed during the flight. We had several of them alive on board the ship for a day or two, where they flew from one leg of the table to another. It was curious to see two animals so widely different in structure, yet provided with so similar means of flight, thus occurring together in the same grove and even on the same tree.

At Malanipa Island, a very small island, not far from Zamboanga, natives had felled a good many large trees to make canoes. The suitable trees are usually at some distance from the water. A straight broad road is cut through the smaller wood direct from the large tree to the sea-shore; and the smaller trees are felled so as to fall across the road. On their prostrate trunks the canoe is hauled to the shore. The open avenues were extremely useful in affording an easy road into the forest for collecting purposes.

Cebu Island, January 18th to 24th, 1875.—The ship was anchored for some days in the harbour of the town of Cebu, in the island of the same name. The special interest of this place lay in its being the locality from which the well-known delicately beautiful silicious sponge, called Venus's Flower Basket (*Euplectella aspergillum*), was first obtained. The sponge is dredged up from a depth of about 100 fathoms in the channel between Cebu and the small island of Mactan.

The fishermen use, to procure the sponge, a light framework, made of split bamboo, with two long straight strips, about eight feet in length, forming its front, and meeting at a wide angle to

MACHINE USED AT CEBU TO DREDGE UP EUPLECTELLA ASPERGILLUM.

form a point which is dragged first in using the machine. The long straight strips have fish-hooks bound to them at intervals all along their length, the points of the hooks being directed towards the angle of the machine.

The whole is very ingeniously strengthened by well-planned cross pieces, and is weighted with stones. It is dragged on the bottom by means of a light Manila hemp cord, not more than ⅛th of an inch in diameter of section, which is attached to the angle. A stone attached to a stick is fastened just in front of the angle to keep the point down on the bottom. The hooks creeping over the bottom and sweeping an area nearly 14 feet wide, catch in the upright sponges and drag their bases out from the mud. These sponges, once so rare and expensive, were a drug in the market at the time of our visit to Cebu.   They were brought off to the ship in washing-baskets full, and sold at two shillings a dozen.

Mactan Island consists of an old coral reef raised a few feet (eight or ten at most) above the present sea level.   At one part of the island, where a convent stands, a low cliff fringes the shore, being the edge of an upper stratum of the upheaved reef, of which the island is composed.   This raised reef is here pre-served, but has over the portion of the island, immediately fronting Cebu, been removed by denudation, with the exception of a few isolated pillar-like blocks, which remain, and which are conspicuous from the anchorage.   These show that the whole island was once of the same height as the distant cliff.

Opposite the town of Cebu, the island of Mactan is bordered by a wide belt of denuded coral flat, partly covered at high tide. The surface is scooped out into irregular basins and sharp projecting pinnacles, and covered in all directions with mud, resulting from the denudation.   Very few living corals are to be found on these flats, but the flats are fringed at their seaward margin by small beds of living corals.

These muddy expanses are the haunt of numerous shore birds.   In the pools a large Sea-Anemone, of the genus *Cerianthus*, expands its tentacles in the full blaze of the sun.   *Cerianthus* is a form which uses its "thread cells," which in all its widely varying allies are apparently only employed as offensive stinging organs, to construct a dwelling.   The cells are shed out in enormous abundance, and with their protruded filaments matted together, form a tough leathery tube with a smooth and glistening inner surface, which is buried upright in the mud.

Within this tube the Anemone lives, expanding its tentacles at the mouth of the tube, on a level with the surface of the mud. It has the power of moving itself with extreme rapidity down its tube, and disappears like a flash when alarmed.   The species at Mactan Island is very large.   The tube measures one foot four inches in length, and is very thick and heavy, though made up almost entirely of thread cells.   The animal itself is six inches in length.

This species of *Cerianthus* lives in shallow water in the full heat and glare of the sun, yet another species, *Cerianthus bathy-metricus,* differing from it in hardly any particular, except that it is of much smaller size, inhabits the deep sea at a depth of three miles, in almost absolute or entire darkness, at a temperature near freezing point, and where the water is at a pressure of roughly, three tons to the square inch.

Camiguin Island, January 26th, 1875.—Camiguin Island lies about 80 miles to the eastward of Cebu Island. "In July 1871 a volcanic eruption of two months' duration took place in the island, and threw up a hill two-thirds of a mile long, and 450 feet in height, destroying the surrounding vegetation and village of Catarman."† A visit was paid to the island in order to see this volcano.

The volcano, a dome-shaped mass standing on the sea-shore, was still red and glowing in cracks at the summit, and smoke was ascending from it. There appeared to be no crater, and Mr. Buchanan, with whom I landed, drew my attention to the fact that the lava of which it was composed was entirely tra-chytic. It recalled in form at once, some of the smaller trachytic domes of the Puy de Dome district, in the Auvergne, concerning the mode of formation of which there has been much doubt.

The mass in this case appeared never to have had any crater. It rose with steep walls directly from the soil formerly covered with vegetation, which it had destroyed. It appeared as if the trachytic lava had issued from a central cavity, and boiled over as it were, till it set into the form of the dome.

The ground around the crater was still almost bare of vegetation, but some plants were beginning to colonize the denuded soil, strongly impregnated as it was with various volcanic chemical products. Three species of ferns, as first colonists, grew as isolated plants here and there: and along the courses of two small streams fed by hot springs, issuing from the base of the volcano, where the poisoned ground was constantly washed,

---

* H. N. Moseley, "On New Forms of Actiniaria dredged in the Deep Sea." Trans. Linn. Soc., 2nd Ser., Vol. I., p. 302.

† "Information received from Francis G. Gray of H.M.S. 'Nassau, Navigating Lieut." Hydrographic Notice, No. 8, 1872, Eyre and Spottis-woode.

a good deal of vegetation was to be found, amongst which were several sedges and grasses, and a rush.

About the mouths of cavities from which hot gases were slowly being exhaled, a moss was found growing in great abundance, with several lowly organised Cryptogams; the whole being confined to the spot occupied by these fumeroles and forming green patches in the midst of the surrounding entirely bare rock.

The hot streams were full of green algæ, and as these streams, being very small, became cooler and cooler from their source downwards, I was able to determine the temperature at which the algæ commenced to flourish.

At the source of one of these streams, as it issued from beneath the volcano, the water had a temperature of 145°·2 F., and was thus too hot to be borne by the hand. Here there were no algæ at all growing in the water. There were, however, small green patches on stones projecting out of the bed of the stream into the air, and also along the margins of the stream where they were not bathed by the hot water itself, but only soaked up the moisture and received the spray occasionally.

At a distance of a few yards lower down, in a little side-pool fed by the stream, abundance of algæ were growing, but the pool had a temperature of only 101°·5 F., though the stream which fed it constantly was at 122° F.

Lower down again, algæ were growing in the middle of the stream, in water at 113°·5 F., and this seems thus to be the limit of temperature at which the particular algæ gathered, will flourish in water impregnated with a certain amount of salts in solution. No doubt the amount of salts present has a limiting effect as well as the temperature.

Oscillatoriæ have been observed growing in water, at a much higher temperature, even 178° to 185° F.* The fact is interesting, as showing that green algæ of some considerable complexity may have commenced life on the earth in its early history, before the water on its surface had anywhere cooled down to a temperature sufficient to be borne by the human hand, and which

* See W. T. Thiselton Dyer, F.L.S., &c., "Proc. Linn. Soc., Bot." Vol. XIV. p. 327. Also pp. 36 and 383 of present work.

may have been strongly impregnated with various volcanic gases and salts.

The upper slopes of the mountains of Camiguin Island were thickly wooded. The lower slopes were cleared and planted with Manila hemp. A Manila hemp plantation is not at all pleasant or easy to traverse. The large trees, a species of Banana (*Musa textilis*) from the stems of which the fibres known as Manila hemp are obtained by maceration, are planted closely together. The plantations are full of fallen stems, which block the way, and are in a half decayed condition, nasty pasty masses which it is very unpleasant to handle and climb over, or crawl beneath.

The ship stopped three days at the town of Ilo Ilo, the head-quarters of the manufacture of a sort of fine muslin, made out of the fibre of pine-apples, and which is known as " piña." This fabric is highly prized by the native Malay and miscellaneous half-caste beauties, but apparently does not find much favour in Europe, because of its always having a dusky tint. A similar fabric is woven in some parts of India.

Manila, November 5th to 12th, 1874, January 11th to 14th, 1875. —As we entered the Bay of Manila, there greeted us the cow-like moan of an American-built steamer, so different from the English whistle, and I felt at once that we had, as it were, turned the corner of the world in our long voyage.

The dress of the Bisayan and Tagalese and half-caste men is very ludicrous. They wear an ordinary shirt without tucking the flaps in. The flaps hang over their trousers, reminding one of the Australian Black's description of a clergyman, as " white fellow belong Sunday, wear shirt over trousers." Men who are well to do wear elaborately embroidered and very transparent shirts of piña.* The shirt is the article of dress on which the wearer prides himself most, and especially is he gratified by the beauty of its front.

The dress of the children at Ilo Ilo and Zamboanga was interesting. It was evidently put on them in many cases by the

---

* The men similarly in Nicaragua wear their shirts over their trousers. See Thos. Belt, F.L.S., " The Naturalist in Nicaragua," p. 63. London, John Murray, 1874.

parents as an ornament or exhibition of wealth, not in the least from any sense of decency. All dress has no doubt been primitively ornamental in origin, and has subsequently come to subserve the functions of increase of warmth or gratification of sense of decency.

A savage begins by painting or tatooing himself for ornament. Then he adopts a moveable appendage, which he hangs on his body, and on which he puts the ornamentation which he formerly marked more or less indelibly on his skin. In this way he is able to gratify his taste for change. No doubt the stripes and patterns on savage dress represent often what were once patterns tatooed on the body.

It is a curious fact that the transverse breast stripes and lateral longitudinal leg stripes worn in some European dresses of ceremony, though quite different in the history of their origin, being, I believe, hypertrophied button-holes and selvages, are exactly similarly disposed to those which the Australian Black paints on his body when he prepares for a Corroboree.

I saw many of the native children in the Philippines playing in the streets, wearing gaudy shirts, which did not reach lower down than six inches or so below their armpits, and practically were nothing more than broad red or blue necklaces.

The Manila natives indulge in a most extraordinary luxury, consisting of ducks' eggs which are brooded until the young are just beginning to be fledged, and are then boiled. It is a sickening sight to see these embryo ducklings swallowed at the roadside stalls, which are common at every street corner, piled high with half-hatched eggs and taking the place of our oyster stalls.

The great business of life in the Philippines, of the men of all the various tame Malay races, the half-castes, and Chinese, is certainly the sport of cock-fighting. The cock-pits in every town are a source of revenue to the Spanish Government. Everyone entering them pays sixpence, and the right of collecting tolls is sub-let by auction, usually to speculative Chinese. Sundays and the numerous Festas and Saints'-days are devoted to cock-fighting.

The galleries are crowded, and the excitement is immense.

It would be hard to say whether the Chinese coolies, who may be seen closely packed aloft, with their legs overhanging the arena, are the more eager spectators, or the darker skinned Malays. The money bet is thrown in a heap at the feet of the judge, in the dust of the arena. There is plenty of gold amongst it, and unless a certain amount is staked, the particular fight arranged is not proceeded with. There are loud shouts of offers on one colour or another, the black cock against the red, the brown against the white, and so on.

The spurs used for fighting are quite different from those formerly used in England, which were conical, and fastened to the natural spurs of the cock, or to the bases of these pared

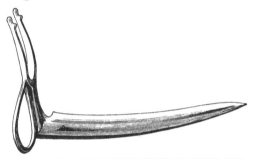

COCK-SPUR USED IN THE PHILIPPINE ISLANDS FOR FIGHTING COCKS.

down. The Philippine spurs are curved blades, like those of penknives, and are fastened by a steel loop over the hind toe of the cock, and secured by means of two prongs, which embrace the base of the natural spur.* Hence the bird deals his blow at the end of a longer lever. A single blow often lays the opponent dead. The spur blades are kept carefully covered with leather sheaths and as sharp as razors. If a cock runs away, as is some-times the case, he is counted beaten. I was told that some of the cocks survive three or four years, and kill twenty or thirty opponents.

When not actually fighting their cocks, on the few days intervening between the festivals, the natives train the birds and teach them to fight, squatting opposite one another, and holding the birds by the tails, and allowing them to strike at each other

* Similar spurs are used in Nicaragua. Thos. Belt, "The Naturalist in Nicaragua," p. 42. London, John Murray, 1874.

without doing injury. The Chinese shopkeepers usually keep a pet cock tied by a string to a peg on the path outside the door, and slip out and have a friendly set-to with a neighbour's cock, in the intervals between the arrivals of customers.

Papal indulgences for sins, and even crimes, are still sold in the Philippines, by the Government, at its offices all over the country, at the same counters with tobacco, brandy, and lottery tickets, and other articles of which the Government retains the monopoly. The perpetual right to sell indulgences in Spain and its colonies, was granted to the Spanish Crown by the Pope in 1750. In 1844–45 the Government received from this source of revenue upwards of £58,000.*

* For the most valuable and exhaustive account of the Philippine Island, See F. Jagor, "Reisen in den Philippinen." Berlin, Wiedmann, 1873. For account of Sale of Indulgences, see s. 108.

CHAPTER XVII.

CHINA   NEW GUINEA.

Hong Kong.   Pigeon English.   Chinese Method of writing compared with European Methods.   Development of Chinese and Japanese Books from Rolls.   Plants colonizing a Pagoda.   Sights of Canton. Chinese and English Examinations, and their subjects compared. The Honam Monastery.   Chinese Floral Decorations.   A Chinese Dinner.   Dragons' Bones and Teeth.   Origin of Mythical Animals. Chinese Account of the Dragon.   The last Dragon seen in England. Use of Unicorn's Horn as Medicine in Europe.   Chinese and English Medicine compared.   Chinese Accounts of the Pigmies and of Monkeys.   English Mythical Animals.   The Sea Serpent.   Owls living with Ground Squirrel in China.   Off the Talaur Islands. Driftwood off the Ambernoh River, New Guinea.   Animals Inhabiting it.   Humboldt Bay.   Signal Fires of the Natives.   Bartering at Night.   Numbers of Canoes.   Relative Prices of Native Property.   Attempts at Thieving.   Modes of Expression.   Mode of Threatening Death by Signs.   Armed Boat Robbed.   Villages of Pile-Dwellings.

Hong Kong, November 17th, 1874, to January 6th, 1875.—The ship was no sooner anchored at Hong Kong, than miserable-looking Chinese came off in small boats, and began dredging round it for refuse of all kinds, carefully washing an old cabbage stalk or beef bone, and preserving it for food.   Such boats, usually worked by a single old man, were at work about the ship during nearly the entire time of our stay in the port, a constant evidence of the desperate nature of the struggle for existence amongst the inhabitants of the country.

We soon began to learn " Pigeon English."   It is not by any means an easy language to learn, that is to really learn it.   A newcomer often mauls his speech in a childish fashion, putting " ey " at the end of every word, and believes he is a master of the language.   But such is not the case, Pigeon English, is a very definite language, as more than one book written on the language has shown, and unless one knows the accepted terms for things in it, one may be entirely at a loss to make oneself understood by the Chinese.

For example, I wanted to visit a Chinese theatre in Hong Kong. I tried the chair coolies with all kinds of explanations and equivalents of "theatre" without success. At last I stopped and got an old resident to explain. He simply said "singsong walkey," and off went the coolies to the theatre at once. As is well known, many of the words in Pigeon English are Portuguese of ancient date, comparatively few are Chinese, though the grammatical construction is all more or less Chinese.

The ordinary visitor using the strange words derived from Portuguese usually imagines that he is employing a Chinese word; but if he asks a Chinaman who can understand him well he will in return tell him to his astonishment that the word is English. The Chinaman using Portuguese thinks he is talking English, and the Englishman using the same thinks he is speaking Chinese.

It is not only the uninstructed who misapprehend the words of the "Business English." I have often been amused in looking at a specimen of a book full of engravings of various Eastern deities, which is exposed amongst the manuscript treasures in the Bodleian Library at Oxford, and labelled in Pigeon English "Pictures of various Josses." Joss is a Chinese corruption of the Portuguese "Deos" (" God "). Most persons suppose it is a sort of Chinese equivalent of the word Idol.

People going from China to Japan usually try to force Pigeon English into the heads of the Japanese. The Japanese language and its construction is of course utterly different from the Chinese. Hence, Pigeon English is probably more difficult for a Japanese to understand than English itself, and the language is really not current in Japan.

I found my servant, on arrival at Japan, attempting to make the washerman understand a series of instructions, in what he rather prided himself as good Pigeon English, though it bore little resemblance to the real article. The Japanese could not understand a word, but he at once comprehended a few words of plain English from me.

The marked feature which renders Chinese and Japanese towns and interiors different from all others, and strikingly peculiar, is due to the vertical method of writing employed. All the flags, all the sign-posts, posters, and shop-signs, and all the tents decorating the walls of the interiors, all the streaks of bright colour in the various views, are drawn out into length vertically, to accommodate the characters, instead of horizontally, as with us.

We are apt to regard the Chinese method of writing as utterly different from our own, because the characters express ideas and not sounds; but in the use of the Arabic numerals in all European languages, there is an exact parallel to the Chinese method. The numerals 1, 2, 3, represent ideas of numbers, and though a Frenchman, German, and Englishman alike understand them when written, when reading them aloud they use different sounds as equivalents, and would not understand one another unless specially instructed.

So it is exactly in the case of Chinese characters, only the system is extended to all ideas, and not confined to numerals. Even in having been derived originally from graphic representations of the numbers themselves, some at least of our numerals, and all the Roman numerals, correspond with Chinese characters.

Though English words are expressed by series of letters strictly representing sounds, yet, nevertheless, when the resulting words are taken as a whole, they are read very differently by the little educated in the various dialects. So much so, that a book read aloud in broad Scotch, would be little understood by an uneducated Englishman at least. Just in the same manner, educated Chinamen, speaking only different dialects, can each read a Chinese book to themselves, with perfect understanding; but neither can comprehend it if it be read aloud to him by the other.

A Chinese book is very interesting in its construction. The back of the book has its edges cut, instead of the front as with us, and the front is left doubled in the condition in which we leave the backs of books. The numbering of the pages and the title of the Chinese book are placed on the front edge of each leaf, where the paper is doubled, so that half of each character is upon one side of the edge, and half on the other; and the folded edge has to be straightened out if the entire characters are required to be seen.

All the leaves in a Chinese book are double, and only one side of the paper is printed on. The back surface of the paper is blank and wasted. The idea of cutting the pages and printing on both sides of the paper seems never to have been attained. Sometimes Japanese picture books, drawing books, and song books, have drawings or printed pictures on both sides of the paper; but even then, the pages are not cut, so that the two sides of each leaf should follow one another consecutively.

Such a book is merely a folded roll. After the folded pages on one side have been looked at, the book must be reversed and opened afresh at what before acted as the back, and thus the opposite sides of the folds are brought into view. If the pages only followed one another in the requisite order, there is no reason why such a folded book should not be at once stitched at the back, and have the leaves cut. The book would thus be rendered far more handy; but the idea seems never to have struck the Japanese.

The folded form of book described, seems to represent a first stage in improvement from the more ancient roll. Japanese paintings and manuscripts are extremely common, executed upon long rolls which are terribly tedious to unroll and roll up again. The folded picture books, such as described, may be pulled out into long strips, on which the pages or drawings follow in regular order, just as on an ordinary roll. Similarly, if ordinary printed Japanese and Chinese books were unstitched, the double leaves might be unfolded, and, if pasted on to a long strip, would follow one another consecutively on the roll.

It seems thus highly probable that the idea of the Chinese and Japanese book arose as an improvement on the roll; and that this is the reason why the leaves are all double, and the paper printed only on one side. The ordinary paper used in printing is possibly too thin to allow of both sides being printed on; but there is plenty of thicker paper available in both countries. Even when very thick paper is used in the folding Japanese books, often one side only of the paper is made use of. I have never seen an example with the front edges cut, even although I possess several folded books made of extremely stout cardboard. The accompanying diagram will serve to illustrate the development of the book from the roll.

DIAGRAM SHOWING THE DEVELOPMENT OF THE CHINESE FOLDED BOOK OUT OF THE ROLL.

Nearly all Chinese and Japanese books are block books, printed from wooden blocks, each of which contains four pages, a pair of pages on each side. All the letters having to be carved out on every wooden block, it is as cheap or cheaper to fill a page with illustrations as to fill it with characters. Hence, no doubt, the profusion of illustration, especially in Japanese books.

I paid the usual visit to Canton from Hong Kong. On the passage of the river the tall pagoda of Whampoa is passed. Pagodas, as is well known, are erected as sanitary precautions for the benefit of the cities near which they are built. They represent sharp peaked mountains, and are intended to preserve the balance of exhalations of the several elements, according to the laws of the mysterious science of Fung Shui, and thus avert pestilence and other ills.

The pagoda interested me, because on every one of the series of balconies or ledges encircling it at successive heights, a large variety of plants had established themselves and were flourishing, in some instances bushes of considerable size. The pagoda stands isolated, and the seeds of all these plants must have been carried up by birds or by the wind. I was told that the Chinese considered it lucky that plants should thus settle on the building.

The strangest sight in Canton is certainly the water-clock,

where a constant attendant watches the sinking of the index attached to the float, as the water slowly runs out; and when an hour is reached, hangs out a board with the hour written upon it on the city wall, and sounds the time on a gong.

The small houses on the ferry-boats on the Canton River, which are the homes of the families which get their living by means of them, are decorated all over inside with prints from illustrated European newspapers, many of them of considerable antiquity. It was amusing to find oneself confronted with "the Funeral of the Late Field Marshal the Duke of Wellington." Pedlars and dealers of all kinds ply their trade in the boat-towns in small boats, with which they traverse the lanes and alleys of water. From one of these pedlars I bought some jewellery, used by the boat population, in which pieces of Kingfishers' feathers are set in a gilt backing, so as to imitate, in appearance, very closely, fine blue enamel. The play of colours on the feathers thus mounted is extremely effective, and the jewellery is very pretty.

One of the places ordinarily visited in Canton by tourists is commonly called the Temple of Horrors. Here the future punishments of the wicked are set forth in a series of groups of modelled figures, representing all horrible tortures conceivable in process of execution. In one of these a man is about to be pounded by demons upon an anvil, but is rescued by the Goddess of Mercy (*Quan Yin*), who, standing on a hill-side at some distance, is represented as letting down a cushion at the end of a string, so that the cushion is interposed between the body of the condemned sinner and the descending mallet. This struck me as a very quaint way of indicating merciful interposition by the Goddess. At this temple some women engaged in some act of religious devotion were pouring libations of some kind of spirit at the foot of one of the pillars.

At the bookshops close by the water-clock, a bookseller, from whom I had bought some books, presented me with an old wood block as a specimen at my request, and refused payment for it. Yet the Chinese are commonly accused of being universally grasping, in their dealing.

The Government competitive examination buildings are

astonishing for the large area which they cover, and the vast accommodation which they afford. It is singular that a similar institution should just now be in course of construction at a vast cost in Oxford. The Chinese examination halls cannot but recall to an English University man the close analogy which exists between Chinese methods of mental training and learned thought, and those in vogue at home. As in our own Universities the main energies of the learned have been devoted to the reiterated translation into English and study of the mouldy and worm-eaten lore of a by-gone age ; so in China successive generations of students have for centuries devoted their lives to the acquisition of the antiquated philosophy of their remote ancestors, for the purposes of display in competitive examination. The reformation of the English Universities proceeds but slowly, and notwithstanding the hopeful movements now in progress in that direction, a period of very màny years must necessarily elapse before all branches of knowledge shall be equally and adequately represented in them.

Like the examination halls, the great monastery at Honam was full of interest from its close resemblances to similar European institutions. We listened awhile to the evening service, intoned and chanted by the monks in their priestly vestments, a gong and a kind of wooden bell giving out a very sharp and short note when struck were used as an accompaniment. We were next shown the refectory; here was a small pulpit for the reading of pious books by one of the monks whilst the others are at dinner, just, for example, as at Tintern Abbey. Close by was the flower-garden of the monastery, where bright flowers were carefully grown, to be used to decorate the holy shrines. The principal flowers in blossom were very fine large red and yellow Cockscombs (*Amaranthus*) of which the gardener of the monastery was very proud and which displayed pyramidal masses of blossom three or four feet in height. Not far from the garden is a fish-pond and near by a small cremation house, where the bodies of monks who die at the monastery are burnt. The whole institution is more or less in decay ; the monks do not act up to the rules of their order.

Chinese are especially tasteful in arranging flower decorations. At a Chinese dinner at which I was present, and which was most hospitably arranged for us by Mr. R. Rowitt, one of the Hong Kong merchants, the entire walls of the room in which the entertainment took place were covered with most beautiful flowers set in tasteful patterns in a backing of moss.

The dining-table was closely packed with dishes of most varied kinds, tastefully ornamented and arranged. There were absolutely no bare spaces, a display of profusion being evidently intended. I was astonished to find as a condiment in the sauce of some stewed pigeons, specimens of the well-known but curious *Cordyceps sinensis*. This is a fungus which attacks and kills the caterpillars of certain moths ; the fungus penetrates the tissues of the living larva, and after the larva has buried itself in the ground in order to assume the pupa state, the fungus throws out above ground a long stem from the dead body of the larva.

The dried dead caterpillar, with the fungus outgrowth attached, is one of the many Chinese delicacies which seem so strange to us, nearly all of which are prized, because, in addition to their gastronomic qualities, they are credited with exercising certain invigorating medicinal effects. The caterpillars are sold tied up in small bundles, and the article is called " the summer grass of the winter worm."

It is the fashion to decry Chinese delicacies as especially nasty, and the well-known eggs, which are pickled and buried for years before being eaten, are always cited as instances of especially disgusting food ; but after all this is more a matter of education and prejudice on the part of the foreign observer, than any real difference of habit in the Chinaman. Englishmen are apt to forget that their countrymen habitually prefer to eat game and cheese in a state of decomposition, and the latter often when swarming with maggots, and in a condition such that it would possibly sicken a Chinaman to look at it. Nearly all races fancy some form of food in a state of decomposition, and no doubt regard that particular food when in that condition as we do cheese, as simply " ripe."

Some of the popular prejudices with regard to Chinese customs are hardly to be comprehended. When I was a child, the one fact I learnt about Chinamen was that they wore pigtails, and I was led to regard that as an extraordinary and peculiar form of hairdressing ; yet the very same fashion had only very shortly gone out of general use amongst Englishmen ; a rudiment of the English pigtail still exists on our court dresses, and foot-men of Royal state carriages, wear a shortened pigtail still, on certain occasions at least.

The women present at Chinese banquets, such as that de-scribed, sit behind the chairs of the men, and receive no share of the luxuries, but are supplied with dried melon seeds, in the cracking and extraction of the kernels of which they occupy their time.

Whilst at Canton, I visited the shop of a Wholesale Chinese Chemist and Druggist, in order to try and select specimens of Dragons' bones which are a highly-prized specific for certain diseases in Chinese Medicine. The wholesale dealer, whose warehouse was very large and full of Chinese medicines in bulk, had no "Dragons' bones and teeth" in stock, but I bought a few specimens from retail druggists who sell them by weight.

The "Dragons' teeth and bones" consist of the fossil teeth and bones of various extinct Mammalia of tertiary age, such as those of *Rhinoceros trichorhinus*, a Mastodon, an Elephant, a Horse, two species of a *Hippotherium*, two of a species of Stag, and the teeth of a large Carnivorous animal.*

The drug is imported into Japan, and I saw samples exposed in a collection of *Materia Medica* at the Kioto Exhibition.

The chief interest in the "Dragons' bones and teeth," seems to me to be that they explain the origin of the Dragon itself, and very possibly of other mythical animals. All mythical animals have a strong foundation in fact and a developmental history. In most instances, no doubt, the mythical animal is derived from a traveller's description, or a description passed on

* For a description of a collection of these objects, by Prof. Owen, see "Quart. Journal of Geological Soc.," 1870, p. 417.

See also D. Hanbury, "On Chinese Materia Medica," p. 40. London, 1862.

Swinhoe refers to a collection of Dragons' bones in "Chinese Zoology," Proc. Zool. Soc., 1870, p. 428.

from mouth to mouth. From this eventually an artist has drawn a picture of the wonderful animal, and this has become the stereotyped representation of the beast, and has been handed down with successive embellishments.

The story of the Argus no doubt arose from a description of the Argus pheasant or peacock. The Dugong (not the Manattee) was long ago shown by Sir Emerson Tennant to have given rise to the story of the Mermaid. No doubt the original Mermaid was a black beauty, and only became white-skinned as the story travelled westwards.

The Unicorn is the Rhinoceros, sketched thus from report; but the Narwhal's tusk having come to hand as the Unicorn's horn, it was placed on the forehead of the animal, in the drawings, and the beast still wears it in our Royal Arms.*   There is the germ of truth in the case of the Narwhal's tusk, that the tusk grows without a fellow on the animal's head ; no doubt it was this fact that led to the blunder.   Marco Polo was astonished to find how different the real Unicorn was from the pictures of it he had been accustomed to see.

The Japanese dealers in carved ivories at Kioto, who speak a few words of English, draw attention to " netskis " cut out of Narwhal ivory, as made from " Unicorn." I suppose this is a survival of an old European term for the tusk, derived from the Portuguese.

The Dragon, however, seems to have had a different mode of origin, and to have sprung from the finding together in a fossil deposit of the bones of various animals, and the inference, that because they were found together they belonged to one animal. An attempt at reconstruction produced the Dragon, and this accounts for the animal possessing stags' horns and carnivorous teeth, and containing in its structure a little of everything.

My friend, Mr. C. V. Creagh, of Hong Kong, kindly trans-

* "The Book of Ser. Marco. Polo," Vol. II, p. 273. Col. H. Yule, C.B. London, Murray, 1875.
The last attempt to resuscitate the heraldic Unicorn, and prove its actual existence as such, was made in 1852, by Baron J. W. von Müller, " Das Einhorn vom geschichtlichen und naturwissenschaftlichen Standpunkte betrachtet." Stuttgart, 1852.

lated for me an account of the Dragons' bones and teeth given in a well known Chinese work, " The Botanical and Medical Works of Li She Chan," sometimes called " Li Poon Woo," Vol. XLIII. I give the account here because it is amusing in many ways as a sample of a Chinese medical work, and seems to bear out the above conjecture as to the origin of the Dragon, or origin of part of the animal's structure at least.

Translation. " Dragon's bones come from the southern part of Shansi, and are found on the mountains. Dr. To Wang King, says that if they are genuine they will adhere to the tongue. He informs us that the bones are cast off by the Dragon. Dr. So Tsung says, that in the autumn a certain fish changes itself into a Dragon, and leaves its original bones, which are of five different colours, and are used by men as medicine. In Shanshi is the Dragon-gate, through which when the fish leaps it becomes a Dragon.

" Dr. Kai Tsung Shik says, that it is well known that the Dragon is invisible to man. If this were the case, how could we see his bones ? I myself have seen a whole skeleton, head, horns and all, in a dilapidated mountain, and have no doubt they come from a dead animal, and have not been cast off by the Dragon.

" Li She Chan, remarks : I believe the above remarks to be inaccurate. In the Tso Chŭne (a history written in the time of Confucius) an official named Wan Lung Shee used to eat spiced Dragons' flesh. A book named Shut Yu Kee (The Record of Curiosities) says that King Wo of Hon Kwok (the old name of China) made soup of a Dragon, which fell into the palace during a heavy rain. He invited all the high officials to partake of the soup. The author of the Pok Mut Chee, says that Cheung Wo got Dragon's flesh, which he steeped in vinegar, and thereby gave to the latter five different colours. As the animal is seen and used in this way, I have no doubt that the bones are those of a dead Dragon, and have not been cast off.

" This medicine is sweet and is not poison. Dr. A. Koon certainly says that it is a little poisonous. Care must be taken not to let it come in contact with fish or iron. It cures heart-ache, stomach-ache, drives away ghosts, cures colds and dysentery, cures fainting in children, irregularities of the digestive organs, heart or stomach, paralysis, nocturnal alarm, &c., and increases the general health."

In the Chinese Repository* is a further quotation from Li She Chan concerning Dragon's bones, as follows: "The bones are found on banks of rivers and in caves of the earth, places where the Dragon died, and can be collected at any time. The bones are found in many places in Szechuen and Shanse, where those of the back and brain are highly prized, being variegated with different streaks on a white ground. The best are known by the tongue slipping lightly over them. The teeth are of little firmness, the horns hard and strong, but if these are taken from damp places, or by women, they are worthless."†

It is possible that other mythical animals besides the Dragon may be, like it, partly of fossil origin, as were, without doubt, numerous races of Giants, which sprung from the discovery of Mammoth bones. Fossil bones from caves, under the name of Dragons' bones, were long used as medicine in Europe. A live Dragon was discovered in Sussex in 1614.‡

It is not so long since all kinds of nastiness, such as powdered Mummy and album græcum were regularly used in English medicine, as now by Chinese doctors. Sir Thomas Browne, in his "Pseudoxia Epidemica," published in 1646, although he explodes many false notions in vogue at his day, as to the Unicorn, yet gravely discusses the power, as antidotes to poisons,

* The Chinese Repository, Canton, 1832–1838, p. 253. Extract from "Pun Tsaou Kang Muh."

† For accounts of Chinese Medicine, see M. P. Dabry de Thiersant, "La Medecine chez les Chinois." Also same author, and Dr. Leon Soubeiran, "La Matière Medicale chez les Chinois," also "Études sur la Matière Medicale des Chinois." Acad. de Medicine, Paris, July 16, 1873.

‡ "True and Wonderful, a Discourse relating to a strange and monstrous Serpent or Dragon lately discovered and yet living, to the great annoyance and divers slaughters both of men and cattell by his strong and violent poison. In Sussex, two miles from Horsham, in a wood called St. Leonard's Forest, and thirtie miles from London, this present month of August, 1614." Printed at London, by John Trundle. In this book a picture of the Dragon is given. It is in the form of a large lizard with protruded barbed tongue and rudimentary wings. The dead victims are strewed in front. The Dragon was nine feet in length. Its principal haunt was at a place called Faygate.

of Unicorns', Elks', and Deers' horns, and their effect on epilepsy when taken as medicine.*

In 1593, a committee of Doctors of Medicine of Augsburg, after a careful examination of a specimen of the very rare drug, the Unicorn's horn (Narwhal's tusk in this instance) in order to confirm their conclusion that the horn was real Monoceros horn and not a forgery, gave an infusion of some of it to a dog poisoned with arsenic, and on the recovery of the animal were thoroughly convinced of the authenticity of the specimen. Their report, duly signed, commences, " Quin etiam visum est nobis, ad experientiam, rerum magistram tanquam κριτήριον descendere."† In the work in which this experiment is recorded, follows an account of another, in which a dram of nux vomica was rendered harmless to a dog, by the action of 12 grains of the precious horn, whilst an exactly similar dog died in half an hour, from the same dose without the antidote.

My friend, Dr. J. F. Payne, has pointed out to me, that Unicorn's horn, and the skull of a man who has died by a violent death, appear as medicines in the Official Pharmacopœia of the College of Physicians of London, of 1678. Unicorn's horn, human fat and human skulls, dogs' dung, toads, vipers and worms, are retained in the same Pharmacopœia for 1724. A Committee revised the Pharmacopœia in 1742. They still retained in the list, centipedes, vipers, and lizards. The use of grated human skull as medicine, by uninstructed persons, survived in England as late as 1858 at least.‡

The idea that Rhinoceros horn acted as an antidote to poisons, was ancient in India. No doubt hence arose the belief that the Narwhal ivory, supposed to be that of the Unicorn, which beast was in reality the Rhinoceros, had the same properties. The story no doubt travelled together with that of the animal. Drinking-cups, elaborately carved out of Rhinoceros horn, were used in the East, and were supposed to detect or

---

* Sir T. Browne's Works, edited by Wilkin, Vol. II, p. 503. London, Pickering, 1836.

† "Museum Wormianum seu Historia Rerum Rariorum," pp. 286–287. Olao Worm, Med. Doct. Amstelodami, 1655.

‡ Rev. T. F. Thiselton Dyer, " English Folk Lore." London, 1878.

neutralize poisons poured into them. The forms of these cups have been largely copied by the Chinese, in ivory-white porcelain.

Rhinoceros horn is still used in Chinese medicine, and is to be seen hanging up, together with Antelopes' and other horns, in every druggist's shop in Canton.

Chinese medical prescriptions are excessively long, containing a vast number of ingredients, most of them inactive. It is only lately that English prescriptions have been shortened, and they still sometimes contain a good deal which is superfluous. A certain air of mystery is still preserved about them. Herbalists still practise upon the uneducated in London, in a style in some respects not very different from that of the Chinese physician.

A large variety of most amusing mythical animals are figured in Chinese works on natural history. Many of them are familiar and classical, such as the Cyclops : and the Pigmies, who are described as going about arm-in-arm for mutual protection, for fear the birds should mistake them for worms and eat them. The story is evidently identical with that of Homer, where the Pigmies are described fighting with the Cranes, on the shores of Oceanos. In Japanese pictures of the Pigmies, the "little men" (sho jin) are represented as walking arm-in-arm on the sea-shore, with the cranes hovering over them ready for the attack. The measured height of the Pigmies is usually given in classical accounts, just as in the Chinese.

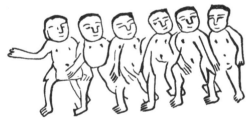

"The Small Men's Country is to the eastward of Tai Tong. The inhabitants are nine inches high."

I give a *fac-simile* of a figure of the Pigmies, and translation of the Chinese explanation of it, taken from the "Shan Hoi King," or Mountain and Ocean Record ; a very ancient work, parts of which were kindly translated for me by Mr. C. V. Creagh. The book is in the preface referred by one commentator to even so early a date as 2205 B.C.

Many of the figures and descriptions in this book are curiously like those which occur in European Natural-History Works of about 250 years ago.

HEN YEUNG KINGDOM.

"The inhabitants of this country have long lips, hairy and dark bodies. They laugh if they see a man laughing, and when they laugh their lips turn over and conceal their eyes."

Some of the strange men figured are evidently monkeys. As for example the men of the Hen Yeung Kingdom, figured and described in the Shan Hoi King. The Chinese figure is given in *fac-simile*. It seems to represent an Ape of the genus *Rhino-pithecus*, and might well be *Rhinopithecus Roxellanæ*, lately discovered by the Abbé David in Eastern Thibet, and figured by A. Milne-Edwards. The prominent nose in this species turned up at the tip just as shown in the Chinese wood-cut. The wide but unscientific distinction, commonly drawn between men and the higher monkeys, is an error of high civilization and comparatively recent. Less civilized races make no such distinction. To the Dyack, the great ape of Borneo, is simply the Man of the Woods, "Orang Utan."

The belief in various mythical animals in England is still very strong. We are probably not far in advance of the Chinese in this matter. So strong is the belief, that several of the animals in question could not be mentioned here without prejudice. The Sea Serpent, however, is always open to criticism. This wonderful animal has hardly ever been seen alike by any two sets of observers. It is nearly always easy to a naturalist to understand the stories told. Sometimes it is a pair of whales that is seen ; sometimes, as when the animal was seen off the Scotch coast, and figured in the "Illustrated London News," a long mass of floating seaweed deceives the distant observer ; sometimes the Serpent has large eyes and a crest behind the head, then it is a Ribbon Fish* (*Gymnetrus*).

I myself am one of the few professed naturalists who have seen the Serpent. It was on a voyage to Rotterdam from the Thames. An old gentleman suddenly started up, shouting, "There's the Sea Serpent!" gesticulating with his umbrella. All the passengers crowded to the ship's side and gazed with astonishment at a black line, undulating with astonishing rapidity along the water at some distance. It was a flock of Cormorants, which was flying in line behind the waves, and which was viewed in the intervals between them with a sort of thaumoscopic effect.

The extremely untrustworthy nature of the descriptions sent home is a constant feature in the natural history of the Sea Serpent. Not long ago he was seen near Singapore (evidently a very large Cuttle fish on this occasion). He was described as with large eyes, spotted with brown, and *without arms* or *legs*, but with a *very long tail*, and was yet said to be like a frog.

Ordinary sailors know nothing about whales or fish, and easily imagine they see wonders. Often, of course, the Sea Serpent stories are entirely without foundation in fact, and

---

* As first, I believe, pointed out by Mr. J. M. Jones, F.L.S., in "An Account of a Ribbon Fish, 16 ft. 7 ins. in length, obtained at Bermuda." Proc. Zool. Soc. 1860, p. 187.

sometimes apparently ships from which they emanate are laden with rum.

Amongst the rough figures in the Shan Hoi Sing, the small book, from which the illustrations already given are taken, is one of a rat-like animal and a bird which lives in the same hole with it. The description of the figures at the margin runs: "The Bird and the Rat live together in the same hole. They come from the mountain of the tailed rats and birds in Wai Une where they may still be seen."

THE BIRD AND THE RAT LIVING TOGETHER IN THE SAME HOLE.

Professor Legge has pointed out to me a reference in "The Chinese Classics" to the mountain called the Neauou-shoo-tung-heŭe, or that of the Bird and the Rat in the same hole; and to a note of his on the subject.* The name of the mountain in "The Classics" certainly dates back as far as 2300 B.C.

No doubt the Rat is the Ground Squirrel (*Spermophilus mongolicus*), and the bird must be an Owl, which is associated with it, just as is the small Ground Owl, *Speotyto cunicularia* of America with the Prairie Dog, and also the Ground Squirrel of California, in the holes of which, as familiarly known, it lives.

The genus *Speotyto*, is, however, peculiar, as far as is known, to America and the West Indies; and the fact that an Owl lives in the holes of the Asiatic Ground Squirrel is not known to naturalists. Mr. R. Bowdler Sharpe, however, tells me that a small owl, *Carine plumipes*, exists in Northern China, which lives in holes in the ground. Possibly this bird has developed

* Rev. James Legge, D.D., &c. "The Chinese Classics," Vol. III., Pt. III. p. 140. London, Trübner, 1865.

the same curious habit of association with a Rodent as the American Ground Owl. If so, the fact is very remarkable.*

Meangis Islands, February 10th, 1875.—The ship left Hong Kong on January 6th, 1875, and after visiting various ports in the Philippine Group as already noted, lay on February 10th between the Meangis and Tulur or Talaur Islands, south of the Philippines. The ship was nearest to the Island of Kakarutan, of the Meangis Group. The large hilly island of the Talaur Group, Karekelang, was seen in the distance, covered with forest, but with numerous patches of cultivation.

A canoe, sharp at both ends and without outriggers, of the Ke Island build, manned by 22 men and boys, came off to the ship. The men wore turbans, like the Lutaos of Zamboanga, and were many of them apparently of the same race, but appeared to be a mongrel lot, and were very dirty-looking. They did not, as far as we could ascertain, understand either Malay, Spanish, or Dutch, but asked for tobacco. They brought mats and very pretty blue and red Lories alive for sale. The birds were secured to sticks by means of rings made of cocoa-nut shell as at Amboina. The men did not chant or use drums as they paddled. They had the Dutch flag flying.

Drift Wood from the Ambernoh River, New Guinea, February 22nd, 1875.—On February 22nd, at noon, the ship was about 70 miles north-east of Point D'Urville, New Guinea, where the great Ambernoh River, the largest river in New Guinea, runs into the sea.† This river probably rises in the Charles Lewis Mountains, on the opposite side of New Guinea; these mountains reach up to the great altitude of 16,700 feet. So large is this river, that even at this great distance from its mouth, we found the sea blocked with the Drift Wood brought down by it.

We passed through long lines of Drift Wood disposed in curves at right angles to the direction in which lay the river's mouth.

---

* An account of Chinese Zoology is given in the "Preussischer Expedition nach Ostasien" Zoologie, Bd. I. s. 169, "Ueber die Thierkunde der Chinesen und unsere Kenntniss Chinesischer Thiere."

† The mouth of the river which is lined with Casuarina-trees, was passed by Rosenberg on his way to Humboldt Bay in 1862. "Nat. Tydsch. voor Neder. Indie." Deel. XXIV. p. 334. Batavia, 1862.

The ship's screw had to be constantly stopped for fear it should be fouled by the wood. The logs had evidently not been very long in the water, being covered only by a few young Barnacles (*Balanus*) and Hydroids. Amongst the logs were many whole uprooted trees. I saw one of these which was two feet in diameter of its stem.

The majority of the pieces were of small wood, branches and small stems. The bark was often floating separately. The midribs of the leaves of some pinnate-leaved palm were abundant and also the stems of a large cane grass, like that so abundant on the shores of the great river (Wai Levu) in Fiji (*Saccharum*). One of these cane stems was 14 feet in length, and from 1½ to 2 inches in diameter.

Various fruits of trees and other fragments were abundant, usually floating confined in the midst of the small aggregations into which the floating timber was almost everywhere gathered. Amongst them were the usual littoral seeds, those of two species of Pandanus, and of the Puzzle-seed (*Heritiera littoralis*), fruits of a *Barringtonia* and of *Ipomœa pes capræ*.

But besides these fruits of littoral plants, there were seeds of 40 or 50 species of more inland plants. Very small seeds were as abundant as large ones, the surface scum being full of them, so that they could be scooped up in quantities with a fine net. With the seeds occurred one or two flowers, or parts of them.

I observed an entire absence of leaves, excepting those of the Palm, on the midribs of which some of the pinnæ were still present. The leaves evidently drop first to the bottom, whilst vegetable drift is floating from a shore. Thus, as the *débris* sinks in the sea-water a deposit abounding in leaves, but with few fruits and little or no wood, will be formed near shore, whilst the wood and fruits will sink to the bottom farther off land.

Much of the wood was floating suspended vertically in the water, and most curiously, logs and short branch pieces thus floating, often occurred in separate groups, apart from the horizontally floating timber. The sunken ends of the wood were not weighted by any attached masses of soil or other load of any kind. Possibly the water penetrates certain kinds of wood more easily in one direction with regard to its growth than the other.

F F

Hence one end becomes water-logged before the other; I could arrive at no other explanation of the circumstance.

It is evident that a wide area of the sea off the mouth of the Ambernoh River is thus constantly covered with drift-wood, for the floating wood is inhabited by various animals, which seem to belong to it as it were. The fruits and wood were covered with the eggs of a Gasteropod Mollusk, and with a Hydroid, and the interstices were filled with Radiolarians washed into them and gathered in masses, just as Diatoms in the Antarctic seas are gathered together in the honeycombed ice. Two species of Crabs inhabit the logs in abundance, and a small *Dendrocœle* Planarian swarms all over the drift matter and on the living crabs also. A *Lepas* was common on the logs.

Enormous quantities of small fish swarmed under the drift-wood, and troops of Dolphins (*Coryphœna*) and small Sharks (*Carcharias*), three or four feet long, were seen feeding on them, dashing in amongst the logs, splashing the water, and showing above the surface, as they darted on their prey. The older wood was bored by a *Pholas*.

A large flock of the very widely spread bird, the Phalarope (*Phaloropus hyperboreus*) was seen flying over the drift-wood. The birds no doubt follow the timber out from shore, and roost on it. In England we consider this bird as one of our visitors from the far north. It seems strange to meet with it at New Guinea. It was previously known from the Aru Islands. Some specimens shot had small surface Crustacea in their stomachs.

The various smaller animals no doubt congregate about the drift-wood because it seems so act as a sort of sieve or screen, and to concentrate amongst it the surface animals on which they feed.

The Charles Lewis Mountains seem to be one of the most promising fields in the world yet remaining unexplored by the naturalist. They no doubt contain an Alpine flora which might prove allied to that of New Zealand, since the great mountain of Kini Ballu in Borneo has southern forms of plants at its top; probably there will be found on these high mountains also allies of the New Zealand Parrots of the genus *Nestor*, one species of which (*Nestor notabile*) is Alpine in its range. There is a *Nestor* in Norfolk Island, and the genus *Dasyptilus* of New Guinea is allied to *Nestor*.

"Talok Lintju" or Humboldt Bay, February 23rd and 24th, 1875.—We sighted the New Guinea Coast as a dark purple line along the horizon, with its upper margin hidden in banks of mist, at about mid-day. On February 23rd, as we approached nearer, in the afternoon, the misty clouds lifted somewhat, and the sharp peak, the highest point of the Cyclops Mountains, 6,200 feet in height, lying just to the north of our destination, Humboldt Bay, showed out isolated and clear above the bank of cloud which concealed all the lower parts of the range.

The opening into Humboldt Bay, between Cape Caillie on the north-west, and Cape Boupland on the south-east, both precipitous and rocky, became gradually well defined. The coast appeared far nearer to us than it was, and its distance was judged at six miles when it in reality was at least 25 miles.

Between 5 and 6 o'clock, the mist lifted almost entirely from the Cyclops Mountains, and they were seen to consist of a series of irregular peaks and sinuous sharp ridges culminating in the one simple terminal peak, which had been seen before above the clouds. The mountain is thickly wooded to the very apex, as could plainly be seen with a telescope. The lines of trees which showed out against the sky along the outline of the mountain and its ridges showed few or no Palms.

The whole coast outside the Bay is steep and rocky, without any sandy beaches, and is thickly wooded with a dark clothing of vegetation with lighter green patches here and there, formed by the cultivated inclosures of the natives, or spaces which have at some time been under cultivation by them.

It was dark when we entered the Bay, steaming slowly to an anchorage. A light flashed from the Cape Caille shore, glimmered and flashed again, then another flashed, then another, and soon a dozen or more lights close together were flashing and moving to and fro. These signal fires were answered from the south side of the Bay, and from another spot higher up on the same side, and we heard the peculiar holloa of warning, "hoa, hoa," coming over the water from many voices, and sounding exactly like the shouts with which the savages at Api in the New Hebrides greeted the ship.

The masses of lights glimmered from the very water level, as could be seen from the mode of reflection of the flashes in the water. The villages of pile-dwellings of Ungrau and Tobaddi were giving the alarm and were being answered by the people of Wawah on the other side of the Bay. We could see the bright lights moving about, and waving to and fro as they were carried by the excited natives along the platforms of the pile-built villages, and could catch a glimpse of the shadows of the natives' bodies as they passed between us and the light.

Just as the anchor was let go in 15 fathoms, a light appeared on the water close to the ship, and a canoe was evidently reconnoitring us, but the natives were shy and wary, and the light disappeared again for some time. Then it was again seen close at hand, being waved up and down; and a native standing up delivered a volley of his language.

Lights were placed at the gangways and were waved as a token of friendship, and all sorts of encouragements were used, but the canoe kept at a distance, paddling to and fro. The only word we caught was "sigor," "sigor!" The canoes had two paddlers, one at either end, apparently boys, and a full-grown savage on the small platform in the centre.

The savage on the platform had his huge mop-like head of hair set off by a radiant halo of feathers stuck into it, and decked with a broad fillet of scarlet *Hibiscus* flowers, placed under the edge of the mop, above his forehead. As he blew up his smouldering fire-stick into a blaze, his dark face glowing in the light and set off by the scarlet blossoms, formed a most striking, but at the same time most savage spectacle.

The canoe at last dropped under the stern, the natives shouting still "sigor" "sigor!" I leaned over the stern boat, and threw down a gaudy handkerchief. It was at once fished out of the water with a four-pronged fish-spear, and examined by the glow of the fire-stick, and then another canoe which was approaching, and which contained four natives, was shouted to in the most excited language, expressive evidently of satisfaction.

Sigor being supposed to mean "tobacco," a cigar was let

down with a line and immediately taken and lighted, and more were shouted for, and two cocoanuts neatly husked and tied together with a part of the husk left attached for the purpose, as in the many islands visited by us, were fastened to the line, to be drawn up in exchange.

Then by cries of " sigor !" which acted as a loadstone, the canoes were drawn up opposite the gangway, and every attempt was made from the bottom of the ladder to invite the natives on board, but without success ; nor would they approach near enough to receive presents from the hand, evidently fearing a trap, but they took a number of cigars, receiving them two at a time, stuck between the prongs of a long fish-spear. The placing of the cigars between the jagged points of the spear was rather trying work, for the ship was rolling somewhat, and the spear was thus prodded to and fro.

Another gaudy handkerchief being given to the boat which had received one already, it was passed over to the other boat at once, either according to some agreement as to division of spoil or because perhaps the occupant of the boat was a chief. The use of ships' biscuit was not understood. One native made signs that he wanted a gun, by pretending to load his bow from some implement picked up from the bottom of his canoe to represent a powder flask, then ramming down in pantomime, drawing the bow as if shooting, and saying " boom."

The natives seemed frightened to some little extent by a " blue light," and shoved off a bit, shouting something as it was lit. At last they left for the shore, using a word very like " to-morrow." At one time they commenced a sort of song in their canoe, as they lay off the ship hesitating to approach.

The canoes hung about the ship nearly all night, and in the morning the ship was surrounded by them, and a brisk barter commenced at daylight. At about 7.30 the ship was moved nearer to the north-west shore of the bay, and to the dwellings of the natives. The canoes paddled alongside, and formed a wide trailing line as they accompanied the ship.

There were then 67 canoes in all present, and this was the greatest number that was seen. Some few of them contained five natives, some four, some three, some only two. In 50

canoes on one side of the vessel there were 148 natives, or
about an average of three to a canoe. In all, therefore, there
must have been 200 natives.

From time to time the shout which was heard the night
before was raised. When heard close by, it is found to commence with a short quick "Wăh Wăh ōh ōh ōh." Some few
natives had perforated Conch shells, both a Triton, and a large
conical Strombus perforated at the apex of the spire, not on the
side of one of the upper whorls, as in the case of the Triton.
These shells they blew, making a booming sound which mingled
with the shouts.

The natives evidently prize these trumpet-shells highly, and
would not part with them, perhaps from the same motives that
prevent them parting with their flutes, as described by the
officers of the "Etna."*

Many of the natives made a sign of drinking, and pointed to
a part of the Bay where water was to be procured, evidently
thinking that the ship required water. This shows that they
are more or less accustomed to ships watering here, and the
fact that the utmost endeavours failed to induce any of the
natives to come on board the ship, and their extreme caution in
their first approach, seemed to show that they must have been
frightened or maltreated in some way by recent visitors to the
Bay. When the Dutch vessel of war, "Etna," came into the
harbour in 1858, the natives clambered on board before the
cable was out.

As soon as the ship anchored again, the natives crowded
round the ship, and barter recommenced most briskly, being
carried on through the main deck ports, the natives passing up
their weapons and ornaments stuck between the points of
their four-pronged spears, and receiving the price in the same
manner.

The constant cry of the natives was "sigor sigor," often repeated (sigōr sigōr, slowly, sĭgōr sĭgōr sĭgōr, quickly). "Sigor"
was found to mean iron; this and "soth," which means more,
were the only words of the language gathered. Iron tub-hoop,
broken into six or eight-inch lengths, was the commonest article

* "Neu Guinea und seine Bewohner." Otto Finsch, S. 144.

of barter, but most prized were small trade hatchets, for which the natives parted with anything they had.

The iron, wherewith to replace the stone blades of their own hatchets, and the miserable ready-made trade hatchets, are to them the most valuable property possible, since they lessen the toil of clearing the rough land for cultivation, and of canoe and house building, which with the stone implements alone to work with, must be arduous indeed.

Hence the natives cared hardly for anything except iron; bright handkerchiefs or Turkey red stuff were seldom taken in exchange, and then for very little value. Beads however were prized. Of their own property, the natives valued most their stone hatchets. Very probably they obtain the stone for making them by barter from a distance, since the rock at Humboldt's Bay is a limestone, and the hatchets are made of jade or greenstone, or of a slate. The labour involved in grinding down a jade hatchet-head to the smooth symmetrical surfaces which these native implements show, must be immense.

Next in value to the stone implements were the breastplate-like ornaments, each of which has as its components, eight or more pairs of Wild Boars' tusks, besides quantities of native beads, of small ground-down Nerita shells. These treasures required a trade hatchet at least to purchase them. All other articles, necklaces, armlets, tortoiseshell ear-rings, combs, paddles, daggers of Cassowary bone and such things, could be bought for plain hoop-iron, as could also bows and arrows in any quantity, and even the wig-like ornaments of Cassowary feathers, which the men wear over their brows, to eke out their mop-like heads of hair.

The natives often attempted, and often succeeded in withdrawing an arrow or two from a bundle purchased, just as it was being handed on board. They understood the laws of barter thoroughly, and stuck to bargains. They attempted once or twice to keep the articles given beforehand in payment without return, but often returned pieces of hoop-iron and other things which had been handed down for inspection and examination, as to whether they were worth the article required for them or no. One or two of these natives tried to fish things out through the lower deck-scuttles from the cabins with their arrows, but were detected and frustrated in their design.

NATIVES OF THE VILLAGE OF TOBADDI, HUMBOLDT BAY.
N.B.—The arrow shown is too short, and should be as long as the bow.

Many of the men wore a pair of Wild Boar's tusks fastened together in the form of a crescent, and passed through a hole in the septum of the nose, so that the two tusks projected up over their dark cheeks as far as their eyes. Most of the men had short pointed beards, apparently cut to that shape ; the old ones had whiskers. One old man who was bald, wore a complete but small wig. None of the men were tattooed, but they had large cicatrized marks on the outer sides of the upper arms, and smaller ones on the shoulders.

The fungoid skin disease was common here as at the Aru and Ke Islands, but only on the adults ; the boys and many of the younger men were free of it.

The men attracted attention to barter by the cries of " urh, urh ! " to express astonishment they struck the top of the outer sides of their thighs with their extended palms. Refusal of barter or negation was combined with an expression of disgust, or rather the two ideas are not apparently separated; the refusal was expressed by an extreme pouting of the lips, accompanied by an expiratory sniff from the nostrils.

The forehead muscles were very little used in expression, though they were slighly knitted in astonishment. In laughing, the corners of the mouth were excessively drawn back, so that four or five deep folds were formed round the angles of the mouth, the head was lolled back, the mouth opened wide, and the whole of the upper teeth uncovered; the whole expression was most ape-like.

I started with a party in a fully armed boat with the intent of landing. As we approached the shore, a native warrior approached, standing as usual on the platform of a small canoe paddled by two boys sitting in the bow and stern; the man held up a yam and made signs that he wished to barter; we halted and made signs of refusal; he then took up one of his arrows, and holding the point to his neck just above the collar bone, made signs of forcing it into his body, and then throwing back his arms and head, and turning up his eyes, pretended to fall backwards by a series of jerks, in imitation of death; then he caught hold of the yam again and proffered it a second time, and on renewed refusal, went through the imaginary killing process again.

We began to move toward shore, when the man ran to the end of the canoe nearest the boat, and fitting an arrow against the string of his bow, drew the bow with his full strength and pointed the arrow full at me; I was standing up at the time with a loaded double-barrelled gun in the stern of the boat.

As he drew the bow he contorted his face into the most hideous expression of rage, with his teeth clenched and exposed, and eyes starting. This expression was evidently assumed to terrify us as an habitual part of the fight, and not because the man was in reality in a rage. In Chinese and Japanese battle-scenes, or hunting-scenes in which attacks upon large animals are depicted, the faces of the combatants are usually represented as horribly contorted with rage. No doubt the grimace is assumed as a menace amongst savages on just the same principle as that on which an animal shows its teeth. The native shifted his aim sometimes on to Von Willemoes Suhm, and sometimes on to Mr. Buchanan, who was nearest to him.

We were in a dilemma; the man evidently did not understand the use of fire-arms, for the whole boat's-crew was fully armed, and we in the stern were all provided with guns. He evidently thought that we were unarmed because we had no bows and arrows; he might have let slip an arrow five feet long into any one of us in an instant.

We of course would not shoot the man in cold blood; if we had fired over his head, he would certainly have let fly one arrow at least, and he was within six yards of the boat. The boys who paddled him were exuberantly delighted at the prowess and success of their warrior.

The canoe was pushed up to the stern of our boat, and the man caught hold of our gunwale. Another canoe joined in to share in the spoil, and closed in at the stern also. The two warriors seized a large tin vasculum of mine from the seat, and immediately began struggling between themselves for it, and taking advantage of the struggle we pulled back to the ship.

The vasculum contained some trade knives and three bottles of soda-water. I expect no savages were ever so thoroughly scared and puzzled as these when they came to open the bottles in the bosoms of their families in their pile-dwellings.

The same man who stopped us had also stopped a boat engaged in surveying, just before in the same manner, and it had also returned to the ship.

All kinds of suggestions were made on our return as to what ought to have been done; we ought to have hit the natives over the knuckles with the stretchers, or run the canoe down, or fired over the natives' heads; but there cannot be the least doubt that in that case some one would have been wounded at least, and one native at least shot.

I cannot understand how it occurred that this native knew nothing of fire-arms, since the Humboldt has often been visited by the Dutch, and many of the natives understood their nature; one man, as has been said, having plainly asked for a gun on our first arrival. Possibly the man had come from a distant part of the Bay either lately or some years before, or had only heard of fire-arms and was a sceptic, or knowing that a gun would kill birds, had thought that special magic, and not comprehended that it would also kill men.

A small party landed with Captain Thomson from the steam pinnace for a short time, and Mr. Murray, led by some natives, shot a few birds. These natives were friendly enough, but when Captain Thomson approached one of the platform villages, the women turned out with bows and arrows, and warned the boat away, using the same signs of death as the man who discomforted us.

A stay of some little time is evidently necessary in order that the natives should become on good terms with visitors in a strange ship, and possibly the natives had been maltreated by the crew of some vessel since the " Etna's " long visit in 1858; no doubt also the natives forget a great deal in the lapse of sixteen years.

As time could not be spared to wait and conciliate the natives, and violent measures were of course out of the question, landing was reluctantly given up, and the ship sailed for the Admiralty Islands in the evening of February 24th.

The bows of the Humboldt Bay natives are cut out of solid palm-wood and have a very hard pull. They taper to a fine point at either end, and in stringing and unstringing them a loop at the end of the string is slipped on and off this point and rests in the extended bow on a boss raised with wicker-work, at some distance from the bow-tip.

The bows are strung quickly by their lower ends being placed between the supports of the canoe outriggers as a fulcrum. If an attempt be made to string a bow, by resting one end on the ground, the tapering end snaps off directly pressure is applied.

The bowstring is a thick flat band of rattan, and the arrows, like all New Guinea arrows, have no notch, but are flat at the ends, and are also without feather. The natives have never learnt the improvement of the notch and feather. The men of Api Island, New Hebrides, have most carefully worked notches to their arrows, but still no feather.   The Aru Islanders have both notch and feather.*

The Humboldt Bay arrows further are excessively long, far too long for the bows, being five feet in length, so that not more than half of their length can be drawn.   They are rather small spears thrown by a clumsy bow for short distances than arrows. They go with immense force for a certain distance, but only fly straight for ten or a dozen yards, wobbling and turning over after that length of flight.

As the anchor was being got up, when the ship's screw was beginning to turn, two natives, who happened to be close to it in a canoe, drew their bows hastily on it as if it were some monster about to attack them from under water.

In the Humboldt Bay stone choppers, the stone blade is mounted in the end of a long wooden socket piece which is fitted into a round hole at the end of the club-like handle.   The socket piece can thus be turned round so that the blade can be set to be used like that of either an axe or an adze.

The handle and socket piece form nearly a right angle with one another, and the socket piece is so long that the whole seems a most clumsy arrangement, and it is most difficult to strike a blow with it with any precision.

The shorter the socket piece the easier it is to direct the blade with certainty in a blow. In Polynesia generally the stone blades are thus fixed close up to the ends of the handles, but in New Guinea this curious long-legged angular handle is in vogue.· It is difficult to understand the reason, unless these natives began with a chisel and mallet; and having got so far in improvement as to join them together, have not yet discovered the advantage to be gained by shortening up the socket piece.

A curious stone implement, similarly mounted to the chopper, was common in most of the Humboldt Bay canoes.   It seems to be a kind of hammer.   The stone head is cylindrical in form tapering to fit the socket at one end, and hollowed slightly on the striking face. . The exact use of the implement is uncertain.

* For the distribution and various forms of bows and arrows, see Gen. Lane Fox, F.R.S., &c., " On Primitive Warfare." Journ. of United Service Inst., 1867-9.

The awkwardness of its method of mounting is at once felt on trying to drive a nail with it.

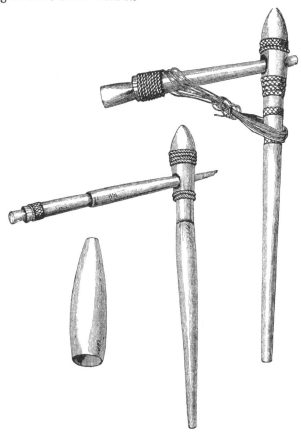

STONE BLADED CHOPPER AND STONE HEADED HAMMER IN USE AT HUMBOLDT BAY, ALSO A LARGER VIEW OF THE STONE HAMMER-HEAD REMOVED FROM ITS SOCKET.

The ethnographical details of the people of Humboldt Bay are, thanks to the investigations of the Dutch commission of the ship " Etna," better known than those of most savages. I extract the following account of the houses from Finsch's compiled account of New Guinea. It is derived from the " Etna " Expedition :

The houses rest on piles which rise three feet above the surface of the water, and are connected with one another by bridges. The walls of each house are not higher than three feet, but the

roof rises as high as 40 feet, is six to eight-sided, and rests on
the central pile of the building, which either stands directly
in the sea bottom, or is built of several trees fastened together.

VILLAGE OF TOBADDI, HUMBOLDT BAY.

Walls and roof are made of bamboos and palm-leaves, and
the interior is separated by partition walls made of palm-

leaves into separate chambers for the men and women and unmarried.

Each house has a fire-place and two small doors, which latter form the only entrance for the light and means of exit for the smoke. The houses are in two rows in each village, with the worst houses at the ends of the rows.

The temples, which are placed in the middle, are mostly

TEMPLE AT TOBADDI.

octagonal, and reach to a height of 60 or 70 feet. Some temples have two roofs, one over the other. There are figures of men, fish, lizards, and other animals at the apex of the roofs, and similar figures at each of the eight angles.

For accounts of Humboldt Bay, see "Dumont D'Urville Voy. de 'l'Astrolabe.'" Paris, 1830. "Voy. au Pôle Sud." Paris, 1841.

"Neu Guinea und seine Bewohner." Otto Finsch. Bremen, Ed. Müller, 1865, s. 132.

"Nieuw Guinea Ethnogr. en Natuururkundig onderzoocht in 1858 door een Nederl. Ind. Commissie." Bijdragen tot de Taal Land en Volkenkunde van Nederlandisch Indie. Amsterdam, F. Müller, 1862, 5th Deel. From this work the three figures given above are copied.

For "Von Rosenberg's Account of the Visit," see Nat. Tydsch. voor. Neder. Indie. Deel XXIV. Batavia, H.M. van Dorp, 1862, p. 333, et seq.

CHAPTER XVIII.

THE ADMIRALTY ISLANDS.

History of Visits to the Island. Eagerness of the Natives for Iron. Trade
Gear. Trading with the Natives. Geological Structure of the
Islands. Orchids and Ferns overhanging the Sea. Fern resembling
a Liverwort. Difficulties in Collecting Words of their Language
from the Natives. Their Methods of Counting. Curious Mode of
Expressing Negation. Physical Characteristics of the Natives.
Hairiness of Races Compared. Possible Signification of Moles.
Clothes, Hair Dressing and Ornaments of the Natives. Tattooing
and Painting. Betel-Chewing and Food. Houses, Temples, and
Canoes of the Natives. Their Implements and Weapons. Artistic
Skill of the Natives. Their Musical Instruments, Dancing and
Singing. Their Polygamy. Fortification of their Villages. Wooden
Gods. Skulls and Hair in their Temples. Their Religion. Dis-
position of the Natives. Their Fear of Goats and Toys. Population
of the Islands. Domestic Animals, Birds and other Animals at the
Islands. Habits of Gar-Fish.

The Admiralty Islands, March 3rd to 10th, 1875.—The Admiralty Islands were sighted on the afternoon of March 3rd.
As we sailed along the north coast of the main island, a Sword-
fish was seen showing its fins above water. It moved rapidly
with a darting motion but sinuous course. It was apparently
about five feet long. The fins showed above water, very differently from those of any other fish. The broad dorsal fin
projected from the water in front, and the upper sickle-shaped
half of the tail fin projected at an interval behind, and seemed
as the fish moved to be chasing the fin in front. The fish was
seen to leap out of the water several times. It was probably a
species of Istiophorus.

The Admiralty Islands are a group, consisting of one large
island and numerous small ones. The group lies between
latitudes 1° 58′ S. and 3° 10′ S., and longitudes 146° E. and

148° 6′ E. between 100 and 200 miles south of the equator. It is distant from New Hanover 130 miles, and from the nearest point of New Guinea about 150 miles.

APPEARANCE OF ISTIOPHORUS WHEN SWIMMING NEAR THE SURFACE OF THE WATER.

The large island of the group which is oblong in form, has an area of about 550 square miles, being thus about twice as large as the Isle of Man. It is mostly low, but contains mountain masses rising to a height of 1,600 feet. Our examination of the group was confined to the extreme north-western portion of the northern coast, and the small outlying islets in the immediate vicinity.

The Admiralty Islands were discovered by Captain Philip Carteret, of H.M. sloop "Swallow," on September 14th, 1767. Captain Carteret lay off small outlying islands to the south of the group. 12 or 14 canoes came off, and the natives at once attacked him by throwing their lances into the midst of his crew. He had to fire on them, and although he made efforts to conciliate them these were entirely unsuccessful. From a statement made by Dentrecasteaux it appears that shortly before 1790 the islands were visited by a frigate commanded by Captain Morelle.

In 1791 the "Recherche" and "Esperance" sailed from France, under the command of Dentrecasteaux, to search for the missing "La Perouse," the "Recherche" having on board of her as one of the naturalists, M. Labillardière.

In the previous year, 1790, the English frigate "Syrius" was wrecked on Norfolk Island, and a Dutch vessel which conveyed her commander, Commodore Hunter, to Batavia, passed by the Admiralty Islands. Whilst she was in sight of the shore, canoes full of natives put off towards the ship, and showed a desire to communicate, and being indistinctly seen in the distance, their white shell ornaments showing against their dark skins were taken for white facings on French naval uniforms, and their reddened bark cloths for European fabrics, and Hunter was persuaded that here were relics of the unfortunate " La Perouse."

Dentrecasteaux received information at the Cape of Good Hope, by a special despatch vessel sent for the purpose from the Isle of France, of what Commodore Hunter had seen, and he in consequence visited the Admiralty Islands with his two ships, arriving off the islands in July, 1792. He visited the outlying islands of Jesus Maria and La Vandola lying to the eastward, and then coasted along the northern shore of the main island to the same spot as that visited by the " Challenger." He communicated with the natives by bartering with them from his ships and from boats, but seeing no trace of any European relics amongst them, he concluded that Commodore Hunter had been mistaken in the manner already described, and set sail without effecting a landing. Two separate accounts were published of Dentrecasteaux's cruise, one by himself, edited by Mr. Rossel, the other by M. Labillardière. Both contain very interesting information concerning the Admiralty Islanders, the account by Labillardière being most complete in this respect, and accompanied by large plates of natives and weapons, and a view of Dentrecasteaux Island.

In 1843 the islands were visited by the American clipper " Margaret Oakley," Captain Morrell. The crew of this ship landed at many points on the coast of the main island, which according to Jacobs's account is called " Marso " by the natives.

They also visited many of the small outlying islands. Jacobs's account* is full of interesting details, but evidently not entirely trustworthy. It will be referred to in the sequel. There is no account extant of the landing of any other Europeans on the Admiralty Islands before the visit of the " Challenger." The well-known explorer Miklucho Maclay has paid a lengthened visit to the islands since our departure.

As the ship approached the anchorage canoes came off

* " Scenes, Incidents and Adventures in the Pacific Ocean," &c., pp. 164 to 182. By T. J. Jacobs. New York, Harper & Bros., 1844.

through openings in the reef to the vessel, though a stiff breeze was blowing, and the natives were evidently in great excitement and eager to reach the ship. Paddles were waved to show friendship, and various articles of barter exhibited to tempt us. The constant cry was "laban, laban!" which sounded to us at first like "tabac tabac," but which we afterwards found out to be, like the Humboldt Bay "sigor," the word for iron. Iron was the wealth they coveted.

Having seen the ship securely anchored, the chief ordered all the canoes away, and we were left alone till the morning. In the morning trade went on briskly, the canoes crowding round the ship, and the natives handing their weapons and ornaments through the main deck ports. The barter we gave in exchange principally was ordinary hoop iron broken up into pieces about six inches in length; but we also disposed of a great quantity of so-called "trade gear."

Trade gear is regularly manufactured for Polynesian trading, and sold by merchants in Sydney and elsewhere. We had bought a stock of about £300 worth for the ship's use. It consisted of a cask of small axes, worthless articles, with soft iron blades, butchers' knives of all sizes, some of them with the blade 12 or 14 inches in length; cotton cloth, Turkey red and navy blue, beads, and other similar articles.

The islanders had possibly traded with Europeans before our visit within tolerably recent time.* They brought off their tortoiseshell ready done up in bundles, and they knew the relative value of various qualities. The chief had a large European axe, which I believe was not procured from the ship; and many natives had hoop iron adzes. Nevertheless they must have had very little experience indeed, otherwise they would not have taken old German newspapers freely as trade as they did at the first, thinking them to be fine cloth, until rain had fallen. They soon took to making trade goods,

---

* There were specimens of Admiralty Island lances and gourds in Gen. Lane Fox's Collection, and in the Christy and British Museum Collections, procured before the "Challenger" visit. These have probably been obtained from Cape York, and no doubt were taken there by tortoiseshell and pearl-shell traders who had visited the Admiralty Group.

shell hatchets, and models of canoes, to sell to us, which were as badly made as the trade gear which we gave in exchange. They understand the rules of barter well, and, as in Labillardière's time, seemed anxious to pay their debts. They pretended, with many expressive grimaces, to be unable to bend pieces of tortoiseshell which they offered for sale, and of the thickness (i.e., fine quality) of which they wish to impress the purchaser. They often thus pretended to try ineffectually to bend very thin pieces indeed, and fully entered into the joke when we did the same with thin bits of hoop iron. They always required to see the hoop-iron tested by bending before accepting it. They must trade with one another regularly. They made signs that the ore of manganese which they use came in canoes from a distance eastwards. The native canoes are so seaworthy, and the natives so enterprising and fearless in going to sea, that possibly articles may pass by barter from island to island here over wide distances, even to New Hanover and New Britain.

The natives took all the hoop-iron from us which they could get, evidently receiving more than they could use, no doubt intending it for future barter. My colleague, the late R. von W. Suhm, believed that the natives on Wild Island recognized the native name of Humboldt Bay (Talok Lintju), and pointed in the direction of New Guinea, having knowledge of the place. Hence he thought that they visited the place to trade. I think, however, that he must have been mistaken. The Admiralty Islanders could never make a stand against a race armed with bows ; they would be cut off at once ; and had they once seen bows and arrows they would surely have adopted them. (The Australians have not done so at Cape York, though the Murray Islanders come to trade there and bring bows and arrows with them, but then they are. far lower in intellect, and have the throwing stick.) Many other circumstances concur against the above hypothesis.

The Islanders were anxious to trade with us to the very last, and followed the ship as she left the anchorage, with that intent. They were in a highly excited state, especially at first, and a man from whom I bought the first obsidian-headed spears I procured, fairly trembled with excitement as I handed him two pieces of tub-hoop.   The natives have no metals of their own, excepting the ore of manganese, with which they blacken their bodies.   This ore they call " laban," and they have adopted the same term for iron.   They appear unable to work iron at all, since they refused any pieces not of a form immediately applicable to use.   They preferred a small piece of hoop iron to a conical mass of iron weighing several pounds.

The natives are quieter than the Humboldt Bay men.   There was comparatively little noise when their canoes were alongside. There was no combined shouting.   The natives are rapacious and greedy, and very jealous of one another.   The chief showed all these traits in the highest degree.   They were ready enough to thieve, but not so constantly on the look-out for plunder, as the Humboldt Bay Papuans.

The men showed no great astonishment at matches or a burning-glass, apparently understanding the latter, and motioning that the operator should wait until the sun came from behind a cloud.   Looking-glasses were not at all understood. They were tried in all positions, as ornaments on the head and breast, for example.   The men seemed to see no advantage in seeing their faces in them.   In Labillardière's time they broke them to look for the picture, or man inside.   Tobacco and pipes were not understood.   Biscuit was eagerly taken and eaten. Great wonder was expressed at the whiteness of our legs and chests by the natives, and the women at Dentrecasteaux Island crowded with great curiosity and astonishment to look at a white arm or chest.   The natives, no doubt, thought our hands and faces only painted white, and took our negroes on board for men who had not got the paint on.

I am convinced that both the Humboldt Bay and Admiralty Island natives believed that we bought their weapons in order to use them as such.   They frequently, when offering spears, and bows, showed by signs how well they would kill.   No doubt they think the Whites are a race which cannot make bows and arrows for themselves.

The ship was anchored in ." Nares Anchorage," which is sheltered by a line of outlying reefs running parallel with the shore, and by numerous small islands.   On the line of

reefs are two larger islands named Wild Island and Dentrecasteaux Island, the latter being nearer to the anchorage.
These were the only two islands seen by us to be inhabited.

The land surface in the vicinity of Nares Bay, consists of a
series of low irregular ridges rising one above another, with
wide flat expanses at the heads of bays on the coast, which are
scarcely or not at all raised above sea-level, and thus are in a
swampy condition. The mountains appear, from their form, to
be volcanic; and it is probable that the obsidian used by the
natives for their spear-heads is procured in them. A trachytic
lava was found to compose one of the outlying islands; and a
similar rock was observed on the mainland where it commenced
to rise. A platform of coral-sand rock forms the coast-line of
the main island in many places; and a similar rock is the only
component of most of the small outlying islands.

From the position of the Admiralty Islands with regard to
the equator, their climate is necessarily an extremely damp one.
A great deal of exceedingly heavy rain fell during the stay of
the "Challenger." Rain fell on five days of the seven, during
which we were at Nares Anchorage, the total fall being 1·66
inch. The temperature of the air ranged between 86° and
75° F., the mean of maximum and minimum observations
being about 80° F.; and the air was loaded with moisture.
Dense clouds of watery vapour hung about the forest-clad
ranges, keeping the mountains most frequently concealed;
and in the evenings clouds of mist hung about the lower
land, looking like smoke rising from between the densely-
packed trees. In a bay some miles to the eastward of the
anchorage of the "Challenger," the mouth of a small river,
apparently the outlet of the drainage of the mountains on this
side, was found, and also a very small brook; but running
water was not elsewhere observed, and the rain probably drains
to a large extent into the swamps.

The main island, as viewed from seawards, is seen to be
densely wooded everywhere. Along the summits of the ridges
Cocoa-nut Palms show out against the sky, accompanied by
Areca Palms, as can be made out on a nearer view. The
general dark-green mass of vegetation on the hill-sides is fes-

tooned with creepers, and the smaller outlying islands dotted about in front of the main island, are all thickly wooded. The inhabited ones are distinguished at once by the large number of cocoanut trees growing upon them and forming the main feature of their vegetation.

I landed twice upon the main land. The trees where the shore is not swampy overhang the sea with immense horizontal branches. The bases of many of the trunks of these trees are constantly washed by the waves; but they nevertheless have large woody Fungi growing upon them, sometimes attached so low down that they are frequently immersed in salt water. The overhanging branches are loaded with a thick growth of epiphytes; and I had to wade up to my middle in the sea in order to collect specimens of orchids and ferns which hung often only a couple of feet above the water.

In other places the shore is swampy, and is either covered with Mangroves, or with a dense growth of high trees with tall straight trunks, so closely set that it was very sensibly dark beneath them. In such a grove near Pigeon Island, a small outlier near the anchorage, whilst the ground beneath is bare and muddy, and beset with the bare roots of the trees, the trunks of the trees and fallen logs are covered with a most luxuriant growth of feathery mosses and Jungermaninas.

On one of these tree trunks I found a very curious and rare Fern, known before only from Samoa and New Caledonia (*Trichomanes peltatum*). The fronds of the fern are circular in form, and, connected by a slender rhizome adhere in rows to the bark. They are pressed absolutely flat against the bark, so as to look like an adherent crust, and have all the appearance of a *Riccia* or some such Liverwort, for which indeed I took them, as I gathered specimens by shaving off the bark. A species of Adders-tongue Fern (*Ophioglossum pendulum*), unlike our humble little English form, grows in abundance, attached to tree stems with long pendulous fronds as much as a yard in length.

Most of my time during our stay was consumed in the collection of plants, since the Botany of the Admiralty Group was entirely unknown. Several of the ferns when examined at Kew, proved, as was to be expected in such a locality, of new species. Amongst the plants was a new Tree-fern; and one Orchid formed a new section of the genus *Dendrobium*.

All my spare time was devoted to studying the habits and language of the natives. I several times visited Wild Island, and roamed about with a native guide. The guides always went armed. The natives were much frightened and astonished at first at the sound of a gun. One of my guides, when I was shooting birds, stopped his ears at first, and bent down trembling every time that I fired. The natives were, however, not much scared by our firing our ship's guns and rockets at night, but came off next day to the ship to trade as if nothing had happened.

I obtained about 55 words and the numerals of the islands, and have published the results elsewhere in a paper, a large part of which is here reprinted, reference to which will be found at the end of this chapter. The difficulty of obtaining correct vocabularies from savages, of whose language the investigator is entirely ignorant, is well known, and has been commented on by many writers on anthropology and philology. I was well aware of these difficulties, and I used great caution, and believe that the words which I obtained are mostly correct.

I met with the following special difficulties with the Admiralty Islanders in obtaining words from them. The natives seemed always ready enough to give the names of particular birds which had been shot, as of two kinds of Pigeons and a Parrot, or of a Cuscus, Hermit-crab, or any such object which they considered was strange and novel to the inquirer, and one for which, as they thought, he wished to learn a name; but immediately they were asked for the word for the nose, or arm, or any such object common to the inquirer and themselves, they seemed to grow puzzled and suspicious, and to wonder why one could want to know the name of a thing for which one must have a name already. Some men were suspicious from the first, and refused sullenly to give any words at all, and prevented others from giving any. Some would give one or two only, and then refuse further information, whilst I came across two who gave me at least ten words each, quickly one after another, but then, like the rest, failed me.

I got a few natives together round some dead birds which I had shot, and gathered small stones and set them to count. The numerals are interesting, because those for 8 and 9 are expressed as 10 minus 2, and 10 minus 1.* In the process of learning the art of counting, a term for the numeral 10 has been reached by the natives, before 8 and 9 have been named. This method of forming the numerals 8 and 9 is known amongst other distant races, such as the Ainos, and some North American races, but apparently does not occur amongst either Polynesians or other Melanesians. It is, however, found in the language of one Micronesian island, Yap in the Caroline Group.

In counting objects, the natives clap their hands, held with the fingers pointed forwards and closed side by side, once when ten is reached, twice when twenty is pronounced, thrice for thirty, and so on. Up to 10 counting is done on the fingers, and after that, 10, 11, &c., are reckoned on the toes.

The idea of counting on the feet as well as the hands still survives in Great Britain. An Irish car-driver in Co. Mayo, a few winters ago, used the expression to me, "as many times as I could count on my fingers and toes" for a score. The use of the toes in counting is apt to seem extraordinary to civilized Europeans, who constantly wear boots and shoes, and sit on chairs. The majority of mankind who squat on the floor or ground, and have their toes generally exposed, and from their posture near to their hands, naturally pass to the toes in counting, after having exhausted the hands.

To express affirmation, the natives jerk the head up, as at Fiji. Negation is expressed by a most extraordinary and peculiar method. The nose is struck on its side by the extended forefinger of the right hand, the motion being as if the tip of the nose were to be cut or knocked off. This sign

* Admiralty Island Numerals. 1, Sip. 2, Huap. 3, Taro. 4, Vavu. 5, Lima. 6, Wono. 7, Hetarop. 8, Anda Huap. 9, Anda Sip. 10, Sangop. Jacobs, *l.c.*, p. 172, gives, See. Maruer, Tollo. Ear. Leme. Ouno. Andru-tollo. Andruruer. Andru-see. Songule. Thus, according to him, the numeral for 7 is formed in the same manner as that for 8 and 9. His numerals are no doubt from a different part of the Admiralty Group, and the method of spelling adopted by him is very different. They still correspond closely with those obtained by me.

was invariably used to express refusal of proffered barter, or that a native had not got some article asked for. It is capable of various modifications. The quick decided negative is given by a smart quick stroke on the nose. In the doubtful, hesitating negative, the finger dwells on its way, and is rubbed slowly across the nose.

The men average about 5 feet 5 inches in height, and the women about 5 feet 1. They contrast at once with the Papuans of Humboldt Bay, in being far thinner and lankier. Three men who were weighed, averaged only nine stone (137 lbs.) in weight. I saw but one native that was at all fleshy, although such were not uncommon at Humboldt Bay. Food is perhaps not so abundant here as on the New Guinea coast, and the natives have not, like the Papuans, the advantage of bows and arrows to kill game with.

The usual colour of the natives is a black-brown, often very dark, and darker than that of the Papuans of Humboldt Bay. The young girls and young boys appear much lighter as a rule than the adults. Some one or two of the young women were of a quite light yellowish-brown, as was also one young man, who came from a distance to the ship to trade. No doubt there is a mixture of blood, and the light coloured natives observed, belonged to the light coloured race described by Jacobs as inhabiting the eastern part of the main island and as constantly made war upon by the dominant black race.*

The hair of the head, which is worn long only by the younger adult males, formed in them a dense mop, projecting in all directions 6 to 8 inches from the head. It appeared less luxuriant in growth than that of the Papuans of Humboldt Bay. The hair is crisp, glossy, and extremely elastic, and every hair rolls itself up into a spiral of small diameter.

In general appearance thus it is fine curly, like that of Fijians. On comparing it with a very small sample of hair of the natives of Humboldt Bay taken from several native combs, the Papuan hair proves to be somewhat coarser, but in other respects the two hairs are closely alike, the diameters of the spirals of the curls being the same. Some hair from a

* Jacobs, *l.c.*, p. 176.

native of Api, New Hebrides, is of about the same coarseness
as the Admiralty Island hair, but the curls are of much smaller
diameter. The hair of the Api Islanders seems to be remarkable
for the fineness of its curls. In Tongan hair the curls are of far
larger diameter than those of the Papuan or Admiralty Island
hair.

The fineness of the curl of the hair in various Polynesian
and Papuan races which I have seen, seems to be pretty constant in each race and characteristic. It might be estimated by
measuring the diameter of the circles formed by the separate
spirally twisted hairs, and taking the average of several measurements. No doubt a certain curve of the hair follicles corresponds
with and produces the curl in the hairs, as in the case of the
hair follicles of the negro as discovered by Mr. Stewart.* But
the amount of curve will be peculiar to each race. The hair of
both head and body of the Admiralty Islanders is naturally
black, that of the head being of a glossy black.

The hair of the men's bodies was not at all abundant, nor by
any means so plentiful as it is often seen to be on the bodies of
Europeans, the hairiness of whom is apt to be underrated. I
lately saw in a travelling show an abnormally hairy Englishman.
His back and chest were covered with a thick growth of coarse
black hair, as thick as that of a gorilla. Unlike most abnormally hairy examples of the human race, the hair was not
continued over the whole body, but ceased at certain lines on
his arms.

The continuous covering ceased abruptly at these lines, but
beyond them were scattered small isolated hairy patches, which
formed a sort of gradation to the ordinary bare skin beyond.
These small patches, the tailing off of the hairy covering, were
regular hairy moles, such as occur so frequently on various parts
of the body in Europeans. It would seem therefore not improbable that such moles are in reality small patches of the
original coating of long hair of the ancestral man, a small spot of
the skin, returning by atavism to its ancestral condition.

Each organ and each histological tissue in animals and plants

* Charles Stewart, F.L.S., "Note on the Scalp of a Negro." Microscopical Journal, 1873, p. 54.

has its own special developmental history. May not many morbid growths and pathological changes in the tissues of higher animals and plants be regarded as instances of reversion in the particular tissues to a condition which was normal in their earlier history? In these the growth of the cells is, as in the embryo, more rapid and less closely restrained by polarity, so that the resulting masses are mostly without definite form.

Eyebrows were generally absent in the Admiralty Islanders, very probably shaved off; the natives made signs when offered razors, that they used obsidian knives for shaving.

I did not notice that the natives seen at Nares Anchorage had excessively large front teeth. This fact was observed by Miklucho Maclay. Figures are given by him of the teeth.

The septum of the nose in all the adults is perforated, and the lower margin of the perforation usually dragged down by the suspension of ornaments, so that in a profile view of the face the large aperture in the septum is looked through by the observer.

Some of the natives, as at Humboldt Bay, have most remarkably long Jewish noses. About 1 in every 15 or 20 has such a nose. I at first imagined that this form of nose was produced to some extent by long action of excessively heavy nose ornaments, but I saw one youth of only 16 or 17 with such a nose very fully developed, and I saw more than one woman with a well-marked arched nose with dependent tip, and the women appear to wear no nose ornaments.

The lobes of the ears of all the men were enlarged, being slit and dragged down into long loops by the weight of suspended ornaments.

The women wear as their only clothes two bunches of grass, one in front, the other behind. The men wear as their only dress, excepting a white cowry shell, occasionally a narrow strip of bark cloth about five feet long and six or eight inches in breadth, which is almost white when new and clean. The cloth is in the form of a long natural sac, open at both ends, being evidently loosened from the cut limb of the tree from which it is made by beating, and then drawn off entire. This cloth is sometimes reddened by being rubbed with a red earth used by the natives for smearing their bodies. No better native cloth was seen; and the natives apparently do not know the method of fusing the fibrous matter from several pieces of bark together, so as to form tappa, like that of Fiji or Tonga.

The hair in the women, young and old, is cut short all over the head, and worn thus simply, without decoration of any kind. In the boys, the hair is short, I believe cut short, as in the women. Only the young men of apparently from 18 to 30, or so, wear the hair long and combed out into a mop or bush. In the older men the hair is always short. There are probably religious ceremonies connected with the cutting of the hair, for the very large quantities of bunches of fresh-looking hair suspended in the temples are probably not all at least, if any, taken from the dead.

The mop of hair in the young men, possibly the warriors (though numbers of adults, still in full vigour, had their hair short), is carefully combed out, often reddened, and greased. A triangular comb is worn in it, also cocks' feathers which are bound together in plumes and fastened on to the ends of short sticks of wood worn as hair pins. Plumes of the Nicobar pigeon, or the Night Heron, are also thus worn.

It must be remembered that Pacific Island native ornaments are all made to show on a dark skin. White shell or tusk ornaments look exceedingly well against the dark skins of natives, although when removed and handled by Whites, they show to little advantage. The young girls amongst the Admiralty Islanders sometimes have a necklace or two on, but they never are decorated to the extent to which the men are. The old women have no ornaments. I saw one girl only with a necklace of the beads procured from the ship. Another girl had one of small unshaped lumps of wood, worn apparently rather as a charm than an ornament.

Amongst the lower races of savages, decoration follows the law which is almost universal amongst other animals. It is the male which is profusely ornamented, whilst the female is deprived of decoration. This condition is almost entirely reversed by civilisation, and the grade of advancement of a race may, to some extent, be measured by the amount of expense which the men are willing to incur in decorating their wives. The males in highly civilised communities revert to the savage condition of profuse decoration only as warriors or officials, and on State occasions.

Amongst the Admiralty Islanders, the decoration is almost entirely confined to the men, and these seem averse to part with any of their finery to the women.

The men wear armlets of *Trochus niloticus* shell, like those of Fiji, the Carolines, and elsewhere. They wear often seven or eight on each arm. The rings are neatly engraved with lines forming lozenge-shaped patterns, and form very effective ornaments indeed.

Circular plates, ground out of *Tridacna gigas* shell, are also worn, either as breastplates or on the front of the head. The discs are faced with plates of thin tortoiseshell, perforated with very elaborated patterns.

Long style-like ornaments of *Tridacna* shell are worn dependent from the nose. They are closely like those which, in the Solomon Islands, are worn stuck transversely through the septum nasi, but are here always worn dependent by a loop of twine. Ear and nose ornaments are also made of the teeth of the *Cuscus* of the islands, and crocodiles' teeth. The ears and nose septa are always perforated. Pieces of rolled-up leaf are worn sometimes in the ear (perhaps those of betel pepper).

Necklaces of native beads of shell or cocoanut wood are also worn. Rings of tortoiseshell are commonly worn in the ears, as at Humboldt Bay. Both waist-belts and armlets of fine plaited work, with patterns in yellow and black, are common. These resemble those of the Aru Islands and Humboldt Bay.

Charms composed of human bones, usually the humerus, bound up with eagles' feathers, are worn suspended round the neck, and hanging down the back between the shoulders.

The body is seldom decorated with green leaves, as at Humboldt Bay. But leaves are occasionally worn, both hanging down the shoulders and on the arms. I saw them once so worn. Flowers, also, are seldom worn, but a single *Hibiscus rosa sinensis* flower is occasionally worn in the hair.

The full-grown men are mostly marked with cicatrizations

on the chest and shoulders. These cicatrizations are in the form of circular spots about the size of half-a-crown.

Tattooing is almost entirely confined to the women, with whom it is universal. Two males, however, were tattooed. One, a small boy, had a simple ring-mark round one eye. The other, an adult, had rings round both eyes. These were, however, exceptional cases. The tattooing is not made up of fine dots or pricks, but of a series of short lines or cuts.* The colour is an indigo blue. The women are tattooed with rings round the eyes and all over the face, and in diagonal lines over the upper part of the front of the body, the lines crossing one another so as to form a series of lozenge-shaped spaces. The tattooing is sparse and scarcely visible at a short distance, and nowhere are the marks placed so close to one another as to form coloured patches on the body, as in Fijian women or Samoan men.

The male natives occasionally had their chests and faces reddened with a burnt red clay. Sometimes one lateral half of the face is reddened, the other being left uncoloured. When vermilion was given to the natives they put it on cleverly and symmetrically in curved lines, leading from the nose under each eye, showing that they understood how to use it with effect. I had expected to find Magenta a popular article of trade, but it was of no use at all. It is too transparent to show on a dark-brown skin, and the natives rejected it directly they tried it. No doubt the reason why they do not tattoo themselves is because the tattooing would show so little. Perhaps it is on account of their dark colour that Melanesians generally have adopted cicatrization as a substitute so largely.

No doubt the natives paint themselves elaborately on festive occasions and in war time. They were fond of being painted, and two natives who were painted on board all over with engine-room oil-paint, yellow and green, in stripes and various facetious designs, were delighted.

The natives were also often coloured black, the colouring matter used being an ore of manganese, which gives their bodies a metallic lustre, like that given by plumbago or boot blacking.

* Probably made with obsidian flakes. I am informed that the Soloman Islanders are tattooed with short cuts thus made.

The blacking was extended over the faces and chests. The old women were often blackened, and a group engaged in singing an incantation were all blackened. A man, who was possibly a priest, was always blackened over the face, arms and chest, and perhaps blackening has here a religious signification.

The natives nearly universally chew betel, using the pepper-leaf, areca-nut and lime together as usual. Some one or two men were observed who did not chew at all, and had no lime-gourds. The lime is carried in gourds of a different form from those used at Humboldt Bay, but perforated in the same manner at one end with a small hole through which the long spoon-stick is inserted. The lime is conveyed to the mouth with the stick. At Humboldt Bay the lime-gourds are not decorated. Here all the lime-gourds are decorated, but all with nearly the same pattern.

The use of kaava and of tobacco is entirely unknown to the natives.

The principal vegetable food of the islanders is cocoanuts and sago. The sago is prepared into a farine, and preserved in hard cylindrical blocks about a foot in height, and six or eight inches in diameter. Specimens of the preparation have been placed in the Kew Museum.

Taro (*Caladium esculentum*) is also eaten. It is cultivated in small enclosures adjoining the houses, but to a very small extent, and there are no large clearings or cultivation of any kind which leaves its mark on the general features of the vegetation of the islands as at Humboldt Bay, or Api, or Fiji. Plantains are grown sparingly round the houses. A Bread-fruit tree also grows about the villages. Several wild fruits, a Hog Plum, (*Spondias*) a small Fig, and the fertile fronds of a Fern are eaten by the natives, and they have a Sugar-cane of better quality than that used at Humboldt Bay. Young cocoanut trees are planted about the houses, and protected from injury carefully by means of neatly-woven cylindrical fences. They are also planted with care on the uninhabited islands.

The natives have no Yams (*Dioscorea*), nor Sweet Potatoes.

The flesh of pigs is roasted by the natives, and served for eating, placed on a quantity of the prepared sago in large wooden bowls, which are often elaborately carved. The Opossum of the islands (*Cuscus*) is also roasted, and is carried about cold, roasted whole with head, tail and legs intact, ready to be torn with the teeth and eaten at any moment. I saw no boiling being done, but the earthenware pots made by the natives were evidently used for that purpose.

There are wells on the inhabited islands; they are at some little distance from the houses. They are shallow holes dug in the coral ground. They are kept covered in with sheets of bark, and at each, cocoanut-shell cups are hung up for drinking.

The houses of the natives are built on the ground,[*] and always close to the shore. They are all of an elongate beehive shape, occupying an oval area of ground. On Wild Island they are built of a continuous wall and thatch of grass and cocoanut-leaves or similar material. They thus look like long haycocks somewhat.

In Dentrecasteaux Island many of the houses have their walls built up neatly of wood cut into billets and piled as fire-wood is in Europe. The roofs are similar to those in Wild Island. They are supported on two stout posts rising from the foci of the oval floor of each house, and by a regular framework of rafters, &c. Shorter posts, placed along the walls at intervals, support the roofs at their periphery and the walls. Very often the ground is excavated to the depth of a foot or so beneath the house, so that the wall is partly of earth, and one has to step down to get into the house.

The dwelling-houses are mostly about 20 to 25 feet long, 10 to 15 feet in height, and about ten feet in breadth. They have a low opening at one or both ends. To the main supporting posts of the roof are secured a series of wide horizontal shelves placed one above another, and on these shelves food, implements and weapons are kept. I saw these shelves in the women's houses. In some of the houses are also bed places, consisting of rough boards fastened against the side posts of the walls on one side, and supported by short special posts on the other. Arms and implements are suspended from the posts and rafters. The

---

* Jacobs, *l.c.*, p. 182, describes, as seen by him, "several large villages built on piles over the water," on the east coast of the Main Island.

dwelling-houses have no further furniture. The posts are sometimes curved and painted, and occasionally a human skull is fastened to a post, or placed under the thatch. Everything about the houses is rough, and there is no neatness as in Fijian buildings.

About the houses in the villages, bright-red *Dracœnas* are commonly planted as ornaments, representing the flower-garden in its most primitive stage. The temples are houses exactly like the dwelling-houses, but larger—about 20 feet long, 15 broad and 20 in height. Some have carved door-posts of wood, the respective carvings representing a male and female figure. The doors are closed by a kind of hurdle.

The canoes are more of the Polynesian than the Papuan form, *i.e.*, they have their bows and sterns low, and simply pointed, and not turned up and built so as to form figure-heads, as at New Guinea and the Aru Islands. The canoes' hulls are formed each of a hollowed trunk of a tree, with a single plank built on above it, and a gunwale-piece as a finish. The hollowed-out portion has slightly and equally rounded sides, and is not flat on one side and rounded on the other, as in the Carolines. The mast is stepped in the bottom of the canoe, just in front of the horizontal outrigger platform. A pole of about similar length, with a natural fork at the top, is stepped against the foremost end of the cross-bar of the horizontal outrigger, and it and the mast being inclined towards one another, the mast is fitted into the fork at the top of the pole, and roused down with a rope-stay so as to remain firm in that position. The bow and stern are ornamented with a simple carved ridge or two, and with *Ovulum ovum* shells, a single row of a dozen or so being fastened on either side. A horizontal outrigger extends from the middle of the canoe on one side, and is connected with a long canoe-shaped float, and opposite to it is an inclined shelf or deck supported on two or three stout projecting beams. A platform is formed with planks on the horizontal outrigger, and on the outer part of this a large store of spears and the mast and sails are kept. On the inner part the natives sit when not paddling, and stow on it some of their gear, food and articles for barter, but most of these are kept on the inclined platform, where also some of the crew often sit. The canoes are from 30 to 40 feet in length.

The sail is nearly square in form. It is hoisted to the top of the mast, and set so that one corner is uppermost. The opposite corner does not nearly reach down to the canoe, hence the square sail being high above the water has a very peculiar look when seen over the sea at a distance. As at all Pacific islands, apparently the outrigger platform is the place of honour, and the seat of the head-man or chief. Oto, the chief of Wild Island, never occupied any other position, and never touched a paddle.

Small canoes with single outriggers, holding one or two persons, are used for paddling about the reefs round the islands. The large canoes are manned by from 10 to 15 men.

The natives swim hand over hand. They never take a header in diving, but jump in after anything upright, sinking feet first with the body inclined forwards.

Long sein-like nets are used for fishing. These nets are probably the property of a community, for they are kept hung up in the temples. I saw one about a fathom in depth and of very considerable length. Hand nets fixed on elbow-shaped frames of wood are also used. Stake nets are used, and lines of stakes are conspicuous objects just off the shore near the villages.

Fish-hooks are used made of Trochus shell, all in one piece. They are of a simple hooked form without barb. The natives did not seem to care for steel fish-hooks, and apparently did not, at first at least, understand their use. It is possible that they have never found out the plan of using bait on a hook. All Polynesian and Melanesian fish-hooks which I have seen are of the nature of artificial baits of bright nacre, imitating small fry in the water. If the natives did not understand the use of baits, it is no wonder that they despised European fish-hooks.

The tool in most constant use by the natives is a small adze, consisting of a natural crook of wood with a *Terebra maculata* shell bound on to it, the shell being ground down until only one lateral half of it remains. Such small shell adzes were abundant enough still, but in most cases the shell had been replaced on the handle by

H H 2

a piece of hoop-iron. Every man almost carried one of these small adzes hung on his left shoulder. From the houses large adze blades made of *Tridacna* and *Hippopus* shell were obtained. They resemble somewhat those of the Carolines, but are very roughly made indeed, only the actual edge being ground. None were seen mounted, and they appeared to have gone out of use. Axes made of hard volcanic rock were also obtained from the houses. They have ground surfaces and are triangular in form, and resemble the stone adzes of the Solomons, but are mounted in an entirely different and very primitive way, as axes, being merely jammed in a slot cut in a club-like billet of hard wood near its end. Only one specimen was obtained mounted. These stone implements did not seem plentiful, and the natives valued them highly and required a high price for them; and when I at first showed them a Humboldt Bay stone axe, to try and explain that I wished to buy such from them, they were immediately anxious to purchase it themselves. The chief had a very fine large one, with which he would not part.

The heads of the obsidian-headed lances serve as knives, being cut off just below the ornamented mounting which acts as a handle.* Long flakes of obsidian are however also mounted specially as knives in short handles. They are excessively sharp, and used to shave with even, but are of course very brittle. Pieces of pearl oyster shell, usually semi-circular in shape, ground down thin to an edge on the rounded border, are used constantly as knives to cut cordage, and for similar purposes. Knives made of the spine of a Sting-ray (*Trygon*) are also used. Large ground pearl oyster shells are used to dig with.

The Admiralty Islanders have no bows, slings, or throwing sticks, ulas (Fiji), nor clubs. Their only weapons are lances of several kinds, which are thrown with the unaided hand, not even with a cord, as in New Caledonia. They have no spears, like the Humboldt Bay men, Fijians, and others, to be used at

---

* This is an interesting instance of the same instruments serving different purposes in a rude condition of the arts, other cases of which have been dwelt on by Colonel Lane Fox, F.R.S., Lecture "On Primitive Warfare," Journal of the Royal United Service Institution, 1867-9.

close quarters, and no shields, though Jacobs mentions shields as in use at other parts of the group.

The principal weapon is a lance formed of a small flexible shaft of tough wood, a natural stem often, with the bark trimmed off, to the thicker end of which is attached a heavy head of obsidian or volcanic glass, which, in size, appears out of proportion with the light shaft. The obsidian lance-head is usually of a conical form, but some of the weapons have a knife-edge in front, and some are irregular. They are shaped by bold wide flaking. The points and edges are often slightly re-chipped in order to sharpen them, but the original faces and angles are never worked up for the sake of symmetry or balance, but remain rough. Many lances have their edges and points sharp and perfect, though formed entirely by the original flaking. The hinder borders of the lance-heads are simply rounded. They are secured in a socket of wood attached to the end of the shaft by means of a cement, and by being bound round with fine twine.

Many of the lance-heads are of most irregular forms, remaining just as they happened to flake out in manufacture.

The heads of the lances are kept covered with a conical sheath of dried plantain leaf made to fit. The natives possess an enormous store of these weapons. They have piles of them lying on the outriggers of the canoes. On shore the men commonly carried two or three in their hands. In a dispute alongside the ship one of the lances was instantly snatched up and made ready. They are used for hunting wild pigs as well as for fighting. The natives pointed to the mountains of the main island as the source of the obsidian. They parted with the lances readily, and the material must be abundant. The lances are thrown in the usual manner, grasped by the naked hand, being first made to quiver by a shaking motion of the hand for some seconds.

Though there is an enormous abundance of Wild Pigeons at the islands the natives have invented no means of shooting them. They can only climb the trees and catch them at roost, or knock them off the nest.

The natives are extremely expert in wood carving, and show

most remarkable taste in their designs. The lance-heads are often carved. The carving taking the form mostly of incised patterns, the effect being heightened and beautified by the use of black, white, and red pigments.

The white coral lime, the red burnt clay, the black, possibly charcoal of some kind. The guardian deities carved on the door-posts of the temples and posts of the houses are ornamented also in the same style. Similar patterns are graved on the ovulum shells and armlets. These patterns are all modifications of the lozenge or diamond, and without curves; but besides this, various patterns are burnt in upon the surfaces of the chunam gourds, and in these the lozenge is combined with various curves.

An entirely different class of carving is that of the large wooden bowls which are used for eating out of. These resemble somewhat those of the Solomon islanders, being, like them, blackened, but in the present case they are most remarkable for their graceful forms and delicately carved handles. The bowls are worked with wonderful precision, considering the tools avail-able, to the circular form, appearing as true as if turned. They are widely open, and are provided with a pair of curved handles, which rise above the level of the tops of the bowls, and are some-times ring-like, sometimes cut in a delicate spiral. They are always ornamented with perforated carving, and often bear a pair of Crocodiles, or roughly executed human figures on their outer margins. The bowls stand always on four short legs, like the Fijian kaava bowls. They never have a circular bottom, no doubt because there are no level surfaces for them to rest upon, and because the idea is derived from a four-legged stool.

A still more remarkable appreciation of symmetry and fer-tility in design is shown in the patterns which are cut upon the circular plates worn sometimes on the forehead, oftener on the breast. These consist of circular white plates ground down out of *Tridacna* shell, with a hole in the centre for suspension. On the front of this white ground is fastened a thin plate of tortoise-shell, which is ornamented with fretwork, so that the white ground shows through the apertures. The patterns are of end-less variety, *no two being alike*, and show all kinds of combinations of circles, triangles, toothing, and radiate patterns. The shell back-ground is often graved also at its margin. Symmetry is evidently striven after, but with the appliances available the execution falls short here and there of the design. Nevertheless these ornaments are very beautiful. Closely similar ornaments are worn in the Solomon Islands, and also in New Hanover, and in the far-off Marquesas Islands, curiously enough.

A regular style of ornamentation is preserved for each class of ornaments, weapons, and implements. Thus I saw no *Ovulum* shells with curved pattern like those on the gourds. Both these and the bracelets bore simple patterns of diagonal lines graved and blacked. The spears, also, never bore curves.

The sticks or spoons with which the chunam is carried from the gourds to the mouth are often richly carved in the handle. The skulls of Turtles suspended in the temples are ornamented with patterns painted in the three usual colours. The human skulls are likewise decorated, and some have eyes of pearl shell inserted into the orbits on a background of black clay.

The musical instruments used are the Conch shells, perforated on the side as usual, a very simple Jews-harp, made of bamboo, of the usual Melanesian pattern, Pan-pipes, of three to five pipes of different lengths (the New Hebrides natives have Pan-pipes with three pipes), and lastly, Drums. These latter are hollowed out cylinders of wood with a narrow longitudinal slit only opening to the exterior. Some of them are small, 1½ foot or so in length, and are carried sometimes in the canoes. The larger drums I saw only in the temples. They are cylinders. 4 feet in height and 1½ foot in diameter, and are fixed upright at the entrances of the temples. There were four such at the four corners of one temple. The slit in these is not more than 4 or 5 inches broad, and I do not understand how the cylinders are hollowed out by the natives. Very similar drums exist at the New Hebrides, at Efate, *e.g.*, where they are stuck upright in the ground in circles.*

The natives seemed to have no idea of tune, they blew the notes on the Pan-pipe hap-hazard. The chief of Wild Island

---

* "A Year in the New Hebrides," by F. A. Campbell. Melbourne, George Robertson, 1873, p. 111, figure Fili Id Efate.

blew a child's tin trumpet with evident satisfaction. He appropriated it from one of his subjects, to whom I had given it, and came off to the ship standing on his canoe platform and blowing it with all his might, with three bright coloured cricket belts which he had purchased, put on one above the other round his middle. The drums were constantly sounded on Wild Island, often in the afternoon.

The women, both old and young, dance, moving round in a ring with a quick step. The men signified that they danced too, but were not seen to do so. I did not see dancing myself.

I saw some old women performing a kind of incantation. They sat on the ground in the yard of one of the houses, four of them sitting facing one another in a circle, whilst two sat outside the circle. They had their faces and bodies blackened. They uttered at regular intervals a chant, " ai aiai aiai aiai aiai umm." The commencement was shrill, in a high key, and the terminal " umm " was sounded low, with the peculiar humming lingering sound, just as in Fijian chants.

Polygamy is practised. Oto, the chief, told R. Von W. Suhm that he had five wives. I do not imagine that the aged are killed. I saw several aged miserably lean hags, one especially emaciated and disgusting to look upon, and also old men. On one occasion amongst a party of 42 natives in nine canoes there were two old men, one with grey hair, the other somewhat infirm. Children are carried by the women generally on the back, but sometimes on the hip astride.

The chief Oto pointed out one youth as his son, and took away presents which were given to him.

The village at Dentrecasteaux Island is fortified. A palisade about ten feet high, stretches right across the corner of the island, where the village lies, shutting this off from the landing-place. The path to the village led through a gate-like opening in the palisade, which seemed in not very good repair. The palisade was without ditch or embankment. The village itself was surrounded by a second wall, low, and crossed by stiles ; at Wild Island there was no fortification. The natives inhabit the small outlying islands, probably for protection from attack. Very few natives were seen living on the main land, and these few at one spot only. Former places of dwelling on the main land appear to have been abandoned. We saw no actual fighting, but in a quarrel about some barter alongside the ship, Oto, the chief, attempted to strike a native in another canoe from a distant small island. He was prevented by his own men, who held him back. The opposite party at once got their spears ready, and threatened him with them.

I saw no traces of Cannibalism, although an anonymous correspondent of the " Times " newspaper, writing from the ship, appears to have thought that he saw evidence of it, and Jacobs relates an instance of the occurrence of what he supposes was a Cannibal Feast.

There are several Temples in Wild Islands; they have already been partially described. One such had as door-posts a male and female figure roughly carved in wood, but elaborately ornamented with incised patterns and colour. Between the legs of the female figure was represented a fish. There are in the same figure black patches with white spots, which appear to mark out the breasts. The hair in both figures is represented as cut short, and thus the mop of hair of the warrior is not represented in the male figure. No clothes, i.e., T-bandage of bark-cloth, bulla shell, nor ornaments, such as ear-rings, nose ornaments, and breast-plates, are indicated on the figures, and the male figure has no weapons. The ears of both figures are, however, slit for ear-rings, and it is possible that a zone of diagonal ornament passing round the body of the male figure represents the plaited waistbelt commonly worn. On the upper part of the chest of the male figure are a series of circular white ring-marks on a black ground, which evidently denote the circular cicatrizations present in all the male natives. In the female figure the tattooing is possibly intended by a wide patch of diagonal ornamentation upon the abdomen, as also by lines drawn round the eyes, and not present in the male figure. In the male figure one lateral half of the face is painted white, and the other red. The arrangement of paint in this way is in vogue amongst the natives here as at Fiji; I saw one Admiralty man with one side only of his face reddened, and in Fiji, at dances, it is common to see natives with one lateral half of the face blue.

and the other red or black. All the ornamentation on the figures is of the common zig-zag pattern, and formed of a series of lozenge and triangular-shaped spaces. The patterns are incised, and coloured of three colours, black, red, and white. The parts coloured white and red are cut in, whilst the patches of original surface left in relief are blackened. Guardian deities, such as these, are common in Melanesia and Papua, as is also their combination with representations of fish ; carefully coloured drawings of the figures were made by Mr. J. J. Wild, artist of the Expedition, and my description of the figures is derived from these drawings.

Another temple had no figures, but the four large drums already mentioned. To the rafters and supports of the roofs of these temples inside are fixed up quantities of skulls of pigs and turtles, all arranged regularly, with the snouts downward. The skulls were decorated with colours. With them were suspended large quantities of balls of human hair, some evidently old, others of recent date : these balls or masses of hair were suspended sometimes in networks of string, sometimes in small receptacles of a very open basket-work. Both the bunches of hair and the skulls appeared often to have regular owners, though dedicated in the temple ; the natives parted with both freely for barter.

The hair is probably cut off as a religious ceremony ; some men had the hair recently cut off. A Dugong's and a Porpoise's skull were produced for barter. The natives evidently treasure skulls of all sorts. Human skulls are likewise kept stuck up in the thatch of the houses. At Dentrecasteaux Island, one having an ornament in the nose was suspended to the front of a house over the doorway by means of a stick thrust through holes in the two squamous parts of the temporal bone. This skull the owner could not be induced to part with, but usually they were sold pretty freely, and they were in considerable abundance about the houses, but often much shattered ; a dozen only were purchased. The natives are very superstitious. When a group was being photographed, the old women put up two long poles transversely between themselves and it in order to protect themselves from its evil influence, and they could not be persuaded

to sit until Captain Thomson seated himself in the centre of
the group, and was taken with them. When I began sounding
the big drums in the temple, my guides hastily drew me out of
the place in terror, and made signs that the people from the
chief's group of houses would come and cut my throat.

NATIVES OF THE ADMIRALTY ISLANDS WITH CAPTAIN F. T. THOMSON, R.N.
(From a photograph.)

A mystery was always made about the principal temple containing the images. Sometimes it was freely open, at others
closed, and I was warned back by the chief on two occasions
when I attempted to enter. The temple with the drums was
used for the suspension of the large fish nets, no doubt common
property.

The charm, made of a human *humerus* wrapped round with
feathers, and worn hung round the neck, was taken in the hand
and flourished about, dashed against the ground, and used
apparently to swear by during a violent harangue of one of the
chief men of Dentrecasteaux Island, who wanted possibly to
incite the natives to attack our boat, or to try and capture a
much coveted bag of trade gear in it. These feather and bone

charms are sometimes made of four human ulnar and radial bones, sometimes of hand bones, and one contained the bones of a large bird, probably the eagle (*Pandion haliœtus*).   It is a curious fact that one such charm which was purchased, contained an imitation head of a human humerus, cut in wood.   Possibly the owner intended to deceive his enemies by this artifice. Some of the officers told me that they made the natives readily understand when they wanted to visit the temple by pointing upwards.   It would appear thus that the gods or religious influence is supposed to reside above.

The only appearance which I saw of a religious ceremony was the chant of the old women.   One man who came off to the ship often, invariably with his body blackened all over with peroxide of manganese, was thought to be a sort of priest; he wore a narrow fillet round his head, with an *ovulum ovum* shell suspended from it on one side.

The dead are buried in the ground.   Two different natives, one on Dentrecasteaux Island, and the other on Wild Island, explained to me by signs in an unmistakeable way, that the skulls put up about the houses were obtained by burying bodies in the earth, and afterwards digging them up again.   The value set upon the skulls and bones as ornaments, and probably also superstitious motives, are no doubt the reason why no marks of burial were seen; no mark is made probably for fear of the bones being stolen.   Two at least of the skulls procured were those of females.

The fact that some of the men restrain themselves and abstain from the use of betel, seems to be a proof of considerable strength of character.   I gave a hatchet to a guide at Dentrecasteaux Island as pay, according to promise.   He seemed grateful, and presented me with his own shell adze in return, unasked, and he made signs that the others had got enough, and that we were not to give more away; that we were being swindled.

The natives delighted in being towed along in their canoes by the steam pinnace, and clapped their hands with delight; but of course did not understand how the boat moved, nor apparently see in the fire the cause of motion.   They came up to the cutter when sailing to get a tow for their canoes, and apparently expected to see the boat go off, head to wind, in the same style.

The inhabitants of each small island appeared to be under a separate chief, and quite independent of each other. The chief's power seemed to depend on his fighting qualities. The chief of Wild Island had considerable power. He ordered all the canoes away from the ship on the first evening of our arrival, on our anchoring. He took articles away from men to whom they were given, and made arrangements for each man of a party getting a hatchet. He never paddled himself, and he pushed canoes out of the way when approaching the ship. He, however, clamoured with the rest for presents and trade. He had no ceremonious respect paid to him at all.

The natives seemed friendly enough, but they were of course excessively excited at our presence. No doubt they were afraid of us. When a party, which landed with Captain Thomson on Dentrecasteaux Island, was putting off from shore in a small boat to reach the pinnace, the inhabitants seemed possibly to be meditating an attack, for they suddenly produced their lances and showed intense excitement; no doubt the sight of a sack full of trade articles in the boat was almost too tempting for them.

We were usually on very good terms with them. On one occasion Mr. R. Richards, Paymaster of the "Challenger," accompanied a number of natives in the chief's canoe, which was guiding a party to Pigeon Island. He took down the names of the whole crew.

The natives were very much frightened at some Goats which were offered to them by Captain Thomson and refused to let them be landed on the inhabited islands. They were very much scared also by a wooden jointed toy Snake which I showed them swaying to and fro; and evidently must be acquainted with poisonous snakes, as they made signs for me to kill the thing or it would injure me. A squeaking Doll, which kicked its legs and arms about, frightened the chief Oto very much, and he and others made signs at once to have the thing put out of their sight.

With regard to the population of the islands, I estimated that the population of Wild Island was about 400 or 500, and that of Dentrecasteaux Island about 250 or 300. This estimate for these two small northern outliers has unfortunately been mistaken* for an estimate of the population of the entire group, which may, perhaps, be conjectured to amount to about as many natives for the same range of coast line all round the main island. Jacobs describes the entire range of outlying islands and part of the coast of the main island as inhabited and in places densely so.

The most remarkable fact about the Admiralty islanders is that of their having no bows and arrows, slings, throwing sticks, or throwing cords for their spears, no ulas, clubs, spears for hand-to-hand fighting, and no shields. Many other Melanesians have no bows and arrows, as the New Caledonian Loyalty Islanders, and apparently the New Britain and New Ireland races, and the same is the case with the natives of the south-east of New Guinea; bows and arrows seeming to commence on the coast only at Humboldt Bay, but all seem to have slings or other additional means of defence.

The only domestic animals possessed by the natives of the Admiralty Islands in any abundance are pigs. These are partly kept in enclosures around the houses, partly run half wild over the inhabited islands. The pigs are small, lean, and black coloured, and appear never to develop large tusks. No ornaments of large pigs' tusks were seen in the possession of the natives. If therefore, as I believe, from signs made by the natives, is the case, there are wild pigs on the main island of the group, they must be unlike the Papuan pigs in this respect, and resemble more the New Hebrides breeds. Two Dogs were seen on Wild Island. I saw one of these a puppy. It was white, smooth haired, like a Fox Terrier in appearance, and very like a dog that was in the possession of the natives at Humboldt Bay. No dogs but these two were seen amongst the natives. No Rats were seen on any of the islands. No Fowls were seen in the possession of the natives, but I obtained a plume of cock's

* Behm und Wagner. "Die Bevölkerung der Erde," V. Petermann Mittheilungen, 1878, s. 48.

feathers worn as a head-dress from one native.   Fowls must
therefore exist in the islands somewhere, but are probably scarce,
as only this one plume was seen.

With regard to the Zoology of the islands, two species of
Fruit-Bats (*Pteropinæ*), and an Opossum (*Cuscus*), were procured.   A Dugon and a Dolphin are also killed by the natives.
Of birds the most abundant are the Fruit-Pigeons (*Carpophaga
Rhodinolæma*), which feed upon the Wild Coffee and Nutmegs,
and roost in vast numbers upon one of the small outlying
islands.   We saw or procured about 28 other species of birds,
including two Eagles, a Lory, and a Kingfisher, many of which
appear nearly allied to, or identical with those of the Echiquier
Islands.   They have been described by Mr. P. L. Sclater, F.R.S.,
who finds several new species amongst them.*

Small Tree-Swifts (*Collocalia*) fly about amongst the Cocoanut-trees, and all day flocks of Terns and Noddies (*Sterna
lunata, Anous*), follow in the still waters within the reefs the
shoals of Skipjacks (*Caranx*), as they pursue the smaller fish.
The shores are inhabited by several species of Shore birds.   I
saw on the main island a scarlet and black Parrot or Cockatoo
of some kind, which flew out of some high trees on the sea-
shore, screaming loudly, like a Cockatoo.   The bird was wary,
and I could not get a shot at it.   It reminded me at the time of the
rare *Dasyptilus pequetti* of New Guinea; it was of about that
size.   Of Reptiles, there are two species of Turtle common here,
Chelone midas and *C. imbricata*, the latter the source of the principal article of barter of the natives, tortoiseshell.   In the swamp
pools is a species of Crocodile, of which the natives are in great
dread.   There are also at least one species of Land and one of
Sea Snakes (*Hydrophidæ*), and the natives showed themselves
acquainted with danger of handling Snakes.   A Gecko and
blue-tailed Lizard (*Euprepes cyanura*) are also present and
abnndant.

I was interested in watching the Skipjacks chase small
shoals of young Garfish (*Belone*).   The little Garfish hotly pursued, dashed out of the water, and by violent lashing of their

* P. L. Sclater, "On the Birds of the Admiralty Islands," Proc.
Zool. Soc., June 19th, 1877.

tails managed to keep themselves above the water in a nearly upright position for a distance of several yards, as they moved swiftly from the danger; their motion seemed a step towards that of the Flying-fish.

The large Gar-fish, when startled, move along the surface of the water by a series of rapid bounds for thirty or forty yards at a time with astonishing rapidity, and are often to be seen dashing thus along when scared by a boat. I was told that in some of the Pacific Islands they not uncommonly cause the death of natives who, when wading in the water, are liable to have their naked bodies dangerously speared by the long sharp bony snouts of these fish. The fish merely bound blindly away from danger and strike such an obstacle hap-hazard, but their weight must render them very formidable to encounter in this manner.

The above account of the inhabitants of the Admiralty Islands is mostly reprinted from the "Journal of the Anthropological Institute" for May, 1877, where, in a paper on the "Admiralty Islanders," further details, and an account of the language is given.

Literature relating to the Admiralty Islands :—"An Account of a Voyage round the World in the years 1766, '67, '68, '69." By Philip Carteret, Esq., Commander of H.M. Sloop "Swallow." Hawksworth's Voyages. London, 1773, Vol. I.

Labillardière, "Relation du Voyage à la Recherche de La Perouse. 1791." Paris, an. VIII. T. I, p. 255.

The above translated by John Stockdale. London, 1800, Vol. I. p. 296.

"Voyage de Dentrecasteaux à la Recherche de La Perouse." Rédigé par M. de Rossel. T. I, p. 131.

Extracts from the above are to be found in general works, such as Waitz "Anthropologie," Meinike "Die Inseln des Stillen Ocean," &c.

"Scenes, Incidents, and Adventures in the Pacific Ocean, during the cruize of the Clipper 'Margaret Oakley' under Capt. Henry Morrell." T. J. Jacobs. New York, Harper Bros., 1844.

My attention was called to the above work by my friend Mr. A. W. Franks, F.R.S. The book is rare in England, but there is a copy in the British Museum Library.

CHAPTER XIX.

JAPAN. THE SANDWICH ISLANDS.

Tedious Voyage to Japan. Jinriksha Coolies. Worship of the White
Horse. Japanese Sight-Seers. Consulting the Oracle. Japanese Pil-
grims. Book Shops and Religious Shops. River Embankments. Rice
Fields. Houses of Wood and Paper. English Bed-room Exhibited at
the Exhibition. Money Boxes. Pilgrims and Priests. Interest
taken by the People in Tojins. Cold Water Cure. Painting of the
Face in China and Japan. Japanese Tattooing. Japanese Modes of
Expression. Japanese Pictures and Theatres. Barren Appearance of
the Sandwich Islands. Honolulu. Supremacy of American over Native
Productions. Principal Trees of Oahu Island. King Kalakaua.
Hawaian Burials. Visit to the Crater of Kilauea. Ponds of Fluid
Lava. Mode of Formation of Peles Hair. Lava Fountains and
Cascades. Recent Eruptions. Hawaian Hook Ornament. Its
Probable Religious Signification. Hawaian Stone Club. Affinities
between New Zealand and Hawaian Art. Inter-breeding on Isolated
Islands.

Japan, April 11th to June 16th, 1875.--The Admiralty Islands
were left behind on March 10th, and a most tedious voyage, of
a month's duration, to Japan ensued. The vastness of the
expanse of water in the Pacific Ocean in proportion to the area
of the dry land, was pressed most strongly upon our attention.
Though the course north lay across a tract, which on the map
appears so crowded with islands that it seems impossible at
first sight that a straight route through them can be marked out
without encountering one of them, the ship nevertheless arrived
at Japan without any land having been sighted during the
whole voyage from the Admiralty Islands.

A fact often brought home to me before, during the " Challenger's " cruise, was tediously forced on our notice on this
voyage to Japan, namely, that the inmates of a sailing ship on
a long voyage, suffer far more from too little than from too
much wind. We were constantly becalmed, and our steam

I I

power being only auxiliary, and coal being short, we had to lie still and wait, or creep along occasionally only at the rate of a mile an hour.

When the ship was about 400 miles distant from the Japanese coast, a flock of about 20 Swallows (*Hirundo rustica*) came to rest on the rigging. They were very tired, and allowed themselves to be caught with the hand. Yokohama was at length reached on April 11th.

At Japan I had the good fortune to become acquainted with Mr. F. V. Dickins, a barrister, practising at Yokohama, who is an accomplished Japanese scholar, and at the same time deeply versed and interested in all branches of science. I am mainly indebted to him for what little knowledge I gained of the country. I travelled with him overland from Kioto to Yokohama.

I have never met with any persons, whether naval officers or members of other professions, or ordinary travellers who have been to Japan, who did not wish to go there again, so charming are the people, and so full of interest to everyone is the country and its belongings.

No traveller can fail to be impressed by the great powers of endurance shown by the Japanese coolies. Two coolies will drag a man in a jinriksha a distance of 30 miles in six hours, along a road anything but good. The same two men dragged me at a fair pace 30 miles on each of two successive days.

When great speed is required, three coolies are taken, and as they run they encourage one another all the way with shouts, "quickly," "quickly," "now pull up," and so on, and when several jinrikshas are travelling together, the shouting reminds one of a pack of hounds in cry. The coolies only get from four to six shillings a piece for such a day's hard work.

I travelled more than 200 miles in this way with Mr. Dickins along the great military road between the two capitals, called the Tokaido (East sea road). The start was from Kobe. Here I was delighted to see a Sacred White Horse kept in a stall at one of the temples. The Japanese came up one after another and uttered a short prayer before the horse, clapping their hands reverently together in the attitude of prayer. Close by an old man sold small measurefuls of boiled maize to be given as offerings. I bought a measureful for the horse, which responded with alacrity to that form of worship, but I could not help going through the other form as well in memory of ancient reverence for the white horse in my own country.

There seems to be a parallel for everything European in Japan, even for the most out-of-the-way customs. At Kama Nisigamo, near Kioto, on the slope of a hill called Daimogiyama, is a huge representation of the written Character " dai " " great." This is cut out on the hill side.

I was told by a Japanese that once in a certain number of years an assemblage of persons collects together and holds a sort of festival, and clears the area of the Character from over-growth ; the ceremony thus exactly corresponding to the " scouring of the white horse." On certain occasions the Character is illuminated with lanterns so as to show out on the hill side at night. I have a Japanese coloured sketch of it thus lighted up.

The Japanese are extremely fond of gadding about, and of sight-seeing, and especially of beautiful scenery. Near Kobe is a very pretty waterfall. It is crowded, wherever a good view is to be obtained, with tea-houses and resting-places for pic-nic parties, and I never saw the place without plenty of holiday-making visitors. When visiting such places the Japanese express their delight, and describe the beauties of the scene in short poems which they write out in the evenings at their inns. A Japanese clerk of Mr. Dickins's, a Mr. Tanaka, who accompanied us on our journey and was a very pleasant companion, often wrote thus short poems about our day's doings.

One of the walks from Kobe is to the Moon Temple, which is perched at the summit of a steep mountain ridge, clad with beautiful woods. The climb to the temple is a severe one, up many hundreds of steps. I was amused to see a Chinaman and a Japanese toiling up together to the top, to consult the Oracle about some matter of business. It seemed extraordinary that a Chinaman, so sharp in business matters, should come so weary a journey to take the opinion of the foreign gods. Yet the two men were evidently equally anxious as to the result of their inquiry. The Oracle was consulted by shaking out a lot from a number of inscribed slips of wood packed in a case. The men received the case of lots from an attendant priest, and hastened off with it to one of the shrines.

From Kobe, the large city Osaka, is reached by rail. As we left the railway station at Osaka, a crowd of pilgrims was just entering it. The pilgrims were clad in white, and carried long staves, and had bottle-gourds of water or saki slung round their necks. They were returning from the holy shrines. A passer-by begged a blessing of one of these pilgrims who was lagging behind the rest. The suppliant crouched down in the street, and the pilgrim blessed him, making passes over him with his wand. This looked strange in front of a brand new railway station.

Pilgrimages are extremely popular in Japan. On the journey along the Tokaido, the road was thronged with pilgrims, going to the ancient shrine of Ise, the oldest temple in Japan of the Shinto religion, the ancient State religion of the country, of which the Mikado, descended from the gods, is the supreme head.

In one large town, which we reached at night, all the inns were full of pilgrims, and we had to journey 10 miles farther to find a resting-place. It was a curious sight to see a string of blind pilgrims on the road, travelling on foot, holding on one behind the other, and led by one man who could see.

In Osaka, I spent much of my time in the booksellers' quarter, where there is nearly a mile of continuous book-shops. I bought here a large collection of illustrated books. The shops of each kind of wares are mostly placed together in the city.

Most interesting are the shops for articles used in religious worship. Here rosaries of the forms proper to the various sects of Buddhism, are manufactured by the gross, religious pictures are sold, and small shrines of the various gods are supplied for domestic worship, with miniature altars, candlesticks, and in-cense-censers. To these also the family god can be sent, when shabby, to be regilt.

Beautiful miniature lacquered shrines are also made at the shops, containing the goddess Kanon or some other popular deity. The shrines close with a pair of small doors, and are sold in great quantities to pilgrims at the temples, which they visit; as, for example, at the Moon Temple near Kobe.

At one temple, that of Tennoji, near Osaka, was a children's shrine, which was hung inside with great quantities of the finest toys of all sorts, and bright holiday clothes, placed there as offerings by children.

From Osaka, the road to Kioto leads all the way along the summit of the great embankment of the Ogawa (great river). These earthworks rather reminded me of the great embankments of the ancient tanks of Ceylon. At intervals, there are sluice-gates to let the water in upon the rice-fields. The sluice-gates are at the bottom of wells, sunk in the centres of the embankments. In the ancient Cingalese embankments, there are similar wells sunk through the middles of the embankments to meet the outflow channels from the tanks which traverse their bases. I was shown such an arrangement at Anuradhapura, by Mr. Rhys Davids, who told me that its use was not understood by engineers.

The land along the road is in the very highest culture. A great deal of it was covered with yellow-blossomed crops of rape, whilst here and there were wheat crops. The straightness of the lines of planting, and the regularity of their distances from one another, was such as I have never seen approached elsewhere in any form of agriculture.

Amongst these crops were the rice-fields, usually small areas surrounded by low narrow banks of mud, made by the laborious process of placing lumps of mud side by side with the hands. These enclosures are turned into shallow ponds by letting water in if the level suit, or by pumping it in by means of a small portable tread-mill or an undershot wheel worked by the stream of the river, if the level is above that of the river. The field surface is worked up by means of a buffalo and plough into a pond of mud, and on this the rice is transplanted. The seed is previously sown broadcast in a small special plot, from which the birds are kept off by a scarecrow, as in England, but here representing the rice-straw rain coat and large mushroom-shaped hat of the Japanese peasant.

The distance to Kioto from Osaka, 32 miles, is run by the jinriksha coolies in from five to six hours. In the hotel at Kioto I had my first experience of a Japanese house. They are all alike in being entirely built of wood and paper. The partition walls are all of light lath lattices, fitted as sliding panels and covered with a tough tissue paper. Even these walls, such as they are, often do not reach up as high as the ceiling, so that everything that goes on or is said over the whole range of rooms upon each floor is plainly heard.

If care is not used, one is apt in stretching oneself at night to push a hand or finger through the wall into the next room. A square of paper and some rice starch put matters all right again, however. One must always take off one's boots in going into a Japanese house, and at theatres and restaurants they are ticketed, and a check is given for them as for umbrellas and coats with us.

The hotel was on the side of a range of hills overlooking the capital. Kioto, the Holy City of Japan, is by far the most beautiful city I have ever seen when thus viewed from the overhanging hills. Everywhere are groves of Cryptomerias surrounding the holy places and monasteries, and above the groves in all directions rise the high temple-roofs and porches.

A great exhibition was going on at the time of our visit. It was amusing in going round this to see the tables completely turned upon the English. One of the exhibits consisted of a couple of rooms with one side removed to show the interior. One of the rooms was fitted up as an English bed-room, and the other as a drawing-room, both completely furnished. These were very popular sights. The Japanese are intensely fond of strange sights, and when the English first settled at Yokohama long journeys were made to look at them and their houses and to watch their strange habits, and guide-books were published for the use of the sight-seers, in which all articles of furniture, all implements and utensils and articles of dress of the Englishmen were figured.

Early every morning in Kioto there is a tremendous clanging and booming of bells from the monasteries, mingled with beating of gongs, to call the monks to matins, and arouse Buddha and Kanon to listen to their prayers. There is a big gong in front of every shrine with a large heavy cord in front of it. As each private worshipper arrives he swings the rope and strikes the gong, to notify the deity that he is about to say his prayers.

The temples of the Holy City are thronged with devout worshippers, and the floors of the shrines strewn with offered cash thrown into them. The receptacles for offerings are not small boxes with a slit, as in England, but large manger-like troughs with mouths many feet long and more than a foot in width, and when a grand service is in progress, I have watched a perpetual rain of cash thrown into such a money-box from the crowd in front.

There is no lack of money-boxes in Japan, every holy tree and holy stone, in however apparently remote a spot, is garnished with one, and even the holy white horse at Kobe solicited offerings, with a box of his own. At one of the temples, we saw a row of country pilgrims who had just arrived, and were having a special service performed for themselves. They evidently knew nothing of the ritual, and a clerk stood by and told them the proper moments in the service at which they were to bow their heads to the ground. But the pilgrims could not fall in with the thing, and were perpetually bowing out of time, much to the excitement of the clerk and their own apparent annoyance.

Mendicant friars sat by the roadsides in groups, perpetually hammering small round flat gongs, and bawling out the oft-repeated prayer, " Namu amida butsu," " Holy Lord Buddha," whilst passers-by threw them coppers. These mendicant priests, with their uplifted hammers and open mouths, are common subjects for caricature in Japanese picture-books.

Other priests perambulate the town with large square-shaped wallets covered with silk hangings, suspended over their chests by a broad band passed round their necks. In these wallets they collect offerings of food. There can be no doubt in the traveller's mind as to the activity and reality of religion in the Holy City, it is impressed on him in some form at every turn.

Very few English travelled along the Tokaido about the time of our journey, because of the existence of the far cheaper and quicker route by sea, by means of a regular line of mail steamers. I was surprised to find that we afforded, towards the middle part of the great road where no open ports were near, in our own persons a gratis exhibition of very great interest.

I was especially worth seeing, since I had a reddish beard of some length. The Japanese consider beards and moustaches excessively ugly, and they even used to put false beards and moustaches, often red in colour, on the face-pieces of their suits of armour, in order to assist the warrior in terrifying his enemies.

It was amusing to watch the faces of the people in some of the towns as they glared at us. I saw one woman look as if taken suddenly ill, on meeting me unexpectedly at a corner. Others burst out into fits of laughter. Everywhere, the idea uppermost in the minds of parents, was, that we were a sight which the children should on no account be allowed to miss. Mothers darted into the back premises and rushed back with their children, and often when we were halting, came and planted them in front of us, and pointed out to the children with their outstretched hands the various points of interest in the Tojins.

I was, as Mr. Dickins said, a first-rate Tojin. "Tojin," originally meaning Chinaman, the only foreigner the Japanese knew, now means foreigner of any kind, and it is also at the same time a term of reproach, like the well-known Chinese "Fan kwai," "Aboriginal Imp." Impudent small boys shout "Tojin Tojin" at an Englishman in the streets.

The Japanese being a race invariably black-haired, and with a tolerably uniform tint of skin, are naturally somewhat astonished at the great diversity in appearance of so mongrel a race as the English, whose hair is of all possible colours, often irrespective of that of the parents, and whose skin varies in colour through so many different shades of brown, red, or milky-white.

The Japanese believe very strongly in the efficacy of natural hot-springs, and also of certain cold-springs. At some springs chapels are erected, and the patient combines the curative effects of prayer with those of the cold douche. I saw a number of bathers near Yokohama, standing one by one under a small intensely cold waterfall, coming direct from a spring. They were shivering and quaking, and half gasping half bellowing out with pain the prayer which had to be repeated a certain number of times before they came from under the spout. A stout healthy priest stood by to direct the ceremony and take the money.

The use of paint as an ornament in China and Japan, seems to me to be of considerable interest. In both countries the women regularly paint their faces when in full dress, of which the paint is a necessary part. The painting is entirely different in principle from that in vogue in Europe. The paint is not put on with any idea of simulating a beauty of complexion which might be present naturally, or which has been lost by age. The painted face is utterly unlike the appearance of any natural beauty.

An even layer of white is put on over the whole face and neck, with the exception in Japan, of two or three angular points of natural brown skin, which are left bare at the back of the neck, as a contrast. After the face is whitened, a dab of red is rubbed in on the cheeks, below each eye. The lips are then coloured pink with magenta, and in Japan this colour is put on so thickly, that it ceases to appear red, but takes on the iridescent metallic green tint of the crystallized aniline colour.

In modern Japanese picture-books the lips of girls will sometimes be seen represented thus green. I suppose the idea is that such thick application of paint shows a meritorious disregard of expense. It is curious that the use of aniline colour should have so rapidly spread in China and Japan. In China at least such was not to be expected; but it seems to have supplanted the old rouge, and it is sold spread on folding cards, with Chinese characters on them, at Canton and in Japan.

This form of painting the face seems to be exactly of the same nature as savage-painting, and possibly is a direct continuation of it. It is like the painting of our clowns in pantomimes. In China, the faces of men seem not to be painted at the present time, either on the stage or elsewhere; but in Japan, actors in certain plays are painted on the face with bright streaks of red paint, put on usually on each side of the eyes. The kind of painting is exactly that of savages.

It is a curious fact that this form of painting, surviving in adults on the stage, is still used elsewhere for the decoration of young children. It is quite common to see children on festive occasions, when elaborately dressed by their parents, further

FACE OF JAPANESE ACTOR.
(To show the mode of painting the face. From a Japanese Theatrical Picture-book.)

adorned with one or two transverse narrow streaks of bright red paint, leading outwards from the outer corners of their eyes, or placed near that position.

Such a form of painting possibly existed in ancient times in China. When a man of distinction was buried in China in former times, a certain number of servants were buried with him. Now, figures made of pasteboard and paper, about 3 feet or so high, are burnt at the funeral service in small furnaces provided for the purpose in the temples, together with cart-loads of similar pasteboard gifts, which are thus sent by the survivors for the use of the dead in the next world. Earthenware figures were similarly buried with great men in old times in Japan.

The pasteboard heads of these funeral servants and retainers are painted with streaks, some of which are put on in almost exactly the same style at the angles of the eyes as those of modern

Japanese actors. It seems a fair conjecture that the streaks on these heads are a direct survival of an actual former savage form of painting, which was once in vogue in China, probably used to make fighting-men hideous.

It is well known that primitive customs survive in connection with funerals all over the world with extreme tenacity. The numerous interesting survivals existing in the case of English funerals are familiar.

The accompanying figure of a Japanese actor's painted face is copied from a Japanese theatrical picture-book. The head of the Chinese servant is drawn from one which I bought at a manufactory of funeral properties in Hong Kong.

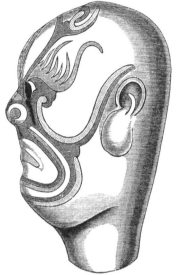

HEAD OF FIGURE BURNT AT CHINESE FUNERALS.
MADE OF PASTE-BOARD.
(To show the mode of painting the face.)

The Chinese are not now tattooed, but in Japan the art of tattooing has reached far greater perfection than anywhere else. Formerly all the coolies were tattooed, often all over the body, but now the practice is forbidden by the Japanese Government as barbarous, and it is a criminal offence to practise the art.

The tattooing was done by special artists, who made this their business. The outline of the subject to be tattooed is first sketched on the skin with great care with the point of a fine brush with Indian ink. The subjects are copied from printed pattern-books, which are very abundant in Japan, suited to all classes of decorative art.

The outline having been drawn, a light wooden handle, like that of a camel's-hair brush, is used, having about five or six fine needles set in its end in a straight line. The needles are dipped in Indian ink, and the fingers of the left hand being used as a guide, the outline is cut in on the skin by a series of punctures with the needles, which in the hands of a skilful operator travel rapidly along the lines, and leave them almost as fine as those sketched with the brush.

For sharp curves, handles with only two needles are used. For shading, handles with needles set in a variety of forms are employed, suited to producing broad flat tints, or, for example, pointed or rounded scales of dragons or fish. For the black parts of the design, Indian ink is exclusively used; it looks blueish when under the skin. Bright red is produced with ver-milion. A madder-colour is also used, and sometimes a yellow.

So rapidly is the work done that an elaborately finished design of a dragon or Japanese girl covering all the front of the forearm will be completed in a couple of hours. Very little pain is caused by the process, and not any or a little scarcely percep-tible bleeding. The area tattooed is slightly inflamed subse-quently, but not so as to cause inconvenience of any kind, and becomes quite healed in eight or ten days.

The results produced are astonishing in their softness, their correctness and delicacy of outline and minuteness of detail; and very far surpass those attained in tattooing by any other race. In a representation of a fish or dragon every scale is separately shaded, often with two strengths of shading, and in birds every feather is separately finished. In some cases large figures on the backs and shoulders of coolies are made to stand out in relief by means of an even dark shading, extending over the whole background. The artists recommend themselves to Europeans, by each asserting that he is the man who tattooed the Duke of Edinburgh.

With regard to gestures and expressions of the Japanese, I was struck by the entire absence of any gesture accompanying affirma-tion. A Japanese says "he," which means "yes," without moving the head at all or making any other sign. In negation the hand is sometimes shaken across the body with the fingers hung down. On receiving a present of money or payment, or a cup of saki, the hand is carried up with it to the forehead as a gesture of thanks. In salutation, or as an expression that a person feels honoured by the condescension of another, a curious half sucking half hissing noise is made by drawing in the breath with the lips partly closed, as if in sipping a fluid.

Large waxwork exhibitions are very popular in Japan. The figures are far better executed than European ones, and photographs of the faces of them would supply most perfect material for studying the facial expressions of the various emotions.

In some of the theatrical books, figures are given of the gestures to be used in declamation and in expressing the various passions.

Japanese picture-books are full of interest. Some of the most striking peculiarities in method of representation are closely paralleled in European art of a few centuries ago. The discharge of a gun or a cannon is represented as a long band of fire stretching from the muzzle to the object hit; and in a picture of a volley from a line of soldiers, a long streak proceeds across the page from every one of the muskets.

In engravings illustrating old Dutch travels, such as Barent's Voyage, a closely similar style is adopted; a line is to be seen drawn from the muzzle of a gun to the body of a Polar Bear, and the bullet is shown in mid-flight. Such a mode of representation survived in cheap European prints till quite recent times. I bought at a stall in London, not long ago, such a print representing the shooting of Marshal Ney, published in London in 1815, within a few days of his execution; in which similar lines are drawn from the muskets of the firing party, and all the bullets are shown on their course.

It is just possible that this method of representing discharges of fire-arms was derived from the Europeans by the Japanese, and is not an instance of the independent commission of a parallel error on their part. One of the most difficult problems in drawing is to separate what is actually seen from what is at the same time mentally present. Many a beginner looking at distant hills infers from their appearance that they are covered with trees, and proceeds to paint them green and cover them with detail, the result being failure. Only after practice does he detect the fact that hills seen at a distance are really blue, and that the details to be made out in a general glance are in reality very slight. No doubt it is from a similar error that the bullet is drawn in a representation of a discharge of fire-arms.

Art is employed largely in Japan in connection with religion.

Lives of the Saints, elaborately illuminated and illustrated, are executed on long rolls, or depicted on sheets arranged for suspension on walls. Similarly pictures of the various deities represented in groups, or singly, are suspended for devotional purposes, and many of them curiously resemble, in general appearance, early European representations of a similar character. Pictures are also suspended in shrines representing the nature of the prayer of the suppliant; as for example, a picture of a mother praying for her child. Pictures representing the pleasures of Heaven and torments of Hell are also common. These various religious pictures are sold in the vicinity of the temples.

The illustrations in many of the Japanese Zoological books are very interesting to a naturalist and remarkably complete. Even Land Planarians (*Bipalium*) are figured in some of them.

In a book in my collection, representing the doings of the Ainos, the Ainos are represented as hunting Seals, or Sea Otters, with bows and arrows from canoes. Some of the men are shown as provided with foxes' brushes tied by strings to the ends of short rods. The foxes' brushes are being caused to dance about on the surface of the water as a lure to the Seals who are following them in a shoal. Seals, or Sea Otters, must be attracted by lures of this kind, though it seems most improbable that they should. The figure almost certainly represents an actual occurrence.

I often visited the Japanese theatres. Besides the ordinary stage there is a second stage, consisting of a narrow platform, which lies on the left side of the audience, and extends from the side of the main stage, the whole length of the theatre, to a point close to the entrance door. Actors go round to the door behind the box seats, and appearing at the end of the long platform, approach the stage along it, acting their parts as they go.

In this way journeys are acted. A man may be represented as on a journey home, and at the same time his family are seen awaiting his return on the main stage, and he may be waylaid and murdered, for example, on the way; two separate but connected scenes being acted at once.

It is a curious fact, which I have not seen mentioned elsewhere, that the customary drink of Japanese women is simple hot water. I imagined that the Japanese were the only race that drink hot water; but I have lately been told, to my surprise, that it is the customary beverage of some old women in England.

The ship left Japan on June 16th for Honolulu. Notwithstanding all that has been written on Japan, the country and its people still remain almost as great a source of interest and field for investigation as does European civilization to the educated Japanese themselves. The English and German Asiatic Societies at Japan, showing as they do, a most remarkable activity, and constantly producing papers of the greatest value and interest in all branches of inquiry, have still probably the most fascinating field of research in the world before them.

The Sandwich Islands, July 27th to August 19th, 1875.—The ship reached Honolulu on July 27th, after an unsuccessful dredging between the Islands of Oahu (pronounced with stress on the penultimate), and Molokai. These islands of the Hawaian group are most remarkable for the extremely barren aspect which they present as viewed from seawards. In this respect they differ from all other Pacific Islands which were visited during the Voyage of the "Challenger"; no trees or shrubs form a feature in the view, but the hill slopes are covered with a scanty clothing of grass and low herbage, which in the summer season is yellow and parched.

Only one scanty grove of Cocoanut-trees is to be seen on the shore of Oahu Island, to the east of the town of Honolulu, whilst westwards the barren plains and distant bare hills recalled almost St. Vincent, Cape Verde Islands, in their sterility. Here are no thick belts of Cocoanut-trees fringing the shores as at Tonga, with littoral vegetation overhanging the very surf; no dense forests clothing the mountains from the summits to the shore as at Fiji, or the Admiralty Islands.

There is little more show of vegetation in the general appearance of the islands, as seen from seawards, than is to be seen on the bleak Marion Island in the Southern Ocean.

The harbour of Honolulu is entered by a narrow channel in a not very extensive fringing reef. The town lies on an almost flat expanse immediately adjoining the shore, and is not very conspicuous from the distance. It is composed of streets of very various widths, laid out at right angles to one another, lined on either side by very irregular rows of houses of all kinds, mostly wooden shanties, the greater part of them occupied as general stores.

There is a large shop of Chinese and Japanese curiosities, and two photographers' shops, where corals, imported mostly from the Marquesas, and spurious imitations of native implements manufactured for sale, are disposed of, at exorbitant prices to passengers by the mail steamers. I was told that a Chinaman is even employed to manufacture *fac-similes* of the stone gods of the ancient Hawaians for sale as genuine curiosities ; the forged deities being represented as having been dug up in taro-fields.

The business streets are very hot and dusty, but around the hotel and villa dwelling-houses on the east side of the town are pretty gardens, filled with the usual imported tropical garden plants, shrubs, and trees, which are maintained alive only by constant irrigation ; hoses from the town supply-pipes being kept playing on them day and night. Twenty years ago, where these gardens now are, there was not a single tree, and now the gardens form only a small oasis in a dry parched desert, which extends along the coast east and west, and which is soon reached on leaving the town in either of these directions.

On this tract, the bare volcanic rock shows out everywhere, and its only conspicuous vegetation is a Prickly Pear (*Opuntia*), introduced from America, which has spread far on either side from the town and multiplied exceedingly, so as in places to form a dense impassable growth, and constitute a most conspicuous feature in the landscape. These barren parts of Oahu reminded me somewhat of the rocky tracts of Teneriffe with their growth of *Euphorbia canariensis*.

The Guava, a second introduced American plant, has spread in all directions, in places forming dense thickets from which it is difficult to drive out the half-wild cattle. The whole town of Honolulu has a thoroughly American aspect. Americans are supplanting the rapidly decreasing native population ; American plants are, as has been said, covering the ground, and American birds have been introduced and bid fair to spread and oust the native avi-fauna, which has no single Land-bird in common with any other Polynesian Island group.

The only vigorous opponents of the Americans in the struggle for existence are the Chinese. The natives speak English commonly with a nasal twang, and I was much amused by a small Hawaian boy from whom I asked the way in the streets, who replied with the strongest twang, but with the utmost readiness, "I don't speak no English, I don't."

Behind Honolulu is a valley, called Nuuanu Valley, with precipitous walls in its upper part, which becomes greener and greener as the ascent is made by the road leading up it. The difference of rainfall in the valley, and in Honolulu, is most remarkable. At Waikiki near Honolulu, at sea level, the rainfall in 1873 was 37·85 inches, whilst in the Nuuanu Valley, 2¾ miles distant inland, and at an elevation of only 550 feet, the fall was in the same year 134·06 inches. Captain Wilkes even remarks that certain streets in the town of Honolulu are said to be more rainy than others.

The leading native trees in the valley, are the Malvaceous *Paritium Tiliaceum, Acacia Koa,* and the Candle Nut (*Aleurites triloba*). The *Paritium* forms curiously tangled impassable thickets. The *Koa* grows only high up on the cliff tops. The Candle Nut, by the peculiar glaucous colour of its foliage, gives a characteristic appearance to the vegetation. Its blue green trees seen in the far distance, appear as rounded bushes, dotted over the high ground above the barren shore region.

At the summit of the valley is the "pali," a narrow cleft in the tops of the mountains, which are on the other side precipitous. A beautiful view of the windward side of the island is here suddenly encountered, and a refreshing breeze blows through the gap. The range of cliffs forming the windward side of the mountain range, is an ancient coast line, and against the foot of the cliffs the sea beat in old time.

The visit of the King of the Sandwich Islands, Kalakaua, to the "Challenger," pleased me very much. The officers of the ship, donned, as in duty bound, full "war paint" to receive him, and even one member of the scientific staff appeared in curious clothes, and was girt with a rudimentary sword for the occasion, yet the Polynesian king arrived in a black frock coat, white waistcoat, and straw hat. To a confirmed "agriologist" the tables seemed completely turned on European civilization.

The king took the liveliest interest in the special work of the "Challenger," and was almost the only distinguished visitor of the many to whom I had exhibited microscopical objects during our voyage, who recognised the well-known anchors in the skin of the Holothurian *Synapta*, and named them at first glance. These anchors stood us in good stead at all the ports visited, and were described in all the colonial newspapers as belonging to the "Admiralty worm," supposed to be the most wonderful of the deep-sea discoveries of the Expedition.

There is a most excellent musical band at Honolulu, composed almost entirely of Hawaians, and numbering 20 or 30 performers, who execute complicated European music with accuracy and most pleasing effect. No one can doubt after listening to this band, that the Polynesian ear is as capable of appreciating the details of music as the European. It will be interesting to observe in the future, whether the Chinese and Japanese, whose music is so very different from that of Europe, and who profess to dislike Western music, and now at least much prefer their own, will develope a similar capacity, and changed appreciation in the future. The Hawaians seem to be ahead of some of our own colonists in the matter of music, and have a better band than existed at the time of our visit to New South Wales, even in Sydney.

Whilst the ship was at Honolulu, I visited the north-east side of the island, and collected at Waimanalo, on the estate of Mr. John Cummins, a series of native skulls from a deserted burial-place. The burials are amongst dunes of calcareous sand, and the bones are exposed by the shifting of the sands by the wind.

The burials are often on the sides of the gullies, between the dunes. They have probably been made in this locality, because of the ease with which the sand is excavated. Similar burials occur at various spots around the coast of Oahu, and I know of no place where so abundant material is ready at hand for the study of the skeletal peculiarities of a savage race, by the examination of long series of crania and skeletons, as here. Other burials occur in caves inland, where the bodies are found in a dried mummy-like condition.

All the bodies at Waimanalo were buried in a doubled-up posture. One which was exhumed with care *in situ*, was buried with the knees bent up to the chest and the head bent forwards, and was placed resting horizontally on the back. Chips and fragments of basalt were found around all the graves, but no implements of stone.

The ship moved to Hilo, in the island of Hawai, in order that a visit might be paid to the crater of Kilauea. A Petrel, possibly *Procellaria rostrata* which occurs at Tahiti, and a Stormy Petrel (*Oceanitis*), were seen about the ship between the two islands. These birds do not seem to be included in lists of the avi-fauna of the group. The appearance of the great volcano of Mauna Loa is most remarkable. The slope of the mountain, as seen from the sea, is so gradual that it seems impossible to believe that it rises to a height of nearly 14,000 feet above sea level. The cause of the peculiar form is the extreme fluidity of the lava, of successive flows, of which the mountain is composed. It has run out almost like water.

Kilauea is a secondary crater on the side of the Mauna Loa, at a height of about 4,000 feet. The island of Hawai is much more fully clothed with verdure than Oahu, and has none of the desert appearance of the latter. The journey to Kilauea is a tedious and monotonous ride. The ascent is so gradual that it is hardly perceived.

The track leads first through a fine belt of forest near the shore, and then emerges on a weary expanse of open country, entirely devoid of any fine trees, and mostly covered with a scanty, low, moorland-looking growth, with Screw-pine trees here and there. The track is scarcely marked on the bare surfaces of the lava flows, which look almost as fresh as if the lava had only set the day before. These surfaces are covered in every direction by ropy projections, curved lines of flow, and small rounded ledges showing where one part of the flow has run over another.

K K 2

The whole looks as if a vast quantity of melted pitch had been turned out of a pot suddenly and allowed to run and set hard.

It was getting dark before the hotel on the verge of Kilauea was reached. During the ascent a globular cloud was seen hanging in the air in the distance, and we were told by the guide that it hung over the summit of Mauna Loa itself, but we could not have told this, for the gradient being so gradual there was no appearance of any mountain at all. As night fell, this cloud, perpetually re-formed by condensation, was lighted up by a brilliant orange glow reflected from the molten lava in the great terminal crater, and the appearance was just as if a fire was raging in the forest in the distance.

With the evening appeared an Owl: I suppose the short-eared Owl (*Otus Brachiotus*), an English, European, Asian and African bird, but which is most curiously found in no other Polynesian group besides the Sandwich Islands. A Duck also rose from a small marsh. A species of Duck is described as visiting the islands from America, a distance of 2,000 miles.* Another species occurring in the islands has been described as peculiar to the group by Mr. Sclater from "Challenger" specimens. Since this latter Duck was formerly supposed to migrate to the islands from America, there may be some mistake also with regard to the other species.

Not far from the crater of Kilauea there are abundant woods of Acacia Koa trees and plenty of herbage, and no doubt Deer which have been turned out will thrive there and multiply rapidly. A few small Sandalwood-trees still remain uncut in the vicinity.

The crater appeared in the dark as a wide abyss filled with gloom, but in the distance were seen three or four glowing spots, reminding one of furnaces seen at night in the Black Country, and every now and again a jet of glowing matter showed itself thrown up from a lava fountain which happened to be playing at the time.

In the morning the crater was seen to be bounded by a

* Finsch und Hartlaub. "Beitrag zur Fauna Central Polynesiens." Halle. H. W. Schmidt, 1867.

range of cliffs all round, and at the bottom was a wide flat expanse of hardened lava which looked as fresh as if it had only just set. The crater has evidently been formed by the sudden falling in of a vast mass of rock resulting from the fusion and flowing away of the supporting rock below. A succession of secondary smaller cliffs round the margin of the crater-bottom inside mark where this process has been repeated several times, as after the crater has been filled to certain levels, and the lava has hardened, the support has given way over the greater part of the area on successive occasions.

The smooth surface of the lava within the crater was closely like that traversed on the journey from Hilo. It was cracked by contraction on cooling in all directions, and in all the cracks, at the depth of a foot or so, was seen to be glowing hot.

The well-known molten lake of Kilauea was at the time of our visit rather to be termed a pond, for a stone could easily be thrown across it. We stood on a low cliff overhanging it on the side from which the wind drifted away the stifling vapours exhaled from it, and threw stones into the pond of melted rock below. A low cliff bounded the expanse nearly all round. At the base of this cliff opposite us, in three places, a violent surging was constantly taking place, the melted rock being thrown up high above the cliff by violent discharges of gas from below.

The melted rock was thrown against the base of the cliff in waves which, as they surged against it, made a noise like that of waves of the sea beating similarly against rocks. There seemed no tenacity in the melted lava, it splashed about just like water. As the waves fell back from the bases of the cliffs, pendent coagulations of lava were formed for an instant, and hung in the glowing cavities like icicles, but were remelted in a moment by the returning waves.

The waves when thrown up were glowing brightly with heat. The lake, itself, was covered with a thin black scum of coagulated lava with red-hot cracks in it, and the whole scum moved slowly round under the influence of the ebullition taking place at one side as described.

Close by was another, but smaller pond, where, however the

churning up of the lava was more violent. It occurred here also as in the other pond, at the bases of the low bounding cliffs only. The waves dashed against the cliffs, threw their spray high into the air above them, and the wind carried part of this spray over the edges of the cliffs, so as to fall on the hard lava platform above.

The spray masses, cooling as they fell, formed in their track the threads known as "Pele's hair," like fine-spun green glass. Many of the threads could be picked up, each with the small mass of hardened lava still attached. These fallen masses are closely like drops thrown out of a pitch-pot. Some were nearly pear-shaped. Others, which had reached the ground before setting, or when only partially set, had coiled up into various forms as they fell, but nearly all showed an upright fine point, where a hair had been attached to them.

Pele's hair, thus formed, drifts away with the wind and hangs in felted masses about the rocks, and the birds sometimes gather it, and make their nests entirely of it.

Between the two ponds was a lava fountain, the one which had been seen playing the night before, but was now quiet. A lava fountain is a tall hollow cone; an extinguisher as it were, with a hole at the summit, which is built up of successive jets of lava thrown out of a hole, and hardened one over the other.

The surface of the cone looks as if built up of small masses of pitch thrown on to it hap-hazard one over another.

As the mouth of the cone contracts, the jet is thrown higher and higher, and the spray falling all around, covers the lava platform around with congealed drops of a lava rain, as it were. Each of these drops forms, like the spray from the waves, a Pele's hair.*

Over one of the ranges of low cliffs in the crater, a cascade of lava had poured, and cooling and setting as it flowed, had been drawn out into long ropes and rounded ridges which were twisted one over another, and formed a curiously gnarled and

---

* Mr. H. C. Sorby, F.R.S., had come to the conclusion from the observations on furnace slag that Pele's hair was probably formed in this manner with globules attached. "Nature," Vol. XVI.

contorted mass. Everywhere were complex ripple marks sharply moulded in the rapidly setting melted mass.

All over the lava surfaces were to be met with bubbles, many of them large, 4 or 5 inches across, blown in the surface of the hot lava by the escaping gases, and now set and covered by convex films of thin transparent lava like thin-blown green bottle-glass.

The following is an account of a great eruption of Mauna Loa, which has occurred since our visit, taken from the "Times" of April 3rd, 1877. "Hawaiian Volcanoes.—The 'Honolulu Gazette,' states that in the last 90 years there have been 10 great eruptions on Hawaii. That of February, 1877, is the eleventh of the series. On the 14th of that month Mauna Loa, which is nearly 14,000 feet high, sent out an immense volume of smoke that rose to a height of 16,000 feet, and spread out, darkening the sky, over an area of 100 square miles, and then a stream of lava started down the mountain sides, but the source dried up at the end of six hours, and the eruption ceased. The sight was grand while it lasted. Mr. C. J. Lyons writes from Wainea that the columns of illuminated smoke shot up with such velocity that the first 5,000 feet were passed inside of a minute. Ten days afterwards, early on the 24th of February, there was a submarine eruption 50 miles from Mauna Loa, near Kealakeakua Bay. Flames were thrown up from the sea, and numerous jets of steam arose on a line about a mile long, where the sea was from 150 feet to 400 feet deep, as if the crust of rock under the sea had been broken in a fissure to let the internal fires out. In many places lumps of lava were thrown up, and it was so porous, somewhat like pumice-stone, that while hot it floated away, but sank as soon as it became cold and saturated with water. Another rupture, doubtless a continuation of the submarine fissure, was traced inland from the shore nearly three miles, varying in width from a few inches to 3 feet. In some places the water was seen pouring down the opening into the abyss below, food for the fiery element. A severe earthquake-shock was felt by those living at Kaawaloa and Keei during the night of the eruption."

The characteristic gods of the Hawaians were not the Sun

and Moon and ocean gods which they had in common with
other Polynesians, but the offspring of the active volcanos, the
Goddess Pele and her train.   I sounded our guides to see
whether they had still any reverence for Pele, the ancient goddess of the mountain, but apparently, according to the teaching
of the missionaries I suppose, Pele and all other Deities of old
Hawai were completely identified in the guides' minds with
the Devil of Scripture.   There are, however, I was told on
good authority, plenty of Hawaians still existing, who have a
lurking reverence for, or fear of the old gods.

It cannot but be a source of regret that more of the old
Hawaian gods were not preserved, and sent to European
Museums, instead of having been burnt and destroyed, a
course which the missionaries found necessary.   Of most of
them, there remain only imperfect drawings.

One of the ornaments of the Hawaians, well-known to ethnologists, is a pendent of a curious shape,
something like that of a fish-hook.   It
is usually cut out of a Sperm-Whale's
tooth, and is worn by both men and
women, suspended round the neck by
means of a necklace composed of small
strands of plaited human hair.   The
reason for the peculiar form of the ornament has not been understood.   I
believe, from the examination of various
drawings extant, representing the
ancient temples of the Sandwich Islands, that the hook represents a sym-
bol for the head of a god.

HOOK-SHAPED HAWAIAN ORNAMENT.
Made of Sperm-Whale's tooth.

In Ellis's account of the Sandwich
Islands, is a figure of the Hare o Keave, or House of Keave, the
sacred depository of the bones of departed kings and princes, at
Honaunau in Hawai.*   Besides the obviously human-like gods,
represented as set up around this building, there are also shown
in the sketch posts of wood, near the tops of which are carved

* " Narrative of a Tour through Hawai, &c.," p. 153.   By William
Ellis. 2nd Ed.   London, Fisher & Son, 1827.

out, crescent-shaped objects surmounted by straight continuations of the posts.

The gods are all shown with widely-open mouths, so that their faces assume a sort of crescent shape, and on comparing

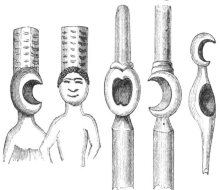

WOODEN GODS, FROM ELLIS'S SKETCH OF THE HARE O KEAVE.
Showing the gradations from the form of the human face to that of the crescent or hook.

them with the posts in question, it seems almost certain that these latter really represented also gods' faces, according to a sort of conventional mode of rendering them, or symbolic representation. Some of the images with well-marked human figures are shown with tall feather crowns on their heads, and together with them are figures with a mere crescent, to represent the face, yet wearing exactly similar crowns. One image has a simple crescent to represent the head, closely like that of the Hook-ornament.

A further figure of a Sandwich Island Deity, also from the writings of Mr. Ellis,* bears out this conclusion, as does also one of the plates of Captain Cook's "Third Voyage,"† in which Cook is shown seated at the base of a wooden idol, in order that he may be worshipped by the sacrifice of a pig. The idol is post-like in appearance, and with a wide crescent-shaped opening for a mouth. No doubt many of these post-like images were, when in use, decorated with ornaments and cloths, and thus, as in Tahiti, made to look more human in appearance.

* "Narrative of a Tour through Hawai, &c.," p. 437. By William Ellis. 2nd Ed. London, Fisher & Son, 1827.

† "A Voyage to the Pacific Ocean." Pl. 60, Vol. III, p. 13. Cook and King. London, G. Nicol, 1785.

The Hawaian gods, made of wicker-work, covered with feathers, show a similar curving inwards of the face. I give a rough sketch of one in the British Museum collection. In one

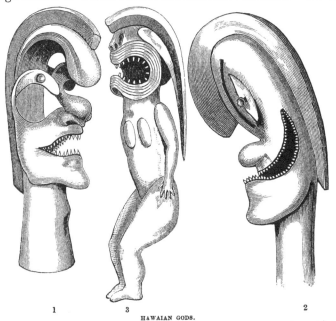

HAWAIAN GODS.

1 and 2 Heads of gods made of wickerwork, covered with feathers. 1 From "Cook's Third Voyage"; 2 Sketch of a specimen in the British Museum; 3 Entire god, copied from Ellis's "Narrative."

figured in "Cook's Voyages,"* the face is entirely hollowed out, and the eyes are borne on small flaps, projecting from the hook-shaped back part of the image, which mainly represents the well-known crested helmet worn by ancient Hawaian warriors.

In some instances, the hollow crescent form, as representative of the face, seems to have been arrived at by an enormous increase in the size of the mouth; in others, as in the case of the wicker image just described, by a hollowing out of the face altogether; the mouth here, though large, not being widened so as to encroach upon the whole area of the face. Since, in the worship of the gods, food was placed in the mouths, the mouths may have been gradually enlarged as the development of the

* "A Voyage to the Pacific Ocean." Pl. 67, fig. 4. Cook and King. London, G. Nicol, 1785.

religion proceeded, in order to contain larger and larger offerings, and the head in the wicker-work image may have been hollowed out for a similar purpose.

All voyagers who saw the Hawaian images, speak of their huge mouths. Lisiansky, evidently describing the same images

MORAI OF TAMEHAMEHA AT HAWAI ISLAND.

as those figured by Ellis, says that some of them bear huge blocks of wood on their heads, and have mouths reaching from ear to ear.* In the accompanying figure of the burial-place of

* "A Voyage Round the World in the years 1803, 4, 5, and 6," pp. 106–107. By Urey Lisiansky. London, 1814.

the Hawaian kings, the god on the left hand shows an extraordinary development of mouth.*

The Hawaians, in common with other Polynesians, recognized a Moon Goddess, " Hina." The crescent-shaped faces figured by Ellis, look almost as if they might possibly have represented such a Moon Goddess ; but there seems to be no evidence in favour of such a conjecture.

With regard to the hook-shaped ornament, Cook writes : " Both sexes adorn themselves with necklaces made of small black cord, like our hat string, often above a hundred-fold, exactly like those of Wateeoo ; only, that, instead of the two little balls, on the middle before, they fix a small bit of wood, stone, or shell, about two inches long, with a broad hook, turning forwards at its lower part, well polished."† " And sometimes a small human image of bone, about three inches long, neatly polished, is hung round the neck."

Captain King writes : " Both sexes wear necklaces made of strings of small variegated shells, and an ornament in the form of the handle of a cup, about two inches long, and half an inch broad, made of wood, stone, or ivory, finely polished, which is hung about the neck, by fine threads of twisted hair, doubled sometimes a hundred-fold. Instead of this ornament some of them wear, on their breast, a small human figure, made of bone, suspended in the same manner.‡

The form of the ornament was thus a matter of invariable usage already in Cook's time. No similarly formed ornament appears to occur in any other Polynesian Island. Nearly all examples of the ornament in museums are of Sperm-Whale ivory. I have seen one wooden one ; but none of stone. They seem all closely alike in form ; but in the British Museum and Christy Collections, there are necklaces made of a number of small Hookornaments strung on the same strands side by side.

From the accounts cited it appears that human figures were

* The figures extant of this Morai vary very much, no doubt partly because taken at different times. The one in " Byron's Voyage," when compared with Ellis's, seems however to be simply excessively badly and carelessly drawn.

† " Cook's Third Voyage," Vol. II, p. 232.

‡ Ibid., Vol. III, pp. 134–135.

worn in the same manner as the Hook-ornament, as if the one
ornament were a substitute for the other.   The Hawaians habitually carried their gods to battle with them, and in the plates
of " Cook's Voyages " several deities are represented as borne in
each fully manned canoe.   Hence it seems probable that the
people would wish to carry a representation of a god constantly
with them, and the comparison of the form of the Hook-ornament
with that of the recent-shaped and hollow-faced images of gods,
seems to leave little doubt that the hook represented the head
of a god; and thus as a religious emblem, suspended round the
neck, corresponded to those in vogue in the case of so many
other religions.   It may thus well be compared to the well-known
jade " Tikis " of New Zealand, similarly worn, which, however,
represented ancestors and tutelary deities rather than gods.*

It must have been a matter of great labour to work hard
ivory or stone into the form of the Hook-ornament.   The curves
in all examples seem to correspond closely; and there is a ridge
on the outer-curved surface of the hook, which appears to
represent the crest of the helmet.   The necklace and ornament
is termed in Hawaian " Lei palaoa," simply " whale's tooth "
necklace.

These speculations as to the meaning of the Hook-ornament
will, I hope, elicit further information on the subject.   General
Lane Fox has rendered familiar to ethnologists the curious tran-
sitions of form which representations of the human faces may
undergo in savage decoration under the process of successive
copyings.   The details of the representation gradually dwindle
away; a mere simple transverse crescent remains to represent
the entire face of a man on some of the paddles of New Ireland.†

Many similar degenerations of form in copying of decoration are well known ; and a well-marked instance is to be seen
in the crockets on the pinnacles of the Bodleian Library at

---

* The origin of tattooing in Polynesia is supposed possibly to have
been from the desire to mark the body permanently with the figure of
the tutelary deity.   Waitz, " Anthropologie der Naturvölker," 6ter Th.
Leipzig, 1872, s. 34–35.

† General A. Lane Fox, F.R.S., " Address to the Department of
Anthropology."   Report of the British Association, 1872.

Oxford. Towards the bases of the pinnacles the crockets are carved in the form of well-defined gurgoyle-like animals, with open mouths ; but in tracing the successive crockets upwards the shape is seen to degenerate gradually in each until towards the tops of the pinnacles the crockets have merely a sort of scroll-form, the origin of which could not possibly be guessed if it were looked at separately.

It seems probable that a very large proportion of what appears, in savage art, to be mere simple pattern ornamentation is in reality derived originally from degeneration of outline drawings representing natural objects. The lowest savages, such as the Australians, excel far more in their drawings of animals and men than in their pattern ornaments on their weapons, and the earliest attempts at art known are drawings of animals, such as the well-known one of the Mammoth cut on its own ivory by contemporaneous man.

At Hilo I obtained from some natives a short stone club,[*] which appears to have been hitherto unknown as a Sandwich Island weapon, and is interesting as approaching in some particulars the New Zealand "Mere." It is made of basalt, with carefully ground surfaces, and is about 10 inches in length. It is cylindrical in form with three sharp edges at the striking end, and was slung to the wrist by a string passed through a hole at one end. It was called "pohaku newa," "stone club."

My attention has been drawn by my friend Mr. A. W. Franks, F.R.S., to the resemblance between the Hawaian images of gods and the New Zealand human images. The accompanying figures are given for comparison. It will be seen that there is in them a similar extraordinary increase in the size of the mouth, which encroaches upon and renders insignificant the remainder of the head. Mr. Franks is of opinion that, as far as regards the special development of art, and forms of implements of use amongst the New Zealanders, that people are nearly allied to the Hawaians, certainly more nearly so than to the Samoans, from colonists of which race Hall supposed that the Maoris were sprung. The stone adzes of the New Zealanders

* H. N. Moseley, "Note on Stone Club." Journal of Anthropological Inst. 1877, p. 52, Pl. XVIII.

are of the same form as those of the Hawaians, and both differ for example from those of Tahiti.

NEW ZEALAND WOOD CARVING OF HUMAN HEAD.
To show the huge size of the mouth, from which the tongue is seen hanging down. (From the stretcher of a canoe in the Ashmolean Museum, Oxford.)

NEW ZEALAND WOOD CARVING OF HUMAN HEAD.
To show the large size of the mouth and concavity of the face. (From a specimen in the British Museum.)

The affinities of the New Zealand language appear to show that the ancestors of the Maoris reached New Zealand from Raratonga, and it appears that Hawaiki, the distant land of which their tradition spoke, is the religious name of the mythical land of origin of the whole Polynesian race, not to be identified with any particular island.*

The well-known posts with images carved on their tops, set up in the fences around New Zealand houses, may well be compared with the somewhat similar posts set up round the temples in the Hawaian group. In many cases, rough blocks of wood on the tops of the New Zealand posts, evidently represent the carved figures with which the other posts associated with them are surmounted, in the same way as the crescent-shaped notches in the Hawaian posts represent heads of gods. In New Zealand, however, images of the actual gods were not made or

* "Die Inseln des Stillen Oceans." C. E. Meinicke. Leipzig, Paul Frohberg, 1875. 1. Th., s. 312.

worshipped ; the images made, represented ancestors or tutelary deities only.

There were, according to the Government census of December, 1872, 438 lepers at the leper establishment in the Island of Molokai. There can be no doubt that the races inhabiting all the isolated Polynesian Islands must have sprung originally from a very small stock, which arrived there probably hap-hazard in canoes, or possibly sometimes in larger vessels. Hence the races must have been produced by close interbreeding, and only very rarely, if at all, can any extraneous blood have been interfused by the arrival of further waifs.

May not this circumstance be connected in some degree with the extreme liability of the Sandwich Islanders to the attacks of leprosy ?

A similar close interbreeding must have occurred in the case of the animals and plants inhabiting isolated islands. No doubt many islands may have been colonized by plants which have sprung from only a single seed transported by birds, or other- wise. Similarly no doubt, all the birds of a species present in an island or group, may have in many cases been the produce of a single pair ; at all events they must certainly have often been the produce of very few pairs ; such interbreeding would be expected to have left its mark on insular floras and faunas.

The Government Library at Honolulu, contains a splendid collection of Voyages and Travels relating to the islands, and also of sumptuous illustrated works on Natural History, mostly from the library of the late Mr. Harper Pease, the Conchologist.

For a Catalogue of various works, including Zoological, Geological and Botanical treatises relating to the Sandwich Islands, see Catalogue d'Ouvrages relatives aux Iles Hawai, par William Martin. Paris, Challamel Ainé, Rue des Boulangers 30, 1867. The List, which forms a somewhat thick octavo volume, is not by any means complete, but contains an im- mense amount of information.

For the Land Shells, see Harper Pease, "On Polynesian Land Shells." Proc. Zool. Soc., 1871, p. 449.

For a detailed account of the Volcanoes and their Geological Pheno- mena, see W. T. Brigham, "Notes on the Volcanic Phenomena of the Hawaian Islands." Memoirs. Boston Soc. Nat. Hist. Vol. I, p. 341 ; ibid., p. 564. Also, J. W. Nichol, F.R.A.S., "Note on the Volcanoes of the Hawaian Islands," Proc. R. Soc. Edin. 1875-76, p. 113.

CHAPTER XX.

TAHITI. JUAN FERNANDEZ.

Death of Rudolph Von Willemoes Suhm. Scientific Papers and Journals left by Him. Papeete. Excursion into the Mountains. Fly-Fishing in a Mountain Stream. Uses of the Wild Banana. Vegetation Composed mainly of Ferns. Camping at Night. Tahitian Mountain Map. Ascent to 4,000 feet Altitude. Petrels Nesting at this Height. Their Possible Influence in Distribution of Plants. Ignorance of the Natives Concerning the Mountains. Mode of Alternation of Generations in the Mushroom Coral. Structure of Millepora. Structure of the Stylasteridæ. Catching Land-Crabs. Tahitian National Air. Juan Fernandez. Preponderance of Ferns. Destruction of Trees. Gunnera Chilensis. Conspicuous Flowers. Humming Birds of the Island. Their Fertilization of Flowers. Smallness of the Island Compared with the Number of Endemic Forms. Endemic Palm. Dendroseris.

Tahiti. Society Islands, September 18th to October 3rd, 1875.— The voyage to Tahiti occupied a month. It was painfully impressed upon the memories of us all by the death of Von Willemoes Suhm, which was caused by a rapid and virulent attack of erysipelas. Rudolph von Willemoes Suhm had been, before he joined the "Challenger" Expedition, assistant to the illustrious Professor von Siebold of Munich. He had distinguished himself by his researches as a naturalist before he joined the "Challenger." A list of his papers published during the voyage will be found at the end of this book, as well as a reference to a collection of his letters published in German after his death.

He left many descriptions of animals and drawings, some complete, others only partly finished. They comprehended about 72 plates of octavo size and a few drawings of larger size. Amongst these there are 13 of Annelids, mostly from the deep sea. About 50 are of Crustacea, including five showing the development of *Euphausia* complete from the *Nauplius* stage ; six

L L

illustrating the development of two species of Sergestes, and three on the development of Amphion. Four are of Pteropods. One of these, labelled by Von Suhm as *Chionider Pteropod*, is a most remarkable form, with large eyes borne on long stalks. Von Suhm was uncertain whether it was to be regarded as the larva of a new form of Cephalopod. It has two arms only, apparently homologous with the tentacular arms of Decapod Cephalopoda.

Besides these drawings Von Suhm left two closely written volumes of zoological journal in German and one volume in English. It is to be hoped that the German journal will be published in due course. It cannot but contain much most valuable matter. Besides this work Von Suhm constantly kept during the voyage the tabular record of the results of the deep-sea dredging in an official book which was called the Station Book.

Von Suhm had been, when a boy, an ardent collector of birds, and some of his first publications were on European birds. He took constant interest in birds during the voyage, and his last excursion on shore was at Hilo, Hawai, in pursuit of the interesting endemic birds of the islands with a native guide. Almost the last notes that he wrote were some on the Sandwich Islands relating especially to the birds.

I sat with him during the whole of the "Challenger" voyage, working day after day with the microscope at the same table. I am very greatly indebted to him for information in all branches of zoology, and especially in the matter of zoological literature, of which he had a most comprehensive knowledge. I also learnt very much from him in the way of method, and I feel that I shall always remain indebted to him for a decided push on in my general scientific training.

He was a most indefatigable worker. He was full of hope for the future, and, no doubt, could he have published his journal himself, would have established a reputation as a man of science, which would have been far greater than that which he most deservedly possessed at the time of his death.

The harbour of Papeete in Tahiti was reached on September 18th.

The beauty of Tahiti, as seen from the sea, is not to be over-rated. It forms a most striking contrast to the barren-looking Oahu. One of the first sights I saw on landing was a party of Frenchmen starting off into the mountains to shoot wild pigs. One of them was laden with long French loaves. Another led a dejected-looking mongrel dog by a large rope tied round its neck, and a third had his body encircled by the usual huge horn, without the assistance of which a Frenchman cannot go out shoot-ing even partridges at home. I little expected that so much of Parisian manners would not have worn off at the distant Tahiti.

The Tahitians appeared, as far as could be judged from so short an acquaintance, to dislike their French rulers, and seemed to like Englishmen all the more by contrast. Some natives grew suspicious and less friendly at once because they found that I could speak French. Possibly if the English were in the position of rulers they would lose their popularity. The natives have remained mostly Protestants, notwithstanding the efforts of Roman Catholic missionaries during the French occupation.

Tahiti is the principal colony of France in the Pacific, and even New Caledonia is under the rule of the head government at Tahiti.

Tahiti is wretchedly supplied with provisions. The Guava bush has overrun all the lower country and covered it with scrub; hence there is scarcely any pasturage. Cattle are procured from the Sandwich Islands, and it depends on the kind of weather which the sailing-vessels that bring them meet with, whether they are worth eating or not when they arrive.

We bought for the use of our mess at Papeete the most miserable specimens of sheep that I have ever seen. They had come from Easter Island which is now principally occupied as a sheep run, the inhabitants having been largely deported to Tahiti, where some of them are employed as household servants, the men waiting very well at the dinner-table in European dress. The sheep had been long on the voyage, and were so miserably poor that one of them only weighed about ten pounds when skinned. Pork is the only animal food which is cheap and plentiful at Tahiti.

One of the greatest treats to the natives is tea and bread-and-butter. A Chinaman keeps a restaurant to which Tahitian girls are taken by their lovers in order to consume these luxuries. Wheaten bread is greatly appreciated by Polynesians, and a baker is one of the first tradesmen who finds a profitable business amongst the natives on any of the islands when in process of civilization. There was an English baker on Tongatabu, he being almost the only White retail dealer established there. He told me he sold a great quantity of bread to the natives.

I made an excursion up into the mountains in search of plants. Some of the mountains rise to a height of over 7,000 feet, and I hoped to be able to reach a considerable altitude in the search of mountain forms. It was settled that at all events I was to reach the head of a valley called Papeno in the interior. I was provided with native guides; one, an old man, supposed to be thoroughly acquainted with the mountains.

I started with Lieut. Channer and F. Pearcey, our excellent bird skinner and factotum. The men carried our little baggage on the ends of poles, resting on their shoulders, like Chinese coolies. The practice of this method of carrying has been remarked upon as one of the many evidences of the Polynesian affinity of the New Zealanders. We traversed the beautiful valley of Fataua, closed at its head to the view by the irregularly peaked outline of the mountain, termed by the French, from its form, the " Diadem."

The stream of the valley pours over a high cliff, which bars the valley across in a very beautiful waterfall. In the cliff beneath the falling water is a wide hollow, overhung by the rock above, and in this Tropic Birds nest, and two or three were constantly to be seen, flying about the cliff and across the deep chasm of the valley, conspicuous against the dense green foliage and dark rocks. Very good strawberries were growing in a garden just above the fall. The plants were mostly in blossom, only a few fruits were ripe. The Mango trees in the island in the same way were mostly now in blossom, or with young green fruit. The orange season was just at its end.

The stream is full of small fish (*Dules Malo*) one of the Perch family. The fish have adapted themselves entirely to a fresh-water life, and rise to a fly like trout. Captain Thomson and the others of us who were fishermen, got out our fly rods and whipped the stream, catching a few dozen. The stream falls over the rocks and stones in small runs and stickles just like a trout stream, and the fish thrive in the rapid water. I carried my salmon and trout rods round the world with me, but the last place at which I should have looked forward to throwing a fly in, was Tahiti.

The first camp was made in the head of Fataua Valley, at a height of about 1,600 feet, amongst the "Fei" or wild Plantain, *Musa uranascopus*, a species which occurs also in Fiji and elsewhere in Polynesia according to Seemann, though I do not know whether the fruit of the wild plant is equally good in other places to that of Tahiti. The plant is closely similar in appearance to an ordinary large Banana tree, but the large bunches of fruit instead of hanging down, stand up erect from the summit of the stem. They are bright yellow when ripe.

A fire is lighted and a bunch of these wild bananas is thrown into it. The outer skin of the fruits becomes blackened and charred, but when it is peeled off with a pointed stick, a yellow floury interior is reached, which is most excellent eating and like a mealy potato. This is one of the very few plants which, growing spontaneously, and in abundance, affords a really good and sufficient source of food to man. Hardly any improvement could be wished for in the fruits by cultivation. It could not but be most advantageous that the plant should be introduced into many other tropical countries. On our way up the valley we had passed numerous natives, going down to Papeete with loads of "Fei."

Rats live in the mountains, and climb up and devour the ripe Bananas, and the groves of the trees are traversed in all directions by the tracks of wild pigs, which likewise feed on the fruit. It is strange that the pig should run wild and thrive, under such widely different conditions as it does, and should be able to exist equally well on wild Plantains in the warm Tahiti, and on Penguins and Petrels in the chilly Crozets. In this power of adaptation it approaches man.

It had been raining heavily during our walk, and was still pouring when we halted, and we were all wet through. The guides soon built a small waterproof hut, with sticks and the huge wild Banana leaves. Then they put up another small roof of leaves, and finding dry dead Banana leaves under the shelter of the freshly fallen ones, soon lighted a fire under the roof, and we dried our clothes in the smoke before nightfall, in the midst of the heavy rain. The Banana leaves afforded further waterproof covers for our clothes and for my botanical drying paper.

We had brought no blankets with us, because I wished to make the utmost attempt to scale the mountains as far as possible, and had therefore reduced the baggage to a minimum. I had not expected that we should suffer from cold as we did. The thermometer showed, at about half an hour before sunset, 75° F., about an hour later, 68°·5, at midnight 63°·0, at daybreak 60°·5, and in about half an hour after daybreak it rose to 61·5°. The main stream of the valley running past the huts, had a temperature at daybreak of 65°·0, having retained throughout the night the heat of the former day, which the air had so rapidly lost. The effect of the stream on the climate here, is thus just the opposite of that of the streams of such an island as Tristan da Cunha.*

From this camp, the way led over several steep minor ridges in the head of the valley, and then up to an elevation of 3,000 feet, which was reached on one of the extremely narrow ridges, characteristic of Tahiti, situate just to the west of the " Diadem." From the ridge, a descent was made into the Punaru Valley by the aid of ropes fastened to the trees. The precipitous side of the valley which we thus descended, was covered at this elevation, from about 3,000 to 2,000 feet altitude, with a dense vegetation, composed almost entirely of ferns. A Tree Fern (*Alsophila Tahitiensis*) formed a sort of forest to the exclusion almost of other trees, and with this were associated huge clumps of the giant fern, *Angiopteris evecta*, and masses of the Birds-nest Fern (*Asplenium nidus*). With these grew a trailing Screwpine and a Dracœna, but the three ferns together formed a greater proportion of the entire vegetation than I have observed to be the case elsewhere.†

---

* See page 111.

† This statement concerning the preponderance of ferns in the vegeta-

The second camp was made at an elevation of about 1,800 feet, at a native hut in the upper part of Punaru Valley. The natives have not forgotten their religion since the time of Darwin's visit.* Our guides said their prayers every evening before sleeping, even when huddled together out of the rain, all repeating the words together, and the native family at the hut did the same. The temperature at this hut sank at daybreak to 59° F. We suffered much from cold in the night, and still more from Mosquitos. We had an old piece of canvas lent us to spread on the ground to sleep on, but we crept together under it for warmth.

In the morning we attempted to cross over a high ridge at the head of Punaru Valley, and so reach our destination, the Papeno Valley, but the attempt failed, the guides, after an elevation of about 3,000 feet had been toiled up to, proving not to know the way at all. One of the guides had been over the pass many years before, but all he seemed to know was that he had been up a stream, so we spent the day in wading through pools and clambering over slippery boulders in the stream beds, creeping along under the overhanging branches. We kept making attempts in various impracticable places, and at last made a hurried descent in the evening into the valley, and had to prepare a camp almost entirely in the dark, and in heavy rain, at a height of 2,500 feet.

This was above the limit of the growth of the wild Banana in any abundance, so the shelter for the night was made of the fronds of the Birds-nest Fern (*Asplenium nidus*). These are tougher and more durable than the leaves of the Banana, and hence are used for permanent thatching, but from their smaller size require much more time in arrangement.

We had to put up with a very small hut, which sheltered our bodies as we lay down, but would not cover our legs, and had to feel in our baggage in the pitch darkness for our food, and eat it by the help of the sense of touch alone. The unfortunate guides who had constructed our hut first, could find

tion of Tahiti is referred to by Mr. Wallace from my MS. " Tropical Nature," p. 269.

* C. Darwin's "Journal of Researches," p. 411.

scarcely any more fern leaves in the dark, and they squatted out the night together, sheltered from the rain by a small extin-guisher-shaped erection, which looked as if one human body could not be forced into it, much less two. The temperature here at daybreak was 60° F., and the morning being cloudy, and the camp lying in a narrow gorge, it remained the same for an hour and a half after daybreak.

In the morning we descended again several hundred feet, and sent back to the hut and procured two young men, supposed to be practised mountaineers, and, as we thought, certain to know the way about every pass within four or five miles of their dwelling. One of them, as a proof of his knowledge, brought with him what I suppose is the most primitive form of a map. It was a thick stick of wood about a foot and a half long, with two short cross pieces on it at some distance from the ends, and on each of these cross pieces were set up three short uprights of wood.

I give a figure of it from memory. The uprights represented moun-tain peaks, and the spaces between, the valleys.

The new guide held his map

TAHITIAN MOUNTAIN MAP.

in his hand and took long con-sultation with his brother, and then explained matters thoroughly to our former guides. He clutched the uprights one after another and dilated upon them, pointing out the peaks to which they corresponded. There seemed no doubt we had got hold of the right man at last.

The guides now lashed our small baggage on their backs, instead of on poles as before, since this mode of carriage was no longer practicable, owing to the steepness of the ascent, and we started up the face of an extremely steep-sided ridge, a spur of Orofena, the highest mountain of Tahiti. At the lower part, we pulled ourselves up by means of the trailing Screw-pine, which covered the ground with a tangled mass of its long serpentine stems so thickly, that as we climbed over it, we did not reach the ground beneath by a yard or more.

Near the summit of the spur, the face of the ridge was almost perpendicular, and one of the men got up by the help of the bushes and let down a rope by which we reached the crest. In order to collect plants, I had to hold my knotted handkerchief in my teeth and fill it. It was impossible to get at a vasculum. The crest of the ridge was nowhere more than a yard wide, often less. There was an almost sheer fall on either hand, and if grass and small bushes had not been growing at the edge on each side, it would have been very difficult to walk along the ridge without becoming giddy. It was as if one was walking along the top of an immensely high wall.

Here and there, small *Metrosideros* trees grew upon the centre of the crest of the ridge, and when these were encountered, we had to climb between the branches, often where they over-hung a sheer drop below, and once we had to swing ourselves along the steep side of the crest for a short distance past one of these trees under its overhanging branches.

We ascended the crest of the ridge, until we had reached an altitude of 4,000 feet, when the guides found the way barred by a precipice and entirely impracticable. The summit of the ridge was covered' with a thick growth of the fern *Gleichenia dichotoma*, and a climbing fern (*Lygodium*), and amongst the bushes on the ridge a Whortleberry (*Vaccinium*) was very abundant, and also two species of *Metrosideros*. The entire vegetation was different from that below. One of the species of *Metrosideros* was however also seen growing much lower down.

Just as the ridge met the face of the mountain, by which we were brought to a halt, its crest widened out, and here there was a damp hollow with mosses and lichens growing in it, in great abundance. Here also grew a tree (*Fitchia nutans*) belonging to the Compositæ, with a large yellow flower. The tree was 20 feet in height, and had a trunk nine inches in diameter. It is allied to the Composite trees of Juan Fernandez, being nearly related to the Chicory.

Here in the soft loose soil, amongst the moss, were numerous burrows of a Petrel, I believe *Procellaria rostrata*. The natives call the bird " Night-bird," just as the inhabitants of Tristan da Cunha call the Burrowing Petrels there "Night-birds." The Tropic Birds also nest far up in the mountains, and in Hawai they nest in the cliffs of the crater of Kilauea at an altitude of

4,000 feet.   Similarly a Puffin (*Puffinus nugax*) nests at the top of the Korovasa Basaga mountain, in Viti Levu Island, Fiji,* and in like manner, a Procellaria breeds in the high mountains in Jamaica.

It seems to me possible that these birds may carry Alpine plants as seeds and spores attached to their feathers from one island to another, for great distances.   They make their holes in the ground where it is densely covered with herbage, and often become covered with vegetable mould.   The *Procellaridæ*, widely wandering as they are, have probably had a great deal to do with the wide distribution of much of the Antarctic flora. Grisebach† lays stress on the range of the Albatross (*Diomedea*) from Cape Horn to the Kurile Islands, as possibly accounting for the occurrence of Northern species of plants amongst the Southern flora, and also the wide range of the Antarctic flora. He supposes the seeds, however, to be swallowed by the Albatross, with its food, after being washed down into the sea by rivers, and perhaps swallowed by fish.

When I mentioned the matter of the birds possibly picking up seeds whilst nesting, and so conveying them, to Mr. Darwin in conversation, he at once objected that at nesting time these birds, like all others, do not wander, and do not fly to a fresh nesting place directly after nesting.   It seems to me, however, that though this objection is almost fatal to the suggestion, occasionally birds may leave an island with mountain seeds attached, and alight in the higher parts of a distant island from habit.   The fact that they do nest amidst the mountain flora, is at all events to be noted.

With regard to the Albatross, it is to be noted that at Tristan da Cunha these birds nest in the terminal crater, at a height of 8,000 feet.   Former Albatrosses may have nested in high tropical mountains ; the plants are possibly very much older than the present species of Albatrosses.   The great Albatross has, on a very few occasions, been found as a straggler, north of the

---

* Finsch und Hartlaub, "Ornithologie der Viti, Samoa, und Tonga Inseln." Halle, 1867. Einleitung, S. XVIII. Peale describes the habit in question of Procellaria rostrata at Tahiti.

† A Grisebach, "Vegetation der Erde," Bd. II., S. 496.

equator in the Atlantic, and has reached Europe. It is most extraordinary that the bird has not established itself permanently in the Northern Atlantic. The genus probably, once extended north in the Atlantic, as it does in the Pacific, for a form possibly ancestral has been described by Prof. Owen as *Cimoliornis Diomedeus*, a fossil bird nearly allied to *Diomedea* which occurs in the lower chalk at Maidstone.* The immense rapidity of birds' flight must always be borne in mind in considering their aid in distribution of plants. A journey of 4,000 miles, at 40 miles an hour is only four days and nights' flight.

As the date of sailing of the ship was uncertain, we were obliged to give up the attempt to reach Papeno Valley, and we therefore returned to the native hut for the night. The sky being remarkably clear, the thermometer sank at daybreak to 55° F. (elevation 1,800 feet). We followed the Punaru Valley down to the sea-shore, and returned to Papeete, along the coast. I am much indebted to Mr. Miller, English Consul at Tahiti, for his kindness in hunting up guides for me, and otherwise assisting me.

Mr. Darwin refers to the fact mentioned by Ellis, that long after the introduction of Christianity into Tahiti, wild men lived in the mountains, whose retreat was unknown. The ignorance of the natives concerning the interior of the island is still, as shown by the failure of our guides, extreme. The guides living on the spot, did not even know on which side of the valley to attempt to scale the ridge at its head. The men can climb extremely well, but they do not seem to have any idea of thinking out a route and judging it as seen from a distance, which is the real art of mountaineering.

The natives are still grateful for favours, as in Mr. Darwin's time. The older of our guides brought me just as the ship was leaving, as a present, a fine stone adze, which he had been at considerable difficulty to procure from Punaru Valley, where it had been found in the earth, he knowing that I wished very much to obtain one. The stone adzes are now scarce, and fetch their full price in Tahiti.

---

* "Trans. Geol. Soc.," 2nd Series VI, Tab. 39, fig. 2. "Quart. Journ.," 1846, II, p. 101.

The orange, lemon, and lime, which grow wild all over Tahiti, do not appear to deteriorate at all in quality, nor in quantity of fruit, although in the ferine condition. The fruit almost appears finer and better for running wild. The oranges we all pronounced the best we had ever eaten. The limes lay in cartloads upon the ground, rotting in the woods. It would pay well to make lime-juice for export in Tahiti. Some native insect must have adapted itself completely to the blossoms of the orange tribe as a fertilizer, so abundant is the fruit. Vanilla, which is cultivated in the island with success, requires, as everywhere else away from its home, to be fertilized by hand.

A Mushroom Coral (*Fungia*) is very common all over the reefs at Tahiti. After much search, I found one of the nurse-stocks from which the disc-shaped free corals are thrown off as buds, as was originally shown by Stutchbury, and confirmed by Semper, who considers the case to be an instance of alternation of generations.*

Though the free corals were so extremely numerous, I could only find the one nurse mass. It, as in Stutchbury's specimen,† consisted of a portion of a very large dead *Fungia*, to which were attached all over numerous nurse stocks in various stages of growth. Some of those in the specimens have only just developed from the attached larva, and have as yet thrown off no buds. A small cup-like Coral is formed, and as it grows the mouth of the cup widens and assumes somewhat the form of the adult disc-shaped free Coral, but is still distinctly cup-shaped. A line of separation forms in the stem of this bud, and the bud falls off; a fresh bud then starts from the centre of the scar left by it on the stock, and the process is repeated. The fresh bud in its growth does not spread its attachment over the whole surface of the old scar, the margins of which persist as a dead zone around its base.

The line of separation of the second bud does not correspond with that of the first, but is beyond it a short distance.

---

* Semper, "Generationswechsel bei Steinkorallen," Zeitschrift für Zoologie, 22. Bd. 1245. Leipzig, Engelmann, 1872, s. 36.

† G. Stutchbury, "An Account of the mode of Growth of Young Corals of the Genus Fungia." Trans. Linn. Soc., Vol. XVII. 1830, p. 493.

Hence, the nurse stem which has thrown off several buds, is transversely jointed in appearance. Some of the stems in the

DIAGRAM REPRESENTING A NURSE STOCK OF THE MUSHROOM CORAL.

a b Successive joints of the stem which have each thrown off a free discoid coral; *d* young mushroom coral still attached to the last joint of the stock; *c* a transverse line marking where the present bud will separate.

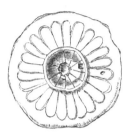

ENLARGED VIEW OF THE SCAR LEFT ON THE END OF THE STOCK WHEN A YOUNG CORAL HAS BECOME DETACHED.

E Fresh discoid coral commencing to bud forth; *e* wide surrounding scar surface.

specimen I found, showed thus three rings. Stutchbury imagined that each mother stock threw off only one bud, and then died; Semper showed that this was not the case, he speaks of three or four generations only being produced by each stock. Apparently the number produced is very limited. None of the stocks in my specimens were branched. A young Coral bud just ripe, 1$\frac{1}{8}$th of an inch in diameter, dropped off one of the stocks as I lifted the specimen from the water. Beneath it on the scar, another very small young *Fungia* had begun to bud out before its predecessor was quite free. The somewhat cup-shaped buds, when set free, become by the direction in which future growth takes place, flat and disc-shaped and develope eggs, from which spring free-swimming larvæ, which start fresh stocks.

The mass of nurse stocks which I found was surrounded on the reef by a group of fully-formed *Fungias* of all sizes, I counted twenty in all. Some six of these were small and still showed the scar of attachment which disappears in the process of subsequent growth.

A species of *Millepora* (*M. nodosa. Esper*), is a very common coral upon the Tahitian reefs. It forms irregular nodular masses usually of small size, and often encrusts dead corals of other species. The tips of the lobes of the living coral are of a bright gamboge-yellow colour, which shades off into a yellowish-brown on either side of the lobes. Mr. Murray succeeded in getting the polyps of the coral to expand under the microscope, and handed them over to me for examination. I found them, as Agassiz had discovered long before, to be Hydroids allied to the *Medusæ* and not to the *Actinozoa* and Sea Anemones, like the majority of modern stony corals; I studied the structure of the coral minutely.[*]

The hard part of the coral or calcareous skeleton is finely porous throughout, being excavated by a complex reticulation of fine and tortuous canals which are in the freest possible communication with one another. Within this porous mass at its surface are excavated cylindrical holes or pores of two sizes.

The canal spaces in the skeleton are, when the coral is living, filled by a network of living tissue made up of a meshwork of branching and communicating tubes, which form a canal system, by means of which a free circulation can pass from one part of the coral to another.

Two kinds of Polyps inhabit the pores described as existing on the surface of the coral. The larger pores are occupied by short stout cylindrical polyps which have each four tentacles and a mouth and stomach, and which are hence termed " *Gastrozooids*," whilst their pores are termed " *Gastropores*." The smaller pores shelter each a very different kind of polyp, which has a long and slender sinuous body provided with numerous tentacles, and devoid of any mouth or stomach; this latter form of polyp, because its function is merely to catch food, is called a " *Dactylozooid*," and its pore a " *Dactylopore*."

The *Dactylozooids* catch food for the colony and deliver it to the *Gastrozooids*, which alone are able to swallow and digest it. All the polyps of the colony are in communication at their bases with the canal system already described, and by means of these canals the nutritive fluids derived by the gastrozooids from the food are distributed to the entire colony and nourish

---

[*] For further details, see H. N. Moseley, " On the Structure of a species of Millepora occurring at Tahiti." Phil. Trans. Roy. Soc., Vol. 167, 1877, p. 117.

it. There is thus a very complete division of labour in the colony.

In all species of *Millepora* the mouth-bearing polyps are much more numerous than the mouth-less ones. In some species the gastropores and dactylopores are scattered irregularly over the surface of the colonies. In the Tahi-tian species, however, they are for the most part gathered into definite groups or systems, each consisting of a centrally placed gastropore surrounded by a ring of five, six, or seven dactylopores, as shown in the accompanying figure, where the circular groups of minute pores are seen scattered over the coral surface.

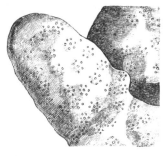

PORTION OF THE HARD CORAL OF MILLE-
PORA NODOSA.
(Twice the natural size.)

The second figure shows, much enlarged, a single system of polyps belonging to one of these pore systems, as it appears

ENLARGED VIEW OF PORTION OF THE SURFACE OF A LIVING MILLEPORA NODOSA, SHOWING THE
EXPANDED ZOOIDS OF A SINGLE SYSTEM.

In the centre is the short mouth bearing Gastrozooid, around are the mouthless Dactylozooids.

when the polyps are fully protruded from their pores and expanded. Beneath is seen shaded dark part of the canal meshwork,
which maintains the general circulation of the colony. From
this stands up in the centre, the short and stout gastrozooid
with its four tentacles, and dark stomach cavity seen through
the walls of its body, and its mouth at its summit. Around are
grouped five dactylozooids, each with many tentacles, but without any mouth or stomach. One of the dactylozooids is seen
bending over to feed the gastrozooid of the system.

Marvellous as is the completeness in the division of labour in
the Millepora Colony, this is far surpassed in the case of the
Stylasteridæ, another family of stony corals, which, as I found to
my astonishment, is also like the family *Milleporidæ*, Hydroid
in structure.

In the *Stylasteridæ* there is a canal network and common
circulation in each colony essentially similar to that in the
Milleporidæ. Two kinds of polyps also, mouth-bearing and
mouthless, are present. The dactylozooids are, however, entirely devoid of tentacles, and are reduced to simple long
tapering bodies, just like the tentacles of Sea Anemones in appearance, and performing the same functions. The gastrozooids
alone bear tentacles round their mouths, and in some genera
even they have lost their tentacles, and the entire colony is thus
devoid of these appendages. In some genera there are two
kinds of dactylozooids, smaller and larger, the latter evidently
intended to be enabled to better catch food by means of their
long reach, the former probably to deliver the food so caught to
the mouth-bearing polyp.

The accompanying woodcut shows the principal living structures as they exist in one of the more simple genera of the
Stylasteridæ, namely, the genus *Errina*. The various component structures are displayed as they are seen when the
calcareous skeleton of the coral has been removed by the action
of acids, and the remaining soft tissues have been cut through
in a direction at right angles to the surface of the coral. The
calcareous style is introduced into the drawing in order to show
its relation to the gastrozooid. In the case of the *Milleporidæ*,
the mode of reproduction is not known ; it is possible that they

produce free-swimming *Medusæ.* In that of the *Stylasteridæ* on the other hand, the process is well understood. Each colony

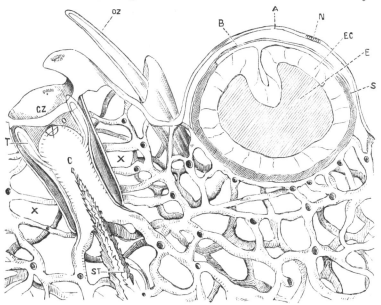

VERTICAL SECTION THROUGH THE LIVING TISSUES OF ERRINA LABIATA GREATLY MAGNIFIED, AND WITH ALL THE CALCAREOUS SKELETON EXCEPT THE STYLE REMOVED.

The mass is seen to be made up of a network of canals, which canals are shown in many places cut across. On the left is a gastrozooid, *c z*, cut through, showing two of its four tentacles, *t*, its stomach cavity, *c*, and its style, *s t*. Large canals pass from the stomach cavity to join the general canal network. The gastrozooid is withdrawn within its sac, which lines the gastropore, the wall of which is removed. To the right of the gastrozooid is seen a single dactylozooid, *d z*, partly protruded from its sac. On the extreme right is seen an embryo or planula doubled up within the ampullar sac and cut through. The planula is mature and nearly ready for escape; *e* Endoderm of the planula; *e c* ectoderm; *s* spadix; *b* layer of ectoderm covering the planula; *a* layer of soft tissue in the wall of the ampulla; *n* nemataphore.

or coral stock is of a separate sex, either male or female. In the female stocks, eggs are developed within special chambers hollowed out in the calcareous skeleton of the stock, and protected by a wall of hard coral, which often projects on the surfaces of the branches, so that the breeding chambers (*ampullæ*) show themselves to the naked eye like small warts on the coral twigs. Each egg is developed within the chamber into a cylindrical larva (*planula*), which is set free when mature, and swimming off fixes itself and develops from itself by growing and budding a new stock.

The nurse structures on which the eggs are developed, represent polyps which have become solely reproductive in function, just as the dactylozooids have become solely tentacular in function. Hence, in these colonies certain members of the community devote themselves to the catching of food, but cannot eat it themselves ; they deliver it to other members of the colony, whose only function is to eat and digest it. These latter nourish the whole colony by supplying blood to it through the common circulation as the products of their digestion ; in several genera, they have become reduced to the condition of mere stomachs, having no tentacles or prehensile organs of their own. Other members again of the colonies neither catch food nor eat it, but are entirely devoted to the production of eggs and larvæ, and have thus become reduced to the condition of mere egg-bags.

In the *Stylasteridæ*, the polyps are lodged within pores of two kinds, just as in the *Milleporidæ*, and, as in these latter, the dactylopores are far more numerous than the gastropores. In some genera of *Stylasteridæ*, the pores are scattered irregularly all over the surfaces of the coral stocks ; but in others they are grouped into systems of very great complexity, and almost all gradations of this complexity are shown in the various genera, so that the successive stages by which natural selection has brought about the development of the systems is clearly to be traced.

This series of stages of development is shown in the set of diagrams on the opposite page. Figure 1 represents the condition existing in the genus *Sporadopora*, the dactylopores shown as the smaller black circles are here irregularly grouped together with a single large gastropore. The gastropore has a white dot in its centre, marked S, indicating the "style," a rod of the calcareous skeleton, which in many genera of *Stylasteridæ* acts as a support to the mouth-bearing polyp within its pore and which by its presence gives the name to the family, *Stylasteridæ*. In *Sporadopora*, the pores of the two kinds are irregularly scattered over the whole coral surface.

In the case of another *Stylasterid*, *Allopora nobilis*, the development of regular systems of polyps is commenced. The arrangement is shown in Figs. 2 and 3. In some parts of the branches of a specimen of this Coral, the dactylopores are to be found simply grouped in rings around a single centrally-placed gastropore, just as in the Tahitian *Millepora* (see Fig. 2). In other parts of the same specimen, a further complication arises,

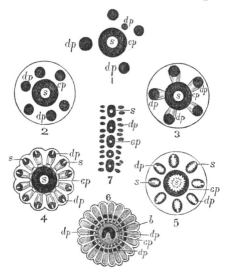

DIAGRAMS ILLUSTRATING THE SUCCESSIVE STAGES IN THE DEVELOPMENT OF THE CYCLOSYSTEMS
OF THE STYLASTERIDÆ.

1 In Sporadopora dichtoma. 2, 3 Allopora nobilis. 4 Allopora profunda. 5 Allopora miniacea.
6 Astylus subviridis. 7 Distichopora coccinea. *s* Style; *dp* dactylopore; *gp* gastropore; *d* in fig. 6, inner horseshoe-shaped mouth of gastropore.

as shown in Fig. 3. A shallow groove leads from each of the dactylopores to join the gastropore cavity, and a radiate figure is produced. No doubt the grooves are developed by the constant bending inwards of the tentacular zooids to feed the mouth-bearing zooid in the gastropore.

A more complete development of radiate systems occurs in another species of Allopora (*A. profunda*), as shown in Fig. 4. Here 12 dactopores surround each gastropore, and the grooves are much deepened. The dactopores in this case have small rudimentary styles, which structures are usually confined to the gastropores. In *Allopora mineacea* (Fig. 5), the styles in the dactopores are large, and have brush-like tips like the styles of the gastropores.

In the genus *Astylus*, neither kind of pore has a style, the radiate arrangement is most complete, and the highest condition of development of the circular systems of zooids (cyclosystems) is arrived at. These radiate cyclosystems in the *Stylasteridæ* so closely simulate in appearance the cups of ordinary Anthozoan corals with their radiate septa, that they were always supposed to be of the same essential nature as these latter, until dissection of the soft structures of the animals of which they are the skeleton, revealed their real significance.

The two radiate calcareous structures are in reality widely different in nature. In the Anthozoan coral, each radiate system is the skeleton of one polyp animal like a single Sea Anemone, and the radiating plates of calcareous matter in the coral cup, are supports developed inside the body of this single polyp. In the cylosystem of the *Stylasterid*, on the other hand, a large number of polyps are lodged, namely a single gastro-zooid and numerous dactozooids.

The radiating plates of the cyclosystem, so like the septa of Anthozoan corals, are formed of the walls of the mouths of the

PORTION OF A SPECIMEN OF THE CORAL OF ASTYLUS SUBVIRIDIS.
Showing the cyclosystems placed at intervals on the branches, each with a central gastropore and zone of slit-like dactylopores.

dactylozooids pressed against one another as they are closely packed together in a ring round the gastropore, and thus flattened out by mutual appressure. The elongate or groove-like form of the mouths of the pores is also to a large extent brought about by the manner in which the dactylozooids are doubled up within them when in the retracted condition.

The figure on the opposite page represents a small part of the skeleton or coral of a stock of *Astylus subviridis*, enlarged to twice the natural size. The cyclosystems, one of which is shown as a diagram in Fig. 6 of the preceding woodcut, are here seen placed at intervals along the branches of the coral.

Still further complexity, however, in the cyclosystems of the genus *Astylus* remains to be described. The figure below shows one of the systems cut through vertically to display the arrangements within. The gastropore has two chambers, an upper and lower. The lower, in which the gastrozooid, which in this genus is a mere flask-shaped stomach sac devoid of tentacles, is lodged, communicates with the upper by a narrowed horseshoe-shaped opening, which is more plainly seen from above in the diagram, Fig. 6, already referred to. The opening is rendered horseshoe-shaped by the projection from one side of it across the aperture of a small tongue-shaped excrescence of hard coral. This projection no doubt serves to protect the polyp from injury.

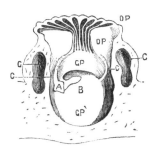

VERTICAL SECTION THROUGH ONE OF THE CYCLOSYSTEMS OF ASTYLUS SUBVIRIDIS.

dp–dp Dactylopore cavities; *gp* upper gastropore chamber; *gp'* lower gastropore chamber; *cc* canals leading from the gastropore to the dactylopore cavities; *b* tongue-shaped projection; *a* its base cut through; *gg* ampullæ cut open.

The lower gastropore chamber further communicates, as seen in the figure, at its margins by means of vertical canals with the bottoms of the dactylopores, which are seen above it. Through these vertical canals in the living coral pass large nutritive vessels from the stomach of the gastrozooid all round to join the dactylozooids and nourish them. The slit-like openings of the dactylopores into the upper chamber of the gastropore, allow the dactylozooids to be bent far down into the gastropore to reach the gastrozooid, and deliver food to it.

Around each cyclosystem are grouped a zone of ampullæ, which contain the reproductive elements, and in which, in the case of female colonies, the young are developed. The *ampullæ* are shown cut open in the figure, and marked GG. Thus each cyclosystem is, in the genus *Astylus*, complete in itself. It contains its single gastropore which placed in the centre nourishes the whole, its zone of dactylopores, and its zone of nurse structures which produce and rear its young. Nevertheless, the numerous cyclosystems of the colony are in communication with one another by a common canal system traversing its branches, and thus each is able to assist the other with nourishment, and any part of the branches thus so perfectly fed is able to increase the size of colony, by growing, and developing on the new twigs fresh cyclosystems as buds. In some other genera of *Stylasteridæ*, various other complications in the grouping and structure of the pores and their zooids occur. In one genus, *Distichopora* (see Diagram 7, *ante*), the gastropores are arranged in regular rows at the edges of the coral branches, and on either side of the row of gastropores is placed a row of dactylopores. The pores are thus everywhere grouped in three parallel rows, and they occur only on the edges of the branches, the rest of the coral surface is devoid of pores altogether.

The *Milleporidæ* and *Stylasteridæ* are so closely allied to one another, that I have grouped them together as a sub-order of Hydroid corals (*Hydrocorallinæ*). The *Milleporidæ* all occur in comparatively shallow water, and are reef corals. The *Stylasteridæ*, on the other hand, although some species occur in quite shallow water on reefs, range also to great depths, some species having been dredged by the " Challenger " from 1,500 fathoms. Specimens of six genera of *Stylasteridæ* were fished up at one haul of the trawl, from 600 fathoms, off the mouth of the Rio de la Plata, and it was from these specimens that the details of the structure of the family were made out, and the *Stylasteridæ* determined to be Hydrozoa.

I have given an account of the *Stylasteridæ* in this place, because any description of them must necessarily follow that of the *Milleporidæ*.* In having in each colony, polyps of several kinds, and of separate functions, with a regular system of division of labour, the *Hydrocorallinæ* resemble the Siphonophora, Hydroids which form similar colonies, which are free-swimming at the ocean surface, instead of fixed to the bottom like the corals.

To return to Tahiti:—The ground just above the shore near Papeete is everywhere burrowed by large Land Crabs. The Crabs are difficult to catch; never, in the daytime at least, going far from their holes, but watching a passer-by from near the mouths of their retreats, and bolting in if suspicious of danger, like rabbits. An old marine, whose name was Leary, who acted as my constant assistant in collecting on shore, invented a plan by which he caught some of the largest and oldest of the crabs. He tied a bit of meat on the end of a string, fastened to a fishing rod, and by dragging the meat slowly enticed the Crabs from their holes, and then made a dash forward and put his foot in the hole, and so caught them. The largest Crabs were far more difficult to catch than the younger ones.

It is curious how little animals seem to be frightened by a long wand like a fishing-rod. I have seen Mr. Thwaites in Ceylon put a noose of Palm Fibre, fastened at the tip of a rod of this kind over the heads of numbers of Lizards, and carry them off thus sniggled to put them into spirit for Dr. Günther. The Lizards sat quite quietly to receive the noose, though if we had moved a foot nearer to them they would have run off at once. Snakes and Lizards are nearly all caught in this manner in the Peradeniya Gardens.

We got up anchor and steamed out of Papeete Harbour to the time of the Tahitian National Air, a quick and lively jig characteristic of the place, and which sets the Tahitians dancing at once. It is popular with the French also : and as we

* For a detailed account of the structure of the Stylasteridæ, see H. N. Moseley, "Preliminary Note on the Structure of the Stylasteridæ." Proc. Roy. Soc., No. 172, 1876, p. 93. Also, "On the Structure of the Stylasteridæ, a Family of Hydroid Stony Corals." The Croonian Lecture. Phil. Trans. Roy. Soc. 1878.

entered Valparaiso Harbour, the band on board a French man-
of-war struck up this tune to greet us and recall the gaiety of
the beautiful island we had left behind us.   I give the air as
written down for me by Mr. T. Bird, the band-master of the
"Challenger," who by his indefatigable efforts, succeeded in training a very creditable brass band, during the voyage, although
only two or three at most of the Blue-jackets composing the
band, had any knowledge of music at all before the voyage
commenced.

MODERN NATIONAL AIR OF TAHITI.

Juan Fernandez, November 13th to 15th, 1875.—The voyage to Juan Fernandez occupied six weeks, as we had the bad fortune to be becalmed for 12 days on the passage. It was with the liveliest interest that we approached the scene of Alexander Selkirk's life of seclusion and hardship, and an island with the existence of which, in the case of most of us, the very fact that we were at sea on a long voyage was more or less distinctly connected. The study of Robinson Crusoe certainly first gave me a desire to go to sea, and " Darwin's Journal " settled the matter. Defoe was obliged to lay the scene of his romance in the West Indies in order to bring in the Carib man, Friday. He thus gained the Parrot, but he lost the Sea-Elephants and Fur-Seals of Juan Fernandez, one of the latter of which would have made a capital pet for Crusoe.

The island is most beautiful in appearance. The dark basaltic cliffs contrast with the bright yellow-green of the abundant verdure ; and the island terminates in fantastic peaks, which rise to a height of about 3,000 feet. Especially conspicuous is a precipitous mass which backs the view from the anchorage at Cumberland Bay, and which is called from its form " El Yunque " (the anvil).

There are upwards of 24 species of Ferns growing in this small island, and in any general view the Ferns form a large proportion of the main mass of vegetation. Amongst them are two Tree Ferns, one of which I only saw amongst the rocks in the distance, but could not reach. The preponderant Ferns, especially the Tree Ferns, give a pleasant yellow tinge to the general foliage. Curiously enough the almost cosmopolitan common Brake Fern (*Pteris aquilina*) does not occur in the island. Four species of the Ferns out of the 24 present are peculiar to the island, and one, *Thyrsopteris elegans*, is of a genus which occurs only here. The appearance of this Fern is very remarkable, for the cup-shaped sori hang down from the fronds in masses, looking just like bunches of millet seed.

Everywhere for the first few hundred feet, trees are absent, the wood having been all felled. In 1830 a large quantity of dry old sandal-wood still remained in the valleys: but even then there were no growing sandal-wood trees remaining.* No doubt the general appearance of the vegetation is very different now from what it was when the island was first visited.

I landed and climbed with a guide a steep path leading directly up from the Bay to Selkirk's Monument. The island is rented from the Chilian Government as a farm by a Chilian who employs a number of labourers and rears cattle, and grows vegetables, doing a very fair trade with passing vessels, the crews of which, like our own, after a voyage from such a port as Tahiti, long for a little wholesome fresh food. A considerable sum is also realized by the sale of the skins of the Fur-Seals. Close to the farm-house at the Bay still remain a row of old caves dug out in the hill-side by the Buccaneers.

In ascending the path the first tree was met with at about 700 feet altitude, all below had been cut down. We passed through a hollow overgrown by a dense growth of the gigantic Rhubarb-like *Gunnera Chilensis*. Darwin remarked on the large size of the leaves of this plant and height of its stalks as seen by him in Chile.† The stalks of the plants he saw were not much more than a yard in height. In this hollow the stalks must have been 7 feet in height. We walked through a narrow passage cut in a thicket of them with the huge circular leaves above our heads. The leaves catch and hold a large quantity of rain-water. The size attained by the *Gunnera* varies with its situation. In many places the leaves are very conspicuous on the hill-slopes, crowding closely as an undergrowth, and not rising high above the ground.

It was now Spring in Juan Fernandez, as at Tahiti. Most excellent Strawberries grow wild about the lower slopes of the

---

* " Narrative of the Surveying Voyages of H.M.S. ' Adventure' and ' Beagle,'" Vol. I, p. 302. London, 1839. Visit of Capt. King, H.M.S. ' Adventure' accompanied by Signor Bertero the Botanist, Feb., 1833."

† C. Darwin, " Journal of Researches," p. 279.

island, and especially well on banks beneath the cliffs close to the sea-shore. The Strawberries are large and fine, but white in colour, being, I believe, a Spanish cultivated variety. If so, they have not at all reverted to the parent wild form, either in colour or size; a few only were just beginning to ripen.

At this time of the year the foliage of the Myrtles, though evergreen, looks half dead, and these trees thus show out conspicuously amongst the rest. Here and there examples of the Magnoliaceous Tree " Winter's Bark " (*Drymis winteri*), a tree common in the Straits of Magellan were covered with showy white flowers, and large patches of a small species of Dock (*Rumex*) in full flower showed out red amongst the general green, whilst a white-flowered Iris, growing socially formed well-marked patches of white. A tall Bignoniaceous Shrub, which was very common, was covered with dark blue tubular flowers.

Hovering over the flowering bushes and trees, were everywhere to be seen two species of Humming-Birds ; one of which (*Eustephanus Fernandensis*) is peculiar to the island, whilst the other (*E. galeritus*) of the same genus occurs also on the main land. A further closely allied but peculiar species occurs in the island named by the Spaniards " Mas-afuera," or farther out, because it lies 90 miles to the westward of Juan Fernandez and so much farther from the Chilian Coast.

The Humming-Birds were extremely abundant, hovering in every bush. In the species peculiar to the Island of Juan Fernandez the male is very different in plumage from the female, being of a chocolate colour, with an iridescent golden-brown patch on the head, whilst the female is green. So different are the two sexes that they were formerly supposed to represent two distinct species, as has happened in the case of so many other birds. This endemic Humming-Bird seemed more abundant than the continental one. Any number of specimens might have been shot.

In skinning some of the birds which I killed, I noticed that the feathers at the base of the bill and on the front of the head were clogged and coloured yellow with pollen. The birds, no doubt, in common with other species of Humming-Birds, and other flower-frequenting birds, such as the *Myzomelidæ*, are active agents in the fertilization of plants. I noticed, as has been already mentioned, pollen attached in a similar manner to a bird at Cape York.* Mr. Wallace concludes that the presence of these birds, as fertilizers, accounts for the abundance of conspicuous flowers in Juan Fernandez.

There are very few insects in the island, according to the observations of Mr. E. C. Reed, and only one very minute species of Bee. Flies, of which there are 20 species, form the most prominent feature of the entomology of the island.† Some fertilizers, either insects or birds, must act on a very comprehensive and effectual scale all over the island, as follows from the abundance of fruit yielded by various introduced plants.

Strawberries, Cherries, Peaches, Apples, and Figs bear well; Strawberries and Peaches at all events very abundantly. The Wild Peaches are spreading everywhere. These, the Cherries and the Apples are possibly fertilized by the birds, but one would hardly suppose that the Strawberries would be also thus pollenized: though at a height of 9,000 feet in the Andes, I have watched Humming-Birds, possibly the same species as that at Juan Fernandez, hovering over the low mountain flowers, quite close to the ground, where nothing like a bush was growing.

It would be very interesting, if it proved to be the case, that Humming-Birds have in this distant island adapted themselves to the fertilization of our common garden fruits. Besides the fruit trees, there are many introduced plants with well-developed flowers which thrive in the island; a Thistle is very abundant and luxuriant, as if eager to remind travellers to what race the world owes the immortal Selkirk, and a Wild Turnip is rapidly spreading. Possibly the abundant flies take some share in the fertilizing work.

It must be remembered, with regard to insular floras, that a plant which had developed showy flowers to attract certain insects on some main land or other place where insects were abundant, might, when transferred to an island devoid of insects suitable to its requirements, nevertheless retain its gaudy flowers little or not at all impaired, for an indefinite period, just as

* See page 354.
† A. R. Wallace, " Tropical Nature," p. 270-271.

animals which have taken to deep-sea life have some of them retained their colours, though living in the dark.*

Selkirk's Monument is placed on the crest of a short sharp ridge in a gap in the mountains at a height of about 1,800 feet above the sea. From this, a steep descent leads down on either side to the shore. Here Selkirk sat and watched the sea on both sides of the island in long-deferred hope of sighting a sail.

Here we rested for some time, enjoying the view. Juan Fernandez is only ten miles in length, and 20 square miles in area, and from this elevated point nearly the whole extent of the island could be overlooked. Yet this tiny spot of land contains birds, land shells, trees, and ferns which occur nowhere else in the vast expanse of the universe, but here or in the neighbouring Mas-afuera. One could almost count the number of trees of the endemic Palm (*Ceroxylon Australe*) and estimate the number of pairs of the endemic Humming-Bird existent, at a bird for every bush. Two of the species of Land-birds, and all the 20 species of Land-shells of the island are endemic.

The temperature at the monument at 11 A.M. was 65° F. A small Bat, possibly disturbed by the sound of the gun, was seen to fly past. The common Sow-thistle (*Sonchus oleraceus*), the ubiquitous weed, has climbed up the pass, and grows by the Monument. The endemic Palm has been almost exterminated, excepting in nearly inaccessible places, as on a rock above the monument, where a group of the trees can be seen, but not reached.

The terminal shoot of the palm, especially when cut just before the tree flowers, is excellent to eat; the developing leaf mass being quite white, and tasting something like a fresh filbert. It seemed to me more delicate than that of the shoot of the Cocoanut. The guide knew where there was a tree remaining in the woods not far above sea-level, and I went with him to it hoping

* See A. R. Wallace, " Tropical Nature," p. 274. London, 1878. Mr. Darwin, " Origin of Species," 6th edition, p. 349, refers to the similar survival of the hooks of hooked seeds in islands where there are no mammals to the fur or wool of which they could cling. Some hooked seeds may, however, surely also be adapted to hang in the feathers of birds, as those of the Uncinia and Acœna of the Southern Islands, possibly, for example, are adapted to those of the Albatross.

to find it in flower.   As it was not, I cut it down for eating, for the guide was only waiting to let it develop further before felling it for that purpose himself.   A few seedling Palms grew near by.   Palms of the same genus occur in the tropical Andes.

Most remarkable in appearance amongst the Composite endemic trees are the species of the genus *Dendroseris*, allied to our Chicory.   The specimens which I saw in flower, were rather large straggling shrubs than trees, but with thick woody stems and branches from 10 to 15 feet in height.   The leaves are very like those of a Dandelion in appearance, and the stem which when split open, has a curiously jointed pith, has just the smell of a Dandelion-root, and would, no doubt yield chicory.   It pours out, like the Dandelion and allied plants, a milky juice when cut.

The flesh of the Wild Goats of the island is most excellent eating, no doubt because of the abundance of the feed.   In some parts of the island, especially to the south-west, there are open stretches covered with long grass.   Pigeons (*Columba œnas*), which are said to have been imported into the island, are common, and feed on the hill-sides in flocks.

Fish are very abundant, and easily caught, as are also Rock-lobsters (*Palinurus frontalis*) which are very large, and especially good to eat.   More than 60 were caught by means of a baited hoop-net put over the ship's side at the anchorage, and hauled up at short intervals.   The meat of the tails of these lobsters is dried at the island for export to Chile.

N.B.—Some of the matter in the above account was sent in MS. to Mr. Wallace, and is quoted by him in "Tropical Nature," pp. 143 ; 270–272.

For an account of the Land-birds of Juan Fernandez, see an article by Mr. P. L. Sclater in "Ibis," 1871, p. 178.

For accounts of the island in old times, see "Anson's Voyage." Account given by Capt. Woods Rogers.   Funnel's (mate to 'Dampier') Voyage. London, 1707.   Shelvocke's "Voyage of the 'Speedwell,' 1719–1722." London, 1726,—and many others.

CHAPTER XXI.

CHILE. MAGELLAN'S STRAITS. FALKLAND ISLANDS. ASCENSION.

Valparaiso. The Andes not Conspicuous. Cattle lassoed in the Streets. Excursion up the Uspallata Pass. Leafless Mistletoe on the Leafless Cactus. An Equestrian Hair Cutter. Dead and Disabled Animals on the Pass. Use of the Lasso in Robbery and Flirtation. Cleverness of a Horse on a Mountain Path. Fjords of the Western Coast of Patagonia. Density of the Forest. An Anchor Broken. Fuegians. Wild Geese at Elizabeth Island. Kitchen Middens. The Falkland Islands. Visit to Port Darwin. Scotchmen turned Gauchos. Chapinas and Tropijes. Wild Horses and their Habits. Various Modes of Handling Cattle in Different Parts of the World. Goose-Bolas made of Knuckle-Bones. Flies and Gnats with Rudimentary Wings. Skeleton of Ziphioid Whale. Fuegian Arrow-heads Scattered in the Islands. Habits of Jackass Penguin. Ascension Island. Land Crabs. The Hatching of Turtles' Eggs. Shooting at Flying Fish. Birds at Boatswain Bird Island.

Valparaiso, Chile. November 19th to December 11th, 1875.— How Valparaiso came to be called the Vale of Paradise I cannot well understand; the voyagers who so named it, must have come from a desert land indeed. The surrounding country has a most barren and inhospitable appearance, the red decomposed granite soil showing bare everywhere, and being only here and there sprinkled over with scanty bushes. Not a tree is to be seen anywhere from the anchorage in the harbour though a wide view is thence obtained of the coast of the Bay.

I had expected the far-famed Andes to show out as a splendid range in the background; but these mountains, though they look close to the coast on the map, lie so far inland that one has to search for them in the view from the Bay in order to see them at all. Even then they are only to be discerned on a clear day, and when seen they look small and not at all imposing. The residents on the hill-slopes know where the higher peaks lie, and point them out to strangers; but there is nothing in their appearance which would lead one to suspect their real grandeur were one not acquainted with it beforehand.

The hill-sides around the town are scored by the straggling tracks of Pack Mules following the crests of the ridges. The earth being so little held together by vegetation is readily cut into by the rain. An excessively heavy rain-storm occurred just before we left Valparaiso. The water poured off the hill-sides, flooding the streets of the town, and carried so much earth with it that it buried the lines of the tramway in some places with two feet of soil, and the line had to be dug out.

One sees the lasso in full use even on the quay of Valparaiso. It is used by the herdsmen who have to assist in shipping the cattle which they drive down from the country. I saw two refractory animals thus thrown down with the lasso on the pavement, and subdued, amongst a crowd of passers-by. It might have been awkward for the crowd if the men had missed their aim; but the matter seemed perfectly safe in their hands.

Amongst the herdsmen was a youth of about 16 years. He made a clumsy shot with his lasso, which interfered with that of one of the other men. The man rode his horse full tilt at that of the boy several times, driving in his spurs and making his horse charge with all its force. The boy returned the charge guiding his horse so that the two met always chest to chest, and eventually the man finding he could not upset him gave up the attempt. I was told that this charging of horses, which corresponds exactly to charging at football, is commonly practised in Chile. It was curious to see it going on in the populous street of a large city.

I went to Santiago, the capital of Chile, and also made an excursion to the summit of the Uspallata Pass, which is traversed by one of the roads leading over the Andes to Mendoza in the Argentine Republic. I started from the town of Sta. Rosa de los Andes. The Pass has been described by Mr. Darwin.[*]

Soon after leaving Sta. Rosa the hill-sides are seen to be covered with the tall Candelabra-like Cactus (*Cereus Quisco*)

* " Journal of Researches," p. 330.

It has a most strange appearance. Other forms of Cacti, each adapted to the climate of a particular altitude, succeed one another as the slope of the Andes is climbed; those that lie highest being dwarf forms scarcely rising above the ground.

On the *Cereus Quisco* grows a Mistletoe (*Loranthus aphyllus*). This Mistletoe is most remarkable, because, like the plant on which it is parasitic, it is entirely devoid of leaves. It is extremely abundant, growing on nearly all the *Cereus* trees, and is very conspicuous, because its short stems are of a bright pink colour. I could not understand what it was at first, as it looked like a pink inflorescence of some kind belonging to the Cactus.

Mr. Thiselton Dyer has examined the mass of parasitic tissue of this Mistletoe which draws the nourishment from the interior of the stem of the Cactus. He finds that having a soft and succulent matter in which to ramify, the basal fibres of the parasite form a large spongy mass of great size within the stem of the Cactus, which curiously simulates a mass of *mycelium*, such as produced by a parasitic fungus.

The fact that the Mistletoe growing on a leafless Cactus has no leaves itself, reminded me of a remark which Sir William MacArthur made to me in New South Wales. He told me that he had noticed that the Mistletoes growing on the various species of Gum-trees (*Eucalyptus*) simulated in their foliage that of the tree on which they grew, so that from that reason they were difficult sometimes to find. He pointed out to me examples.

The leaves of one Australian species of Mistletoe, *Loranthus celastroides*, which grows on species of *Eucalyptus*, are so like those of the *Eucalyptus* itself, that the varieties of the species have been termed *L. eucalyptoides* and *L. eucalyptifolius*. The Australian species of *Loranthus* have commonly two very different forms of leaves, broad and narrow. In the case of *L. celastroides* the broad-leaved varieties grow on *Banksias* mostly, and the narrow-leaved on *Eucalypti;* but both forms occur on species of *Casuarina*, which is a tree with narrow needle-like leaves; all gradations occur between the two varieties of this Mistletoe.*

* "Flora Australiensis," Vol. III, pp. 388, 390. Bentham and Müller. London, 1866.

Loranthus aphyllus is the only *Loranthus* without leaves. It grows only upon the *Cereus Quisco*. There are, however, species of the genus *Misodendron* of the Mistletoe family, which are leafless, and yet grow on trees with well-developed leaves, such as the Fuegian Beech Trees.

Probably the leafless Mistletoe on the Cactus has got rid of its leaves for the same reason as the Cactus, viz., to minimise loss of moisture by evaporation in an arid climate. The Australian Mistletoes possibly are adapting their leaves to the forms of those of the Gum-trees, in order to benefit the trees, and thus themselves, by interfering as little as possible with the vegetation at the roots of their host. They can hardly be supposed to gain by being inconspicuous, but must rather be certain to lose thereby.

After accompanying me for about half the distance up the Pass, my companion, Lieut. G. R. Bethell, had to return and left me to proceed with a Chilian rustic guide. As a substitute a travelling barber joined us and attached himself to me to my great amusement. It was curious to meet with an equestrian hair-cutter. He had his scissors slung to his saddle. He was a most useful man to me, for, true to his trade, he persisted in talking to me and telling me long stories, riding beside me all day until at last I really began to understand part of what he said, and made rapid progress in Spanish. His great wish was that we should reach the new house that he was building, that I might see it. At last he led me off the road in a turn of the valley which was excessively barren-looking, like the rest of the landscape at this altitude, 7,000 or 8,000 feet. I could see no house, but he led me to a large square block of fallen rock. Here, against the rock on one side, was a sort of pen enclosed on three sides by a wall of roughly piled stones about a yard high and by the rock on the other.

There was no roof of any kind, but this was the "casa." It measured about six feet square. A hole excavated under the rock at the back was the store-room. My friend motioned me with most elaborate politeness to enter, and offered refreshment. He pressed especially coffee, so I agreed to that, whereupon his servant or assistant, a lad whom we found at the "new house," produced, after a long delay, some hot water slightly tinged brown by about half a dozen coffee beans.

The hair-cutter had turned a rill from the river over the dry and dusty soil near by, and grass was beginning to spring. He insisted on riding farther with me to an inn at the bottom of the final steep climb to the summit of the Pass, and having slept a night and waited at the inn till my return from the summit, accompanied me back to his house. He ceased not to talk to me all the time, and though I was becoming comparatively proficient in the language, I got tired of him at last, and treacherously gave him the slip whilst he rode off into a side valley to find some wonderful plant for me of which he only knew the locality.

It pleased me very much to find amongst the Alpine vegetation, at 7,500 feet elevation, a plant of the genus *Azorella* (*A. trifoliata*), a genus with which I had become so familiar in the far-off Kerguelen's Land.* A plant, *Chevreulia Thouarsii*, which occurs in the isolated and distant Tristan da Cunha, is common all over Chile; the species found on the continent being identical with that of the island.

Near the summit of the Pass the slopes are almost absolutely barren.

The line of the track is strewn with the skeletons of mules and cattle which have perished on the journey. Very large numbers of cattle are constantly driven over the Pass, though it is 12,500 feet in height, from the Argentine Republic, and the Chilians, in exchange for this meat, supply corn to the Argentiners, which, however, of course goes round by sea.

The cattle can find little or no food on the journey over the Pass, and many die on the way; many others are obliged to be killed, and men occupying houses on the route buy the disabled ones, and make a profit by drying the meat.

At one spot an unfortunate mule had fallen from a zigzag path down a steep slope, and lay at the bottom with one of its legs broken, and the bone protruding for six inches. My guide went up and kicked the poor beast, which was lying down, till it got up on three legs, but only to see if it was of any good, and he

* See page 166.

laughed at it without the slightest feeling of compassion. I would have given a great deal to have been able to put it out of its misery, but I did not want the man to see that I had no pistol with me, and I was, therefore, obliged to let the animal lie.

There was absolutely no food, yet the man said the mule would live eight days. There were plenty of Condors wheeling about in different directions, but they took not the slightest notice of the beast. I was told that they never approach until an animal is actually dead. The drover who took the pack off the mule had, no doubt, never given a thought to taking the trouble to kill the animal.

There were several patches of snow which were crossed by the track close to the summit (*Cumbre*), but there was no snow on the track at the actual summit itself.

I was told that when highway murders were committed on the Pass, the traveller attacked was usually lassoed and dragged off his horse, and some way away from the track; the assailant, as soon as his man is noosed, putting spurs to his horse; a very unpleasant mode of death. The lasso is, however, used on human beings occasionally with far different intent. I saw a young girl going out on foot to milk the cows, at a farm at some distance down the Pass, playfully lasso a young man with whom she had been flirting, catching him round the neck as neatly as possible, just as he was going away.

I rode a horse on the journey whilst my guide rode a mule. We made a *détour* on our return journey in order that I should see a remarkable chasm in the rock called "El salto del soldado" (The Soldier's Leap). We had to traverse an old and neglected route for some distance. In one place the hill-side had slipped somewhat, and the track was gone, but steep slopes of loose stones had to be crossed between short lengths of the remaining path. There was a deep drop into the river below. My horse halted a second or two before each of these slopes, evidently well knowing their treacherous nature and also the best way of crossing them, and then went across with a quick run as fast as he could make his way.

Just so I should have crossed them myself on foot; the

momentum helps one across the sliding stones, and there is no time for stones to roll down from above. I certainly thought that the horse managed his feet better than the mule on this occasion, and as far as my experience goes, a horse that is thoroughly accustomed to mountain work is better than a mule to ride in difficult places, and is certainly quicker, though the mule has secured the credit of being the better mountaineer.

Messier Channel and the Straits of Magellan, December 31st, 1875, to January 20th, 1876.—The ship entered the Gulf of Peñas on the coast of Patagonia, south of the Chonos Archipelago, on December 31st, and for a fortnight steamed through the wonderful series of sounds or fjords into which the south-west coast of South America has, like the coasts of British Columbia, of Greenland, Norway, and other countries, been slowly engraved by the prolonged action of glaciers. Such an indented coast-line occurs only in those regions in high latitudes where there is a constant precipitation of moisture, since glaciers can only be fed, and perform the eroding work where there is an abundant snowfall.*

The Western Patagonian fjords are very beautiful. The route led through narrow channels, between successive ranges of mountains, capped here and there by snow and glaciers, the dwindled representatives of those that scooped out the main features of the scenery. The fjords remind one somewhat of those of Norway. They branch and send off offsets on either hand perpetually. Thus, as these long sounds are traversed, constant glimpses are obtained down the communicating channels, which show themselves bounded by successions of mountain ridges, fading gradually out of sight, one behind the other in the distance.

In the upper part of the Messier Channel, near the Gulf of Peñas, the mountains are covered from top to bottom by a dense forest of small trees, and one of the chief peculiarities of the scenery is caused by the fact that these forests come right down to the sea-shore, and overhang the beds of mussels growing on the rocks. The channels are full of Fur Seals, which were to be

* O. Peschel, "Neue Probleme der Vergleichenden Erdkunde." Leipzig, Duncker und Humboldt, 1876. Die Fjordbildungen, s. 9.

seen progressing through the water alongside the ship in troops, by series of bounds, just like porpoises.*

The anchor was dropped every night, it being impossible to proceed without daylight, because of the intricacy of the channel. Every evening I went on shore at some wild harbour, to wade through swamps and crawl through the dense undergrowth, in pursuit of wild geese, ducks, snipes, and woodcocks. In some of the harbours it was impossible to get away from the sea-shore, so dense was the barrier of forest everywhere. The ground is encumbered with prostrate trees and logs, which are overgrown with the most delicate and beautiful ferns, mostly *Hymeno-phyllums*, which thrive in the constantly moist atmosphere.

At one place we fired the forest. The fire spread rapidly for miles, covering the mountains with clouds of smoke, and somewhat endangering Mr. J. J. Wild, one of the members of the scientific staff, who was on shore alone. After an anxious hue-and-cry he was found safe on a rocky promontory, and brought back to the ship in one of the boats in triumph.

About Sandy Point there is more open country, and wide stretches of grass-land, on which we found abundance of mushrooms. A curious accident happened at Port Churruca, in Desolation Island. The ship's anchor was let go in a glassy calm, and apparently the ship was safely anchored. A short time later, however, a slight breeze sprung up, and the officer of the watch found that the ship was drifting freely before it. He had just time to let go another anchor and save the ship from drifting on shore, which was a very short distance off in the narrow fjord. It was found that the anchor, falling heavily on the rocks when let go, had broken in two short off, so that the remnant did not hold at all, a fact which had not been apparent during the calm.

Many deserted huts of the Fuegians were seen at the various harbours; but to my great disappointment we met with no natives themselves. Only one day, as we steamed along the middle of the main Strait of Magellan, near the southernmost point of America, Cape Froward, in a bitterly cold blast, we saw on the shore, in the distance, three fires, with their smoke

* See page 265.

streaming out before the gale, and we could make out through the rain the forms of the natives around them.

At Sandy Point there were two Fuegian girls and a boy, who had been picked up in a canoe by a Chilian war-vessel. I was struck by the ruddy colour of the cheeks of the girls, which closely resembles that of Japanese women, especially that of many older ones. Two Fuegian men who belonged to a Mission schooner at the Falkland Islands did not show any ruddy colour, but were of a uniform light-yellowish brown.

The girls and the boy slept huddled together in a heap, and curled up for warmth. The girls were photographed by the "Challenger" photographer. They were very shy and suspicious, and both put one of their fingers in their mouths during the process, on three successive occasions, that being evidently with them the natural mode of expressing shyness.

There were no Patagonians at the Sandy Point settlement at the time of the ship's visit. We were told that they visit the settlement at intervals to sell their Guanacho robes. When the tribe arrives at a short distance from the settlement, a messenger is sent forward to tell the Chilian Governor that the tribe is coming on a certain day, and expects a salute to be fired. As they approach accordingly, a salute is fired from the fort, and they ride in, making their horses caper, and showing off their horsemanship.

When they have stayed some time in the settlement, and have sold their robes and spent the money, mostly in drink, they send word that they are going, and require another salute, and as everyone is very glad to get rid of them, and they will not go without it, they are again saluted, and depart to hunt the Guanaco again.

After leaving Punta Arenas, we landed at Elizabeth Island, which is without trees, but covered with grass, and is likely soon to be occupied as a sheep-run. The island is the breeding-place of large numbers of Wild Geese (*Chloephaga Patagonicha*). The geese were very abundant, and a wild-goose-chase in Elizabeth Island is a very different matter from one at home. When I had shot nine geese I found that I had no light task before me in carrying them to the boat at the end of the island, over the soft and yielding soil. Goose-shooting in the Falkland

Islands similarly soon satiates the sportsman, who finds himself early in the day with a heavier bag than he can stagger under.

The geese at Elizabeth Island showed some wariness, and some little trouble had to be taken in order to get within shot of them, unless they were met with in long grass. When on the alert, they settled on the summits of the hillocks and ridges, in order to have a wide view of the enemy. One had to creep up under cover of the hill-slopes, and make a final rush forwards towards the flock. The birds are startled by this, and it is some time before they make up their minds to fly.

No doubt the wariness of these geese is due to their progenitors having been hunted for generations by natives in old times. Elizabeth Island is fringed with Kitchen-middens of large extent, which are full of vast quantities of bones of the Sea Lion (*Otaria jubata*). Mr. Murray excavated some of these mounds, and found some stone arrow-heads and stone fishing-net sinkers. The island was inhabited at the time of the early Dutch Voyages.

Besides the middens there are plenty of small shallow circular excavations with the thrown-out earth heaped around, which mark the site of Fuegian huts. The human *débris* is evidently of all ages, and I even found a sardine tin amongst it, perhaps left there by Cunningham.

The geese at the Falkland Islands are far tamer than those at Elizabeth Island, and seem not to understand a gun, though they have been shot at now for a long period. The Falkland Islands, however, were never inhabited by any savage race, and the birds have not had time to learn. The other birds in Magellan's Straits, which also occur at the Falklands, as for example the Loggerhead Ducks, show the same contrast in their wildness. They have been hunted for generations by the hungry Fuegians.

The young wild geese at Elizabeth Island, whilst still covered with black down, run amongst the grass with astonishing quickness, and are as difficult to shoot as rabbits. It is no easy task to catch them by running. A brood when met with separates, every gosling running off in a different direction. The young birds dodge behind a tuft of grass, and squatting closely under it are at once safe. It is quite impossible to find them, and a brood of ten or twelve goslings, as large almost as full-grown fowls, disappears as if by magic. The goslings can only be caught by the pursuer keeping his eye on one bird only, and running after it at the utmost possible speed. I had no idea that goslings would be able to secure their safety so completely. No doubt a terrier would find them one after another. They are far better to eat than the full-grown geese.

The ship was anchored in about 16 successive harbours in the passage through the long Patagonian Channels and Magellan's Straits. The run across from the eastern mouth of the Straits, to the eastern extremity of the Falkland Islands, consumed only three days. The sea crossed over is extremely shallow, varying from 50 to 20 and 110 fathoms only in depth.

For the Natural History of the Straits of Magellan, see R. O. Cunningham, M.D., "Notes on the Natural History of the Straits of Magellan." Edinburgh, 1871.

For Accounts of the Patagonians. G. C. Musters, R.N., "At Home with the Patagonians." London, Murray, 1873.

The Falkland Islands, January 23rd to February 7th, 1876.— The ship reached Stanley Harbour in the Falkland Group, on January 23rd. The Falklands are a treeless expanse of moorland and bog, and bare and barren rock. Though it was summer, and the Islands are in about a corresponding latitude to London, a bitterly cold hail-storm pelted in my face as I was rowed to the shore. The islands are occupied as sheep and cattle-runs, and since sheep are found to pay best, they are supplanting the cattle, formerly so numerous, to a large extent.

The mutton is most excellent, but the supply is so far in excess of the small demand, that the Falkland Island Company have a large boiling-down establishment, where their sheep are boiled down for tallow.

I rode with Lieut. Channer 60 miles across the large island, on which the town of Stanley is situate, to Port Darwin, in order to examine some reported coal-beds, at the request of the Governor. The route lay over the dreary moorland, and wound and turned about in order to avoid the treacherous bogs. A " Pass " in the Falkland Islands means, not a practicable cleft in the mountains, but a track by which it is possible to ride across a bog. The horses born and bred in the island, know full well when they are approaching dangerous ground, and tremble all over when forced to step upon it.

At every ten miles or so a shepherd's cottage was met with. Usually the shepherd was a Scotchman in the employ of the Falkland Company. Otherwise the entire route was uninhabited. Some of the shepherds are married. They seem well off and were very hospitable. These Scotchmen have almost entirely supplanted the "gauchos" from the mainland, who did all the cattle work at the time of Darwin's visit to the islands. They come out from home usually entirely unaccustomed to riding, but very soon become most expert with the lasso and bolas, and can ride and break the wildest horses. There were only two Spanish gauchos in the employ of the Company at the time of our visit.

The Company's shepherds are each allowed eight horses, a fresh one for every day of the week, and a pack-horse. The horses feed together on the moorland near the shepherd's cottage, and keep together in a band though quite free. An old broken-down mare, which cannot roam far, is usually kept with each band.

Generally, the mare is one in which the hoofs, as occurs quite commonly in the Falklands from the softness of the soil, are grown out and turned up, somewhat like ram's horns.[*] Though the gauchos themselves are a thing of the past in the Falklands, their Spanish terms for all matters connected with cattle and horses survive, and are in full use among the Scotch shepherds. Such a maimed animal as above described is accordingly called a "Chapina" (chapina, a woman's clog). The band of horses, which is called the "Tropija," never deserts the "Chapina."

A man, after riding 30 or 40 miles and about to change horses, merely takes the saddle off his horse, gives the animal's back a rub with his fingers, to set the hair free where the saddle-cloth pressed, and lets the horse go. The horse never fails to

---

[*] The hoofs of cattle in the islands grow out in a similar manner. "Proc. Zool. Soc.," 1861, p. 44, 1869, p. 59. See also C. Darwin's "Journal of Researches," p. 192.

return to its "tropija" and feeding-ground. We changed horses several times on the route, since we were the guests of the Company, and were treated most hospitably. We always simply turned our tired horses loose, to find their own way back for 20 miles or so.

An experienced guide is required, in order to traverse the Falkland Island wastes and find the Passes. To a stranger, every hill and mountain appears alike, and many persons have lost their way and their lives on the moors. The most experienced "camp" men (Spanish campo) get lost sometimes, especially when a thick fog comes on, and then they trust entirely to their horses, which make their way when left to themselves back to their accustomed feeding-ground.

Mr. Fell, the head man of the Company at Darwin Harbour, told me that a band of horses will always stay with a mare that has a foal. Mr. Darwin has described a degeneration in the size and strength of the horses which have run wild in the Falkland Islands,* ascribing the degeneration to the action of the climate on successive generations. Mr. Fell, and other persons brought in constant relation with the horses, hold the opinion that it is only the wild horses, occupying a particular district in the neighbourhood of Port Stanley, which are small and pony-like.

Further, they believe that the reason why these particular wild horses are small, is that they are sprung from a stock originally inferior in size when imported. The wild horses which are abundant in the large peninsula, known as Lafonia, were said to be of full size and vigour, and to show no signs of degeneration, and to be preferred for all purposes to those bred in domestication. I saw several of these horses which had been wild ones, and rode one. They were not at all undersized. My guide rode a sturdy pony, which he said was one of the smaller wild breed. I give these opinions merely as a suggestion for further inquiry.

Mr. Fell has watched the habits of the Wild Horses in Lafonia closely. The strong and active horses each guard their own herd

---

* "Journal of Researches," p. 192. "Animals and Plants under Domestication," Vol. I.

of mares. They keep the closest watch over them, and if one strays at all, drive her back into the herd by kicking her. The younger horses live in herds apart, but the more vigorous ones are always on the look-out to pick up a mare from the herds of the older ones, and drive her off with them, and they sometimes gather a few mares and hold them for a short time, till they are recaptured from them. When they think they are strong enough, they try the strength of the old horses in battle, and eventually each old horse is beaten by some rival and displaced. The fighting is done mainly with the tusks, and front to front, not with the heels. Thus the most active and strongest males are constantly selected naturally for the continuation of the herds.

The wild horses, as well as others, are often broken in by tying them with a raw hide halter to a post, and leaving them for several days without food or water. After long ineffectual struggles to break loose, the animals become convinced of the absolute power over them of the halter, and in future become cowed and docile directly a halter or lasso is over their heads. The wild horses when broken in, are very tame and· quiet to ride.

I was astonished at the facility with which the Falkland Island horses obey the rein. There is no necessity, as a rule, to make them feel the bit at all, in order to turn them. Merely laying the part of the reins close to the hand against that side of the neck from which they are wanted to turn is sufficient. Well-broken horses can be turned round and round in a circle by this means, by a gentle touch on the neck only. Our horses in England are certainly not half so well broken.

Our progress on our ride was mostly slow, because of the bogginess of the ground, and it was dark by the time that we reached the end of our 60 miles ride. Mr. Fell gave us an opportunity of seeing an assembly of the herd of tame cattle belonging to the Company in the part of Lafonia near Darwin Harbour. The Company has imported some first-class Bulls of the hornless polled breed.

The wild cattle in Lafonia will probably all be killed off in order that sheep may be substituted. At present, the Company pays men to kill these wild cattle for their hides. The cattle are thrown by means of the lasso or bolas, and ham-strung, or " cut down," as the term is, and then killed and skinned at leisure. 2,000 had been thus killed in Lafonia in the year of our visit.

It seems remarkable that such very different means of handling wild or half wild cattle should be adopted in different countries, and that one method should not long ago have been found the best. The bolas is used in the Argentine Republic and the Falklands, but not, I believe, in Chile. The lasso is always used with it. In California the lasso only is used, as also in the Sandwich Islands, the inhabitants of which derive their methods of cattle herding from the former country.

In Brazil the cattle, as I have described,* are brought into subjection by being tailed; the lasso is used, but not the bolas. In Australia and New Zealand, neither of these appliances are used, but only the stock-whip. An experienced owner of large herds of cattle in Australia, tells me that he considers that these various appliances are really not wanted, and that the great art in driving cattle is to get them to move quite slowly, and never to excite or terrify them, and that he can tell a good manager at once by observing whether his cattle are quietly and easily driven. There seem to be no differences in the condition of the country in the various regions which should render the lasso or bolas more necessary in some than in the others.

The bolas,† as is well-known, is an apparatus consisting of heavy balls of stone, metal, or wood fastened at the ends of long thongs of raw hide. In the Patagonian Ostrich bolas, only two balls are used ; for cattle and horses, three, one ball being smaller than the others. The three thongs are brought together at one knot. The bolas is held by the smaller ball, and whirled round the head, and then thrown so as to entangle the legs of the animal aimed at.

* See pp. 99–101.

† Mr. Darwin's " Journal of Researches," pp. 44 and 111, in his accounts of the bolas, calls it by this name as also other authors, Musters included. A hunter, however, from whom I bought one at Sandy Point, and also the Falkland Islanders, said the name was not bolas, but "boleaderos," or some word closely similar, and they considered the word bolas incorrect. Possibly the name has changed.

The boys at the Falkland Islands have invented a small bolas in which the large knuckle-bones of cattle are used as the larger balls, and a smaller bone from the foreleg, as the small ball for the hand. They use the bone bolas for catching wild geese, creeping up to a flock and throwing the bolas at the birds on the wing as they rise. They constantly succeed in thus entangling them, and bringing them to the ground, and their mothers always send out their boys when they want a goose, so that the birds are seldom shot at around Darwin Harbour.

Flocks of the geese were to be seen there feeding on the grass close to the houses, looking just like farm-yard geese. The birds take no notice of a gun, but I soon found that they were very quick at seeing a bolas when I carried one, well-knowing that they were going to be molested. I could not catch one with the bone bolas, though I came very near it, and should have succeeded with a little practice. The bone bolas comes curiously near that of the Esquimaux in structure. The Esquimaux bolas, used also for catching birds, has more than three balls, and these are made of ivory.

Near Darwin Harbour, I found some Dipterous insects with rudimentary wings, a species of Fly (*Muscidæ*) and a species of Gnat (*Tipulidæ*), which are of especial interest because similar. *Diptera* incapable of flight, occur, as already described,* at Kerguelen's Land, and the Fly at least appears to be of the same genus as one of the Kerguelen Flies ; a genus which has been hitherto found nowhere else but in Kerguelen's Land and Marion Island. It is of importance to find further connections between Fuegia and the distant Kerguelen's Land, the connections between which regions in the matter of the flora, were so long ago demonstrated by Sir Joseph Hooker.

The Fly has small rudiments of wings. It appears closely allied to *Amalopteryx maritima* (Eaton) of Kerguelen's Land, and corresponds closely to that insect in its habits. The flies were found near Darwin Harbour, only on the sea-coast, in hollows under overhanging slabs of the sandstone rocks, and sheltering in crevices. They spring nimbly like fleas or small

* See pages 192–193.

grasshoppers, and are a little difficult to catch. They cannot fly at all.

The Rev. H. C. Lory, late Colonial Chaplain in the Falkland Islands, writes to me that these flies inhabit in immense numbers dried matted seaweed which is to be found on the sea-beaches. He says that they escape in hundreds from the seaweed masses when they are broken up, and that the masses are full of the chrysalises of the flies.

The Gnats which I found, also cannot fly, having even smaller rudiments of wings than the flies. They were found crawling on rocks, on the shore in sheltered places, and also on the sunny sheltered face of a peat-bank, which formed the cattle fence across the narrow neck of the promontory of Lafonia. The gnats run quickly, and when in danger, draw up their legs and drop amongst the grass in order to escape. A Gnat with rudimentary wings, occurs also in Kerguelen's Land. Some species of flies and gnats with rudimentary wings, are known in Europe and elsewhere, and Prof. Westwood has shown me an apterous fly which occurs in England (*Borborus apterus*). I found besides a wingless Beetle, and one also with perfect wings, near Darwin Harbour.

From the head of Port Sussex, not far off, I obtained the skeleton of a Ziphioid Whale, complete except the paddles which had been dragged away tied to the ends of lassos, in order to get the oil out of them. The skull was given to me by Mr. John Bonner, a farmer in the neighbourhood. The Whale measured exactly 14 feet in length. It ran on shore in accordance with the usual unaccountable propensity of Ziphioid Whales.*

We lashed the skeleton on a pack-horse, by no means an easy matter in the case of so unusual a load. We rode at a good pace, but during the long ride the lashings were constantly getting loose, and we had to dismount at least 30 times. We led the first pack-horse, and hunted and drove along before us the second for which we changed it; but night overtook us before we reached Stanley with the skeleton, and we almost lost our way near the end of the journey.

* See page 158.

Many of the seamen living at Stanley constantly visit the Straits of Magellan, and bring back with them very often Fuegian bows and arrows for their children to play with. The boys shoot at a mark with the stone-tipped arrows, and the tips are soon broken off and lost. The stone arrow-heads thus become scattered about the moorland anywhere near a habitation, and before long they are sure to be picked up, being indestructible. It must then be remembered that they are not proofs that the Falkland Islands were once inhabited by a savage race. Difficulties of this kind are constantly occurring; for example, part of a New Zealand jade Mere has been found in Yorkshire; ancient Chinese Seals turn up in the ground in Ireland, and I lately had a New Zealand fish-hook sent to me by a Canadian, as found on the shores of a Canadian Lake and the work of North American Indians.

I wished very much to taste the luxury which Darwin partook of when travelling, the Falklands meat roasted with the hide on "Carne con cuero,"* but on my asking for it everyone spoke of the practice of so cooking food with horror, as only fit for savages and almost with as much disgust as if I had suggested cannibalism. No doubt this notion has been fostered by the cattle owners, because of the great value of the hides, which are necessarily spoilt by the process.

Not far from Stanley Harbour there are rookeries of the Magellan Jackass Penguin (*Spheniscus Magellanicus*). The birds make large and deep burrows in the peat banks on the sea-shores, and large numbers make their burrows together, so that the ground is hollowed out in all directions.

Round the mouths of their burrows and on the even surface of the banks, between the holes, the birds lay out pebbles which they must carry up from the sea-shore for the purpose. The pebbles are of various colours, and the birds seem to collect them from curiosity, at least there appears to be no other explanation of the fact.† The edges of the birds' bills are excessively sharp, and one of them bit me as I was trying to secure it, and cut a strip out of my finger as clean as if it had been done with a razor.

* " Journal of Researches," p. 190.   † See pages 156–159.

Ascension Island. March 27th to April 3rd, 1876.—After a stay of ten days at Monte Video, during which time was afforded for a visit to Buenos Ayres, the ship reached Ascension Island on March 27th. Land Crabs swarm all over this barren and parched volcanic islet. They climb up to the very top of Green Mountain, and the larger ones steal the young rabbits from their holes and devour them. They all go down to the sea in the breeding season.

It always seems strange to me to see Crabs walking about at their ease high up in the mountains, although the occurrence is common enough and not confined to the Tropics. In Japan a Crab is to be met with walking about on the mountain high-roads far inland, at a height of several thousand feet, as much at home there as a beetle or a spider. Crabs of the same genus, *Telphusa*, live inland on the borders of streams in Greece and Italy.

The sea is usually so rough around Ascension that a sort of crane is provided at the landing steps with a hanging rope, by which one can swing oneself on shore from a boat when it is too rough for the boat to come close to the steps.

Close to the Dockyard is the Turtle Pond, in which there were over a hundred Turtles at the time of our visit. At the side of the pond an enclosed area of sand is provided, in which the Turtles dig great holes, large enough to bury themselves in, laying their eggs at the bottom of them. Some Turtles were still laying, but a good many lots of eggs were beginning to hatch out.

The Turtle eggs have a flexible leathery shell, and are rather smaller than a billiard ball, and of the same shape. The fresh-laid egg is never quite full, so that there is always a slight fold or wrinkle in the yielding shell, and the seamen sometimes puzzle themselves by trying to squeeze the egg so as to get the dint out, but it always forms in a fresh place notwithstanding their efforts. When the eggs are near the time of hatching they are perfectly filled out, the shell being tense, no doubt from the development of small quantities of gas within it.

The sand in which the eggs are hatched does not feel warm

O O

to the hand, but rather, in the daytime at least, cool, and it is always moist. I gathered several sets of eggs, placed them in large vessels full of sand, and took them on board the ship, thinking that I should easily succeed in hatching them artificially. I wished to obtain eggs in all stages of development. I found, however, that all my eggs perished within a couple of days. No doubt a certain definite amount of moisture must necessarily be maintained in the sand as well as a certain constant temperature in order to keep the eggs alive and develop them. I exposed the sand in which my eggs were to the sun in the daytime and covered it up at night.

I used to imagine, from what I had read, that Turtles' eggs were hatched by the direct daily heating by the sun of the sand in which they were buried. What appears to be the case is, however, that the eggs are buried at such a depth that the sand there maintains a constant mean temperature, never hot and never cold. The eggs of a species of Mound Bird (*Megapodius*) are hatched under closely similar conditions in the Philippine Islands.*

The young Turtles fresh from the eggs are kept as pets by the seamen at Ascension in buckets of sea-water. They eat chopped-up raw meat ravenously, using their fore-fins to assist their beak-like jaws in tearing the morsels. Turtle-meat is served out twice a-week as rations to the inhabitants of Ascension, who are all naval employés. The island is commanded by a captain, and is treated by the Admiralty as a man-of-war, a sort of tender to the "Flora," the Guardship stationed at the Cape of Good Hope, to which the Ascension officers theoretically belong.

I paid a visit in the small steam-vessel which is employed in collecting Turtles from the various bays of the island to Boatswain-Bird Island, a breeding-place of various Sea Birds. As we steamed along the shore of the main island large Flying Gurnets (*Dactylopterus*) rose, scared by the vessel, and skimmed rapidly away in front of the bows. I stood in the bows with my gun and tried to shoot Flying Fish on the wing, a novel experience, but quite without success. The flight was rapid and the boat was in constant motion, pitching and rolling; no doubt in calm weather the thing might be done.

* See page 403.

Boatswain-Bird Island is a high rock separated from the main island by a narrow channel. The sides of the rock are precipitous, but some sailor had managed to climb up and fix a rope at the summit, so that it hung down the cliff. The cliff surface was covered with guano, hanging everywhere upon it in large projecting masses and stalactite-like formations. We clambered up the cliff by means of the rope, being half blinded and choked by the guano dust on the way.

In holes on the sides of the cliff, burrowed in the accumulated guano, nest two kinds of Tropic Birds (*Phaethon æthereus* and *P. flavirostris*). In bracket-like nests, as at St. Paul's Rocks, fixed against the lower parts of the cliffs, breeds a species of Noddy (*Anous*), and together with these birds, a beautiful small snow-white Tern with black eyes (*Gygis candida*), called by the seamen the White Noddy, to distinguish it from the Black Noddy.

The summit of the rock is flat, and the plateau is covered with guano, in hollows on which nest the Booby (*Sula leucogaster*) and a Gannet (*S. piscatrix*), and the Frigate Bird (*Tachypetes aquila*). The throat of the Frigate Bird hangs in the form of a sort of pouch in front. This pouch is bare of feathers and coloured of a brilliant vermilion, looking as if rubbed over with some bright red powder. The bird is thus very handsome.

All the birds allowed themselves to be knocked over with sticks on their nests or when near them on our first reaching the plateau, but they soon became generally alarmed and took to flight. The Frigate Birds were on the look out whenever the Gannets were molested, and snatched the small fish which they disgorged, profiting thus by the general disaster. A single "Wideawake," the name given to the Tern (*Sterna fuliginosa*), which breeds in millions gregariously at "Wideawake fair" on the main island, was found on the plateau. The bird was nesting all alone amongst the Gannets for some reason or other.

It was striking to find breeding thus in the middle of the Atlantic, on the top of a steep volcanic rock, the same assemblage of birds which we had seen breeding together on a coral island at sea-level off the north-east coast of Australia. At this latter island, namely Raine Island,* there is a third

* See page 348.

species of Gannet and no *Gygis;* but a Frigate Bird, the same Noddy, the same two Gannets, and the "Wideawake" breed there together as at Ascension, and also one of the species of "Tropic Birds" of Ascension.

After a halt at Porto Praya, and St. Vincent Cape Verde Islands, the ship was steered for England, but being long delayed by contrary winds, had to put into Vigo for more coals before it reached the Channel, and anchored at Spithead in the evening of May 24th, 1876.

CHAPTER XXII.

LIFE ON THE OCEAN SURFACE AND IN THE DEEP SEA. ZOOLOGY AND BOTANY OF THE SHIP. CONCLUSION.

Plants of the Ocean Surface. Fauna of the Sargasso Sea. Protective Colouring of Pelagic Animals. Variety of Pelagic Animals. Flight of the Albatross. Flight of Flying-fish. A Pelagic Insect. Pelagonemertes described. Phosphorescence of Pelagic Animals. Giant Pyrosoma. Uncertainty as to Range in Depth of Pelagic Animals. The Depth of the Oceans and Depression on the Earth's Surface. Deep-Sea Dredging. Vast Pressure existing in the Deep Sea. Experiment showing this made by Mr. Buchanan. Conditions under which Life Exists in the Deep Sea. Range of Plants in Depths. Food of Deep-Sea Animals. Experiment on Rate of Sinking of a Salpa. Vegetable and Animal Débris Dredged from Great Depths. The Deep Sea, a High Road for Distribution of Animals. Deep-Sea Faunas and Alpine Floras Compared. Nature of Deep-Sea Fauna a source of Disappointment. Remarkable Deep-Sea Ascidian. Localities specially Rich in Deep-Sea Forms. Relations of Deep-Sea Animals to One Another. Phosphorescent Light in the Deep-Sea. Colours of Deep-Sea Animals. Cockroaches, Moths, Mosquitos, House-flies, Crickets, Centipedes and Rats on board the "Challenger." Plants on Board the Ship. Pet Parrot, Cassowary, Ostriches, Tortoises, Spiders, Fur-Seal, and Goat on Board. Adaptation to Sea Life. Smallness of the Earth's Surface. Slow Rate of Travelling. Man and possibly Protoplasm existent on the Earth alone. Necessity for Immediate Scientific Investigation of Oceanic Islands.

Plants and Animals of the Ocean Surface.—The three-fourths of the surface of the earth which is covered with sea is thickly tenanted by its own peculiar forms of vegetable and animal life. These forms of life are termed " Pelagic," to distinguish them from the Marine animals and plants which inhabit the shores and sea-bottoms; they inhabit the surface waters of the open ocean and reach the shores only when washed thither accidentally by the waves and currents. Some of these forms, such as the *Pteropods, Ctenophora* and *Siphonophora,* belong to groups peculiar to the sea surface, and have, no doubt, a most ancient connection with it, whilst others are forms, the more immediate progenitors of which lived a terrestrial or littoral existence, and which, having taken to Pelagic habits, have become modified only in less important particulars of their structure to suit their new habits of life.

The surface water of the open ocean is full of vegetable life. Diatoms are to be found with the surface net everywhere, and in high northern and southern latitudes* they abound extremely, so as to colour the ice with their *débris,* change the tint of the water, fill the towing net up with slimy masses and cover the deep-sea bottom with a silicious deposit of their skeletons.

In tropical seas, other lowly organized algæ especially abound : mainly *Oscillatoriæ,* of the genus *Trichodesmium.* These algæ occur in the water as small brown faggots of minute threads, resembling, as Mr. Berkeley says, minute fragments of chopped hay. Together with these forms others often occur in which the threads are gathered into small globular masses with the ends of the threads all directed outwards. When tracts of the sea are passed through, which are full of this *Trichodesmium,* the water lighted up by sunlight, when looked down into, appears as if full of small particles of mica, or some such substance, so strongly is the light reflected from the minute bundles of the algæ.

We met with this alga in greatest abundance in the Arafura Sea, between Torres Straits and the Aru Islands. Here it was at first encountered discolouring the sea-surface in bands and streaks ; as the ship moved farther on, it became thicker, and at length the whole sea, far and wide, was discoloured with it. It remained still, however, denser in long streaks, and within these again it was massed in small patches. There was a strong smell from these patches, as from a pond covered with vegetation. So abundant is *Trichodesmium* in some seas, that one of the explanations of the name of the Red Sea is that the term was derived from the discolouration of the water by vast quantities of *Trichodesmium Erythrœum.*

* See page 249.

On the voyage from Ternate to the Philippine Islands, the sea was again seen to be full of minute algæ. In this case there were several other forms beside *Trichodesmium*, and they were embedded together in small masses of a jelly-like substance, which also contained Diatoms. The water was perfectly full of these masses, and tinted by them of a light brownish colour.

Besides these smaller algæ living in the open ocean, there are abundance of several species of larger seaweeds which are Pelagic in habit. The Gulf Weed, *Sargassum bacciferum*, of the Sargasso Sea in the Atlantic, is well-known. It is brown when dried or preserved, but when living is of a very bright yellow colour, which contrasts pleasingly with the deep blue of the open Atlantic. Another seaweed (*Fucus vesiculosus*) is to be found also living free in the Atlantic, and the Giant Kelp (*Macrocystis pirifera*), in the floating condition, ranges over a wide belt of the Southern Ocean, as proved by Sir Joseph Hooker.*

All these seaweeds grow attached to rocks on various shores as well as free, but they all produce spores, only when attached. The Pelagic varieties multiply only by simple growth and sub-division. A wide area covered with seaweeds corresponding to the Sargasso Sea occurs in the North Pacific Ocean.

Were it not for the existence of this vast Pelagic vegetation the Pelagic fauna would be but a scanty one, since the *débris* derived from the land could only support a small amount of animals. Plants are as necessary in the open sea as on land to form the starting-point of the organic cycle by building up those compounds required by animals as food. The algæ, though brown in appearance, contain and build up *Chlorophyll*, the same green colouring matter as that which tinges the leaves of our trees and plants on land, and which is now the only starting-point and foundation-stone of life.

The Sargasso Sea has its own fauna of animals specially adapted to life amongst the Gulf Weed. Amongst these there is a small fish, *Antennarius*, allied to the Angler, which has long arm-like fore-fins with which it clings on to the bunches of Weed. The fish makes a nest of the Weed, binding together a globular mass of it, as big as a Dutch cheese, by means of long

* " Flora Antarctica," Vol. I, pp. 464–465.

sticky gelatinous strings, which it forms for the purpose.   In
the centre of the nest are deposited the eggs.

The Weed is much encrusted by a Bryozoon (*Membranipora*),
which makes conspicuous white patches upon its surface.
Numbers of the detached air-vessels of the Weed are to be seen
floating about amongst the living Weed-beds, coated entirely
with the white *Membranipora,* and they look at first like small
globular Pelagic animals.

All the inhabitants of the Gulf Weed are most remarkably
coloured, for purposes of protection and concealment, exactly like
the Weed itself.   The Shrimps and Crabs which swarm in the
Weed are of exactly the same shade of yellow as the Weed, and
have white markings upon their bodies to represent the patches of
Membranipora.   The largest shrimp occurring has a dark-brown
colour with brilliant-white sharply defined areas upon its surface,
thus closely resembling the older darker-coloured pieces of Weed,
which are also most thickly covered with *Membranipora.*

The small fish (*Antennarius*) is in the same way coloured
Weed-colour with white spots.   Even a Planarian worm, which
lives in the Weed, is similarly yellow-coloured, and also a Mollusc
(*Scyllæa pelagica*).   The white patches on some of the Crabs, no
doubt, represent also, to some extent, the white shells of
Barnacles, though these are not very abundant in the Weed.   A
small Crab, *Nautilograpsus minutus,* which varies very much in
colour, very abundant amongst the Weed, is constantly to be
found also in large numbers hanging on to floating logs and
similar objects elsewhere, and in these cases the white patches
on its body correspond closely with the barnacles by which the
logs are covered.   These little crabs vary extremely in the
arrangement and forms of the white patterns on their backs,
and we found a number of them once (I believe of the same
species) which were clinging to the floats of the blue-shelled
Pelagic Mollusc *Ianthina,* and these were all coloured, for concealment, of a corresponding blue.

Pelagic animals generally seem to be either colourless or
specially coloured, with a view to protection from enemies both
above and below the surface of the water.   Probably the blue
colour of *Ianthina* and *Velella* is protective as resembling that

of the ocean water. *Velella* has serious enemies in the oceanic birds and in turtles. We caught a small turtle (*Chelone imbricata*) which had its stomach full of *Velellas*. There are numerous other Pelagic animals thus coloured blue for protection, such as the Mollusc *Glaucus, Porpita* allied to *Velella*, and some *Salpœ* in which the nucleus is blue. There are also blue *Medusœ*.

The dark red-brown colour of the nucleus of most *Salpœ* is probably an imitation of that of floating seaweed, and it occurs in several other Pelagic animals, as, for example, *Pelagonemertes*. The extraordinary transparency of most Pelagic animals is, no doubt, a protective contrivance. In both *Salpa* and *Pelagonemertes*, above referred to, almost the entire body, with the exception of the smaller parts coloured brown, as described, are colourless and transparent, like glass. It is extremely difficult to see these transparent animals, when one attempts to collect them from a boat.

Almost all classes of land or shore animals seem to have contributed to the Pelagic fauna forms which have become in most cases extremely modified to suit their changed mode of existence. Amongst Mammals there are the Whales and Porpoises, the ancestors of which, no doubt, long after they had deserted the land and had taken to a Pelagic existence, came on shore regularly, like the Seal, at certain seasons to breed, but at length acquired the power of even rearing their young in the open sea.

Amongst birds the Petrels are Pelagic in habit, the largest amongst them being the Albatross. Of the various kinds of Petrels we necessarily saw a great deal. They were our constant companions in the Southern Ocean, following the ship day after day, dropping behind at night to roost on the water and tracing the ship up again in the early morning by the trail of *débris* left in its wake.

The Oceanic Petrels have reduced the science of flight to the condition of a fine art. The flight of the Albatross has always excited wonder and admiration, nevertheless, some of the smaller Petrels fly quite as well. There are almost all gradations to be observed in the powers of flight of different birds, in the various stages of perfection in the shaping of the wings, and the skill of the use of them shown by the birds. Refinement in the art of the use of the wings by birds seems to run in two different directions. The flight of the Albatross, regarded as the perfection of one mode, the soaring method, performed by aid of great length of wing, may be contrasted with that of the Humming Bird, equally perfect in its way and far more rapid, but performed by the use of short wings and excessively rapid motion of them.

The movement of the Albatross may be compared to that of a skilful skater on the outside edge; the Humming Bird's flight is just like that of an insect. The Albatross ekes out to the utmost the momentum derived from a few powerful strokes, and uses it up slowly in gliding, making all possible use at the same time of the force of the wind.

I believe that Albatrosses move their wings much oftener than is suspected. They often have the appearance of soaring for long periods after a ship without flapping their wings at all, but if they be very closely watched, very short but extremely quick motions of the wings may be detected. The appearance is rather as if the body of the bird dropped a very short distance and rose again. The movements cannot be seen at all unless the bird is exactly on a level with the eye. A very quick stroke, carried even through a very short arc, can of course supply a large store of fresh momentum. In perfectly calm weather, Albatrosses flap heavily.

The Great White Albatrosses which are seen behind ships, are usually by no means beautiful objects. The long wings look far too long for the body, and being so narrow, the body looks heavy and out of proportion to them. Further, five out of six of the birds seen are young ones, in immature brown plumage, and look dirty and draggled. The old birds when in their best breeding plumage, as seen on their nests, are handsome enough.

Whilst on the subject of flight, I would say a few words about the flight of the Flying-fish. Dr. Möbius has lately produced an elaborate paper* on the much vexed question as to

* K. Möbius, Die Bewegungen der fliegenden Fische durch die Luft z. für. Wiss. Zool. 1878, s. 343.

whether Flying-fish move their wings in flight or not, and after examination of the muscular apparatus, and watching the living fish, has come to the conclusion that they do not do so at all. There are two widely different genera of fish, which have developed long wing-like fins for support in progress through the air, the ordinary Flying-fish, the various species of *Exocœtus* allied to the Gar-fish, and the Flying Gurnets, species of the genus *Dactylopterus*.

I have never seen any species of *Exocœtus* flap its wings at all during its flight. These fish merely make a bound from the water, and skim supported by their extended fins, the tips of which meanwhile quiver in the air somewhat occasionally, from the action of air-currents against them, and sometimes from the shifting a little of their inclination by the fish.

I believe, however, that I cannot be mistaken in my conviction, that I have distinctly seen species of Flying Gurnets move their wings rapidly during their flight. I noticed the phenomenon especially in the case of a small species of *Dactylopterus* with beautifully coloured wings, which inhabits the Sargasso Sea. Whilst out in a boat collecting animals amongst the Gulf Weed, these small Flying Gurnets were constantly startled by the boat and flew away before it, and as they did so, appeared to me to buzz their wings very rapidly.

Their mode of flight seemed to me to be closely similar to that of many forms of Grasshoppers, which cannot fly for any great distance, but raise themselves from the ground with a spring, and eking out their momentum as much as they can by buzzing their wings, fall to the ground after a short flight.

I watched these little Flying-fish fly along before the boat, at a height of about a foot above the water, for distances of 15 or 20 yards, and I chased them and caught one or two with a hand net amongst the Weed. Dr. Möbius who similarly watched the flight of a species of Flying Gurnet maintains that neither forms of Flying-fish flap their wings at all during flight. I do not consider the question as yet set at rest. Of course no Flying-fish can raise themselves in the air at all by means of their wings alone.

There are even Pelagic insects. One of these (*Halobates*) was constantly caught during our voyage in the towing net in the open ocean. The Atlantic species differs from the Pacific one. The insect is one of the Bug family, with a small round wingless body and long legs, and is black coloured. It is closely allied to the long-legged insects (*Gerrys*) which are so commonly to be seen resting on the surface of ponds and ditches in England, moving along by a series of jerks, and casting curious looking shadows on the bottoms of shallows when the sun is overhead. The *Halobates* lives entirely at sea, and carries its eggs about attached to its body.

Most fish live about the coasts, and comparatively few are met with far away from land, but there are regular Pelagic fish. There are Pelagic Mollusca of all kinds, including perfectly transparent Cuttle-fish, transparent Pelagic Crustaceans, transparent Pelagic Annelids, and Pelagic Planarian worms.

There are even Pelagic Sea Anemones (*Nautactis* and its allies) which have their bases, by means of which shore-inhabiting Sea Anemones cling to the rocks, so modified as to form chambers containing air, and thus acting as floats. Many Pelagic animals form highly complex colonies, which float about in the surface water, combined in one mass. Such are Chain-Salpæ and *Pyrosoma*. In some of these compound organisms, such as the *Siphonophora*, there is a complex combination of variously modified zooids, with a complex division of labour amongst the members composing the colony, just as amongst the closely allied *Stylasteridæ*. The *Siphonophora* like the *Stylasteridæ* are Hydrozoa, but the compound organisms they form, are soft, hyaline, and free-swimming, whilst the stocks formed by the *Stylasteridæ* are stony, hard, opaque, and firmly rooted to the sea bottom.

I have described a Land Nemertine worm,* which exists in Bermuda. Nemertines however, though like Planarians normally shore inhabiting animals, have adapted themselves not only to terrestrial, but also to Pelagic existence. One of the most remarkable animals discovered by the "Challenger" Expedition, is a Pelagic Nemertine, which I have called *Pelagonemertes Rollestoni*, after my friend Prof. Rolleston of Oxford.

The body of the animal is leaf-shaped and gelatinous, and perfectly transparent, with the exception of the digestive tract,

* See page 27.

which is branched as in Planarians,* and is of a burnt-sienna colour. The worm is provided with a proboscis like that of other Nemertines, which may be compared with that shown in the figure of the Land Nemertine, but it is not armed with stylets as in the latter animal. *Pelagonemertes* is devoid of eyes and apparently of any other special sense organs. It constitutes a special family of Nemertines, the *Pelagonemertidœ*.†

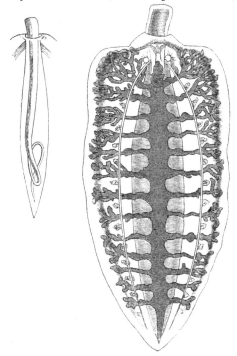

PELAGONEMERTES ROLLESTONI.

The branched digestive tract is shaded dark; behind its central tube is seen the wide sac of the proboscis. The proboscis is seen at the upper extremity of the body, partly protruded. Beneath it is the mouth, with a folded opening shaded dark. On either side of the mouth are the nervous ganglia, giving off each a long nerve tract which passes to the extremity of the body. Just exterior to the nerve tracts on each side is seen a row of ovaries.

The smaller figure shows the proboscis sheath and coiled proboscis, as seen from the hinder surface of the animal.

* Prof. Giard has lately described a gigantic Nemertine (Avenardia Priei) a yard and a half in length, which has a similarly ramified intestine, otherwise this arrangement does not occur amongst Nemertians. Ann. and Mag. Nat. Hist., Sep., 1878.

† For a detailed description of Pelagonemertes, see H. N. Moseley, "On Pelagonemertes Rollestoni." Ann. and Mag. Nat. Hist., March, 1875; Ibid., Dec. 1875.

Most important to the student of deep-sea phenomena, are the Foraminifera with calcareous shells, covered with long delicate tubular calcareous spines, such as *Globigerina* and its allies, which float everywhere on the surface, and the dead shells of which form the vast calcareous deposits on the deep sea bottom of *Globigerina* mud.

At night, the Pelagic animals render themselves conspicuous by their phosphorescence. The kind of light emitted, and the manner of its appearance, varies according to the nature of the animal causing it. Sometimes the sea far and wide, as far as the eye can see, is lighted up with sheets of a curious weird-looking light, and wherever the water breaks a little on the surface before the breeze, the white foam is brilliantly illuminated. This particular kind of illumination is due to *Noctiluca*. One night, when we were between the Cape Verde Islands and St. Paul's Rocks, the sea was thus illuminated by myriads of *Noctiluca* and the lower sails of the ship were seen to be distinctly lighted up by the light given off from the broken water thrown up by the hull of the vessel.

At other times, the water where disturbed is seen to be full of small luminous scintillating specks. This is the commonest form of phosphorescence, and is due to various small animals, principally small Crustacea, which give out their light thus by flashes. Some Crustacea certainly derive their phosphorescence from containing in their stomachs phosphorescent food, and their excrement is phosphorescent, as first pointed out to me by my friend, Captain Tupman, R.M.A. When large fish, or porpoises or penguins, dash through water full of luminous Crustaceans or *Noctiluca*, their bodies are brilliantly lit up, and their track marked as a trail of light.

The most beautiful kind of phosphorescence is however that of the Ascidian colony *Pyrosoma*. This, when stimulated by a touch, or shake, or swirl of the water, gives out a bright globe of blueish light, which lasts for several seconds, as the animal drifts past several feet beneath the surface and then suddenly goes out.

A giant *Pyrosoma* was caught by us in the deep-sea trawl. It was like a great sac, with its walls of jelly about an inch in thickness. It was four feet in length, and ten inches in diameter. When a *Pyrosoma* is stimulated by having its surface touched, the phosphorescent light breaks out at first at the spot stimulated, and then spreads over the surface of the colony as the stimulus is transmitted to the surrounding animals. I wrote my name with my finger on the surface of the giant *Pyrosoma*, as it lay on deck in a tub at night, and my name came out in a few seconds in letters of fire.

Pelagic animals range through a considerable depth of water, near the surface of the sea, ascending to the surface at times, especially at night when safe from enemies, and again descending. It is quite uncertain to what depth they extend their range, and whether there is a zone of water intermediate between that near the bottom and that near the surface, which is devoid or nearly devoid of life, as is believed by Sir Wyville Thomson to be probably the case.

The trawl net used on board the "Challenger" swept, in going down to the deep-sea bottom and in coming up, the entire depth of the sea, and animals were constantly being found in the net, about which it was quite uncertain as to what depth they came from. Amongst these were, for example, some *Medusæ*, which have been found by Prof. Haeckel to be of peculiar structure, and which may possibly be deep-sea forms ; they may, however, also have come from a few fathoms depth only.

A net of some kind is required to settle this question which shall be capable of being sent down completely closed to any required depth, then opened and towed for some time, and then again closed before it is raised. It is by no means an easy matter to devise such a net which will be practically available. There are numbers of animals, fish, *Medusæ*, and *Actinias* for example, which are found in the deep-sea trawl, and about which it is a matter of speculation only as to the depth from which they came.

Mr. Murray hit upon the expedient of using the ordinary towing net at considerable depths* and with great success, since

---

* A. Baur was, I believe, the first to use the towing-net at considerable depths. "Beiträge zur Naturgeschichte der Synapta digitata." Verhandl. der K.L.C.D. Akad. 1864. Mr. Murray, however, invented the method independently.

he constantly obtained large catches of Pelagic animals, when very few were obtainable·at the surface.

Pelagic animals are most widely-spread, closely similar forms occurring in widely distant oceans. In this particular, the Pelagic fauna resembles that of the deep sea. In the case of the sea surface winds and currents are present both to aid or limit the range of species, and the variety of climate acts as a barrier. In the deep sea all these forms of restriction are, however, absent.

The Deep Sea and its Fauna.—I have above briefly described the vegetation and fauna of the ocean surface, because, did these not exist, life would be impossible, or only extremely scanty, in the deep-sea bottom. Before referring to the fauna of the deep sea, it will be well to consider briefly the conditions under which it exists.

If a globe, 40 feet in diameter, be taken to represent the earth, this will be on the scale of 1 foot to 200 miles, or 1 inch to $16\frac{2}{3}$ miles, or 88,000 feet.* Thus on such a globe the highest mountain and the deepest sea would be on true proportional scale represented severally by an elevation or depression of $\frac{1}{3}^{rd}$ of an inch. Were the land surfaces and sea beds sculptured in due proportion on the face of this globe, the surface would at a little distance hardly appear roughened, so insignificant is the altitude of the highest mountains and the depth of even the deepest seas in proportion with the dimensions of the earth itself. The oceans in relation to their superficial area are as shallow as a sheet of water one hundred yards in diameter, and only an inch in depth.†

We are apt to form an erroneous impression as to the actual shapes and distributions of the elevations and depressions on the earth's surface, because only the very tops of the elevations stand above water. The outlines of the various continents and islands with which we are familiar on maps, are merely lines marking the height to which the water reaches up. A very

---

* Lieut. Gen. R. Strachey, R.E., F.R.S., "Lecture on Scientific Geography." Proc. Geogr. Soc., 1877, p. 191.

† James Croll, "Climate and Time," p. 135. London, Daldy & Co., 1875.

small proportion of the elevated masses projects above water, hence from an ordinary map we gain no truer impression of the form of the sculpturing of the surface of the earth itself than we should of the shape of a range of mountains if we viewed it when all but its summits were hidden by a flood.

So small a proportion does the mass of dry land elevated above sea-level bear to the hollows on the earth's surface beneath this level, that the cavities now occupied by the sea would contain three times the volume of the earth existing above the sea surface. If the surface of the land and the sea bottom were brought to a complete level, the waters of the sea covering its even face would still have a depth of about 1,700 fathoms, being reduced in depth by the process only about 800 fathoms.*

We should obtain a more correct idea of what are the real elevations and what the depressions on the earth's surface, if we drew on the map or globe a contour line marking the level at which the mass of the earth raised above this line is just equal to the excavations beneath it, and would just fill up these hollows if the surface of the earth were rendered even and smooth.

Although the depth of the ocean is so small in proportion to the vastness of its expanse, the depth is, nevertheless, so great as to be difficult of adequate realisation. The greatest depth as yet ascertained by sounding occurs, as will be seen from the map at the commencement of this work, in the North West Pacific Ocean; it amounts to about five miles and a quarter.

In order to realize such a depth, the reader should think of a spot distant several miles from his actual position, and then attempt to project the distant point downwards, until it lies vertically beneath him. The average depth of the ocean between lats. 60° N. and 60° S.† is about three miles or 2,500 fathoms. The great depth of five miles occurs only exceptionally over very small areas.

The vastness of the depth of the Ocean was constantly brought home to us on board the "Challenger" by the tedious

* O. Peschel, "Neue Probleme der Vergleichenden Erdkunde." Leipzig, 1876, s. 82.

† J. J. Wild, "Thalassa," pp. 14–15. London, Marcus Ward & Co., 1877.

length of time required for the operations of sounding and dredging in it. When the heavy sounding weight is dropped overboard, with the line attached, it takes about an hour and a quarter to fall to the depth of 4,500 fathoms, and thirty-five minutes to reach the bottom in the average depth of 2,500 fathoms.

The winding in of the line again, is a much slower process. It used to take us all day to dredge or trawl in any considerable depth, and the net usually was got in only at nightfall, which was a serious inconvenience, since we could not then, in the absence of daylight, make with success the necessary examinations of the structure of perishable animals.

The ship, when deep-sea operations were going on, used to lie rolling about all day, drifting along with the wind, and dragging the dredge over the bottom. From daybreak to night the winding-in engine was heard grinding away with a painful noise, as the sounding-line and thermometers were being reeled in.

At last, in the afternoon, the dredge-rope was placed on the drum, and wound in for three or four hours, sometimes longer. Often the rope or net, heavily weighted with mud, hung on the bottom, and there was great excitement as the strain gradually increased on the line. On several occasions the rope broke, and the end disappeared overboard ; three or four miles of rope and the dredge being thus lost.

At first, when the dredge came up, every man and boy in the ship who could possibly slip away, crowded round it, to see what had been fished up. Gradually, as the novelty of the thing wore off, the crowd became smaller and smaller, until at last only the scientific staff, and usually Staff Surgeon Crosbie, and perhaps one or two other officers besides the one on duty, awaited the arrival of the net on the dredging bridge, and as the same tedious animals kept appearing from the depths in all parts of the world, the ardour of the scientific staff even, abated somewhat, and on some occasions the members were not all present at the critical moment, especially when this occurred in the middle of dinner-time, as it had an unfortunate propensity of doing. It is possible even for a naturalist to get weary even of deep-sea dredging. Sir Wyville Thomson's enthusiasm never flagged, and I do not think he ever missed the arrival of the net at the surface.

Often when the dredge or trawl appeared, there was nothing in it at all, and then frequently a somewhat warm debate ensued between the members of the scientific staff and the naval officers as to whether the instrument had ever been on the bottom or no, the scientific view being that it had not.

Sometimes there would be only a bright red Shrimp in the net; and this fact, on the one side, would be held as proof that the bottom had been reached, whilst, on the other, it was maintained that the Shrimp probably inhabited a region lying at some distance above the bottom. The sledge irons of the trawl-net were carefully examined as evidence in the matter, to test whether they had been polished by friction on the bottom or no, or whether they had any mud adhering to them. In future dredging operations, it would be well to have a small cup with a valve to it attached to the dredge or trawl, so that it shall always retain a little of the bottom, and prevent the possibility of the occurrence of such doubts.

The conditions under which life exists in the deep sea, are very remarkable. The pressure exerted by the water at great depths is enormous, and almost beyond comprehension. It amounts roughly to a ton weight on the square inch for every 1,000 fathoms of depth, so that at the depth of 2,500 fathoms, there is a pressure of two tons and a-half per square inch of surface, which may be contrasted with the 15 pounds per square inch pressure to which we are accustomed at the level of the sea surface.

An experiment made by Mr. Buchanan enabled us to realize the vastness of the deep-sea pressure more fully than any other facts. Mr. Buchanan hermetically sealed up at both ends, a thick glass tube full of air, several inches in length. He wrapped this sealed tube, in flannel, and placed it, so wrapped up, in a wide copper tube, which was one of those used to protect the deep-sea thermometers when sent down with the sounding apparatus.

This copper tube was closed by a lid fitting loosely, and with holes in it, and the copper bottom of the tube similarly had holes bored through it. The water thus had very free access to the interior of the tube when it was lowered into the sea, and the tube was necessarily constructed with that object in view, in order that in its ordinary use the water should freely reach the contained thermometer.

The copper case containing the sealed glass tube was sent down to a depth of 2,000 fathoms, and drawn up again. It was then found that the copper wall of the case was bulged and bent inwards opposite the place where the glass tube lay, just as if it had been crumpled inwards by being violently squeezed. The glass tube itself, within its flannel wrapper, was found when withdrawn, reduced to a fine powder, like snow almost.

What had happened was that the sealed glass tube, when sinking to gradually increasing depths, had held out long against the pressure, but this at last had become too great for the glass to sustain, and the tube had suddenly given way and been crushed in the violence of the action to a fine powder. So violent and rapid had been the collapse that the water had not had time to rush in by means of the holes at both ends of the copper cylinder, and thus fill the empty space left behind by the collapse of the glass tube, but had instead crushed in the copper wall, and brought about equilibrium in that manner.

The process is exactly the converse of an explosion, and is termed by Sir Wyville Thomson an "implosion." Gunpowder exploded in the centre of a similar copper tube would in a corresponding manner, have bulged the sides of the tube outwards, notwithstanding the existence of the openings at its ends.

Marine animals, no doubt, easily accommodate themselves to these enormous pressures in the deep sea. Their tissues being entirely permeated by fluids, the pressure has little or no effect upon them. Moreover amongst all the various animals dredged up from great depths, it is only some fish which show any marked effects of the alteration of pressure to which they are subjected in being brought to the surface. Fish with swimming bladders come up in the deep-sea dredge in a horribly distorted condition, with their eyes forced out of their heads, their body tense and expanded, and often all their scales forced off.

No sun-light penetrates the deep sea; probably all is dark below 200 fathoms, at least excepting in so far as light is given out by phophorescent animals. At depths 2,000 fathoms and upwards the temperature of the water is never many degrees above the freezing point.

The nature of the food of deep-sea animals has been a matter of some considerable speculation.* Owing to the lack of sunlight in the depths, there is an entire absence there of vegetable life, such as could build up the necessary food of the animals living there, and thus render the cycle of life in those regions self-supporting and complete as it is on land and in the shallow seas.

Dr. Carpenter tells me he dredged living calcareous algæ (*Corallinaceæ*) in the Mediterranean Sea at a depth of 150 fathoms. As far as I observed, the "Challenger" dredgings did not on any occasion yield algæ from a depth so great. The greatest depth from which seaweeds were dredged by us in any quantity was, I believe, 30 fathoms. It is a curious fact that a species of *Halophila*, one of the Sea Grasses, which are flowering-plants which have become modified to a marine existence, was obtained by us in abundance off Tonga Tabu from so great depth as 18 fathoms. At this depth it was, when we obtained it, in full flower.

The only plants which extend their range to any great depth are certain lowly organized parasitic *Thallophytes*, which infest corals and bore for themselves branching tubular cavities in the hard skeletons of their hosts. These parasites have been found by Prof. Martin Duncan in corals which have been dredged from a depth of 1,095 fathoms.† These plants, nourished on the tissues of their hosts, are able to thrive without the aid of sunlight, just as do fungi in dark cellars and mines.

In the absence of plants amongst them, the deep-sea animals have to derive their food entirely from the *débris* of animals and plants falling to the bottom from the waters above them. This *débris* is no doubt mainly derived from the surface Pelagic flora and fauna, but also to a large extent composed of refuse of

---

* See K. Möbius, "Wo kommt denn die Nahrung von den Tiefsee-thieren her." Z. f. Wiss., Zool. 21. Bd. Heft 2.

† P. M. Duncan, F.R.S., &c., "On some Thallophytes parasitic within recent Madreporaria." Proc. Roy. Soc., 1876, p. 538.

various kinds washed down by rivers, or floated out to sea from shores and sunken to the bottom when water-logged.

The dead Pelagic animals must fall as a constant rain of food upon the habitation of their deep-sea dependants. Maury, speaking of the surface Foraminifera, wrote, "the sea, like the snow-cloud, with its flakes in a calm, is always letting fall upon its bed showers of microscopic shells."*

It might be supposed that these shells and other surface animals would consume so long a time in dropping to the bottom in great depths that their soft tissues would be decomposed, and they would have ceased to be serviceable as food by the time they reached the ocean bed. Such is, however, not the case, partly because the salt water of the sea exercises a strongly preservative effect on animal tissues, partly because the time required for sinking is in reality not very great.

In order to test the matter for myself I made the following experiment. I took a dead Salpa, of about 2 inches in length, and placed it in a glass cylinder full of water, and 3 inches in diameter. I allowed the Salpa to fall from the surface of the water in the cylinder to the bottom a number of times and noted carefully the time which it took to traverse this distance, which was about 8 inches. I found that on an average it took 20 seconds to fall the 8 inches. This gives at the same rate, without allowance for acceleration, a distance of a fathom to be traversed in three minutes, or 2,000 fathoms in four days four hours.

I allowed the Salpa to remain in the sea water in the cylinder for a long time. It was still not greatly decomposed after having remained in the same water for a month, whilst the ship was in the tropics; the nucleus was after this interval still undestroyed. The dead animal might have thus sunk to the bottom in the greatest depths almost six times over without having become so much decomposed as to be unserviceable for food to deep-sea animals.

We obtained by our dredgings several interesting proofs of the feeding of deep-sea animals on *débris* derived from neighbouring shores. Thus, off the coast of New South Wales we dredged from

* M. F. Maury, LL.D., "The Physical Geography of the Sea," 15th Ed., p. 322. London, Sampson Low and Marston, 1874.

400 fathoms a large Sea-Urchin which had its stomach full of pieces of a Sea Grass (*Zostera*) derived from the coast above.

Again, we dredged between the New Hebrides and Australia from 1,400 fathoms, a piece of wood and half-a-dozen examples of a large palm fruit as large as an orange. In one of these fruits which had hard woody external coats, the albumen of the fruit was still preserved and perfectly fresh in appearance, and white, like that of a ripe cocoanut. The hollows of the fruits were occupied by a small Lamellibranch Mollusc and a Gasteropod, and the husks and albumen were bored by a small Teredo or allied Mollusc. The fibres of the husks of the fruits had amongst them small Nematoid Worms.

We dredged up similar land vegetable *debris* on many other occasions, of which I will cite some, because they are interesting, not only as showing that deep-sea animals must derive food largely from such sources, but because they are necessarily of great geological importance as showing how specimens of land vegetation are becoming imbedded in deposits which are being formed at very great depths.

Between the Fiji Group and the New Hebrides we dredged from 1,450 fathoms a piece of a branch of a tree, 3 feet in length. Off the Island of Palma, one of the Azores, we dredged from 1,135 fathoms, the leaf of a Shrub, possibly a Holly-leaf which was still green and firm, though water-logged. With this leaf were numerous fish otoliths and eye-lenses. We constantly dredged bones of whales and fish from great depths. Off the coast of Nova Scotia we dredged a quantity of glacially striated stones.

The deep-sea animals of course prey upon one another just as do shallow-water species. We dredged once a fish from 2,500 fathoms which had a deep-sea Shrimp in its stomach. A Cerianthus dredged from 2,175 fathoms had a small Crustacean in its stomach.

The waters of the deep sea being everywhere dark and always cold, the conditions of life in them are the same all over the world. The temperature of the deep sea is practically the same, as far as effect on life is concerned, under the Equator and at the Poles. Hence there are absolutely no barriers to the migrations of animals in the deep sea. Time only is required for any deep-sea animal to roam from any distant part of the earth to another.

It is only in the strata of water, comparatively near the ocean surface, that there is any great difference in range of temperature in various latitudes. Up to a depth of 1,000 fathoms, even from the greatest existing depths, the range amounts only to a few degrees Fahrenheit ; and at 1,000 fathoms everywhere the water is cold and dark, and the conditions of life practically the same as those in the greatest depths ; even at a depth of 500 fathoms the water is almost everywhere as cold as 40° F.· The effects of difference of pressure may be neglected, since, when encountered gradually, they would be of no injury to migrating animals.

Hence, even the ridges, which project up from the ocean floor and separate areas of great depths from one another by intervening expanses, over which the depth is only 1,000 fathoms or somewhat less, do not oppose any obstacle to the migration of deep-sea animals. Such ridges will be seen, by reference to the map at the commencement of this work, to exist in the Atlantic and Pacific Oceans.

In the Atlantic, a long sinuous ridge, with a depth of only 1,000 fathoms over it, separates the two deep troughs on either side of the Atlantic from one another, and were the conditions existing in 1,000 fathoms very different from those obtaining in depths of 2,000 and 3,000 fathoms, it might well be conceived that the Western Atlantic deep-sea animals might be isolated from those of the Eastern Atlantic, and very greatly different from them. As will be seen from the map, there is only one narrow channel, lying just north of Tristan da Cunha, in the South Atlantic, where a depth of 2,000 fathoms extends over from one side of the Atlantic to another, and by which thus migration in the supposed case would be possible.

Similarly in the case of the Pacific, there is only a narrow channel, situate between the Fiji Group and Tahiti, by which the deep waters of the Southern Pacific communicate directly with those of the Northern.

The deep-sea animals are however not restricted by these ridges, and the shallows of 1,000 fathoms depth do not act as barriers. Were there any marked isolation by great depth, we might have hoped to have met with animals of great antiquity in the deepest holes, since these must possibly be regarded as occupying the sites of very old depressions on the earth's surface.

Dr. Wallich, in his celebrated work, "The North Atlantic Sea Bed," unfortunately never completed, though so full of most important discussions of deep-sea phenomena, speaks almost prophetically of the migrations of animals which "must take place along the deep homothermal sea; that great highway, extending from Pole to Pole, which is for ever closed to human gaze, but may nevertheless be penetrated by human intelligence."*

Marine animals may throughout all time have migrated in the course of generations across the equator, from north to south, by way of the deep sea, and on reaching temperate or cold latitudes, may have worked their way up into shallow water and taken to coast life, and assumed forms more or less like those of their ancestors who started on the journey.

Regarded as a high-road for migration across the equator, the deep sea may well be compared with the summits of those mountain chains which, in a similar manner, have acted as bridges across the tropics for the passage of non-tropical plants. The deep-sea animals themselves also, considered as a group, may be well compared to Alpine floras, there being many points of analogy between the two assemblages.

As in the case of Alpine floras, plants which occur at sea-level in cold or arctic regions, are found on high mountains in temperate or tropical latitudes; so in the case of the deep sea, certain animals which in high northern or southern latitudes exist in comparatively shallow water, occur at great depths near the equator. Again, just as Alpine floras consist to a considerable extent of modifications of forms growing at lower levels in other regions of the earth, altered somewhat in non-essentials to suit an Alpine existence, rather than of ancient and isolated forms greatly differing from those of the lowlands; so in the case of the deep-sea fauna, hardly any of the animals dis-

* G. C. Wallich, M.D., F.L.S., F.G.S., Surgeon-Major on the Retired List, H.M. Indian Army, "The Atlantic Sea Bed," Pt. 1, p. 105. London, Van Voorst, 1862.

covered as composing it are of any very important or widely aberrant zoological structure.

Just as some members of Alpine floras are dwarfed by the climate to which they are exposed, so does it occur in the case of some of the deep-sea animals : but by no means in that of all, for some forms seem even to increase in size, through their existence in the great depths. A deep-sea Cerianthus a Sea Anemone living in a tube, already described in this work,* may be cited as an instance of dwarfing. *Pycnogonids* may be referred to as examples of increase of size in great depths. We dredged in deep water gigantic examples of these latter animals, measuring more than a foot between the tips of the legs. Nearly all Crustacea seem to increase in size in the deep sea; we dredged large specimens of *Serolis* and other large Isopods, and large *Scalpellums;* the Decapod Crustacea obtained were however none of them as large as the larger shallow-water forms.

One coral, *Bathyactis (Fungia) symmetrica,* ranges from a depth of 30 fathoms to one of 2,900 fathoms, and varies very much in size. No very large specimens were obtained in small depths; but very small adult specimens were found in great depths, and no direct connection between increase of depth and increase in dimensions was able to be determined in this case, though the great number of specimens obtained rendered the case a good one for examination with regard to the question under consideration.

In many respects, the zoological results of the deep-sea dredgings were rather disappointing. Most enthusiastic expectations were held by many naturalists, and such were especially put forward by the late Prof. Agassiz,† who had hopes of finding almost all important fossil forms existing in life and vigour at great depths. Such hopes were doomed to disappointment, but even to the last, every Cuttlefish which came up in our deep-sea net was squeezed to see if it had a Belemnite's bone in its back, and Trilobites were eagerly looked out for.

---

* See p. 408.
† " A Letter concerning Deep-Sea Dredging, addressed to Prof. Benjamin Pierce, Superintendent of the U.S. Coast Survey." Ann. & Mag. Nat. Hist. 1872, p. 169.

A certain number of animal forms have been obtained in the living condition from the deep sea, which were supposed, until thus found, to be extinct, and to exist only as fossils; but there are a considerable number of shallow-water and terrestrial forms which have similarly survived for long periods, and exist in the fossil condition as well as in the living one. The exploration of any vast hitherto uninvestigated area must necessarily add from amongst the numerous animal forms discovered in it, some to the list of those which are both fossil and recent. It has yet to be shown, that in the case of the deep-sea fauna, the numbers of such comparatively long-lived forms, are greater proportionately than in that of shallow water faunas.

Large numbers of interesting new genera and species of well-known families of animals were obtained by the dredge, but very few which were widely different in their essential anatomical structure from hitherto known forms, and thus of first-rate zoological importance. We picked up no missing links to fill up the gaps in the great zoological family tree. The results of the " Challenger's " voyage have gone to prove that the missing links are to be sought out rather by more careful investigation of the structure of animals already partially known, than by hunting for entirely new ones in the deep sea.

The excessively wide area of the floors of the oceans in the matter of production of species contrasts markedly with wide areas upon the land surface, which are, as has been shown by Mr. Darwin,* specially favourable to the development of variations and development of new forms.

The deep-sea animals obtained by the ship are now in the hands of various specialists for description, and are as yet only partially reported on. As far as I can judge from cursory examination of what was dredged, I believe that the most aberrant and important new animal obtained by the " Challenger's " deep-sea dredgings is an Ascidian, which I have described under the name of *Octacnemus Bythius*.†

---

* "Origin of Species," 10th Ed., p. 83.

† H. N. Moseley, "On Two New Forms of Deep-Sea Ascidians obtained during the Voyage of H.M.S. 'Challenger.'" Trans. Linn. Soc. 2nd Ser. Zoology, Vol. I, p. 287.

The animal, of which a figure of one-half the natural size is here given, is of a most remarkable form for an Ascidian, having eight conical radially disposed lobes. The walls of the body are perfectly transparent. The animal is provided with a small pedicle for attaching itself to the sea bottom; but the greater part of its under surface is free and unattached. The usual exha-

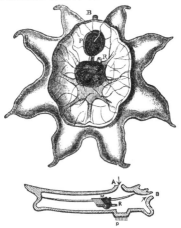

DEEP-SEA ASCIDIAN, OCTACNEMUS BYTHIUS.

Above. The animal viewed from below of one-half the natural size; the nucleus is seen in the centre through the transparent base of the animal. *P* Pedicle of attachment; *B* Exhalant orifice; *R* Rectum.

Beneath. Diagrammatic section through the middle line of the animal's body; *A* Inhalant orifice; *M* Muscle attached to nucleus, other letters as in the figure above.

lant and inhalant apertures are present, as will be best understood by reference to the diagrammatic section shown in the woodcut.

There appears to be no gill network present, but the respiratory sac is flattened out so as to be stretched as a horizontal membrane across the cavity of the body between the inhalant and exhalant apertures, as shown in the section. The principal viscera are gathered together into a compact nuclear mass, just as in Salpa, and this nucleus is attached to the under surface of the horizontal membrane.

The nerve ganglion lies on the nucleus, and there is a globular sense organ in connection with it. Special muscular slips are present on the surface of the nucleus, and there are elaborate muscular arrangements within the conical processes of the body of the animal, and in connection with the horizontal membrane. The animal seems to be entirely without immediate affinities amongst other Ascidians and must be placed in a special Family, *Octacnemidæ*.

I cannot here enter into descriptions of the many deep-sea forms of animals which we dredged. For accounts of these and most beautiful figures, I refer the reader to Sir C. Wyville Thomson's " Depths of the Sea " and " The Atlantic."

We obtained the same animals from the depths in the most widely separated places over and over again, with tedious reiteration. There were, however, one or two localities which we hit upon which are worth referring to, because they are especially rich in deep-sea forms, and because these occur at them in comparatively shallow water.

The first of these localities lies off the Island of Sombrero, in the Danish West Indies. Here, within sight of the lighthouse, in from 450 to 490 fathoms, the dredge yielded a very rich harvest of deep-sea Blind Crustacea, Corals, Echinoderms, Sponges, &c. Another very rich spot lies off the Kermadec Islands. Here, from 630 fathoms, a marvellously rich collection was brought up by the trawls, including very curious new blind deep-sea fish. Ascidians, Cuttle-fish, Crustaceans (*Polycheles, Cystisoma*), many specimens of *Pentacrinus*, large vitreous Sponges (*Poliopogon, Euplectella, Ventriculites*), and many other very valuable specimens. This is probably the richest ground dredged by us at all.

Another rich locality lies between the Aru and Ke Islands, and a further one, almost or quite as rich as that off the Kermadecs, lies between the Meangis Islands and the Talour Islands. Here, from 500 fathoms, more than thirty specimens of living *Pentacrinus* were obtained at one haul of the net, and with them all kinds of other deep-sea forms, very many of the same species as dredged at all the other three localities mentioned. Any yachtsman or collector wishing to obtain, with the least trouble and most certainty rare deep-sea animals, would do well to put his dredge overboard at one of these four above-mentioned localities.

The deep-sea animals are, as I have said, mostly closely allied to shallow-water forms. They appear also to live associated together in closely the same manner as their shallow-water representatives. Some are confined to the sea-bottom, and can only crawl upon it; others, such as the fish and shrimps, have a power of extending their range vertically, but some of the fish at least never rise to more than a very small height above the bottom on which they live.

Lophioid fishes, like the Angler their close ally in shallow water, dangle out in the great depths their lures from above their huge mouths, to attract their prey. Hermit-crabs in the deep sea, crawl about protected by a borrowed shell, and on this, lives an animal allied to a Sea Anemone (*Epizoanthus parasiticus*), so that the combination is closely similar to that so familiar in shallow seas. *Pycnogonid* larvæ rear themselves as parasites within Hydroid colonies in the depths, just as in the shallows.

The depths of the sea being mostly dark, many of the animals inhabiting them are blind, like cave animals, and have their eyes reduced to mere rudiments. Many of these, such as some blind fish and Crustacea, are provided with enormously long and delicate feelers or hairs, in order that they may feel their way about with these, just as a blind man does with the aid of his stick.

Other deep-sea animals have their eyes enormously enlarged, and thus make the best of the little light there is in the depths. This light is, no doubt, as suggested in the early days of deep-sea dredging by Dr. Carpenter, Sir Wyville Thomson, and Mr. Gwyn Jeffreys,[*] that emitted by phosphorescent animals, especially Alcyonarians.

All the Alcyonarians dredged by the "Challenger" in deep water, were found to be brilliantly phosphorescent when brought to the surface, and their phosphorescence was found to agree in its manner of exhibition with that observed in the case of shallow-water forms. There seems no reason why these animals should not emit light when living in deep water, just as do their shallow-water relatives.

The light emitted by phosphorescent animals is quite possibly in some instances to be regarded only as an accidental product, and of no use to the animal producing it, although of course, in some cases, it has been turned to account for sexual

[*] " Proc. Roy. Soc., 1869," p. 431.

purposes, and may have other uses occasionally. There is no reason why a constant emission of light should be more beneficial than a constant emission of heat, such as takes place in the case of our own bodies, and it is quite conceivable that animals might exist to which obscure heat-rays might be visible, and to which men and Mammals generally, would appear constantly luminous.

However, be the light beneficial to them or not, it seems certain that the deep sea must be lighted here and there by greater or smaller patches of these luminous Alcyonarians, with wide intervals, probably, of total darkness intervening; very possibly the animals with eyes congregate round these sources of light.

The nature of the light existing in the depths, has an important bearing on the question of the colouring of deep-sea animals. I examined the phosphorescent light emitted by three species of deep-sea Alcyonarians with the spectroscope, and found it to consist of red, yellow, and green rays only. Hence, were the light in the deep sea derived from this source alone, in the absence of blue and violet light, only red, yellow, and green colours in animals could be effective; no blue animals were obtained in deep water, but blue animals are not common elsewhere.

It is remarkable that almost all the deep-sea shrimps and Schizopods, which were obtained in very great abundance, are of an intense bright scarlet colour, differing markedly in their intensity of colouring from shallow-water forms, and having, apparently for some purpose, developed an unusually large quality of the same red pigment matter which colours small surface Crustacea.

Dr. Wallich refers at length in his work, cited above,* to the absence of light in the deep sea, and explains the possibility of persistence of colouring in deep sea animals, even though they live in absolute darkness. Many deep-sea Holothurians are coloured of a deep purple; no doubt the colouring is useless in their case, and is merely due to the persistence of a colouring developed originally in shallow-water ancestors.

* "The Atlantic Sea Bed," p. 108.

The same purple colouring matter, which is easily distinguished by means of the spectroscope, occurs in a shallow-water (nine fathoms) Comatula at Cape York, in the tropics, and in a Holothurian, found in 1,955 fathoms, near the Antarctic Sea. Many deep-sea Corals have their soft structures tinged with a madder colouring matter which occurs also in surface swimming Medusæ of various kinds.*

No doubt, in the case of many deep-sea possessors of complex colouring matters, these pigments never exercise their peculiar action on light during the whole life of the animals, but remain in darkness, never showing their colour at all. Just so in the case of many Mammalia, with thick or fur-clad skins, the bright red colouring matter of the blood never sees the light or appears as a red colour. It is only in a few Mammals, that this red colouring matter is turned to account, as, for example, in the white races of man, in which case sexual selection has brought about a tinging of the cheeks by its aid.

Most deep-sea fish are of a dull black colour, some are white as if bleached. The majority of deep-sea animals are coloured in some way or another, many brightly so.

Zoology and Botany of the Ship.—The zoology of "Challenger" itself was rather interesting. At the time that England was left the ship seemed nearly free of animals, other than men, dogs, and live stock required for food. The first Cockroaches apparently came on board at St. Vincent, Cape Verdes, for a large one of these insects was caught by one of the lieutenants on his bed, soon after we left that port. Cockroaches soon became plentiful on board, and showed themselves whenever the ship was in a warm climate. A special haunt of a swarm of them was behind the books in the chemical laboratory, from which Mr. Buchanan in vain attempted to evict them.

At one period of the voyage, a number of these insects established themselves in my cabin, and devoured parts of my boots, nibbling off all the margins of leather projecting beyond

* For observations on the Colouring of Deep Sea Animals, see H. N. Moseley, "On the Colouring Matters of various Animals, especially of Deep-Sea Forms dredged by H.M.S. 'Challenger.'" Quart. Journ. Micro. Sci., Vol. XVII, New Ser., p. 1.

the seams on the upper leathers.   One huge winged Cockroach baffled me in my attempts to get rid of him for a long time.   I could not discover his retreat.   At night he came out and rested on my book-shelf, at the foot of my bed, swaying his antennæ to and fro, and watching me closely.   If I reached out my hand from bed, to get a stick, or raised my book to throw it at him, he dropped at once on the deck, and was forthwith out of harm's way.

He bothered me much, because when my light was out, he had a familiar habit of coming to sip the moisture from my face and lips, which was decidedly unpleasant, and awoke me often from a doze.   I believe it was with this object, that he watched me before I went to sleep.   I often had a shot at him with a book or other missile, as he sat on the book-shelf, but he always dodged and escaped.   His quickness and agility astonished me. At last I triumphed, by adopting the advice of Captain Maclear, and shooting him with a pellet of paper from my air-gun, a mode of attack for which he was evidently unprepared; but I was taken to task for discharging the air-gun in my cabin, because it made a noise just like the sharp crack of a spar when broken by the force of the breeze, and created some excitement on the upper deck, where the sound was plainly heard.

In the zoological laboratory on board, small red ants established themselves, and used to follow trails up the legs of the tables, and find out anything eatable.   Clothes-moths were a terrible pest, and destroyed several garments for me in my cabin. Mosquitos swarmed in the ship at some ports, as well as house-flies, but these both disappeared when we had been at sea for a few days in a breeze.

Once, when we were becalmed three days out from Teneriffe, on the voyage to St. Thomas, I went out in a boat to collect surface animals.   Some of the house-flies, which were swarming in the ship, accompanied the boat on the excursion in sufficient numbers to be a pest, I suppose in expectation of reaching the shore.

House-crickets appeared in the ship towards the end of the voyage, and two of them established themselves in Staff Commander Tizard's cabin, to his great annoyance, as they were as noisy as at home.   They were, however, caught with some difficulty. Centipedes, of two kinds at least, were also amongst the navifauna, and many species of spiders. Some of these latter were, however, deliberately imported on board by the navigating officers, in order that they might use their webs, if wanted, as cross-wires in their theodolites.

When the ship was moored at Bermuda, alongside the wharf in the dockyard, boards were placed on all the mooring chains as a fence against rats. Rats nevertheless appeared in the ship, and were all curiously enough of the old species, the Black Rat (*Mus rattus*). One night, as we were sitting at whist, Mr. J. Hynes, the Assistant Paymaster, suddenly started up with a yell, and danced about as if gone mad, clutching one of his legs with both hands. A rat had mistaken his trousers for a pipe or wind-sail, and had gone up.

The only plants which made their spontaneous appearance on the ship were Moulds. Whenever the ship entered damp latitudes everything in our cabins on the lower deck became moist, and mould grew thickly over boots and all other leathern articles. I grew mustard and cress with great success in my wardian cases before these were required for other purposes. I failed, however, entirely with onions and radishes, which I also tried to grow. The plant most commonly grown on board ship in the tropics is the Sweet Potato. It can be grown in water and made to climb up the wall of a cabin and afford a pleasant green.

Besides Dogs and Cats we had many different pets on board the ship at various times. First amongst these must be placed " Robert " the Parrot. The bird belonged to Von Willemoes Suhm. He and I bought a young Grey Parrot each at Madeira, from a ship bound from the Bight of Benin to Liverpool, with a cargo of these birds on board.* One of the Parrots flew into a dish full of boiling caustic potash solution in the laboratory and perished, and we had to draw lots for the remaining bird, and I lost.

" Robert " survived all the extremes of the heat and cold of the voyage and perils of all kinds, from heavy tumbles, driving gales of wind, and the falling about of books and furniture. He

* See page 41.

had one of his legs crippled, and his feathers never grew properly, but he was perfectly happy, and from his perch, which
was one of the wardroom hat-pegs, he talked away and amused
us during the whole voyage.   His great triumph, constantly
repeated, was " What! two thousand fathoms and no bottom ?
Ah Doctor Carpenter, F.R.S."   He knew his own name perfectly,
and I have known him climb over the ledge in at the door of
the cabin of Dr. Maclean, his chief friend, when I have been
sitting there on a dark rough night, after he had come to grief
and tumbled off his perch with a thump, plaintively appealing
with " Robert," " Robert."

    After leaving the Aru Islands a young Cassowary roamed
about the decks for some time, but was soon killed as a nuisance.
No doubt, had it not been killed, it would soon have committed
suicide, like an Ostrich on board one of the men-of-war at the
Cape, which stole a piece of hot iron put down by the black-
smith beside his forge, and swallowed it hastily with fatal effect.

    At Monte Video some very young South American Ostriches
(*Rhea Americana*) were brought on board the ship.   It was
amusing to see them chasing flies on the upper deck, and, as
they darted forwards, instinctively spreading their little wings
as sails to catch the tiny draughts reflected from the bulwarks.
Mr. Darwin has described the use of the wings as sails by the
adult birds on the plains of Patagonia.*

    At the Sandwich Islands, two large living Tortoises from the
Galapagos Archipelago were received on board from Captain
Cookson, R.N., who had visited the group with the special object
of collecting the very curious Tortoises found there.   The Tortoises were fed a good deal on pine-apples, a number of which
were hung up in the Paymaster's office.   The animals used to
prop themselves up against a board put across the door of the
office to keep out dogs, unable to surmount the obstacle, and
used to glare and sniff longingly at the fruit.   They also learned
to know their way along the deck to the Captain's cabin, where
there was another store of Pine-apples, and where they were
often fed.

    At Madeira, I had given to me some living specimens of the

---

* " Journal of Researches," pp. 43, 89.

huge Spiders (*Lycosa*), which inhabit the "Desertas," small out-liers of the island, and which feed on Lizards, which they hunt and kill. I fed the Spiders on Cockroaches. One of them escaped, but it was brought back to me after a week by Captain Maclear, rather crushed, he having discovered it with his toe in the extremity of one of his boots.

At Juan Fernandez a living young Fur-Seal, about two feet in length, was taken on board. It followed us about, crying like a child to be fed, and was never happy unless it was being nursed and petted. I tried to feed it with condensed milk, but it soon died. When it was hungry, if blandishments did not succeed in drawing attention at once to its wants, the animal, though so young, became at once enraged and made determined efforts to snarl and bite, with a view of enforcing its demands.

At the same island a Kid, one of the direct descendants of Alexander Selkirk's Goats, also came on board, and learnt all kinds of tricks on the homeward voyage. We should have liked to have had a pet Monkey with us, but Monkeys are strictly forbidden, by a special Admiralty regulation, on survey-ing ships, because one once destroyed a valuable chart which had just been completed with great labour. Even a Marmoset, which I bought at Bahia, was considered to come under the regulation and perished in consequence.

Concluding Remarks.—I did not suffer at all from the confine-ment of ship-life. It is wonderful how completely practice enables a man so to modify his movements as to perform with success, in a ship constantly in motion, even the most delicate operations. The adjustments of the body to the motion of the ship in ordinary weather, become, after a time, so much a matter of habit as to be quite unconscious. I found no difficulty in working with the microscope with the highest powers (1,100 diameters), even when the ship was rolling heavily.

There are many worries and distractions, such as letters and newspapers, which are escaped in life on board ship, and the con-stant leisure available for work and reading is extremely enjoy-able. I felt almost sorry to leave, at Spithead, my small cabin, which measured only six feet by six, and return to the more complicated relations of "shore-going" life, as the sailors term it. I had lived in the cabin three years and a half and had got to look upon it as a home.

After a voyage all over the world, there is nothing which is so much impressed upon the mind as the smallness of the earth's surface. We are apt to regard certain animals as fixed and stationary, and to contrast strongly with their condition that of forms possessing powers of active locomotion. In reality we are as securely fixed by the force of gravity as is the Sea Anemone by its base; we can only revolve as it were at the end of our stalk, which we can lengthen or shorten only for a few miles' distance. We live in the depths of the atmosphere as deep-sea animals live in the depths of the sea. We can, like these, crawl up into the shallows or we can occasionally mount at peril in a balloon; but the utmost extent of our vertical range is a distance no greater than that which we can walk in a couple of hours horizontally on the earth's surface.

The " Challenger " travelled on the voyage from Portsmouth and back to the same port, 68,690 miles, and this distance, taking into consideration the time consumed from port to port, was traversed at the average pace of only four miles an hour, or fast walking pace. In an express train on land the entire distance could be conceived of as being accomplished in eight weeks, and at the rate at which a Swallow can fly in about half that time.

If there were land all along the equator it would be possible to run round the world in a train in less than three weeks. I used to wonder how the main mass of the inhabitants of America could have peopled the entire country down to Cape Horn, from so remote a starting-point as Behring's Straits; but a walk of four miles a-day would bring a man from Behring's Straits to Cape Horn in about seven years, and a move of a quarter of a mile a-day would bring a tribe the same distance in a little over a century.

The earth, considered as a comparatively insignificant component particle of the universe, may be justly compared to a small isolated island on its own surface. As, in the course of ages, such an island developes its own peculiar insular fauna and flora, so probably on the surface of the earth alone has the peculiarly complex development of the element Nitrogen occurred which has resulted in the various forms of animal and vegetable life.

On the theory of evolution, it is impossible that plants or animals of any advanced complexity, at all resembling those existing on the earth, should exist on other planets or in other solar systems. It is conceivable that very low forms of vegetable life may exist on other planets and may have been by some means transported to the earth : the idea is conceivable, though highly improbable. But it is quite impossible that that infinitely complex series of circumstances which on the earth has conspired to produce from the lowest living forms a Crustacean for example, should have occurred elsewhere ; still less is it possible that a bird or a Mammal should exist elsewhere ; still more impossible again that there should be elsewhere a monkey or a man.

All these forms are quite certainly terrestrial, and terrestrial only, as surely as is the Apteryx a peculiar development of New Zealand alone, or the Dodo a production of the Mascarene Islands only. It is even probable that protoplasm, itself, the basis of all life, is a production entirely confined to our small planet.

That the " Challenger " Expedition has been a great scientific success has been fully acknowledged, and all praise is due to the Government which promoted it, and to the present Government which has supplied funds for the publication of the results. The highest praise is, however, due to those naturalists, especially Sir Wyville Thomson and Dr. Carpenter, who, by their energy and perseverance, actually originated the Expedition.

With regard to any future scientific expeditions, it would, however, be well to bear in mind that the deep sea, its physical features and its fauna, will remain for an indefinite period in the condition in which they now exist and as they have existed for ages past, with little or no change, to be investigated at leisure at any future time. On the surface of the earth, however, animals and plants and races of men are perishing rapidly day by day, and will soon be, like the Dodo, things of the past. The history of these things once gone can never be recovered, but must remain for ever a gap in the knowledge of mankind.

The loss will be most deeply felt in the province of Anthropology, a science which is of higher importance to us than any other, as treating of the developmental history of our own species. The languages of Polynesia are being rapidly destroyed or mutilated, and the opportunity of obtaining accurate information concerning these and the native habits of culture will soon have passed away.

The urgent necessity of the present day is a scientific circumnavigating expedition which shall visit the least-known inhabited islands of the Pacific, and at the same time explore the series of islands and island groups which yet remain almost or entirely unknown as regards their botany and zoology. These promise to yield results of the highest interest if only the matter be taken in hand in time, before introduced weeds and goats have destroyed their natural vegetation; dogs, cats and pigs, their animals, and their human inhabitants have been swept away, or have had their individuality merged in the onward press of European enterprise. There is still, to the disgrace of British enterprise, even in the Atlantic Ocean, an island, the fauna and flora of which are as yet absolutely unknown. The past history of the deep sea, of the changes of depression and elevation of its bottom, is to be sought to a large extent in the study of the animals and plants inhabiting the islands which rear their summits above its surface. These insular floras and faunas will soon pass away, but the deep-sea animals will very possibly remain unchanged from their present condition long after man has died out.

LIST OF BOOKS AND PAPERS RELATING TO THE "CHALLENGER" EXPEDITION.

BOOKS AND PAPERS BY NAVAL OFFICERS OF THE "CHALLENGER" EXPEDITION.

Captain Sir G. S. Nares, R.N., K.C.B., F.R.S.
  Series of Reports to the Hydrographer of the Admiralty. Nos. 1, 2, 3, 1873–74.

Captain F. T. Thomson, R.N.
  Continuation of the above Reports. Nos. 4, 5, 6, 7, 1875–76.

Staff-Commander T. H. Tizard, R.N.
  Remarks on Deep-sea Temperatures, &c., embodied in the above Reports.
  The "Challenger" Expedition. On the Methods adopted in Sounding and Dredging : Naval Science. London, Lockwood and Co., 1873, p. 409.

The above Reports are mainly reprinted with a reproduction of the Section Maps of Deep-sea Temperatures in Petermanns Mittheilungen, 1873–76, where also will be found references to various papers on Deep-sea Physics resulting from the " Challenger " Expedition.

Lieut. Lord George Campbell, R.N.
  Log-letters from the "Challenger." London, Macmillan and Co., 1876.

W. J. J. Spry, R.N.
  The Cruise of Her Majesty's Ship "Challenger." London, Sampson Low and Co., 1876.

BOOKS AND PAPERS BY MEMBERS OF THE CIVILIAN STAFF OF THE SHIP.

Sir C. Wyville Thomson, Knt., LL.D., F.R.S., &c.
  The Voyage of the "Challenger." The Atlantic, 2 Vols. London, Macmillan and Co., 1877.
  Series of Reports to the Hydrographer of the Admiralty in the Proc. Roy. Soc., 1874–76.

602   LIST OF BOOKS AND PAPERS.

Preliminary Notes on the Nature of the Sea-bottom, &c. Proc. Roy. Soc., 1874, p. 32.

Notice of New Living Crinoids belonging to the Apiocrinidæ. Linn. Soc. Journ. Zoology, Vol. XIII, p. 47.

Notes on some Peculiarities in the Mode of Propagation of certain Echinoderms of the Southern Sea. Ibid., p. 55.

On the Structure and Relations of the genus Holopus. Proc. Roy. Soc., Edin., 1876–77, p. 405.

J. J. WILD, Ph.D.

Thalassa. An Essay on the Depth, Temperature and Currents of the Ocean. London, Marcus Ward and Co., 1877.

At Anchor. London, Marcus Ward and Co., 1878.

J. Y. BUCHANAN, M.A., F.R.S.E., &c.

On the Absorption of Carbonic Acid by Saline Solution. Proc. Roy. Soc., 1874, p. 483.

Some Observations on Sea-water Ice. Proc. Roy. Soc., 1874, p. 431.

On the Vertical Distribution of Temperature in the Ocean. Proc. Roy. Soc., 1875, p. 123.

Preliminary Report, Chemical and Geological, on work done on board Her Majesty's Ship "Challenger." Proc. Roy. Soc., 1876, p. 593.

Preliminary Note on the Use of the Piezometer in Deep-sea Sounding. Proc. Roy. Soc., 1876, p. 161.

Note on the Specific Gravity of Ocean Water. Proc. Roy. Soc., Edin., 1876–77, p. 283.

Note on Manganese Nodules found on the Bed of the Ocean. Proc. Roy. Soc., Edin., 1876–77, p. 287.

On the Distribution of Salt in the Ocean. Journ. Roy. Geogr. Soc., Vol. XLII, 1878, p. 72.

Laboratory Experiences on board the "Challenger." Journ. of Chem. Soc., Oct., 1878.

On the Use of the Piezometer in Deep-sea Sounding. Proc. Roy. Soc., 1877, p. 161.

RUDOLPH v. WILLEMOES SUHM, Dr. Phil., &c.

"Challenger" Briefe. Nach dem Tode des Verfassers, herausgegeben von Seiner Mutter. Leipzig, W. Engelmann, 1877.

Briefe an C. Th. E. v. Siebold. I–VII, Z. für Wiss. Zoologie, 1873–77.

Observations made during the earlier part of the Voyage of Her Majesty's Ship "Challenger." The Atlantic. Surface of the Atlantic. Islands of the Atlantic. Proc. Roy. Soc., 1876, p. 569.

On a Land Nemertean found at the Bermudas. Ann. and Mag. Nat. Hist., Vol. XIII, 1874, p. 209.

On some Atlantic Crustacea from the "Challenger" Expedition. Trans. Linn. Soc., 2 Ser. Zoology, Pt. 1, 1875, p. 23.

On Crustacea observed during the Cruise of Her Majesty's Ship "Challenger" in the Southern Sea. Proc. Roy. Soc., 1876, p. 585.

On the Development of Lepas Fascicularis and the Archizoëa of Cirrhipedia. Phil. Trans., 1876, p. 131.

Preliminary Note on the Development of some Pelagic Decapods. Proc. Roy. Soc., 1876, p. 132.

Notes on some Young Stages of Umbellularia and on its Geographical Distribution. Ann. and Mag. Nat. Hist., 1876.

JOHN MURRAY, F.R.S.E.

On Oceanic Deposits examined on board Her Majesty's Ship "Challenger." Proc. Roy. Soc., 1876, p. 471.

On Surface Organisms and their relation to Ocean Deposits. Ibid., p. 532.

Preliminary Report on Vertebrates. Ibid., p. 537.

On the Distribution of Volcanic Débris over the Floor of the Ocean, &c. Proc. Roy. Soc., Edin., 1876-77, p. 247.

H. N. MOSELEY, M.A., F.R.S.

On the Structure and Development of Peripatus Capensis. Phil. Trans., Vol. CLXIV, 1874, p. 757.

On Peripatus Novæ Zealandiæ. Ann. and Mag. Nat. Hist., Jan. 1877.

On Pelagonemertes Rollestoni. Ann. and Mag. Nat. Hist., March, 1875.

On a Young Specimen of Pelagonemertes Rollestoni. Ann. and Mag. Nat. Hist., Dec. 1875.

On Stylochus Pelagicus and other Oceanic Planarians, &c. Quart. Journ. Micro. Sci., Vol. XVII, New Ser., p. 23.

Notes on the Structure of several forms of Land Planarians, &c., with a list of all Species at present known. Ibid., p. 273.

On the Structure and Relations of the Alcyonarian, Heliopora Cærulea, &c., Phil. Trans. Vol. CLXVI, 1876, p. 92.

On the Structure of a species of Millepora occurring at Tahiti, Society Islands. Ibid., Vol. CLXVII, 1877, p. 117.

Preliminary Note on the Structure of the Stylasteridæ. Proc. Roy. Soc., 1876, p. 93.

On the Structure of the Stylasteridæ, a Family of Hydroid Stony Corals. The Croonian Lecture. Phil. Trans., 1878, p. 425.

Preliminary Report on the True Corals dredged by Her Majesty's Ship "Challenger" in deep water. Proc. Roy. Soc., 1876, p. 543.

On New Forms of Actiniaria dredged in the deep sea, &c. Trans. Linn. Soc., Zoology, Vol. I, p. 295.

On two new forms of Deep-sea Ascidians obtained during the Voyage of Her Majesty's Ship "Challenger." Ibid., p. 287.

On the Colouring Matters of Various Animals, especially of Deep-sea Forms, dredged by Her Majesty's Ship "Challenger." Quart. Journ. Micro. Sci., Vol. XVII, New Ser., p. 1.

On the Inhabitants of the Admiralty Islands. Journ. Anthropological Inst., May, 1877.

Botanical Notes in the Journal Linn. Soc., Vol. XIV, 1874 :—On the

Vegetation of Bermuda, p. 317. On Fresh-water Algæ obtained at
the Boiling Springs at Furnas, St. Michael's, Azores, p. 321. On
Plants collected at St. Vincent, Cape Verdes, p. 340 ; at St. Paul's
Rocks, p. 354 ; at Fernando do Norhona, p. 359 ; in the Islands of
the Tristan da Cunha Group, p. 377. On the Botany of Marion
Island, Kerguelen's Land, and Young Island of the Heard Group,
p. 387.

Journ. Linn. Soc., Vol. XV. Further Notes on the Plants of Kerguelen, with some remarks on the Insects, p. 53. Notes on Plants
collected and observed at the Admiralty Islands, p. 93. Notes on
the Flora of Marion Island, p. 481.

PAPERS BY AUTHORS NOT MEMBERS OF THE EXPEDITION.

Reports on the Collection of Birds made during the Voyage of Her
Majesty's Ship "Challenger." Proc. Zool. Soc., Nos. I–XII.

No. I. General Report, by P. L. SCLATER, F.R.S., 1877, p. 534.

No. II. Birds of the Philippine Islands, by the Marquis of TWEED-
DALE, F.R.S. Ibid., p. 535.

No. III. Birds of the Admiralty Islands, by P. L. SCLATER. Ibid.,
p. 551.

No. IV. Birds of Tongatabu, Fiji, Api and Tahiti, by Dr. O. FINSCH.
Ibid., p. 723.

No. V. On the Laridæ, by HOWARD SAUNDERS, F.Z.S., &c. Ibid.,
p. 794.

No. VI. Birds of Ternate, Amboina, Banda, Ke and the Aru
Islands, by Count T. SALVADORI, 1878, p. 78.

No. VII. Birds of Cape York and Raine, Wednesday and Booby
Islands, by A. W. FORBES, F.Z.S. Ibid., p. 120.

No. VIII. Birds of the Sandwich Islands, by P. L. SCLATER. Ibid.,
p. 346.

No. IX. Birds of Antarctic America, by P. L. SCLATER and
O. SALVIN, F.R.S. Ibid., p. 431.

No. X. Birds of the Atlantic Islands and Kerguelen's Land, and on
the Miscellaneous Collections, by P. L. SCLATER. Ibid., 567.

No. XI. On the Steganopedes and Impennes, by P. L. SCLATER
and O. SALVIN. Ibid., p. 650.

No. XII. On the Procellariidæ, by O. SALVIN, p. 735.

By means of the above Papers the scientific names of birds mentioned
in this book have been, as far as possible, corrected. Certain of the
Papers were not available for use in the earlier sheets of the book.

ALBERT C. L. G. GUNTHER, M.D., F.R.S., &c.

Notice of Deep-sea Fishes collected during the Voyage of Her Majesty's
Ship "Challenger." Ann. and Mag. Nat. Hist., 1878, Pt. 1, p. 13,
Pt. II, p. 179, Pt. III, p. 248.

T. Spence Bate, F.R.S.
 On the Willemosia Group of Crustacea. Ann. and Mag. Nat. Hist., 1878, p. 273, Ibid., p. 484.

REPORTS ON BOTANICAL COLLECTIONS MADE BY H. N. MOSELEY DURING THE VOYAGE OF HER MAJESTY'S SHIP "CHALLENGER."

In the Journal of the Linnean Society, Botany, Vols. XIV, XV, XVI, XVIII.

Prof. Oliver, F.R.S., &c.
 List of Plants collected by H. N. Moseley, M.A. Kerguelen's Land, Marion Island and Young Island. Vol. XIV, p. 389.

Prof. G. Dickie, M.D., F.L.S.
 On the Marine Algæ of St. Thomas and the Bermudas. Vol. XV, p. 311 ; of the Cape Verde Islands, p. 344 ; of St. Paul's Rocks, p. 355 ; of Fernando do Norhona, p. 366. Algæ from 30 fathoms off Pernambuco, Brazil, p. 375. Algæ from Bahia, 377 ; from Tristan da Cunha, 384 ; from Inaccessible Island, p. 386 ; from Simon's Bay, Vol. XV, p. 40 ; from Seal Island, p. 41 ; from Marion Island, p. 42 ; from Kerguelen's Land, p. 43 ; from Heard Island, p. 47. Algæ, chiefly Polynesian, p. 235 ; from Torres Straits, Japan and Juan Fernandez, p. 446 ; from various localities, p. 486.

Rev. M. J. Berkeley, M.A., F.L.S.
 Enumeration of Fungi, Vol. XIV, p. 350.
 Ibid., Vol. XV, p. 48. Ibid., Vol. XVI, p. 38, Pl. II.

W. T. Thistleton Dyer, M.A., B.S.C., F.L.S.
 Note on Algæ in Hot Springs, Vol. XIV, p. 326.

W. Archer, F.R.S., &c.
 Notes on Collections made from Furnas, Azores, containing Algæ, &c., Vol. XIV, p. 328.
 Notes on Fresh-water Algæ collected by H. N. Moseley, M.A. Vol. XV, p. 445.

Dr. J. Stirton.
 Enumeration of Lichens from the Islands of the Atlantic. Vol. XIV, p. 366. Ibid., Vol. XVII, p. 152.

Rev. J. M. Crombie, F.L.S., &c.
 Lichens of the "Challenger" Expedition. Vol. XVI, p. 211.

The Rev. E. O'Meara, M.A.
    On the Diatomaceous Gatherings made at Kerguelen's Land by H. N.
    Moseley, M.A.   Vol. XV, p. 55, Pl. I.

J. G. Baker, Esq., F.R.S.
    On the Polynesian Ferns of the "Challenger" Expedition.

Prof. H. G. Reichenbach.
    On some Orchidaceæ collected by Mr. Moseley of the "Challenger"
    Expedition in the Admiralty Islands, Ternate and Cape York.
    Vol. XV, p. 112.

William Mitten, A.L.S., &c.
    The Musci and Hepaticæ collected by H. N. Moseley, M.A., Vol. XV,
    p. 59.

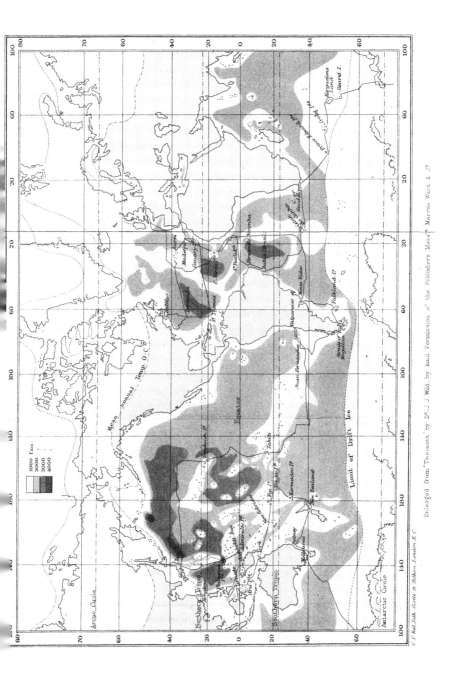

The material originally positioned here is too large for reproduction in this reissue. A PDF can be downloaded from the web address given on page iv of this book, by clicking on 'Resources Available'.

GENERAL INDEX.

A.

ABBOTT, Mr. W. J., R.N., shooting Bird of Paradise, 378.
Acæna ascendens, 116, 165, 167, 191.
Acacia Koa, 497.
Acrocladia Mamillata, 307.
ADMIRALTY ISLANDS, the, 448; Fruit-pigeons of the, 386; natives of the, 451–80.
Æolian formation, at Bermuda, 21; in icebergs, 240.
AGASSIZ, Prof., on deep-sea animals, 586.
Agaricus phylicigena, 136.
Agouti at St. Thomas, 14.
Agrostis Antarctica, 194.
Ainos, 496.
Aizoon canariense, 46.
Albatross, origin of word, 129; Great White, the, 171, 180, 254; nesting of the, 130, 172; Sooty, the, 180, 183, 254; flight of the, 570; diffusion of plants by, 522; range of, 522.
Albino Fijian, 335.
Alcyonarians, phosphorescent, in deep sea, 590; various at PHILIPPINES, 404.
Aleurites triloba, 497.
Algæ, at St. Paul's Rocks, 76; in hot water, 36, 383, 416; parasitic in *Foraminifera*, 293; parasitic in corals, 581.
Allopora, miniacea, 531; *nobilis*, 530; *profunda*, 531.
Alpine, floras and deep-sea faunas compared, 585; plants, probable, of NEW GUINEA, 434.
Alsophila Tahitiensis, 518.
Amalopteryx maritima, 181, 192, 558.
AMBERNOH RIVER, the, 432.
AMBOINA, 387.
Amphioxus lanceolatus, 361.
AMSTERDAM ISLAND, 135, 170.
Anas superciliosa, 372.
Anchor, the ship's broken, 550.
ANDES, the, not well seen from the coast, 543.
Angiopteris evecta, 518.
Anous stolidus, 68, 123, 349, 479; *melanogenys*, 68, 123.
Antelopes, habits of, 150.
Antennarius, nest of, 567.
ANTIPODES ISLAND, parroquets at, 211.
Ant-lion, 50.
Ants, Leaf-cutting, habits of, 104; curious relation of, to plants, 389; on board the "Challenger," 593.

API ISLAND, 342.
Apium australe, 111, 115.
Aplysia, 48.
Aptenodytes longirostris, 176, 197.
Apteryx Australis, attitude in sleeping, 125.
ARAFURA SEA, 366.
Areca, 351.
ARCHER, Mr. W., on algæ in hot springs, 36.
Arctocephalus, 204.
Arenga saccharifera, 392.
Argus, origin of story of, 424.
Argynnis, 134.
Armadillo, the Pigmy, 146.
Arrows, leaf, at KE ISLANDS, 381; poisoned, at API, 346; at the ARU ISLANDS, 374; of HUMBOLDT BAY, 444.
Art, native, in the ADMIRALTY ISLANDS, 470.
Artamus leucopygialis, 353.
ARU ISLANDS, 366; bows and arrows of, 374; houses of, 370, 374; ornaments of, 371; dredgings off, 589.
Arundo donax, hedges of, 33.
ASCENSION ISLAND, 561; migration of turtles at, 133.
Ascidian, remarkable deep-sea, 587.
Aspidium mohrioides, 167.
Asplenium, Nidus, 308, 518, 519; *obtusatum*, 110.
Asthenosoma, 13.
Astrœidæ, 307.
Astrœa, 360, 362, 385.
Astylus subviridis, 531, 533.
ATLANTIC OCEAN, form of the sea-bed of the, 584.
Atya sulcatipes, 61.
Australian Blacks, 353; camp of, 354; caves middens, and drawings of, New South Wales, 273; description of clergyman by, 411; English of, 359; food of, 357; at Government Reserve, 261; habits and utensils of, 355; ideas of after life, 359; method of smoking of, 356; nakedness of, 359; tracks of, on trees, 258.
Azorella, selago, 165, 167, 168, 191, 224; *trifoliata*, 547.
AZORES, The, 29.

B.

Baboons, habits of, 113.
BAKER, Mr. J. G., on ferns, 393.
Balanus, 433.

608 INDEX.

BALFOUR, A. F., Lieut. R.N., ascent of Ternate, 392.
Bulistes, habits of, 74; noise made by, 51.
BAHIA, appearance of town, 85.
Bamboo Jews-harp, 401; spears, 402.
BANDA ISLAND, 382; zones of vegetation at, 45.
Bandicoots, hunting, 269.
Banksia, 266; *Loranthus* on, 545.
Barracuda, 74, 156.
BARRINGTON, Hon. DAINES, experiments on song of birds, 377.
Barringtonia, 387, 433.
BASILAN ISLAND, 405.
Bat, Fruit, feeding on flowers, 291; habits of, 268; at ARU, 375.
Bat, pouch-winged, 103.
Bathyactis symmetrica, size of, in deep water, 586.
Bathyergus suilus, habits of, 145.
Bdellostoma, 156.
Bee Eater, 353.
Belemnites, search for, 586.
Bell, origin of the, from wooden drum, 322.
Belone, 58, 83, 88; habits of, 479.
BELT, Mr. THOMAS, on Ants, 104; on cockfighting, 413; on dress in NICARAGUA, 411; on sharks, 325.
BENDIGO, 259.
BENNETT, Dr. G., on *Nautilus*, 299; on *Ornithorynchus*, 263; on PORTO PRAYA, 65.
BERKELEY, Rev. M. J., on fungi of TRISTAN DA CUNHA, 136.
BERMUDA, 18.
Betel, chewing of, 464.
BETHELL, G. R., Lieut. R.N., excursion with, 546.
Betsey Cove, 196.
Bible, the, of a Boer, 150; King THACKOMBAU'S, 319.
Bipalium, 494.
Birds, at API, 345; burrowing, 123, 125, 131, 207, 560; change of habits in, 125; migrations of, in TORRES STRAITS, 364; learning by experience, 74; Land, met with at sea, 482; notes of, not necessarily a test of race, 377; of HEARD ISLAND, 230; of prey at ST. JAGO ISLAND, 59; Sea, nesting with Land, 83; Sea resting on drift-wood, 367; tameness of, 122, 210, 552; young, pugnacity of, 72; young, power of concealment of, 552.
Birds of Paradise, the King, 377; arrows used to kill, 374; for sale, 391, 392; the Great, cry of, 376; the Great, pursuit of, 375.
Birgus latro, 304, 403.
BLEEK, Dr. W. H., on Bushmen, 148.
Bladder-nose Seal, 129.
Bligh's Cap, 183.
Boatswain Birds, 25, 282, 516, 563.
BOATSWAIN-BIRD ISLAND, 562.
Bodleian Library, Pigeon-English at the, 416.
Bolas, the, for cattle, &c., 557; for wild geese, 558.
Bolax glebaria, 166.
Bombinator igneus, 93.

Bonito, mode of catching, 53.
Boobies, 68, 83, 363, 563; nests of, 72, 83, 349, 563; young of, 72.
BOOBY ISLAND, 83, 363; a resting place for migrating birds, 364.
Books, Chinese, development of, from the roll, 417.
Borborus apterus, 559.
BOTANY BAY, 266.
Botryococus in hot springs, 36.
Bows and arrows, of Api Islanders, 346; gradual development of, 443; at HUMBOLDT BAY, 443; not known at the ADMIRALTY ISLANDS, 468, 478.
Bradypus tridactylus, 104.
Brazil, convict settlement of, 78; slavery in, 106; excursions in, 85.
BRINE, CAPT. LINDESAY, R.N., visit to CROZETS, 183.
Bromeliaceous epiphytes, 15, 90, 278.
BROWERA CREEK, 270.
BROWN, Mr. R., on colour of muscles of seals, 204; on discolouration of Arctic seas, 249; on food of seals, 189.
BROWNE, SIR THOMAS, on Unicorns' horn, 426.
BUCHANAN, Mr. J. Y., on natural sandblast, 222; on rocks of KERGUELEN'S LAND, 195; on ice, 233, 250; on stones dropped by ice, 219; on ST. PAUL'S ROCKS, 75; at HUMBOLDT BAY, 442; experiment on deep-sea pressure by, 579.
BUCKHOLTZ, Dr., on food of seals, 189.
Buffaloes, 395.
BURETA, 310.
Burials, of Admiralty Islanders, 476; of Australian Blacks, 360; of Hawaians, 498; native, at the CAPE OF GOOD HOPE, 149; in TORRES STRAITS, 363.
Bushmen, middens of, &c., 149.
Butterfly, clicking, 89; at TRISTAN DA CUNHA, 134; Bird-winged, 373.

C.

Cacatua galerita, 264, 352.
Cactus, 16, 141, 544.
Caladium esculentum, 35, 464.
Calcareous, algæ, 12, 76, 581, at St. VINCENT ISLAND, 46; sand, pipes formed by, 149; sandstone, with volcanic intermixture, 78; sandstone, weathering of, 78, 83.
Caldeira des Sette Cidades, 37.
Callitriche Verna, 191, 224.
Calophyllum inophyllum, 305, 387.
Calornis metallica, 372, 382.
Calotragus melanotis, 151.
Calycopterix Moseleyi, habits of, 192.
CAMERON, S. L., Commander R.N., on piledwellings, 400.
CAMIGUIN ISLAND, volcano at, 409.
CAMPBELL and AUCKLAND ISLANDS, the, 189.
Cannibalism, at FIJI, 318, 320, 325, 327; at FIJI and NEW ZEALAND, 339; of white men, 341; at ADMIRALTY ISLANDS, 473; cannibal convicts interviewed, 309.

Canoes, at Caxobira, 93; building, 326 felling trees for, 406; Fijian double, 340; voyage in a, 314; of Admiralty Islanders, 466.
Canton, 419.
Cap worn by Madeira peasants, 40.
Cape flats, the, 139, 148.
Cape of Good Hope, 151, 152.
Cape Town, 143.
Cape Verde Islands, 41.
Cape York, 350.
Caranx, 53, 71, 479.
Carcharias, 434; fishing for, 71; large, at St. Jago Island, 57; brachiurus, 8, 281; gangeticus, 325.
Cardisoma, 26, 64.
Carine plumipes, 431.
Carmichael, Captain, experiment on the Albatross by, 131.
Carpenter, Dr. W. B., 595; on algæ, 581; originator of the expedition, 598; on phosphorescence, 590.
Carpophaga, Concinna, 382, 386; pacifica, 292, 304; rhodinolœma, 386, 479.
Caryota, 351.
Cassava, preparation of, 102.
Cassowary, 372, 390; tame on board the "Challenger," 595.
Casuarina, 290; Loranthus on, 545.
Catamaran, 81.
Cattle, driven over the Andes, 547; mode of driving in Brazil, 98, 101; at Tristan da Cunha, 113; various modes of handling, 557; wild, of the Falkland Islands, 556.
Caulerpa clavifera, 76.
Cavalli, fishing for, 71, 74.
Caves, corals growing in, 27; in Inaccessible Island, 128; in sandstone rock, 22; at Kerguelen's Land, 196.
Caxobira, 92.
Cebu Island, 407.
Cephaloptera, 65.
Centrococcyx viridis, 404.
Centropus phasianus, 352.
Centipedes on board the "Challenger," 594.
Cereus, 16, 141; Quisco, 544.
Cerianthus, Bathymetricus, 409, 583; very large, 408.
Cerozylon Australe, 541.
Cestracion Philippi, 276.
Chameleon, tameness of, 154.
Chamisso, on drift wood at sea, 368.
Channer, A., Lieut., R.N., excursion with, 516, 553; killing a Sea-elephant, 201; pet spaniel of, 132; sketch by, 355.
Chant, of Fijian pilot, 321; Fijian, used in dances, 329; of the Ke Islanders, 379; of the Admiralty Islanders, 472.
Chara, 200.
Charles Lewis Mountains, The, 434.
Chatham Islanders, 340.
Chaves, Sr. J. M. Q., Crustacea from, 62.
Chelifer, 78.
Chelone, imbricata, 479; food of, 568; midas 479, 561.
Chenopodium tomentosum, 111.

Chinese, dinner, 422; and English delicacies compared, 422; and Japanese books, 418; loss of language by; graves of, 391; fighting cocks of the, 414; writing and European compared, 417; in the Sandwich Islands, the, 497; examinations, 420; floral decorations, 421.
Chionis, alba, 209; minor, 171, 179, 209.
Chirodota, 402.
Chloephaga Patagonicha, 551.
Chlorophyll, importance of, 567.
Chevreulia, 136; Thouarsii, 547.
Cimoliornis Diomedeus, 523.
Clapmatch, origin of name, 129.
Climate of Antarctic Ocean, 255; of Heard Island, 227; of Kerguelen's Land, 214; of Tristan da Cunha, 112.
Clock, water, at Canton, 419.
Clotho arietans, 153.
Clubs, decorative and processional in Fiji and England, 332; Fijian, 329.
Coal at Kerguelen's Land, 199.
Cobra, the 153.
Cochineal fields, curious appearance of, 2.
Cockatoos, 264, 352; and parrots, 372; Black, 352, 479.
Cock fighting, 95, 412; spurs for, 413.
Cockroaches on board the "Challenger," 592.
Cocoanut palm scarce inland at Fiji, 324; does not thrive at Cape York, 351.
Cænobita, 304.
Cænoplana, 279.
Cænopsammia Ehrenbergiana, 52.
Coffee, wild, berries eaten by pigeons, 387; at Bermuda, 24.
Cold water cure in Japan, 488.
Collocalia, 479; spodiopygia, 292.
Colobanthus Kerguelensis, 193, 224.
Colours, protective, of animals, 13, 349, 567-8; of birds, experiments proposed on, 373; of deep-sea animals, 591; of muscles of shark, 281; of muscles of seals, 204.
Colouring matters of Turacou, 161; of deep-sea animals, 592.
Columba, livea, 60; œnas, 542.
Condor, the, 548.
Cook, Captain, observations of, 386, 508; worshipped by Hawaians, 505.
Cookson, Commander R.N., 595.
Coolies, Japanese, endurance of, 482-5.
Corals, 306; at Bermuda, 27; at St. Vincent Island, colours of, 52; Blue and Organ corals, 404; Brain and Mushroom corals, 385; dying at the top, 290, 343, 385; growing in caves, 27; living exposed at low tide, 385; living detached, 344; large size of certain, 17; mud flats formed by, 360-62; Mushroom, life history of, 524; Precious, fishery of, 65; coral rock at Aru Islands, 378; stinging of, 388; parasites in, 581, 590; deep-sea, 586.
Corallinaceæ, 65, 74; at Fiji, 306; at St. Vincent Island, 46; in great depths, 581.
Coranderrk, 261.
Cordyceps sinensis, 422.
Coriphilus fringillaceus, 292.

Cormorants, 152, 155, 171, 212, 229.
Costumes, Fijian fancy, 331; of women at
  Fayal, 31.
Counting, of Gudangs no higher than three,
  358; method of, by Admiralty Islanders, 456.
Courtship, of Great White Albatross, 174; of
  Mollymauk,, 131.
CORVO ISLAND, 54.
Coryphœna, 434.
Cotula plumosa, 191.
Cox, Mr., R.N., boatswain H.M.S. "Chal-
  lenger," use of trammel net, 51; seining
  at ST. JAGO ISLAND, 57.
Crabs, habits of, 59, 70; King, 402; at
  BERMUDA, habits of, 26; at ST. VINCENT
  ISLAND, 48; Land and Birgus, breathing
  organs of, 305; Hermit, terrestrial, 17, 304;
  Hermit, deep-sea, 590.
Crater, broken-down at HEARD ISLAND,
  223; at FAYAL, 30; of KILAUEA, 500; of
  MATUKU ISLAND, 293; of TENERIFFE, 6;
  of TERNATE, 393.
CREAGH, Mr. C. V., translation by, from
  Chinese, 424, 428.
Crickets, House, on board the "Challenger,"
  593.
Crithagra insularis, 122 (note).
Crocodile, 479.
Crotophaga ani, 13.
Crow, Piping, the, 257.
CROZET ISLANDS, 181.
CROZIER, MOUNT, 185.
CROSBIE, A., Staff Surgeon R.N., 380, 578.
CRUSOE, ROBINSON, 537.
Cryptogamia of TRISTAN DA CUNHA, 136;
  absent at RAINE ISLAND, 348.
Cry, of the Api Islanders, 343; used in
  mountains by Fijians, 326; of the Gibbon,
  337; of Papuans, 438.
Ctenomys, 146.
Cuajiro Indians, houses of, 399.
Cuckoo, Golden, the, 404; Pheasant, the,
  352.
Currents, oceanic, 68; at KERGUELEN'S
  LAND, 185; seeds transported by, 17, 135,
  164.
Cuscus, 384, 465, 479.
CYCLOPS MOUNTAINS, the, 435.
Cystisoma, 589.

D.

Dactylopterus, 562; flight of, 571.
DANA, Prof. J. D., on the area of the FIJI
  GROUP, 302; on basalt at FIJI, 317; on
  corals, 290, 306, 307; on coral rock, 21;
  on the geology of EUA ISLAND, 282; on
  the genus Atya, 61.
Dance, club, 331; fan, 332; development of
  the, 333; of Admiralty Islanders, 472;
  of Fijians, 314; grand, at FIJI, 327; of Ke
  Islanders, 380; Mahommedan, 387; of
  Lutaos, 400; waltz by Malays, 387.
Daption capensis, 134, 183, 229.
DARWIN, Mr. C., on Aplysia, 48; on the bolas,
  557; on "carne con cuero," 560; on

continental areas, 587; on Diodon antennatus, 52; on expression, 284; on flora of
  KERGUELEN'S LAND, 169; on the Tucutuco,
  146; on getting fire by friction, 289; on
  Gunnera, 538; on hooked seeds, 541; on
  horses of the FALKLAND ISLANDS, 555; on
  icebergs, 243; journal of, 537; on limestone bed at ST. JAGO ISLAND, 55, 65; on
  ostriches, use of wings by, 595; on peat,·
  23; on Pelecanoides urinatrix, 209; on
  Petrels as carriers of seeds, 522; on sexual·
  selection in birds, 373; on spiders' webs,
  382; on ST. PAUL'S ROCKS, 73; at TAHITI,
  523; on tree in FERNANDO DO NORHONA,
  78; on the USPALLATA PASS, 544.
Dasyptilus, 435; pequetti, 392, 479.
Dasyurus viverrinus, 268.
DAVIDS, Mr. RHYS, on embankments in
  CEYLON, 485.
Deep-sea animals, food of, 581; colours of,
  591; relations of, 590; fossil forms
  amongst, 587; range of, 584; rich locali-
  ties for, 589.
Deep-sea, faunas and Alpine floras com-
  pared, 585; absence of sunlight in, 580;
  pressure in, 579.
Deer, at AMBOINA, 390; at the ARU ISLANDS,
  379; introduced into the SANDWICH
  ISLANDS, 500.
Delphinus, 82.
Demiegretta sacra, 291, 322.
Dendrobium, 456.
Dendroseris, 542.
Denudation, by rain in NEW SOUTH WALES,
  267; of TRISTAN DA CUNHA group, 136.
Depth, average, of the ocean, 577; relation
  of, to area in the oceans, 576.
Diadema Antillarum, 12.
Diatoms, Pelagic, 216, 566; staining ice, 249.
DICKENS, Mr. F. V., trip with, 482.
DICKIE, Prof. G., on algæ, 65; on seaweeds
  of HEARD ISLAND, 227.
Digging sticks of Gudangs, 357; of Hotten-
  tots, 148.
Dicksonia culcita, 34.
Diodon, 271; antennatus, 51; hystrix, 52.
Diomedea, exulans, 134, 171, 180, 183, 254;
  fuliginosa, 180, 183, 254; culminata, 129,
  183, 254; Melanophrys, 254.
Diptera, of the FALKLAND ISLANDS, 558;
  of KERGUELEN'S LAND, 192.
Dispersion, of insects, 384; of plants, 17,
  135, 368, 419; of plants by birds, 164, 386,
  522.
Distichopora coccinea, 531, 534.
Distribution, of genus Atya, 61; of Marsu-
  pials, 142; of penguins, 119.
DOBBO, 367.
Dogs, run wild in penguin rookery, 132; of
  the ADMIRALTY ISLANDS, 478.
Dodo, the, 598.
Domicella solitaria, 295, 309.
Draco volans, habits of, 406.
Dracænas, planted by savages, 466.
Dragon, the, last seen in England, 426;
  bones and teeth of, 423; Chinese account
  of, 425; origin of, 423.

Dredging, deep-sea, process of, 578.
Dress, origin of, 412; peculiar, of Bisayans, 411.
Drift-wood, large masses of, 367; off New Guinea, 432; on deep-sea bottom, 583.
Drums, 379; log, 309; wooden, of Admiralty Islanders, 471; wooden, Fijian, 321.
Drymis Winteri, 539.
Duck, loggerhead, the, tameness of, 552.
Dugong, the, 424; bones of, used as decoration, 363; skulls of, placed in temples, 474.
Dules Malo, 516.
Duncan, Prof. Martin, on parasites in corals, 581.
D'Urvillœa utilis, 165, 171, 227.
Dyer, Mr. W. T. Thiselton, on algæ, 36, 410; on *Loranthus Aphyllus*, 545.
Dykes, of Basalt, cleavage of, 46; volcanic, 109, 115, 196.

E.

Easter Island, a sheep run, 515.
Eaton, Rev. A. E., on cry of petrels, 181; on Diptera of Kerguelen's Land, 193.
Echidna, 266.
Echinometra, 47.
Echineis Remora, habits and colour of, 8.
Echium stenosiphon, 44.
Eggs, half-hatched, eaten, 412; of birds hatched in sand, 403; of turtles, 561.
Emberiza Braziliensis, 122 (note).
Empetrum nigrum, 110.
Equatorial current, 68.
Enhalus, 361.
Entada scandens, 357.
Epeira clavipes, 382.
Epizoanthus parasiticus, 590.
Erica, arborea, 5, 7; *Azorica*, 34.
Erythrina Indica, 295, 352.
"Etna," the, Dutch ship of war, 438, 443, 445.
Errina labiata, 528.
Eua Island, 282.
Eubalœna Australis, 197.
Eucalyptus Amygdalina, 261.
Eudyptes, saltator, 117, 175, 195; *chrysolophus* described, 195, 229.
Euphorbia, Canariensis, 2; *tuckeyana*, 42.
Euphyllia, 362.
Euplectella, 589; *aspergillum*, dredge used for, 407.
Euprepes cyanura, 479.
Eustephanus Fernandensis and *E. galeritus*, 539.
Examination halls, Chinese, 421.
Exhibition at Kioto, 486.
Expressions and gestures, at the Admiralty Islands, 457; of the Api natives, 346; at Fiji, 336; of Fuegians, 551; of Gudangs, 360; of Japanese, 492; of Papuans, 440; of rage used in fighting, 441.

F.

Falkland Islands, the, 135, 553.
Fayal Island, 29.
Feathers, remarkable, of young ostriches, 152; of *Chrysœna*, 303.

Feira St. Anna, 93, 95.
Fernando do Norhona, 77.
Ferns, in the Azores, 33; preponderance of, in vegetation, at Juan Fernandez, 537, at Kermadecs, 280, at Tahiti, 518, of Ternate, 393.
Fertilization of flowers, by bats, 291; by birds, 354, 539; of introduced plants at Juan Fernandez, 540; of introduced plants at Tahiti, 524.
Festuca Cookii, 191.
Ficus Norhonœ, 82.
Fig tree choking itself, 371.
Fijian, chief, domestic life of, 328; "ula," 338; convicts, 309; group, 293, 301; mountaineers, 309, 321; native's ignorance of his age, 325.
Fiji, reality of rank at, 336.
Fire, getting of, by friction, 288; signal, of Papuans, 435.
Fish, deep-sea, at Madeira, 38; living on land, 295.
Fishing, for Snook, 156; mode of, at St. Vincent Island, 52; at St. Paul's Rocks, 71.
Fish-hooks of Admiralty Islanders, 467.
Fitchia nutans, 521.
Fjords, formation of, 185, 199, 549.
Flies, with rudimentary wings, habits of, 170, 181, 192, 558; House, on board the "Challenger," 593.
Flight, of bats, 291; of birds, 13, 206; of *Draco volans*, 406; of flying fish, 570; of petrels and other birds, 569; loss of, at breeding season, by albatross, 131, 173.
Flower, Prof. W. H., on tusks of Ziphioids, 158 (note).
Flowers, variety of, at the Cape of Good Hope, 153; conspicuous, at Juan Fernandez, 539.
Flowering season, Marion Island, 169; at Tristan da Cunha, 134.
Flying-fish, flight of, 571; development of, flight of, 479; hooked, 51; shooting at, 562.
Fogo Island, 54.
Fox, Gen. Lane, collection of, 451; on development of weapons, 468; on savage decoration, 509.
Foraminifera, at St. Vincent Island, 47; large, 292.
Fortifications at Admiralty Islands, 472; Fijian, 326; origin of idea of, 355.
Fossil, animals found in the deep-sea, 587; wood at Kerguelen's Land, 195.
Francolinus, 145.
Frigate bird, 11, 82, 349, 563.
Fritz Muller on *Ocypoda*, 49.
Frogs, noise made by, in Brazil, 93.
Fringilla Teydeanu, 26.
Fruit pigeon, 292, 304, 479.
Fuegian natives, 550.
Fucus vesiculosus. 567.
Fulmarus glacialis, 206.
Funchal, 38.
Fungi, growing, washed by sea-water, 455.
Fungia, 385; alternation of generations in, 524.

Fung Shui, 419.
FURNAS, hot springs at, 35.
Fur Seal, 538 ; eating penguins, 189 ; killed, 187 ; pet, on board the " Challenger," 596.

G.

Galeopithecus Philippensis, habits of, 405.
Galinis at St. JAGO ISLAND, 57, 59.
Galinula nesiotis, 122.
Garfish, 53, 479.
Gauchos, Scotch, 554.
Gammarus, Arcticus and Themisto, 189.
" Gazelle," ship of war, sounding by, 170,281.
Gecko at St. VINCENT, 49, at TENERIFFE, 7.
Geese, wild, caught with the bolas, 558 ; shooting, 551.
Genetta felina, 153.
Geology, of BERMUDA, 18; of BLUE MOUN-
TAINS, 267 ; of KERGUELEN'S LAND, 185,
194, 195, 197 ; of MARION ISLAND, 164,
180; of MBAU ISLAND, 315 ; of RAT
ISLAND, 83 ; of RAINE ISLAND, 347; of
St. PAUL'S ROCKS, 75 ; of TERNATE, 393 ; of
TRISTAN DA CUNHA, 127 ; deposits of land
organic remains in the deep sea, 583.
Geoplana, 279.
Geopelia, 353.
Gerrys, 571.
Georychus Capensis, 145.
Geyser, formation at the AZORES, 35.
Gibbon, cries of the, 337.
Gilbert Islanders, 400.
GILOLO, 394.
Glaciers, at HEARD ISLAND, 217 ; descent of, 244.
Glacial epoch in HEARD ISLAND, 223.
Glaciation in KERGUELEN'S LAND, 185, 197.
Glaciated stones dredged in deep water, 583.
Gladiolus, 153.
Glaucus, 569.
Gleichenia dichotoma, 393, 521.
Globigerina, 574.
Gnat, wingless, 193, 559.
Goats, wild, at JUAN FERNANDEZ, 542 ; wild,
colour of, at St. VINCENT ISLAND, 54;
at TRISTAN DA CUNHA, 124 ; feared by
savages, 477.
Goat, pet, on board the " Challenger," 596.
GODDEFROY, BROTHERS, 283.
Gods, of Admiralty Islanders, 473 ; Chinese,
420 ; of HAWAII, 503-509 ; Japanese, 484.
Gonostomyus, 325.
GOOD HOPE, CAPE OF, 138, 150, 152.
GOODRIDGE, CHARLES, on King Penguins,
178 ; nesting of albatross, 173; on the
Sea-elephant, 228 ; on tree trunks in the
CROZETS, 182.
GOUGH ISLAND, plants of, 136.
GOULD, Mr., on birds from BOOBY ISLAND,
365 ; on eggs of Chionis, 210.
Grampus, a, 253.
Grapsus strigosus, habits of, 48, 59, 70, 83 ;
climbing trees, 26.
GREEN MOUNTAIN, St. VINCENT ISLAND, 43.
GRISEBACH, A., on diffusion of plants by

birds, 522 ; on Eucalypti and Acacias, 264 ;
on vegetation, of BERMUDA, 24, of the
AZORES, 35.
Ground Squirrel and owl in same hole in
CHINA, 431.
Grysbök, 151.
Gryllus, 79.
Guava, the, encroachment of, at the SAND-
WICH ISLANDS, 496 ; at TAHITI, 515.
Guilandina bonduc, 17, 135.
Gulf-weed, fauna of the, 567.
Gymnetrus, 430.
Gum trees, 33, 141, 258 ; giant, 260.
Gunnera Chilensis, 538.
GUNTHER, Dr. A. C. L. G., on Chloroscartes,
304 ; on deep-sea fish, 38 ; on Periophthal-
mus, 296 ; size of sharks, 10.
Gurnet, Flying, 51, 562, 571.
Gygis candida, 563.

H.

Haastia, 169.
Haidahs, houses of the, 399.
HAECKEL, Prof. E., on Medusæ, 575.
Hæmoglobin, in fish, 281 ; in Mammalia, 592.
Hair, of Admiralty Islanders, 458 ; curl of, in
various races, 459 ; forcing growth of, by
Fijians, 326 ; dressing at FIJI, 310, 328 ; at
TONGA, 284 ; on moles, significance of, 459.
Haircutter, equestrian, 546.
Halcyon, Erythrogastra, 56 ; sacra, 292 ; sancta,
278, 365.
Halimeda opuntia, 12.
Haliotis, 147 ; mode of cooking, 148.
HALMAHERA, 394.
Halmaturus ualabatus, 261, 269.
Halobates, 571.
Halobæna cærulea, 181.
Halophila, 322, 361 ; flowering in 18 fathoms, 581.
Halyritus amphibius, 193.
Hammer-headed shark, 52.
Hands clapped, during dancing, 329 ; to ex-
press astonishment, 337 ; to express respect,
in Japan, 337 ; in FIJI, 328.
Hands, gestures of, in dancing, 401.
Hand-marks made by Australian blacks, 275.
HAWAII ISLAND, 499.
HEARD ISLAND, 196, 216.
Heat, toleration of, by plants, 36, 384, 410.
Helichrysum, 153.
Heliconiæ, 90.
Heliconia narcea, 85.
HELIGOLAND, migratory birds at, 364.
Heliopora cærulea, 404.
Hemiramphus, 58.
Hemp, Manila, 411.
Heritiera littoralis, 433.
HERMIT ISLAND, 224.
Herpestes, 152.
Herpetolitha limax, 386.
Hibiscus tiliaceus, 288.
Hippotherium, 423.
Hirundo, tahitica, 295 ; rustica, 482.

Holothurians, abundance of, at BERMUDA, 28; deep-sea, 591.
HONAM, Monastery of, 421.
HONG KONG, 415.
HONOLULU, 495; rainfall at, 497; scientific library at, 512.
HOOKER, Sir J. D., on the Big Trees of CALIFORNIA, 260; on the flora of AUSTRALIA, 142; on the flora of KERGUELEN'S LAND, 169, 200; on the flora of TRISTAN DA CUNHA, 135; on ice stained by diatoms, 249; on vegetation of the KERMADEC ISLANDS, 280; on vegetation of POSSESSION ISLAND, HEARD ISLAND, &c., 225.
HOOKER MOUNT, 185.
Horses, charging with, 544; dealing in, in BRAZIL, 100; with deformed hoofs, 554; FALKLAND ISLAND, domestic and wild, 554; learning not to trip in mole-holes, 145; and mules compared, 549; white, worshipped, 482.
HORTA, Town of, costume of women at, 31.
Hot springs, AZORES, 35; at CAMIGUIN, 409.
Hottentots, middens of, 147.
Houses, of the Admiralty Islanders, 465, of natives of ARU, 370; at HUMBOLDT BAY, 445; at KE DULAN, 381; at the PHILIPPINES 396; of Tongans, 286; origin of first story in, 399.
Hoya, 345.
HUMBOLDT BAY, 435.
Humming birds, of JUAN FERNANDEZ, 539; flight of, 13, 570; shooting, 91.
Hura crepitans, 15.
HUTTON, Capt. F. W., on *Peripatus*, 161; on Land Planarians, 279.
Hybernation of animals on the PEAK OF TENERIFFE, 7.
Hydrocotyle, 135.
Hydnophytum formicarum, 389.
Hydrocorallinæ, 534.
Hydrophidæ, 479.
Hydrosaurus marmoratus, 405.
Hymenophyllum, 550; *tunbridgense*, 167.
HYNES, Mr. J., R.N., adventure with a rat, 594.
Hypsiprimnus, 269.
Hyrax capensis, habits of, 144.

I.

ISABELLA, PORT, houses of Lutaos at, 397.
Ianthina, 568.
Icebergs at CROZETS, 183; bi-tabular, 236; colour of, 245; cleavage of ice of, 241; dimensions of, 242; first sighted, 232; foreign matter on, 243; height of, 238; run into by ship, 251; stratified structure of, 239; sunset effects on, 247; typical form of, and immersion of, 233; wash lines on, 235.
Ice, pack, 248; stained by diatoms, 249; stream, 249.
Ichneumon, 153.
Indulgences, Papal, sold in PHILIPPINES, 414.

ILO ILO, pile-dwellings at, 398.
INACCESSIBLE ISLAND, 114; position, appearance, and vegetation, 115; German settlers at, 116; penguins at, 117; other birds at, 122; wild goats and pigs at, 124.
Insects, ancestor of, 159; of HEARD ISLAND, 230; Pelagic, 571; at summits of volcanoes, 384; of ST. PAUL'S ROCKS, 73; swarms of, blown off land, 85.
Inter-breeding in islands, 512.
Invocation of winds at Fiji, 321.
Ipomœa pes Capræ, 18, 56, 79, 304, 345, 433.
Iron, clamoured for by savages, 438, 451.
Islands, oceanic, necessity for investigation of, 599.
Istiophorus, 448.
Itch, Vegetable, the, 379.

J.

Jack fruit, 88, 392.
JAPAN, 481.
Japanese, picture books, 493; sight seeing, 483; women, drink of, 495.
Jatropha, curcas, 57, 63; *gossypifolia*, 78; *manihot*, 102; *urens*, 79, 83.
JEFFREYS, Mr. GWYN, on phosphorescence, 590.
Jew's-harp, of Admiralty Islanders, 471; of Lutaos, 401.
JUAN FERNANDEZ, 537; Sea-elephant at, 201; Fur Seal of, goat of, 596.
JUKES, Mr., on calcareous rock, 21; on RAINE ISLAND, 347.
Juniperus barbadensis, 23.

K.

Kaava, at TONGA, 287; drinking, 320; mode of preparing, effects of, &c., 311.
KANDAVU ISLAND, 301.
Kathartes pernicopterus, 53.
KE ISLANDS, 379.
Kentia, 323; *exorhiza*, 294.
KERGUELEN'S LAND, 45, 135, 169, 184.
Kerguelen Cabbage, described, 167, 191; seeds of, eaten by teal, 190.
KERGUELEN PLATEAU, 170.
KERMADEC ISLANDS, 280; dredgings off, 589.
KIDDER, J. H., M.D., on *Chionis*, 210; on the flora of the CROZETS, 183.
KILAUEA, crater of, 500.
Kingfisher, at ST. JAGO ISLAND, 56; jewels made of feathers of the, 420; marine habits of a, 278.
King, KALAKAUA, 497; THACKOMBAU, 309, 319; GEORGE, of TONGA, 320.
KIOTO, 486.
KIRK, Mr. T., on the Rata, 278.
Kitchen middens, at the CAPE OF GOOD HOPE, 147; Australian, 273, 354; at FIJI, 316, 326; at MAGELLAN'S STRAITS, 552.
Kites, mode of feeding of, 55.
KNYSNA FOREST, the, 161.

L.

LABILLARDIERE, M., 450.
Land Crabs, catching, 536; killing rabbits, 561; young of, 64.
Land Nemertine, 26.
Land Planarians, 89, 154, 279, 494.
Language, loss of by all races, but English, 391; Malay and Spanish mixed, 405; Malay, simplicity of, 369; savage, changes of, 358.
LANKESTER, Prof. E. RAY, on *Hæmoglobin*, 281; on Terrestrial Annelida of KERGUELEN'S LAND, 215.
Larus,. dominicanus 155, 212, 230, 278; *Novæ Hollandiæ*, 266, 348; *scopulinus*, 266.
Lasso, the, used in the streets, 544; use of in robbery and flirtation, 548.
Laughing Jackass, 257.
Laughter, of Fijians, 337; remarks on, 337.
Lava, ponds of fluid, at KILAUEA, 501; flow in INACCESSIBLE ISLAND, 127.
Lavandula rotundifolia, 42.
Leaves, verticality of Australian, explained, 264.
Lecanora esculenta, 344.
LEFROY, Gen. Sir J. H., self-sown grapes found by, 24; assistance to Expedition by, 28.
LEGGE, Rev. JAMES, on Chinese natural history, 431.
Leprosy at the HAWAIAN ISLANDS, 512.
Leptocephalus, 281.
Leucadendron argenteum, 142.
Libations, pouring of, 420.
Lichens at KERGUELEN'S LAND, 193.
Limestone bed at ST. JAGO ISLAND, 64.
Limosella, 194, 227.
Limnlus rotundicaudatus, 402.
Lingula, 402.
LI SHI CHAN, medical works of, 425.
Lithodomus caudigerus, 47.
Lithothammion, 47, 65; *polymorphum*, 76.
LITTLE SABA ISLAND, 16.
LIVONI, visit to, 308.
Lomaria, Alpina, 111, 167; *Botyana*, 113.
Loranthus, Aphyllus, 545; *celastroides*, 545; *Eucalyptifolius*, 545; *eucalptoides*, 545.
Lutaos, the, 396,. 402.
Lutra inunguis, habits of, 154.
LORY, Rev. H. C., on Diptera, 559.
Lumbriculus, 194.
Lyallia Kerguelensis, 167, 169.
Lycopodium, 393.
Lycosa, habits of, 13; on board the "Challenger," 596.
LYELL, Sir C., on diffusion of plants, 386.
Lygodium reticulatum, 295.
Lyre Birds, 261, 270.

M.

Mæandrina, 385.
MBAU ISLAND, excursion to, 314.
McARTHUR, Sir W., seat of, in NEW SOUTH WALES, 267; on *Loranthus*, 545.
MACDONALD ISLAND, 216.

MACDONALD, Dr. J. D., R.N., on fish at FIJI, 325.
MACGILLIVRAY, Mr., 280; on Gudangs, 354, 358.
MACLEAN, Staff-Surgeon G., R.N., 595; curious medical fee to, 361.
MACLEAR, Capt. J. P. L. P., R.N., use of air gun, 593.
Macrocystis pirifera, 116, 205, 227, 567.
MACTAN ISLAND, 408.
MADEIRA, 38; cap worn by peasants of, 40; streets of, GRAND CURAL, 39; sunset effect at, 61; wine, 40.
Madrepora, 17, 306.
MAGELLAN STRAITS, 549.
Magenta, useless on a dark skin, 463; in China and Japan, 489.
Majaquens æquinoctialis, 137, 208, 254.
MALAMOUI ISLAND, 405.
MALANIPA ISLAND, 406.
Malays, at the CAPE OF GOOD HOPE, 141; pets kept by, 384.
Manchineel, 16.
Manganese, used as paint, 463.
MANILLA, 411.
MAUNA LOA, form of, 499; glow from crater of, 500.
Maoris, cannibalism of, 339.
Maorioris, 340.
Map, Tahitian mountain, 520.
MARCO POLO, on the Unicorn, 424.
MARION ISLAND, 163.
Marsupials, distribution of, 142.
MAS-AFUERA, birds of, 539.
Materia Medica, ancient, in ENGLAND, 427; Chinese and Japanese, 423.
MATUKU ISLAND, 293.
McKELLAR, Mr., ostrich farm of, 151.
MEANGIS ISLANDS, 432; dredgings off, 589.
Medicine, Chinese, 425; old English, 426, 427; flesh of strange animals used as, 97.
Medusæ in fresh water, 272; upturned, 404.
Megapodius, eggs of, 403; *tumulus*, 353, 365.
Melastomaceæ, 95.
Melbourne, 256.
MELVILLE ISLAND, 225.
Membranipora, 567.
Merops ornatus, 353, 365.
Mesoplodon Layardii, 157.
MESSIER CHANNEL, the, 549.
Metrosideros, 521; *Robusta* and *Florida*, 278.
Migration, of animals in the deep sea, 583, 585; of birds to Bermuda, 24; of birds of the SANDWICH ISLANDS, 500; of birds, TORRES STRAITS, 364; of penguins, turtles, and seals, 133.
Mice, at FERNANDO DO NORHONA, 79; at TRISTAN DA CUNHA, 114; at MARION and ST. PAUL'S ISLANDS, 181.
MIKLOUCHO MACLAY, 392; at ADMIRALTY ISLANDS, 450, 460.
Microglossum aterrimum, 352.
MIERS, Mr. E. J., Crustacea of KERGUELEN'S LAND, 215.
Mill, women at the, in the AZORES, 34.
Millepora, nodosa, structure of, 525; stinging of, 388.

Milleporidæ, 307.
Milvus korschum, 55.
MILNE EDWARDS, M. A., on *Rhinopithecus*,429.
MINDONAO, 395.
Misodendron, 546.
Missionaries, Dutch, 369; at FIJI, 315, 320, 327, 330, 335; at TONGA, 285, 287.
MITTEN, Mr. W., mosses of MARION ISLAND, 168; mosses of TRISTAN DA CUNHA, 136.
Mistletoe, 90; leafless on cactus, 545; variable on Australian trees, 545.
MOBIUS, Dr. K., on the flight of Flying-fish, 570, 571; on deep-sea animals, 581.
Mole, the Golden, 147; the Sand, habits of, 145; true, and Rodent compared, 145.
Moles, hairy, signification of, 459.
Mollusca, shells of, composing rock, 19.
Mollymauk, 183, 254; nests of, 129, 130.
Monasteries, Chinese, 421.
Monkeys, at ST. JAGO ISLAND, 60; figures of in Chinese books, 429.
Montia fontana, 191.
Moraines at HEARD ISLAND, 217; at KERGUELEN'S LAND, 198.
Morbid growths, due to reversion, 460.
Mormon fratercula, Celtic name of, 129.
Morunga elephantina, 114, 171, 187, 200, 222, 227.
Moros, the, 396.
MOSELEY, Rev. CANON, on the descent of glaciers, 244.
MOSELEY, MOUNT, 185.
Moss, Staff-Surgeon R.N., on Arctic ice, 239.
Mosses at HEARD ISLAND, 225.
Mound-birds, 353, 365; eggs of, 403.
Mud spring, AZORES, 37.
Mule, and horse as mountaineers, compared, 548; load of, in BRAZIL, 97; riding in BRAZIL, 94; wounded, 547.
MÜLLER, Baron von, on *Eucalyptus*, 260.
Mullet, Gray, 271.
MURRAY, Mr. JOHN, at ELIZABETH ISLAND, 552; at HUMBOLDT BAY, 443; on sharks, 9; use of tow-net in deep water, by, 575.
Musa textilis, 411; *uranoscopus*, 517.
Museums, origin and development of, 3.
MUSSCHENBROEK, Mr. S. C. J. W. VAN, on *Trigonia*, 276; at TERNATE, 390.
Music, origin of, 333; instruments of, in the ADMIRALTY ISLANDS, 471; knowledge of, by modern Hawaians, 498.
Mus rattus, 594.
Mya truncata, eaten by seals, 205.
Myrmecodia armata, 389.
Myrmeleon, 50.
Mythical animals, 428, 430.
Myzomela jugularis, 295.

N.

Naja haje, 153.
Names, vernacular, of southern animals, 129.
NARES, Sir G. S., excursion with, 52, 79, 84; manœuvres amongst icebergs, 252; landing with, at HEARD ISLAND, 217.
Narwhal, the, 424.

Natural selection amongst wild horses, 556.
Naucrates, 9, 281.
Nautactis, 572.
Nautilograpsus minutus, 568.
Nautilus, ornament made of shell of, 345; *Macromphalus*, 300; *Pompilius*, living, caught, 297.
NAVUSA, 323.
Na Vatani Tawaki, 317.
Nectarinia frenata, 352.
Nectarinidæ, 145.
Nelson, flag captured from, 8.
NELSON, Major-Gen., R.E., on geology of BERMUDA, 21.
Nemertine, Land, 26; Pelagic, 572.
Nephila, web of, 50.
Nesospiza Acuhnæ, 122.
Neuopogon Taylori, 193.
Nertera depressa, 111, 116.
Nesocichla eremita, 122.
Nestor, 434; *Meridionalis*, 279.
Nettle cells, tube composed of, 408; of corals, 388.
NEW GUINEA, 432,435; drift wood from,432.
NEWPORT, Mr., on the genus Atya, 61.
NEW HEBRIDES, 340.
NEW ZEALAND, 277; art of, related to Hawaian, 510; tikis of, 509.
NIGHTINGALE ISLAND, 126.
Ninox boobook, 352.
Nitella Antarctica, 194; *flexilis*, 290.
NUKUALOFA, 286.

O.

OAHU ISLAND, 495.
Oasis, miniature, 51.
Obsidian, lance-heads of, 468.
Oceanitis, 183, 208, 499.
Octacnemus Bythius, described, 587.
Octopus, 125.
Ocyponda, 26, 403; *ippeus*, habits of, 48.
Œstrelata Lessoni, habits of, 208.
OLIVER, Prof., on plants of MARION ISLAND, 170.
Olfersia, 78.
ONSLOW, Capt., R.N., on chasm formed by rain, 267.
Ophioglossum pendulum, 455.
Opuntia, plantations of, at TENERIFFE, 2; at OAHU, 496.
Opossums, Australian, shooting, 267; traps for, 257.
Oracles, Japanese, consulted, 483.
Orotavu, 4.
Orang Utan, significance of name, 430.
Orange and lemon, feral in TAHITI, 524.
Orbitolites, *algæ* parasitic in, 292.
Orca amongst the ice, 253.
Oreaster, very large, 363.
Ornament, Hawaian Hook, origin of, 504.
Ornithorynchus paradoxus, 262.
Ornithoptera poseidon, 373.
OSAKA, 484.
Oscillatoriæ in hot water, 410.
Ossifraga gigantea, 107, 134, 180, 183, 205, 254.

Ostrich farm, habits of the birds, at, &c., 151.
Ostriches, young, on board the "Challenger," 595.
Otaria jubata, 189, 552.
Otariadæ, 188, 204.
O'MEARA, Rev. E., on Diatoms, 216.
Otosaurus microblepis, 292.
Otters, the sea, 494; the Clawless, habits of, 154.
Otus Brachiotus, 500.
Ovulum ovum shells, 466.
OWEN, Prof., on fossil Mammalia, 423; on *Cimoliornis*, 523; on tusks of Ziphioids, 158 (note).
Owl and rat in same hole, 431.
Oxalis, 135.

P.

PACIFIC OCEAN, shallows in, 584.
Pagoda, the Whampoa, 419.
Pagodroma nivea, 253.
Paguridæ, terrestrial, 17, 304; deep-sea, 590.
Painting of the body at FIJI, 331; at the ADMIRALTY ISLANDS, 463; of face in JAPAN and CHINA, 489.
Palæmon in fresh water, 60.
PALMA ISLAND, seen from the PEAK OF TENERIFFE, 5.
Palinurus, 74; *frontalis*, 542.
Palythoa, 48, 73.
Pandanus, 294, 370, 499.
Pandarus with *Lepas* attached, 281.
Pandion haliætus, 476.
PANSCH, Dr., on Arctic vegetation, 226.
PAPEETE, town of, 514.
Papilio feronia, 89.
Paradisea Apoda, 375; *papuana*, 392; *rubra*, 392.
Paritium Tiliaceum, 497.
Parroquet, 211; brush-tongued, 352.
Parrots, African, ship full of, 41; pet, on board the "Challenger," 594; at TONGA, 292; at FIJI, 304, 322, 325; at ARU, 372.
PATAGONIA, fjords of, 549.
Patagonians, saluting, 551.
Patella, 148.
Peak, of TENERIFFE, ascent of, 2; of TERNATE, ascent of, 392; the, of TRISTAN DA CUNHA, 109.
Peat, at BERMUDA, 23; at MARION ISLAND, 165.
Pelagic, animals, range in depth of, 575; animals, protective colouring of, 567, 569; plants and animals, 565, 576.
Pelargonium, 153; at TRISTAN DA CUNHA, 134, 135.
Pelagonemertes Rollestoni, 569, 572.
Pelamys bicolor, 292.
Pelea capreola, 151.
Pelacanoides urinatrix, 129, 171, 208.
Pele's hair, formation of, 502.
Pelicanus fuscus, habits of, 11, 15.
Penguins, 113, 115; origin of word, 129; eaten by seals, 189; Jackass of the FALK-LAND ISLANDS, 560; Jackass at the CAPE

OF GOOD HOPE, 155; Johnny, 175, 189; King, 176, 197; nesting in caves, 196; used as fuel, 229; Rock-hopper, 175, 195; Rock-hopper, mode of swimming, 117, 125; sensitiveness of iris of, 118; Rock-hopper, numbers of, at INACCESSIBLE ISLAND, 133; migrations of 133; Rock-hopper, rocks worn by feet of, 128; Rock-hopper, rookery of, 120, 127, 132.
Pentacrinus, many dredged, 589.
Parameles nasuta, hunting, 269.
Periophthalmus, 295, 322.
Peripatus, at ST. THOMAS, 14; *capensis*, structure and habits of, 159; *Novæ Zealandiæ*, 279.
Peristera geoffroyi, 79.
PERUAGUACU RIVER, 92.
PERON, on drawings of Australian Blacks, 275; on the Sea-elephant, 228; figure of Sea-elephant, 202.
Petrel, diving, habits of, 208; Giant, 171, 180, 254; nesting of, 207, 521; Snow-white, 253; flight of, 569.
Phaethon, 282, 516, 521; *æthereus*, 563; *flavirostris*, 25, 563.
Phalacrocorax, *verrucosus*, nesting of, 212; at HEARD ISLAND, 230; *Capensis*, 152, 155; *imperialis*, 230.
Phalangista vulpina, shooting, 267; young of, 268; traps for, 267.
Phalanger, Woolly, 384, 465, 479.
Phalaropus hyperboreus, 434.
Phascogale penicillata, 268.
Phascolarctos cinereus, 259.
Pheasants at the CAPE OF GOOD HOPE, 145.
PHILIPPINE ISLANDS, 395.
Phoca Greenlandica, 189.
Phosphorescence, in deep sea, 590; of Pelagic animals, 574.
Phylica arborea, 110, 111, 115, 126, 135.
Physalus Australis, 252.
Picnics, Ministerial, 266.
PICO ISLAND, 32.
Pigeon, with aberrant plumage, 303; Fruit, 292, 304, 364,382, 386, 479; Fruit, diffusion of plants by, 386; Ground, 353; Nutmeg, 382; at ST. JAGO ISLAND, 60; at JUAN FERNANDEZ, 542.
Pigeon English, 415.
Pigmies, the, 428.
Pigs, of the ADMIRALTY ISLANDS, 478; wild, 125, 183; wild, destroying crabs, 403; at TAHITI, 517.
Pigtails, Chinese and English, 423.
Pile dwellings, Papuan, 445; modifications and origin of, 396-400.
Pilgrims, Japanese, 484.
Piper methysticum, 311.
Pilot fish, 281; habits of, 9, 10.
Pillows of Tongans, Japanese, &c., 286.
Pine-apple, fabric made from, 411.
Pityriasis versicolor, 380.
Planarians, large, marine, 402; Land, 89, 154, 279.
Plants, introduced, supremacy of, 496; diffusion of, by flotation on sea, 135, 368, 419, 433; diffusion of, by *Procellaridæ*, 522;

diffusion of by winds and birds, 24, 164, 386; on board the "Challenger," 594; range of, in depth, 581; Pelagic, 216.
Platycerium, 270, 371.
Platycercus splendens, 304, 322, 324; *tabuensis*, 292.
Pleuronectids, Pelagic, 281.
Plumage, aberrant, of pigeon, 303; change of, by heron, 291; of young ostriches, 152.
Poa cookii, 165, 167, 224.
Poetry, origin of, 333.
Polyopogon, 589.
Polycheles, 297, 589.
Polypodium Australe, 167.
Population, of HEARD ISLAND, 229; of TONGA, 287; of ADMIRALTY ISLANDS, 478.
Porcupine, hole of, 153.
Porites, 48, 344.
Porpita, 569.
Porpoises, 30, 271.
PORTO GRANDE, 41.
PORT JACKSON, 266.
PORTO PRAYA, 55.
POSSESSION ISLAND, CROZETS, 181; in ANTARCTIC SEA, plants of, 224.
POTTS, Mr. T. H., on habits of the Kaka, 279; on Apteryx, 125.
PONTA DELEGADA, 32, 37.
Prawn, fresh water, at CAPE VERDES, 60; at FIJI, 324.
Pressure, the, in the deep sea, 579; effect of, on animals, 580.
PRINCE EDWARD ISLANDS, 163.
Pringlea antiscorbutica, 167, 169, 191, 224.
Prion, 123, 181, 229, 253; *desolatus*, habits of, 207.
Procellaridæ, 206.
Procellaria rostrata, 521, 499.
Procession, at PONTA DELEGADA, 37; at BAHIA, 86.
Property, retained by natives of API, 345; relative value of Papuan, 439; hidden by savages, 357.
Proteaceæ, 142.
Protococcus affinis, 76.
PSAMMETICHUS, King, experiment of, 377.
Ptilinopus, *porphyraceus*, 292, 324; *superbus*, 364.
Ptilotis, *carunculata*, 292; *crysotis* and *filigera*, 351; *procerior*, 309.
Ptilorhis Alberti, 352.
Pteris aquilina, 33, 537.
Pteropus, 323, 375, 479; *keraudrenii*, 291; *poliocephalus*, 268.
Pteropod with eyes on stalks, 514.
Puff-adder, 153.
Puffinus, 16, 123; *nugax*, 522.
Pycnogonids, giant, 586.
Pygosceles tæniata, 175, 189.
Pyrosoma, giant, 574.

Q

Quail, painted, the, 364; at ST. JAGO ISLAND, 56; at ST. VINCENT ISLAND, 54.
Quan Yin, 420.

Quedius, 73.
Querquedula Eatoni, habits of, 190.

R.

Rabbits of TENERIFFE, 6; of CROZETS, 183.
Races, congregation of, at FIJI, 336.
Radiolarians, yellow cells of, 293.
Rain, belt of excessive, 66; fall of, HONOLULU, 497; effect of on distribution of ferns, 280.
RAINE ISLAND, 347; birds of, 348; insects of, 350.
Rallus pectoralis, 348.
Ramphastos ariel, 92.
Raoulia, 169.
Ranunculus, *biternatus*, 168; *Moseleyi*, 194.
RAT ISLAND, 83.
Rats, Black, on board the "Challenger," 594; at FIJI, 308, 324; Kangaroo, 269; at TAHITI, 517.
Rattans, 351, 372.
Reef, coral, encircling, 294; barrier, visit to, at FIJI, 306; at API, 343; raised, 282, 344, 389, 408.
Religion, in the ADMIRALTY ISLANDS, 473; in CHINA, 420; in JAPAN, 487, 493.
Remipes, 26.
Remora, the, 8, 11.
Restiaceæ, 142.
Retama, 5.
Rhea Americana, 595.
Rheebök, 151.
Rhizomorpha girdles, 330.
Rhinopithecus Roxellanæ, 429.
Rhinoceros horn, an antidote, 427; horn of, cups of, 427; *trichorhinus*, 423.
Rhynchodemus, 154, 279.
RIBEIRA GRANDE, 34.
RICHARDS, MOUNT, 185.
Rice fields, 395, 485.
Richardia Æthiopica, 153.
RICHARDS, Mr. R., R.N., on habits of birds, &c., at TRISTAN DA CUNHA, 116 (note).
Rifle Bird, habits of the, 352.
River, great, of FIJI, 322; Ambernoh, the, 432.
Roches Moutonnes, 197.
ROLLESTON, Prof. G., Nemertine named after, 572.
ROSS, Sir J., on icebergs, 242.
ROSS, MOUNT, 185, 214.
ROWITT, Mr. R., 422.
ROYAL SOUND, excavation of, by ice, 198.
Rumex, 110, 115.
RUMPHIUS, his account of *Nautilus*, 298.
Rusa moluccensis, 379, 390.

S.

Saccharum, 323, 433.
Saccopteryx canina, 103.
Salpæ, 569: experiment on sinking of, 582.
Saluting, the, Patagonians, 551; remarks on, 387.
Samolus Valerandi, 51.

SAN DOMINGO VALLEY, 59.
SANDHURST, Town of, 259.
Sand-box tree, 15.
Sand, calcareous and silicious associated, 362; shifting of, 18; rock, calcareous, 347; rock, calcareous, absent at TONGA, 290; tracks of animals on, 152.
SANDWICH ISLANDS, the, 495.
SANTA CRUZ MAJOR ISLAND, excursion to, 402.
Sarcostemma Daltoni, 44, 63.
SARGASSO SEA, 567.
Sargassum, 362, 367; bacciferum, 567.
SAUNDERS, Mr. HOWARD, on Laridæ, 69, 266.
Savages, counting of, 358, 457; flower gardens of, 466; forgetfulness of, 325, 358; difficulties in learning language from, 456; mistakes of, as to white men, 453; relative decoration of sexes amongst, 461.
Sawfish in fresh-water, 325.
SCLATER, Mr. P. L., on duck of the SANDWICH ISLANDS, 500; and Mr. SALVIN on Phalacrocorax, 230.
Scyllæa pelagica, 568.
Screw pines, 283, 294, 370; in HAWAII, 499.
Sea-anemone, living in tube, 408; deep-sea, 409, 586; Pelagic, 572.
Sea Beans, 17, 135.
Sea-elephants, 114, 171, 187; bones of, 222; food of, 205; herd of, 200; trunk of, 201; mode of hunting, habits of, 227.
Seal, Bladdernose, 203; food of, 189, 205; Fur, 128, 188, 204; Fur, young, killed, 207; swimming of, 265; pet, 596.
Scalpellum, large, 586.
Sea-leopard, 200; bones of, 222.
Sea Serpent, the, 430.
Sea urchins, with poisonous spines, 12; borings of, 42; at FIJI, 307.
SEEMANN, Dr., on Fijian calendar, 295; on Musa, 517.
Seining at ST. JAGO ISLAND, 57.
SELKIRK, ALEXANDER, 537, 538, 541.
SEMPER, Prof., on Fungia, 524; on lungs of Land Crabs, 305.
Sequoia gigantea, 260.
Serolis, large, 586.
Sexual selection in butterflies, 373; experiments proposed on, 373.
Sharks, at FIJI, 308; catching, 8, 71; colouring of muscles of, 281; freshwater, at FIJI and elsewhere, 325; large, caught, 57; Port Jackson, 276; size of, and largest known, 10; treatment of, by sailors, 58.
SHARPE, Mr. R. BOWDLER, on birds of KERGUELEN'S LAND, 214; on Ground Owls, 431.
Sheathbill, habits of, 171, 179, 196, 206, 209.
Sheep, from EASTER ISLAND, 515; at the FALKLAND ISLANDS, 553.
Signs, Papuan, expressing a gun, 437; expressive of killing a man, 441.
Silk-cotton trees, 15.
Silver tree, 142.
SIMON'S BAY, 139.
Sinapidendron Vogelli, 51
Siphonaceæ, 12.
Sipunculids, larvæ of, 402.

STUTCHBURY, Mr. G., on fungia, 524.
Skua, 123, 131, 174, 191, 206, 254.
Slaves, condition of, in Brazil, 105; property of Europeans in Brazil, 102.
Snakes, at CAPE OF GOOD HOPE, 153; feared by savages, 477; Sea, 292.
Snipe, 395.
Solanum anthropophagorum, 339.
SOMBRERO ISLAND, 8, 589.
SOMERSET, CAPE YORK, 350.
Sonchus oleraceus, 115, 541.
SORBY, Mr. H. C., on Pele's Hair, 502.
Sounding, deep sea, time occupied in, 578.
Spalacini, 146.
Spartina arundinacea, 110, 115, 117.
Spartocytisus nubigenus, 5.
Speotyto cunicularia, 431.
Spermophilus mongolicus, 431.
Spheniscus, Magellanicus, 119, 125, 156, 560; demersus, 119, 155; minor, 125, 196.
Sphyræna barracuda, 74.
Spider, Ground, 13; large, on board the "Challenger," 596; making horizontal web, 8.
Spider's web, bird caught in, 382; strong, 50.
Spondylus shells at BERMUDA, 21.
Sporadopora, 530.
ST. ANTONIO ISLAND, 41, 42, 50.
Stalactite deposit in streams, 378.
Starfish, giant, 363.
Starling, glossy, 382; nests of, 372.
Statice Jovis barba, 44.
Stenorynchus Leptonyx, 200.
STEPHENSON, Mr., excursion with, 256.
Stercorarius Antarcticus, 123, 131, 174, 191, 206, 254.
Sterna, fuliginosa, 349, 363, 563; lunata, 479; virgata, 171, 211.
ST. JAGO ISLAND, 55, 56.
ST. MICHAEL'S ISLAND, 32.
Stone, sacrificial, at FIJI, 318.
Stone implements, at the ADMIRALTY ISLANDS, 468; at the CAPE OF GOOD HOPE, 147; grindstones for, 302, 319; used by Gudangs, 357; at FIJI, 313; improved modifications of, 326, 467; Papuan, 444; Tahitian adze, 523; at TONGA, 289.
Stone, mounds surrounded by slabs of, 317, 327.
Stone weapons, of the ADMIRALTY ISLANDS, 468; scattered at the FALKLAND ISLANDS, 560; Hawaian stone club, 510.
STORCK, Mr., at FIJI, 323.
ST. PAUL'S ISLAND, 170, 189.
ST. PAUL'S ROCKS, 67; seaweeds of, 76; Noddies and Boobies at, 68, 72; smallness of, 74.
Strawberries, at JUAN FERNANDEZ, 538; at TAHITI, 516.
Struggle for existence amongst birds, 213.
ST. THOMAS, 11.
ST. VINCENT ISLAND, 41.
Stylasteridæ, structure of the, 527–35.
SUCKLING, Lieut. R.N., trip with, 308.
Sugar-cane, grown for chewing, 11; plantations at FIJI, 324; at ADMIRALTY ISLANDS, 464; wild, 328, 433.

SUHM, R. VON WILLEMOES, 26; on Admiralty Islanders, 452; at HUMBOLDT BAY, 442; bird caught in spider's web, 382; on *Cardisoma*, 64; birds and butterflies of TRISTAN DA CUNHA, 122, 134; death of, 513; excursion with, 402; papers and drawings left by, 513; wingless fly, 192.

Sula, capensis, 155; *cyanops*, nest of, 349; *leucogaster*, 68, 349, 563; *piscatrix*, 349, 563.

SULU ISLANDS, 396.

Survivals in Chinese and Japanese funerals, 490.

Swiss chalets, origin of, 399.

Swallows, nesting on rocks, 145; seen at sea, 65, 482; on mountain tops, 295, 384.

Swifts, Tree, 292.

SWIRE, H., Lieut. R.N., sketch by, 81.

Swordfish, 253, 448.

Synapta, the Admiralty worm, 498; large, at API, 344.

SYDNEY, 266.

Sylvia, 79.

T.

TABLE BAY, 138.

TABLE MOUNTAIN, 138; KERGUELEN'S LAND, 187.

Tachypetes aquila, 11, 82, 563; *minor*, nests of, 349.

TAHITI, 513.

Tahitians, ignorance of mountains by, 519, 523; national air of, 536.

TALAUR ISLANDS, and natives of, 432.

Talegalla Lathami, 353.

Talpa, 146.

Tamandua, roasted, 97.

Tameness of birds in islands, &c., 122, 190, 209, 552.

Tarentola, 7; *Delalandii*, 49.

Taro, 35, 464.

Tattooing of Admiralty Islanders, 463; at FIJI, 319; Japanese, described, 491; relation of, to dress, 412; origin of, 509 (note).

TAYLOR, Rev. RICHARD, on NEW ZEALAND Lake Pas, 400.

Teal at KERGUELEN'S LAND, 190.

Telphusa, 561.

Temperature, of air at MARION ISLAND, 170; in the deep sea, 584; high, of sea water, 368; limit of, at which algæ grow, 410; maintenance of, by seals, 204; of mountain pools at MARION ISLAND, 168; in the mountains at TAHITI, 518; at TRISTAN DA CUNHA, 111.

Temples, at ADMIRALTY ISLANDS, 473; at FIJI, 316; of HORRORS, Canton, 420; Papuan, 447; of HAWAII, 504.

TENERIFFE, 2.

TENNANT, SIR EMERSON, on the Mermaid, 424.

Terebra maculata, adze blades made from, 467.

Termites, nests of, 14, 350.

TERNATE, 390; Peak of, ascended, 392.

Testudo geometrica, 152.

Tetrastemma agricola, 26.

THACKOMBAU, King, burns a town, 309; interview with, 319.

Thalassia, 361.

Thalassæca glacialoides, 134, 253.

Theatres, Japanese, 494.

THOMSON, F. T., Capt. R.N., amongst Admiralty Islanders, 475, 477; at HUMBOLDT BAY, 443.

THOMSON, Sir C. WYVILLE, 166, 185; Director of Scientific Staff, 1; on deep-sea phenomena, 589; at the dredge, 578; on "Implosion," 580; originator of the Expedition, 598; on range of Pelagic animals, 575; on phosphorescence, 590.

Three-toed Sloth, roasted, 97; habits of, 104.

Threshing floor in the AZORES, 33.

THWAITES, Mr. G. H. K., in CEYLON, 535.

Thynnus argentivittatus, 53.

Thyrsopteris elegans, 537.

Thysanodactylus bilineatus, 84.

Thysanozoon, 402.

TIDORE, 390.

Tillæa moschata, 165.

TIZARD, T. H., Staff-Com. R.N., 369, 378; 593; temperatures of MARION ISLAND, 170.

Toads, mewing of, 93.

Tobacco, introduction of, to NEW GUINEA, 357; market in BRAZIL, 97; pipes of Gudangs, 356; smoked by Papuans, 437.

Tokelau race, 336.

Tombs, of chiefs at FIJI, 327; of Chinese, at TERNATE, 391.

TONGATABU, 282.

TONGA, King George of, 320.

TORRES STRAITS, 361.

Tortoises, 152; large, on board the "Challenger," 595.

Toucans, shooting of, 91.

Tracks, of animals in sand, 152; of Australian Blacks on trees, 258; of *Ornithorynchus*, 263.

Trade, of FIJIANS, 324; of Papuans, 439; of Admiralty Islanders, 451; gear, 451; Wind, climatic effect of the, 45, 50.

Trap for Opossum, 258; for wild swine, 381.

TRAVERS, Mr. H. H., on Maorioris, 340.

TRAVERS, Mr. T. W. LOCKE, present of specimens by, 279; on Maoris, 340.

Trees, destruction of, at JUAN FERNANDEZ, 538; high at ARU, 371; highest existing, 260; Composite, at TAHITI, 542; with plank-like roots, 405; shedding leaves in dry season, 94; transplanted by the waves. 368; trunks of, found in CROZETS, 182; fossil at KERGUELEN'S LAND, 195.

Tribulus cistoides, 43.

Trichodesmium, 566.

Trichoglossus Swainsonii, 352.

Trichomanes peltatum, 455.

Tridacna, ornaments made from shell of, 470; living, appearance of, 362.

Trigger fish, 51, 74.

Trigonia, 276, 361.

TRISTAN DA CUNHA, 108; insects at, 134; relations of flora of, 135.

W.

Wai Levu, the, 322.
Wallabies, Bush, 261, 269.
Wallace, Mr. A. R., on the Agouti at St. Thomas, 14; on ferns at Tahiti, 518; on Juan Fernandez, 540-41; on Ke Island boats, 379; on moulting of Great Bird of Paradise, 376; on roots of trees, 405; at Wanumbai, 373.
Wallich, Surgeon Major, G. C., on distribution of deep-sea animals, 585; on colours of deep-sea animals, 591; on icebergs, 243.
Walrus, food of the, 205.
Wamma Island, 367.
Wanumbai, 374.
Water clock at Canton, 419.
Waterhouse, Mr. C. O., on the *Coleoptera* of Kerguelen's Land, 193.
Waxworks in Japan, 493.
Weapons, native, of the Admiralty Islanders, 468; of Api, 346; of Fiji, 338; of Humboldt Bay, 443; of Aru Islanders, 370, 374; mart for, at Cape York, 361; spurious, 496.
Wednesday Island, 361; birds of, 363.
Wellington, New Zealand, 277.
Wells, at Admiralty Islands, 465; at Tonga, 290; at St. Vincent Island, 51.
Wesleyans at Fiji, 327, 331.
West Indies, 11.
Whales, 569; origin of tail fin of, 265; remarkable, with tusks, 157; southern, mode of killing, 213; southern Fin-back, blowing of, &c., 252; southern Whalebone, 197; Ziphioid at the Falklands, 559; following ships, 11.
Whisky Bay, 217.
White ants, nests of, 14, 350.
Wideawake, the, 349, 363, 563.
Wild, Dr. J. J., 550; on Kerguelen Plateau, 170; sketches of gods, 474.
Williams, Mr. Thomas, on Fiji, 317.
Wilkes, Commodore, on Fiji, 318; on icebergs, 242, 250; on the Lutaos, 398 (note); on Fiji, 318, 341.
Wokan Island, 367.
Wood, drift, 367; from New Guinea Coast, 432.
Wood-lice, at St. Vincent Island, 51.
Woodwardia radicans, 33, 34.
Wynberg, 140, 141.

Z.

Zamboanga, 395.
Ziphioid whale, 157, 559.
Zones of vegetation, on mountains generally, 45; at Kerguelen's Land, 194; at Marion Island, 168; at St. Jago Island, 63; at St. Vincent Island, 44; at Ternate, 392; in Arctic and Antarctic regions, 225.
Zostera eaten by deep-sea animal, 583.
Zosterops, luteus and flaviceps, 365.
Zygæna malleus, 9, 52.

Trygon, 271; knives of spine of, 468.
Tubipora musica, 404.
Tucotuco, 146.
Tumuli, under temples, 317; over graves, 327.
Tunny, the, 38.
Tupman, Capt. R.M.A., on phosphorescence, 574.
Turacus albocristatus, 161.
Turbo operculum, 307; pica, 17.
Turkey Brush, the, 353.
Turnix melanonotus, 364.
Turnstone, the, 348.
Turtles, 350, 479; eggs of, hatching of, 561; food of, 568.
Tussock grass, Tristan da Cunha, 115, 117, 126.

U.

Ula, the, of Fiji, 338.
Unicorn, the, 424; horn of, experiments on, as an antidote, 427; the origin of, 424.
Unio, 325, 326.
Uspallata Pass, excursion to, 544.

V.

Vaccinium, 393, 521.
Vaginulus, 89.
Valparaiso, 543.
Vanessa, 134.
Vanilla at Tahiti, 524.
Vaquerios of Brazil, 98.
Vegetable itch, the, 379, 440.
Vegetation, Arctic and Antarctic, compared, 225; débris of, on deep sea bottom, 583; of the Admiralty Islands, 454; at Api, 345; at Aru, 366; of the Azores, 34; of the Cape of Good Hope, 140, 142, introduced at the Cape of Good Hope, 141; of Cape York, mixed aspect of, 351; at Hawaii Island, 499; of Heard Island, 224; of Kermadec Islands, 280; of Matuku Island, 294; at Tahiti, 518, 521; of Tonga, 283, 290; of banks of the Wai Levu, 323; limit in altitude of, in Kerguelen's Land, 194; in Marion Island, 168; in the Southern Islands, 225.
Velella, 568.
Ventriculites, 589.
Verreaux, M. Jules, on turacou, 162.
Virgin Islands, 11.
Visibility of islands at a distance, 54.
Viti, 323.
Viti Levu, 315.
Volcano, of Banda, ascent of, 382; active, at Kerguelen's Land, 186; active, Trachytic, 409; Hawaian, eruptions of, 503; zones of vegetation on, 45.
Voyage, objects and duration of the, 1; slowness of, 597.
Vultures at St. Vincent Island, 53.

Printed in the United States
By Bookmasters

/